Experimental Approaches of NMR Spectroscopy II

The Nuclear Magnetic Resonance Society of Japan
Editor

Experimental Approaches of NMR Spectroscopy II

Application to Life Science and Materials Science

Second Edition

Springer

Editor
The Nuclear Magnetic Resonance Society
of Japan
Kobe, Japan

ISBN 978-981-97-6829-5 ISBN 978-981-97-6830-1 (eBook)
https://doi.org/10.1007/978-981-97-6830-1

This Springer imprint is published by the registered company Springer Nature Singapore Pte Ltd.
The registered company address is: 152 Beach Road, #21-01/04 Gateway East, Singapore 189721, Singapore

Preface

This book is the second edition of the first edition that described advanced developments in methodology and applications of NMR spectroscopy to life science and materials science. Experts who are leaders in the development of new methods and applications of life and materials science have contributed an exciting range of topics that cover recent advances in the structural determination of biological and material molecules, dynamic aspects of biological and material molecules, and the development of novel NMR techniques, including resolution and sensitivity enhancement. First, this book particularly emphasizes the experimental details for new researchers to use NMR spectroscopy and pick up the potential of NMR spectroscopy. Second, the book is designed for those who are involved in either developing the NMR technique or expanding the NMR application fields by applying them to specific samples. Third, the Nuclear Magnetic Resonance Society of Japan has organized this book not only for NMR researchers in Japan but also for readers worldwide who are interested in using NMR spectroscopy extensively.

In Volume II, first, the application of solid-state NMR spectroscopy to silk materials, polymer materials, functional materials, and organic light-emitting diodes (New chapter) is extensively described. Second, the application of advanced solution NMR techniques to structure and dynamics determinations of natural products, glycoproteins and ribonucleic acid (RNA), and paramagnetic compounds are comprehensively accounted. Third, analyses of molecularly complex compounds are described using NMR data science. Fourth, NMR spectroscopic studies of quadrupole nuclei in organic compounds and inorganic materials are described in detail. Fifth, advanced solution NMR methods are focused on the characterization of protein-ligand interactions and protein structure and dynamics determined by residual anisotropic spin interactions.

Kobe, Japan
June 2024

The Nuclear Magnetic Resonance
Society of Japan
Editorial Board

Akira Naito
(Editor in Chief)

Tetsuo Asakura
Toshimichi Fujiwara
Koichi Kato
Ichio Shimada
Kiyonori Takegoshi
Yasuhiko Yamamoto

Acknowledgements

We would like to acknowledge the authors of this book who are members of the NMR Society of Japan for their excellent contribution to writing exciting topics on NMR spectroscopy.

We would also like to thank Mr. Shinichi Koizumi of Springer Japan and Mr. Rajasekar Ganesan of Springer Nature for their encouragement and assistance in editing the book.

Member of the Editorial Board

Contents

Chapter 1
NMR Studies on the Structure of *Bombyx mori* Silk Fibroin

Tetsuo Asakura

Abstract The structural analysis of *Bombyx mori* silk fibroin (SF) has received considerable attention because of its supreme physical and medical properties. These properties originate from the structure of silk fibroin, making structural analysis key to the further development of silk in wider applications. Two distinct structures of SF in the solid state have been reported, i.e., Silk I as the solid-state SF structure stored in the middle silk glands after drying without any external forces, and Silk II as the structure after spinning. Here, detailed structures of Silk I and Silk II were determined using mainly solid-state NMR. Silk I was proposed to be a repeated type II β-turn structure with the torsion angles $(\varphi, \psi) = (-62°, 125°)$ for Ala residue and $(\varphi, \psi) = (77°, 10°)$ for Gly residue. The intra- and intermolecular hydrogen bonding was formed alternatively along the chain. On the other hand, Silk II was proposed to be a lamella structure, i.e., two kinds of Ala residues in β-sheet structure, plus one distorted β-turn formed by repetitive folding using β-turns every eighth amino acid in an antipolar arrangement.

Keywords Silk fibroin · Silk I structure before spinning · Silk II structure after spinning · Solid-state NMR

1.1 Introduction

Bombyx mori (*B. mori*) silk fibroin continues to attract the attention of many researchers in textile technology, biochemistry, biology, polymer chemistry, and biomaterials [1]. In particular, the application of silk fibroin in the biomaterial field is a very active area due to its excellent properties, such as high strength, high toughness, and excellent biocompatibility [2–5]. These properties originate from the structure of silk fibroin, making structural analysis key to the further development of silk in wider applications.

T. Asakura (✉)
Tokyo University of Agriculture and Technology, Tokyo, Japan
e-mail: asakura@cc.tuat.ac.jp

The NMR Society of Japan (ed.), *Experimental Approaches of NMR Spectroscopy II*,
https://doi.org/10.1007/978-981-97-6830-1_1

The silk fiber from *B. mori* cocoon is a twin fiber consisting of two proteins: silk fibroin and silk sericin. The fibroin consists of a heavy (H) chain of 390 kDa and a light (L) chain of 26 kDa connected by a disulfide bond, as well as a glycoprotein (P25 (30 kDa)) [6–9], which is secreted into the posterior silk gland. The H-chain, L-chain, and P25 are thought to be assembled in a ratio of 6:6:1. The silk fibroin is then stored in the middle silk gland and coated by silk sericin. The proteins are spun out through the anterior silk gland and converted into silk fibers. For use as a biomaterial, the coated silk sericin is usually removed by a degumming process [10]. The H chain occupies most of silk fibroin and its amino acid composition is Gly (46%), Ala (30%), Ser (12%), Tyr (5.3%), Val (1.8%), etc. [11]. The primary structure of the H chain has been reported by Zhou et al. [11, 12]. The primary structure contains a large number of repeated sequences organized into 12 domains, each with related subdomains as shown in Fig. 1.1I. The amino acid sequence of the silk fibroin H chain is roughly divided into four motifs: (i), (ii), (iii), and (iv) that alternate along the chain. Motif (i) consists of a highly repetitive AGSGAG sequence and constitutes the crystalline regions of the fiber. It has a total of 433 AGSGAG sequence repeats, which is equivalent to 2598 amino acid residues as summarized in Fig. 1.1II [13]. Since the total number of amino acid residues in the H chain is 5263, almost half of it is made up of the AGSGAG sequence. Motif (ii) is a relatively less repetitive sequence that contains hydrophobic and/or aromatic residues such as GAGAGY, GAGAGV, and GAGAGVGY. Motif (iii) is very similar to Motif (i) except for the presence of an AAS motif. Finally, Motif (iv) contains negatively charged, polar, bulky hydrophobic, and/or aromatic residues such as TGSSGFGPYVANGGYSGYEYAWSSESDFGT [11, 12] and makes up the amorphous regions of the silk fibroin H chain.

It is well known that *B. mori* silk fibroin has two crystalline forms, Silk I and Silk II [14]. Silk I is the structure before spinning, while Silk II is the structure after spinning in the solid state. The structure of Silk I has been a topic of controversy for a long time because X-ray diffraction or electron-diffraction provides only limited structural information [14–17]. To make matters more complicated, the structure of silk fibroin before spinning was initially called the α form to distinguish it from the β-form (Silk II). However, Silk I was later renamed from the α-form to avoid confusion with the α-helix of Pauling and Corey [14]. Despite this, many papers still refer to the structure of Silk I as an α-helix or partially α-helix based on IR and Raman spectra [18–23]. Therefore, the author will first introduce the process that led to the Silk I structure being identified as a type II β-turn using in vivo NMR and several solid-state NMR methods together with selectively stable isotope-labeled model peptides in detail.

On the other hand, the structure of Silk II was first characterized by Marsh, Corey, and Pauling [24] using fiber X-ray diffraction as shown in Fig. 1.2. However, subsequent studies using X-ray diffraction [14, 15, 25, 26] showed a less regular structure than the model proposed by Marsh et al. [24] although the general features of the AP β-sheet structure were accepted. Thus, the structure of SF fibers is still not conclusive. For example, Takahashi et al. [26] conducted a detailed study using superior X-ray diffraction data based on 35 quantified intensities. The Takahashi model consists of structures that are statistically stacked with different orientations, occupying the

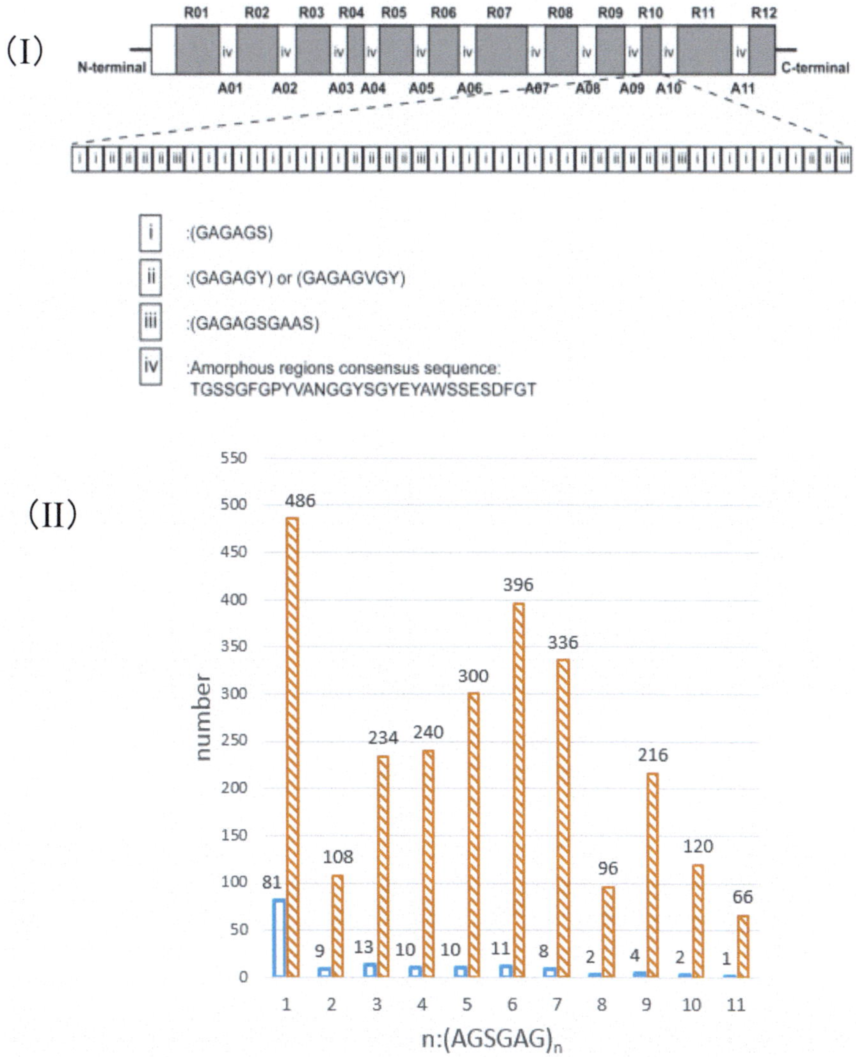

Fig. 1.1 **I** A schematic representation of the organization of *B. mori* silk fibroin heavy chain gene. R01····R12 and A01····A11 represent the predicted 12 repetitive and 11 amorphous regions, respectively, in the polypeptide chain. An approximate amino acid sequence of the R10 region is illustrated by i–iv. Reprinted with permission from Ref. [53]. **II** Histograms of the number of occurrences of each $(AGSGAG)_n$ sequence for n = 1–11 (blue), and the total number of amino acid residues of each $(AGSGAG)_n$ in the primary structure of silk fibroin heavy chain (orange stripes). Reprinted with permission from Ref. [13]. Copyright 2020 Elsevier

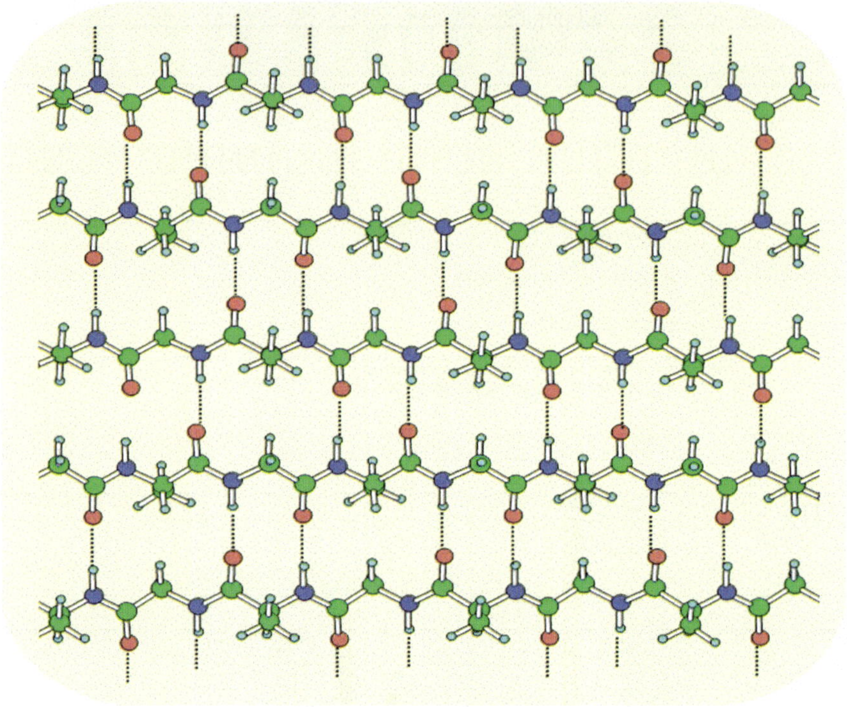

Fig. 1.2 Marsh, Corey, and Pauling model of Silk II (*B.mori* structure after spinning). Reprinted with permission from Ref. [24]. Modified from Ref. [24]. Copyright 1955 Elsevier

crystal site with a ratio of 1:2. Their model is a better fit to the experimental data than the Marsh model, but is not consistent with the distances of the intermolecular hydrogen bonds between the NH···OC groups of Ala and Gly. Therefore, the Takahashi model was energetically unstable, and a more energetically favorable model was required. Most recently, the authors proposed that Silk II structure was a lamella structure using ^{13}C selectively labeled model peptides and solid-state NMR method [13, 27]. Namely, two kinds of Ala residues in β-sheet structure, plus one distorted β-turn formed by repetitive folding using β-turns every eighth amino acid in an antipolar arrangement.

1.2 Structure of Silk Stored in the Middle Silk Gland of *B. Mori* Silkworm

The authors first reported the ^{13}C solution NMR spectra of liquid silks stored in live silk moths, *Samia cynthia ricini* (*S. c. ricini*) and *B. mori*, in 1983, together with the spectrum of pupa as shown in Fig. 1.3 [28, 29]. The shapes of these 5th larval stage silk moths and pupa are suitable for direct placement into the ^{13}C NMR sample tube with a 10 mm outer diameter. The ^{13}C NMR peaks of the liquid silks could be obtained and easily assigned because these peaks should not be observed in the spectrum of pupa (Fig. 1.3b) after making cocoons. From the comparison of the spectra between two kinds of silkworms, it is important to emphasize that the conformation of liquid silk from *B. mori* is different from an α-helix. There are mainly two kinds of Ala residues in the liquid silk of *S. c. ricini* stored in the middle silk glands: Ala residues in the polyalanine sequences (A)n (n = 12,13) and isolated Ala residues (Fig. 1.3a) [30]. The polyalanine sequences take an α-helical conformation, whereas the isolated Ala residues take a random coil conformation in the liquid silk [28, 29, 31–36]. Therefore, the chemical shifts of the Ala main peaks of *S. c. ricini* liquid silk are a good chemical shift reference for α-helix, and the small peaks are a good chemical shift reference for the random coil.

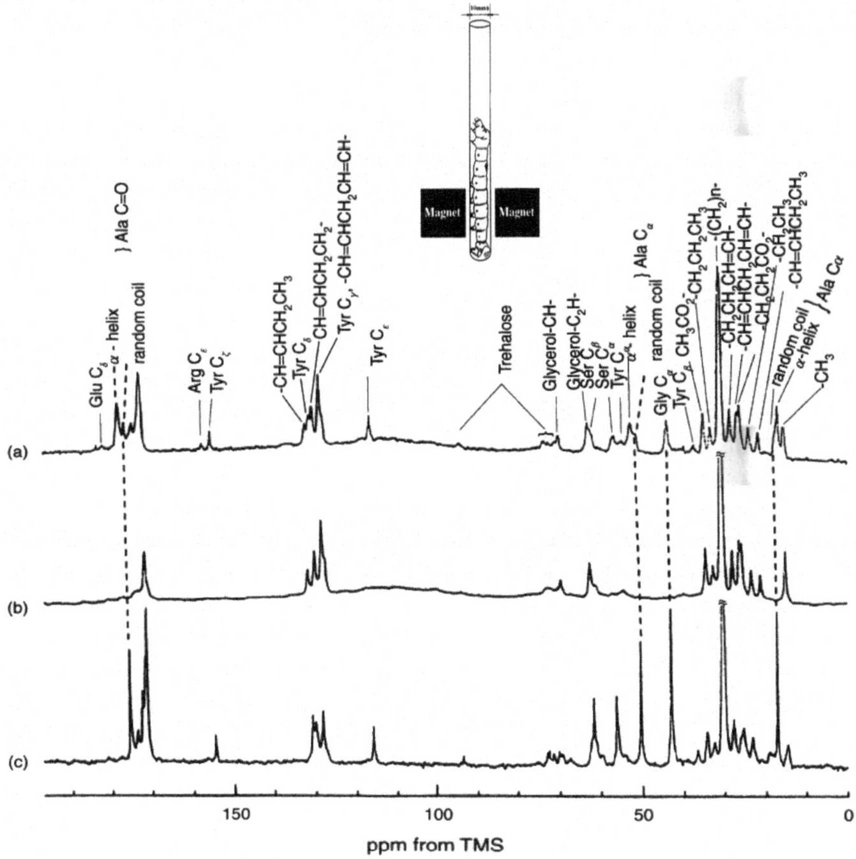

Fig. 1.3 ^{13}C solution NMR spectra of the middle silk gland portions of living **a** *S. c. ricini* and **c** *B. mori* mature larvae together with **b** abdomen of *S. c. ricini* pupa together with the assignment [28]. Reprinted with permission from Ref. [28]. Modified from Ref. [28]

As shown in Fig. 1.3c, the chemical shifts of the single sharp peaks from Ala Cα, Cβ, and C = O carbons of *B. mori* liquid silk are quite different from the corresponding α-helical Ala peaks of *S. c. ricini* liquid silk and close to the random coil peaks (Fig. 1.3a). Thus, the spectral comparison of Figs. 1.3a and c provides clear evidence against the presence of an α-helix in the liquid silk from *B. mori* silkworm. In fact, X-ray diffraction data of liquid silk dried without using any external force also clearly deny the presence of an α-helix [14, 15, 17]. However, the in vivo NMR study does not mean that all of the *B. mori* liquid silk stored in the middle silk glands have a simple random coil structure. It is expected to be difficult for silkworms to prepare silk fibroin fiber with high strength entirely from a random coil structure. Namely, some orderly structure, Silk I, that is not an α-helix, is speculated to be present in the liquid silk.

1.3 Structure of Silk I in the Solid State

In order to determine the structure of Silk I, the authors used the peptide model $(AG)_{15}$. This is because the number of Ala, Gly, and Ser residues in the repeated AGSGAG sequences (crystalline region) make up almost half of the total amino acid residues of silk fibroin as mentioned above [13]. In fact, the alternating copolypeptide (AG)n has been used as a structural model for silk fibroin by X-ray diffraction analysis [14, 15, 25, 37–39]. The $(AG)_{15}$ peptide model was obtained after dialysis from a 9 M LiBr aqueous solution and was used as a model for Silk I. This is because the IR spectra and ^{13}C chemical shifts of its main peaks from Ala and Gly residues are the same as those from *B. mori* silk fibroin. The latter was obtained by slowly drying liquid silk taken from the silk gland at room temperature [10, 16, 17, 40–43]. In order to determine the atomic coordinates of the Silk I structure, the authors used 2D spin-diffusion solid-state NMR and rotational-echo double-resonance (REDOR) NMR spectroscopies together with usual ^{13}C cross-polarization/magic angle spinning (CP/MAS) NMR spectroscopy. The 2D spin-diffusion solid-state NMR is a powerful method for obtaining the relative orientation of two chemical shift tensors of ^{13}C labeled sites in the local molecular framework [44–48]. This solid-state NMR technique was used to determine the torsion angles of both silk fibroin and $(AG)_{15}$ with Silk I form [16, 17, 49–53]. The 2D spin-diffusion NMR spectra of $(AG)_6A[1-^{13}C]G^{14}[1-^{13}C]A^{15}G(AG)_7$ and $(AG)_7[1-^{13}C]A^{15}[1-^{13}C]G^{16}(AG)_7$ were observed under off magic angle spinning [16, 49] and are shown together with the calculated spectra in Fig. 1.4. The errors (RMSD) between the observed and calculated 2D spin-diffusion NMR spectra were mapped in Fig. 1.5. There are two minima for the Ala^{15} residue, $(\varphi, \psi) = (-60°, 120°)$ and $(-30°, 100°)$, and four minima for the Gly^{16} residue, $(\varphi, \psi) = (70°, 10°)$, $(100°, 0°)$, $(120°, -50°)$, and $(100°, -60°)$ (red X marks). To further determine the torsion angles of Silk I, the REDOR method was used [16, 51–59].

REDOR is an excellent method for determining inter-atomic distances from dipolar couplings in solids. The atomic distance between the C = O carbon of the

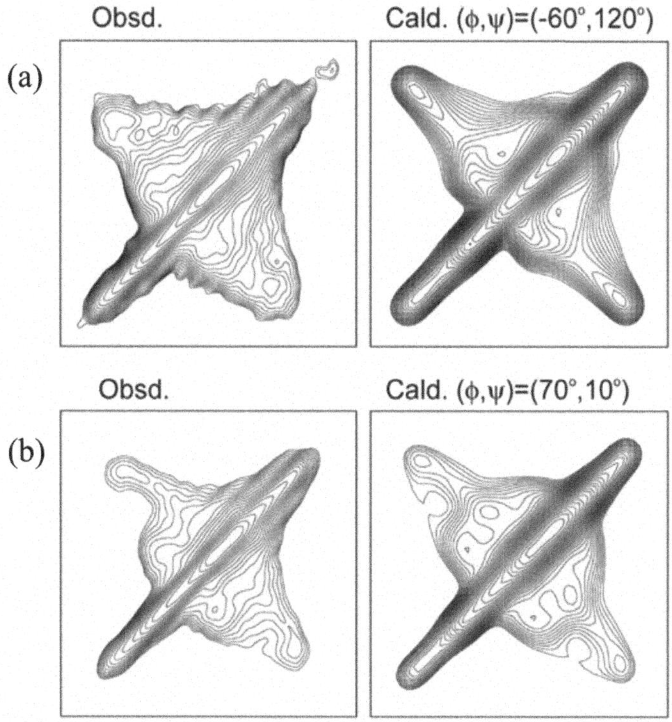

Fig. 1.4 Observed 2D spin-diffusion NMR spectra of **a** $(AG)_6A[1-^{13}C]G^{14}[1^{13}C]A^{15}G(AG)_7$ and **b** $(AG)_7[1-^{13}C]A^{15}[1-^{13}C]G^{16}(AG)_7$ with Silk I structures for determination of the torsion angles (ϕ, Ψ) of Ala^{15} and Gly^{16} residues, respectively. The calculated spectra assuming the torsion angles of Ala^{15} (ϕ, Ψ) = = (−60°, 120°) and Gly^{16} (ϕ, Ψ) = (70°, 10°) were also shown for comparison [17]. Reprinted with permission from Ref. [17]

Gly^{14} residue and the NH nitrogen of the Gly^{16} residue changes significantly when the torsion angles (φ, ψ) of the Ala^{15} residue are changed. Therefore, this atomic distance information can be used as a further structural constraint. The atomic distance between $[1-^{13}C]Gly^{14}$ and $[^{15}N]Gly^{16}$ in $(AG)_6A[1-^{13}C]G^{14}A^{15}[^{15}N]G^{16}(AG)_7$ was determined to be 3.8 ± 0.1 Å. Thus, the contour lines of the atomic distances, including experimental errors, are shown in Fig. 1.5, and the angles (φ, ψ) = (−60°, 120°) were selected for the Ala^{15} residue. A good agreement between the observed and calculated 2D spin-diffusion NMR spectra was obtained when the angles (φ, ψ) = (−60°, 120°), as shown in Fig. 1.4. Next, the atomic distance between $[1-^{13}C]Ala^{15}$ and $[^{15}N]Ala^{17}$ in $(AG)_7[1-^{13}C]A^{15}G^{16}[^{15}N]A^{17}G(AG)_6$ was determined to be 3.2 ± 0.1 Å from the REDOR experiment. Based on the observed atomic distance constraint, the angle (φ, ψ) = (120°, −50°) was ruled out in Fig. 1.5. Then, the 2D spin-diffusion NMR spectrum of $(AG)_6A[1-^{13}C]G^{14}A^{15}[1-^{13}C]G^{16}(AG)_7$ was used to select one of three possible torsion angles (φ, ψ) marked by X. The spectral pattern should depend on both torsion angles of the Ala^{15} and Gly^{16} residues.

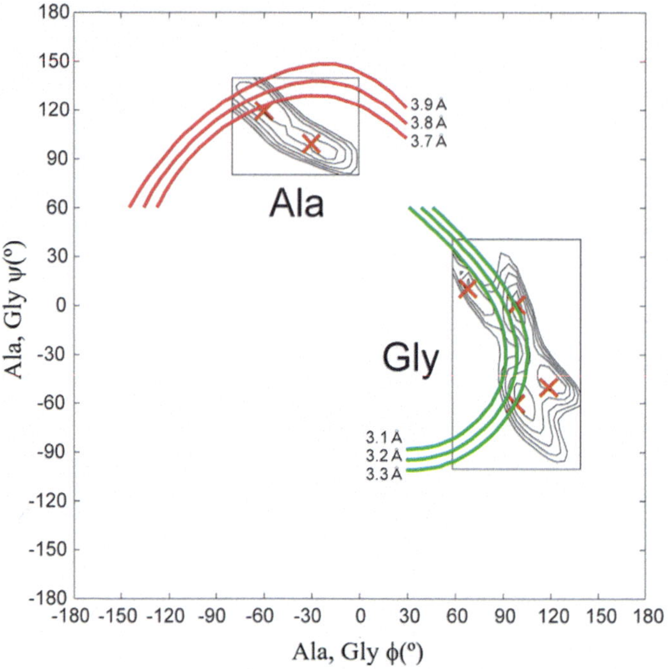

Fig. 1.5 Determination of the torsion angles, (ϕ, Ψ) for Ala^{15} and Gly^{16} residues in $(AG)_{15}$ with silk I structure from both 2D spin-diffusion NMR and REDOR experiments. The candidates of the torsion angles (red X marks) were determined from the observed 2D spin-diffusion NMR spectra in Fig. 1.4. The atomic distances between ^{13}C and ^{15}N nuclei were determined to be 3.8 ± 0.1 Å for $[1-^{13}C]G^{14}A^{15}[^{15}N]G^{16}$ and 3.2 ± 0.1 Å for $[1-^{13}C]A^{15}G^{16}[^{15}N]A^{17}$, respectively from the observed REDOR data [17]. Reprinted with permission from Ref. [17]. Modified from Ref. [17]

However, the torsion angles of the Ala^{15} residue were determined to be (φ, ψ) = (−60°, 120°), as mentioned above. Therefore, the spectral pattern depends only on the torsion angle of the Gly^{16} residue. Three spectral patterns were calculated for the torsion angles (φ, ψ) = (70°, 10°), (φ, ψ) = (100°, 0°), and (φ, ψ) = (100°, −60°). By comparing the observed and calculated spectra, the torsion angles (φ, ψ) = (100°, −60°) could be excluded. However, it is difficult to select between the torsion angles (φ, ψ) = (70°, 10°) or (φ, ψ) = (100°, 0°), because both calculated patterns can reproduce the observed spectrum well. Therefore, X-ray diffraction data of the crystalline fraction of *B. mori* silk fibroin with Silk I structure reported by Lotz et al. [14, 60] and Okuyama et al. [37, 61] were used to determine the torsion angles of the Gly^{16} residue. The X-ray diffraction data of Silk I were reported as follows: the unit cell is orthorhombic and the space group is $P2_12_12_1$. The lattice constants are a = 4.65 Å, b = 14.24 Å, and c = 8.88 Å, with α = β = γ = 90°. In a unit cell, there are four repeats of Ala-Gly and two 2_1-helix chains in an anti-parallel manner. The structure of Silk I was then calculated, including the inter-molecular chain arrangement [14, 17, 37, 60–62]. The initial torsion angle for the Linked-Atom Least-Squares

(LALS) calculation [17] was (φ, ψ) = (−60°, 120°) for the Ala residue, but there were two candidates for the Gly residue: (φ, ψ) = (70°, 10°) and (φ, ψ) = (100°, 0°), as mentioned above. Although the LALS calculations were performed for two cases, the same set of torsion angles was obtained: (φ, ψ) = (−62°, 125°) for the Ala residue and (φ, ψ) = (77°, 10°) for the Gly residue. These torsion angles were obtained with an R factor of 10%, independent of the initial sets of torsion angles for the Gly residues.

The final Silk I structure of (AG)n with intra-(green) and inter-(red) hydrogen bonds is shown in Fig. 1.6 [16, 17, 53]. The most definite evidence of a type II β-turn structure is the presence of an intra-molecular hydrogen bond formation (green) between the carbonyl oxygen atom of the ith Gly residue and the amide hydrogen atom of the i + 3-th Ala residue in Fig. 1.6. The Silk I structure was also supported from the solution NMR analysis of liquid silk fibroin [63]. This type II β-turn structure can easily form a lamella structure (Silk II) with a distorted β-turn formed by repetitive folding using β-turns every eighth amino acid in an antipolar arrangement, as will be mentioned below [13, 27].

1.4 Verification of Silk I Structure (Type II β-Turn)

First, the proposed type II β-turn structure of Silk I was verified by comparing it with previous Silk I models. So far, the Silk I structure has been studied using a combination of X-ray diffraction studies and conformational energy calculations by Lotz and Keith [14, 60], Okuyama et al. [37, 61], and Fossey et al. [25]. In the Lotz and Keith model [14, 60], Ala residues are in a β-sheet structure and Gly residues are in either a left-handed or right-handed α-helical conformation, i.e., Ala (φ = −105° and ψ = 112°) and Gly (φ = 80° and ψ = 50°) or Ala (φ = −125° and ψ = 88°) and Gly (φ = −50° and ψ = −76°). The Fossey model has right-handed and left-handed twisting of sheets with approximately equal magnitudes of twist, Ala (φ = −80° and ψ = 150°) and Gly (φ = −150° and ψ = 80°). The Okuyama model is Ala (φ = −110° and ψ = -2°) and Gly (φ = 73° and ψ = -102°). Therefore, 2D spin diffusion NMR spectra were calculated using the torsion angles of each model and compared with the observed spectra shown in Fig. 1.7. The agreement between the calculated spectral pattern and the observed spectrum were poor for all these models [17], and only the type II β-turn model proposed by the authors could reproduce the observed spectrum very well (Fig. 1.4).

Next, the structure of Silk I, including the intermolecular arrangement proposed in Fig. 1.6, was examined by determining the accurate positions of ^{1}H atoms in the structure and calculating the theoretical chemical shifts of ^{1}H, ^{13}C, and ^{15}N nuclei. The coordinates of heteroatoms such as C, N, and O atoms in the type II β-turn structure of Silk I were determined as shown in Fig. 1.6. However, the coordinates of H atoms were uncertain. Since H atoms are located on the surface of the silk fibroin molecule, their coordinates are sensitive to intra- and intermolecular arrangements. Therefore, it is important to determine the position of H atoms. However, ^{1}H NMR spectra in solid-state are generally very broad compared to ^{1}H solution

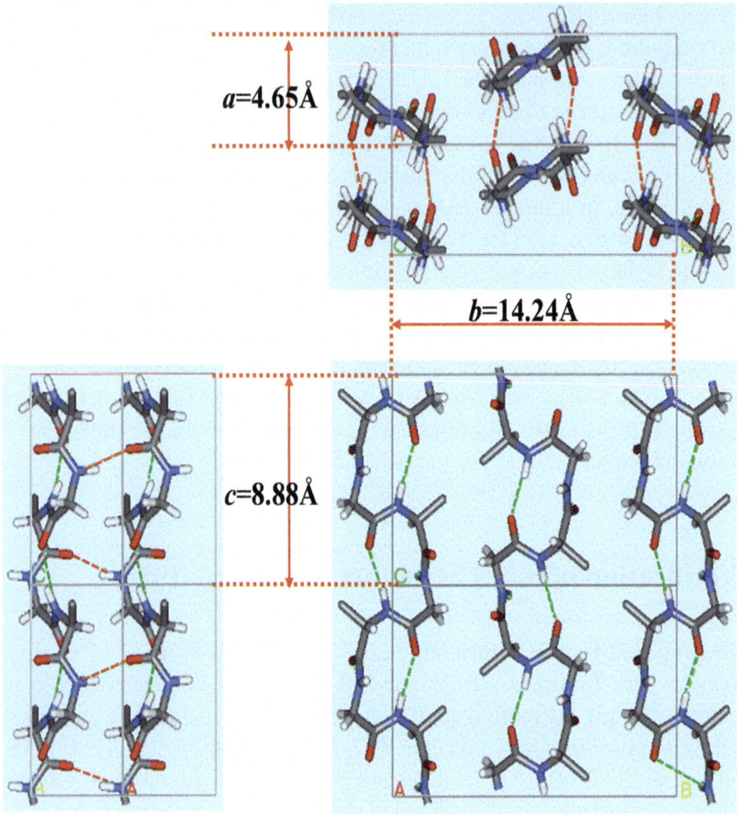

Fig. 1.6 Packing structure of poly(AG) chains with type II β-turn structure as a model for silk I. Dotted lines denote intra-(green) and inter-(red) molecular hydrogen bonds. The unit lattice values, *a*, *b*, and *c* were obtained from X-ray diffraction data [17]. Reprinted with permission from Ref. [17]. Modified from Ref. [17]

NMR spectra due to strong dipolar coupling. The authors developed a 1 mm micro-coil MAS NMR probe-head with high-speed spinning [64]. This micro-coil MAS NMR probe-head with a high spinning rate of 70 kHz was used to observe a well-resolved ^{1}H NMR spectrum of $(AG)_{15}$ using an ultrahigh-field NMR spectrometer at 920 MHz. Figure 1.8 shows the ^{1}H DQMAS (double quantum magic angle spinning) NMR spectrum of $(AG)_{15}$ with Silk I structure, along with Double-Quantum (DQ) correlations to obtain information on intra- and intermolecular ^{1}H-^{1}H distances in $(AG)_{15}$ [65–74]. As a result, the ^{1}H chemical shifts of $(AG)_{15}$ with Silk I structure could be obtained in the solid state from Fig. 1.8 with high accuracy. To calculate the coordinates of ^{1}H atoms, geometry optimization was applied only to protons under periodic boundary conditions after fixing the atomic coordinates of heteroatoms. All calculations were carried out using the NMR-CASTEP program. The intra-and inter-molecular ^{1}H−^{1}H distances less than 4 Å were calculated from the coordinates of ^{1}H

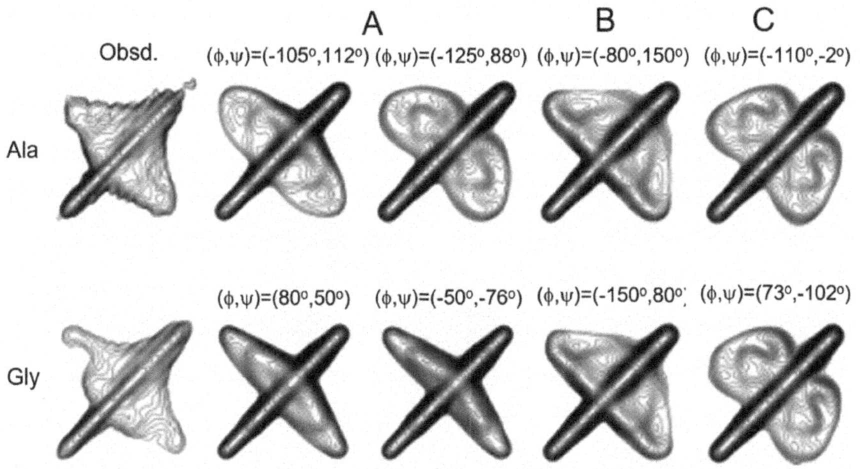

Fig. 1.7 Observed and calculated 2D spin-diffusion NMR spectra of $(AG)_6A[1-^{13}C]G^{14}[1-^{13}C]A^{15}G(AG)_7$ and $(AG)_7[1-^{13}C]$ $A^{15}[1-^{13}C]G^{16}(AG)_7$ for determinations of the torsion angles $Ala^{15}(\varphi, \psi)$ and $Gly^{16}(\varphi, \psi)$ in $(AG)_{15}$, respectively. The calculations were performed for the structural models reported previously. **a**: Two models by Lotz and Keith [14, 60], **b**: Model by Fossey et al. [25], and **c**: Model by Okuyama et al. [37, 61]. Reprinted with permission from Ref. [17]

atoms of the packed poly(AG) chains determined here, as listed in Table 1.1. Here, the intra-and intermolecular chain arrangements of $(AG)_{15}$ molecules with type II β-turn structure are used to calculate the ^1H-^1H distances for the DQ correlations ①–⑮. For example, in column ①, DQ correlation (① G-HN A-HN) and Intramolecular (2.83 (5 3)) means that the intramolecular ^1H-^1H distance between HN Gly and HN Ala residues is calculated to be 2.83 Å in chain number 5 (bold) in group (ac) and the number of Ala-Gly segment 3 (bold) in group (bc). Although the authors did not obtain ^1H DQ build-up curves for quantitative simulation, the relationship between the observed DQ correlation and relative proton-proton proximities from the coordinates of $(AG)_{15}$ with Silk I form could be obtained qualitatively. In fact, a detailed analysis of the DQ correlations showed that relative DQ peak intensities are a reliable measure of relative distances [56, 57]. The ^1H DQMAS spectrum indicates that all of the ^1H-^1H distances predicted in Table 1.1 could be observed in Fig. 1.8.

Next, ^1H, ^{13}C, and ^{15}N chemical shift calculations were performed using GIPAW methods [75]. The observed ^1H, ^{13}C, and ^{15}N chemical shifts of $(AG)_{15}$ with Silk I structure [53, 65, 76] were compared with those calculated by theoretical chemical shift calculation using all coordinates of heteroatoms such as C, N, and O atoms (Fig. 1.6) and those of H atoms determined here. The agreement between the calculated and observed chemical shifts of all ^1H, ^{13}C, and ^{15}N nuclei of $(AG)_{15}$ with Silk I structure was excellent, as shown in Fig. 1.9. For example, the correlation coefficient was 0.99745 for ^1H and 0.99998 for ^{13}C. This excellent agreement strongly suggests that the coordinates of H atoms of $(AG)_{15}$ in the Silk I form after CASTEP

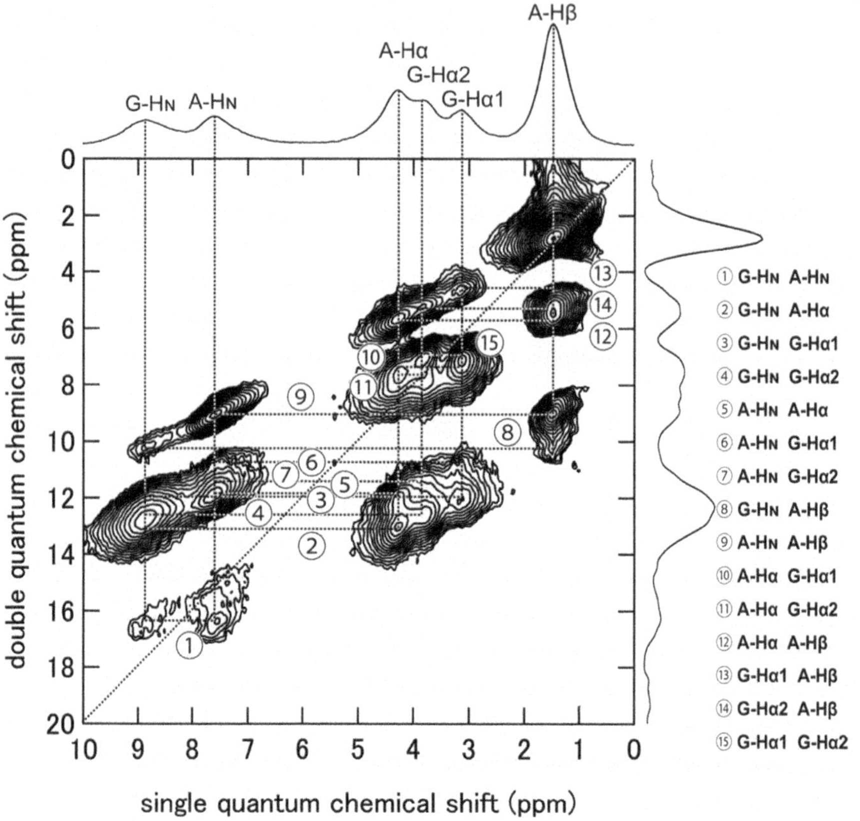

Fig. 1.8 2D ^{1}H (920 MHz) DQMAS NMR spectrum and the skyline projection of $(AG)_{15}$ with Silk I structure together with assignments. The DQ correlations ①–⑮ are shown on the right side [65]. Reprinted with permission from Ref. [65]. Modified from Ref. [65]

calculation could be determined with a high degree of accuracy, as well as supporting the coordinates of heteroatoms such as C, N, and O atoms in Fig. 1.6.

The geometry of the hydrogen bonding arrangements was examined using the coordinates of ^{1}H atoms in the Silk I structure. The intramolecular hydrogen bonding distance was determined to be 2.01 Å between the NH of the (i + 3)th Ala residue and the CO of the (i)-th Gly residue. Additionally, the intermolecular hydrogen bonding distance was determined to be 1.85 Å between the NH of the (i + 2)-th Gly residue in one chain and the CO of the (i + 1)th Ala residue in another chain. On the other hand, the angles between NH and CO bonds that contribute to direct hydrogen bond formation were determined to be 123.2° and 168.1°, respectively. In particular, the angle of 123.2° is far from the preferred linearity for a hydrogen bond. Thus, it can be easily speculated that intermolecular hydrogen bonding is significantly stronger than intramolecular hydrogen bonding based on structural information about both the distance and angle of hydrogen bonding. The planes of the NH-CO peptide and

Table 1.1 The intra- and inter-molecular $^1H-^1H$ distances (in Å) of $(AG)_{15}$ molecules with type II β-turn structure calculated using the coordinates of ^1H atoms obtained from the CASTEP calculation

	DQ correlation		Intra-molecular		Inter-molecular	
①	G-HN	A-HN	2.83	(5 3)	3.43	(4 3)
②	G-HN	A-Hα	2.12	(5 2)	3.46	(4 3)
③	G-HN	G-Hα1	2.30	(5 3)		
④	G-HN	G-Hα2	2.88	(5 3)	2.76, 3.42	(2 5), (4 3)
⑤	A-HN	A-Hα	2.86, 3.52	(5 3), (5 2)	2.96	(4 3)
⑥	A-HN	G-Hα1	3.28, 3.99	(5 3), (5 2)		
⑦	A-HN	G-Hα2	3.47	(5 3)		
⑧	G-HN	A-Hβ	3.89	(5 2)		
⑨	A-HN	A-Hβ	2.87	(5 3)		
⑩	A-Hα	G-Hα1			3.23	(6 3)
⑪	A-Hα	G-Hα2			3.66	(9 4)
⑫	A-Hα	A-Hβ	2.70	(5 3)		
⑬	G-Hα1	A-Hβ	3.65	(5 4)	3.56	(2 4)
⑭	G-Hα2	A-Hβ			3.60	(3 4)
⑮	G-Hα1	G-Hα2	1.77	(5 3)		

The number of the DQ correlations ①–⑮ is the same as those in Fig. 1.8

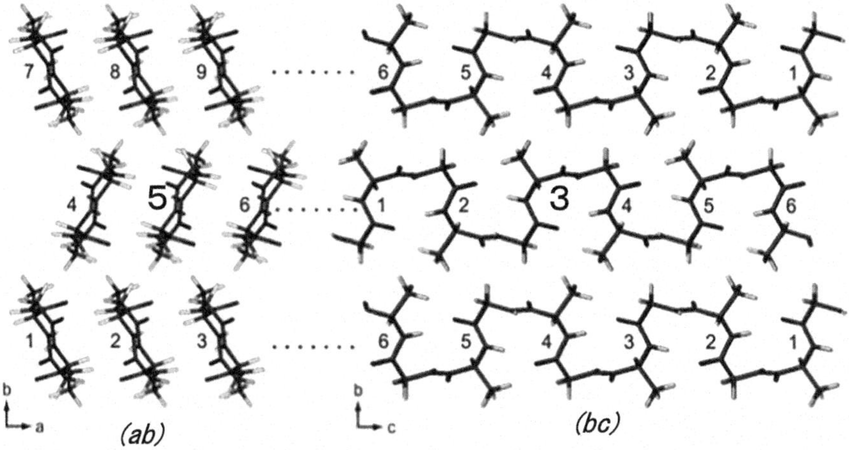

Intra-and inter-molecular chain arrangements of $(AG)_{15}$ molecules with type II β-turn structure (Fig. 1.6) in calculating the $^1H-^1H$ distances for the DQ correlations ①–⑮ in Table 1.1 using coordinates of the ^1H atoms. For example, in column ①, G-H_N A-H_N 2.83 (5 3) in Table 1.1 means that the $^1H-^1H$ distance is calculated to be 2.83 Å between H_N Gly and H_N Ala residues in the chain number 5(bold) in the group (*ab*) and the number of AG segment 3(bold) in the group (*bc*). In addition, G-H_N A-H_N 3.43 (4 3) in Table 1.1 means that the $^1H-^1H$ distance is calculated to be 3.43 Å between H_N Gly and H_N Ala residues in the chain number 4 in the group (*ab*) and the number of AG segment 3 in the group (*bc*). Reprinted with permission from Ref. [65]. Modified from Ref. [65]

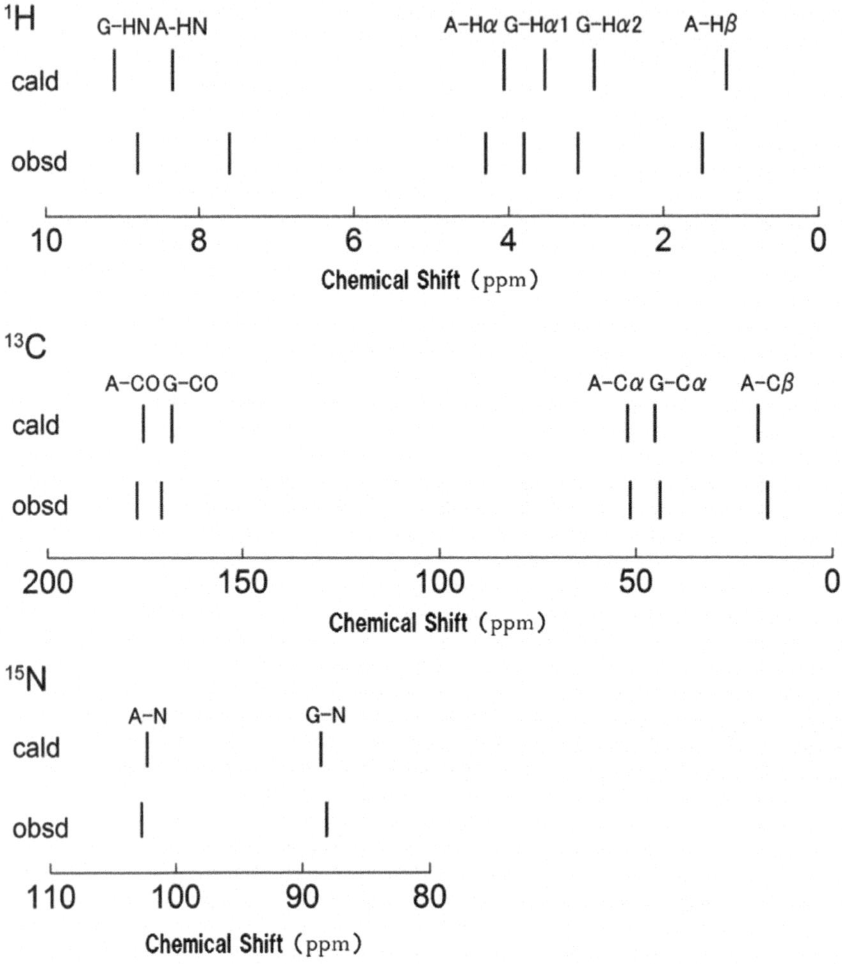

Fig. 1.9 The stick spectra for the ^{1}H, ^{13}C, and ^{15}N chemical shifts calculated for $(AG)_{15}$ with Silk I structure together with the observed chemical shifts [65]. The reference chemical shift was shown to be the mean of the calculated and observed chemical shifts for all peaks. The values were 30.51 ppm, 171.31 ppm, and 197.22 ppm for ^{1}H, ^{13}C, and ^{15}N nuclei, respectively. Reprinted with permission from Ref. [65]

intra- and intermolecular hydrogen bonding appear alternately and are perpendicular to each other. Therefore, it is possible to speculate that intramolecular hydrogen bonding in the Silk I structure can be easily destroyed by external forces such as stretching.

1.5 ^{13}C CP/MAS NMR Spectra of *B. Mori* Silk Fibroin Fiber (Silk II)

The ^{13}C CP/MAS NMR spectrum of the Ala Cβ carbon of native [3−^{13}C] Ala-SF fibers is shown in Fig. 1.10 [77, 78]. The spectrum indicates a heterogeneous structure of SF fibers, which is inconsistent with the homogeneous structure reported by Marsh et al. using X-ray diffraction data [24]. A sample of the crystalline (Cp) fraction (56%) was obtained as a precipitated fraction after chymotrypsin digestion of all SF [79], and the Ala Cβ peak was also observed. The Cp fraction consists of 18% distorted β-turns (**t**), 25% β-sheets (**A**), and 13% β-sheets (**B**), with **t**: **A**: **B** accounting for approximately 1:2:1. In contrast, the fraction of the non-crystalline domain is 44% and consists of 22% each of distorted β-turns and distorted β-sheets [78]. The assignment of each peak to each conformation was performed based on conformation-dependent ^{13}C NMR chemical shifts [10, 16, 42, 76, 80–83].

Next, we concentrated on the secondary structure of the Cp fraction because the fraction is more than half of the total SF, and the primary structure of the sequence of the fraction $(AGSGAG)_n$ is a characteristic of SF. The expanded Ala Cβ peaks in the ^{13}C CP/MAS NMR spectra of $(AGSGAG)_n$ (n = 4, 5, and 8) and the Cp fraction are shown in Fig. 1.11, together with the spectra of $(AG)_m$ (m = 12, 15, and 25) [13, 27]. The ratios of each are 23.3–24.0% at 21.8 ppm for peak **B**, 44.0–45.4% at 19.5 ppm for peak **A**, and 30.9–32.7% at 16.5 ppm for peak **t**. Namely, the fractions of **t**, **A**, and **B** are roughly 1:2:1, which is almost the same for the $(AGSGAG)_n$ and

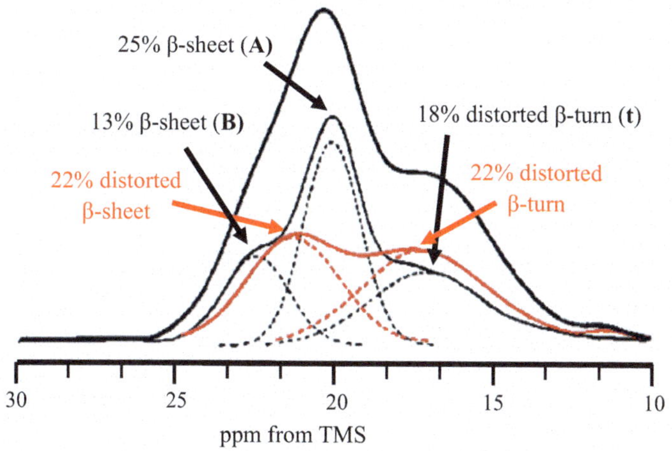

Fig. 1.10 Expanded Ala Cβ peak in the ^{13}C solid-state NMR spectrum of the [3−^{13}C]Ala silk fibroin fiber (thick solid, black) along with the assignment. The crystalline domain (56%; thin solid, black) with the sequence, $(AGSGAG)_n$ consists of 18% distorted β-turns (**t**), 25% β-sheet (**A**), and 13% β-sheet (**B**). The non-crystalline domain (44%) indicated in orange consists of 22% distorted β-turns and 22% distorted β-sheets. Reprinted with permission from Ref. [77]. Modified from Ref. [77]

Cp fractions. Moreover, the relative intensities of **t**, **A**, and **B** are also the same in the expanded Ala Cβ peaks of $(AG)_m$ (m = 12, 15, and 25) as in $(AGSGAG)_n$ and Cp fractions. This indicates that these molecules are considered to have a lamellar structure which consists of repeated smaller units.

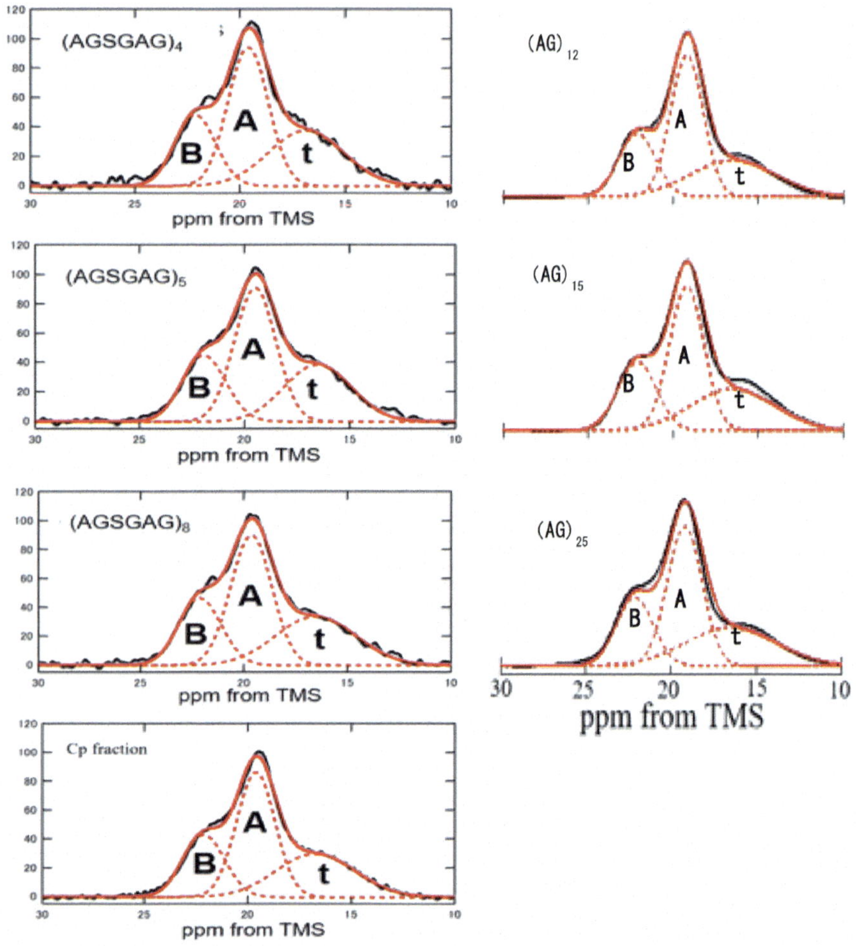

Fig. 1.11 (Left) Three deconvoluted Ala Cβ peaks: **t** (distorted β-turn), **A** (antiparallel-β sheet), and **B** (antiparallel-β sheet) of $(AGSGAG)_n$ (n = 4, 5, and 8) and Cp fractions in the Silk II form (bottom). The deconvoluted Ala Cβ peaks: **t**, **a**, and **b** of $(AG)_m$ (m = 12, 15, and 25) were also shown. Reprinted with permission from Ref. [13]. Copyright 2020 Elsevier and Reprinted with permission from Ref. [27]. Modified from refs. [13] and [27]

1.6 Antipolar-Antiparallel β-Sheet Structure of Silk II Form

Before considering the lamellar structure model in detail, we will consider the packing structure of the lamellar structure, that is, polar or antipolar as shown in Fig. 1.12. The polar arrangement was proposed by Marsh et al. [24], but Takahashi et al. [26] concluded that the crystallographic data fit an antipolar model better.

To examine these models, we obtained the $^1H-^1H$ distance information observed from the double-quantum magic angle spinning (DQMAS) ^1H NMR spectra. Nine $^1H-^1H$ correlation signals in ^1H DQMAS NMR spectrum of $(AG)_{15}$ are indicated in Fig. 1.13I:(a) [84]. We examined the $^1H-^1H$ distances underlying these observed $^1H-^1H$ correlations by inspecting the $^1H-^1H$ distances predicted according to the polar and antipolar models. Figure 1.13II indicates that the Gly $H\alpha_2$ protons in adjacent β-strands are very close to one another (about 2 Å) in the polar model and therefore a diagonal peak for Gly $H\alpha_2$ should be observed. However, this feature is difficult to judge from Fig. 1.13I(a) because the Gly $H\alpha_2$ and Ala Hα peaks overlap in the relevant spectral region. We therefore synthesized deuterium-labeled $([2-d]AG)_{15}$ and acquired another DQMAS ^1H NMR spectrum shown in (Fig. 1.13I(b). We could say clearly that no Gly $H\alpha_2$ peak was observed on the diagonal because the Ala Hα peak of $(AG)_{15}$ at around 5.0 ppm has been removed for the $([2-d]\ AG)_{15}$ sample. This observation provides very strong evidence that the polar model cannot be correct. Actually, the $^1H-^1H$ distances between two Gly $H\alpha_2$ protons in adjacent β- strands are more than 4 Å in the antipolar model, and therefore no observation of the Gly $H\alpha_2$ peak is reasonable. Other experimental $^1H-^1H$ distance data were compared with the distances predicted between the polar and antipolar models. Namely, the observed cross-peak (v) is between Gly $H\alpha_2$ and Ala Hα which is longer than 4 Å in the polar model, but about 2.4 Å in the antipolar model. In addition, observed cross-peaks (viii) and (ix) are from Ala Hβ to Gly $H\alpha_1$ and Hα2 which are longer than 4 Å in the polar model, but about 2.3 and 2.6 Å, respectively, in the antipolar model. Thus, the $^1H-^1H$ distance information obtained from the DQMAS ^1H NMR method gives a

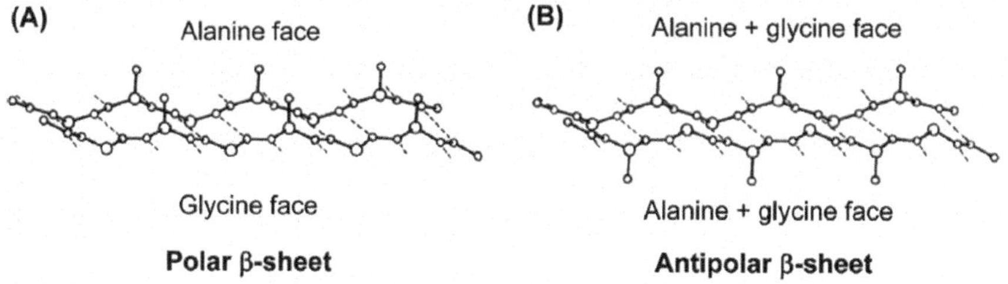

Fig. 1.12 Schematic illustration of **a** polar and **b** antipolar poly(GA) β-sheets obtained by conformational energy calculations. The polar β-sheet has an Ala face and a Gly face, while the antipolar β-sheet has two identical faces. Reprinted with permission from Ref. [25]. Copyright 1991 John Wiley & Sons Inc

clear conclusion that Silk II form takes an antipolar AP β-sheet packing structure, not a polar packing structure.

1.7 Lamellar Structure of Silk II Form Using $(\text{Ala-Gly})_{15}$ as Model Peptides

To obtain further structural information, we synthesized 15 types of $[3-^{13}C]$Ala-$(AG)_{15}$ by ^{13}C labeling different $[3-^{13}C]$Ala positions. The $[3-^{13}C]$Ala Cβ peaks are shown in Fig. 1.14I. We could determine the fractions of all three components, **t**, **A**, and **B** from the deconvoluted ^{13}C-labeled Ala Cβ peaks of 15 kinds of $[3-^{13}C]$Ala-$(AG)_{15}$. The fractions of three Ala Cβ peaks (**t**, **A**, and **B**) were plotted from higher to lower fields against the ^{13}C-labeled position in Fig. 1.14II [27]. At first, we concentrated on changes in the fraction of the peak **t**. The fraction decreased rapidly towards the inner part of the chain and then increased at positions 9 and 11. Beyond that, the fraction decreased slightly and then increased to a maximum at position 19. The fraction then decreased again till the position 25 and remained almost constant at the positions 27 and 29. Thus, there are two maxima at positions 11 and 19 as reported previously [85, 86]. On the other hand, the fraction of the peak **A** assigned to AP β-sheet in Fig. 1.14I was two minima at positions 11 and 19. These data suggested a possible lamellar structure with repetitive folding predominantly through turns at positions 11 and 19, i.e., every eighth amino acid. If we average only for the central parts of $(AG)_{15}$, that is, positions 9 to 23, we get the ratio (%) of the averaged fractions: 24.9: 49.5: 25.6 for **t**: **A**: **B**, respectively. Namely, the fractions were 1: 2:1. Thus, we proposed a lamellar structure for the crystalline domain of SF fiber as shown in Fig. 1.15. It consists of two β-sheet geometries **B** and **A**, and one β-turn **t** with the relative peak intensities 1: 2: 1. These peaks correspond to three Ala Cβ carbons with different spatial arrangements. In the extended part of the lamellar structure, two Ala Cβ carbons with the same arrangement are assigned to **A**, while the other one with a different arrangement is assigned to **B**. This suggests an antipolar lamellar structure with repetitive folding through β-turns every eighth amino acid.

Similarly, we determined the fractions of several conformations of Ala residues using ten kinds of $[3-^{13}C]$Ala-$(AGSGAG)_5$ peptides with different ^{13}C labeling positions (data not shown). There are two maxima at positions 11 and 19 for the fraction of the peak **t**, which is the same tendency as that of $(AG)_{15}$ peptides mentioned above. Thus, a lamellar structure of $(AGSGAG)_5$ with repetitive folding predominantly through turns at positions 11 and 19, i.e. every eighth amino acid, is the same as that of $(AG)_{15}$.

Tirrell's group has reported the use of genetic engineering to produce artificial proteins that adopt predictably, chain-folded AP β-sheet structures in the solid-state [87–89]. The chain-folded lamellar architecture is a kinetic trap in the crystallization of all linear polymers sufficiently flexible to form hairpin-like folds, and it is known that protein chains in the β-form can make tight turns, with either one or two

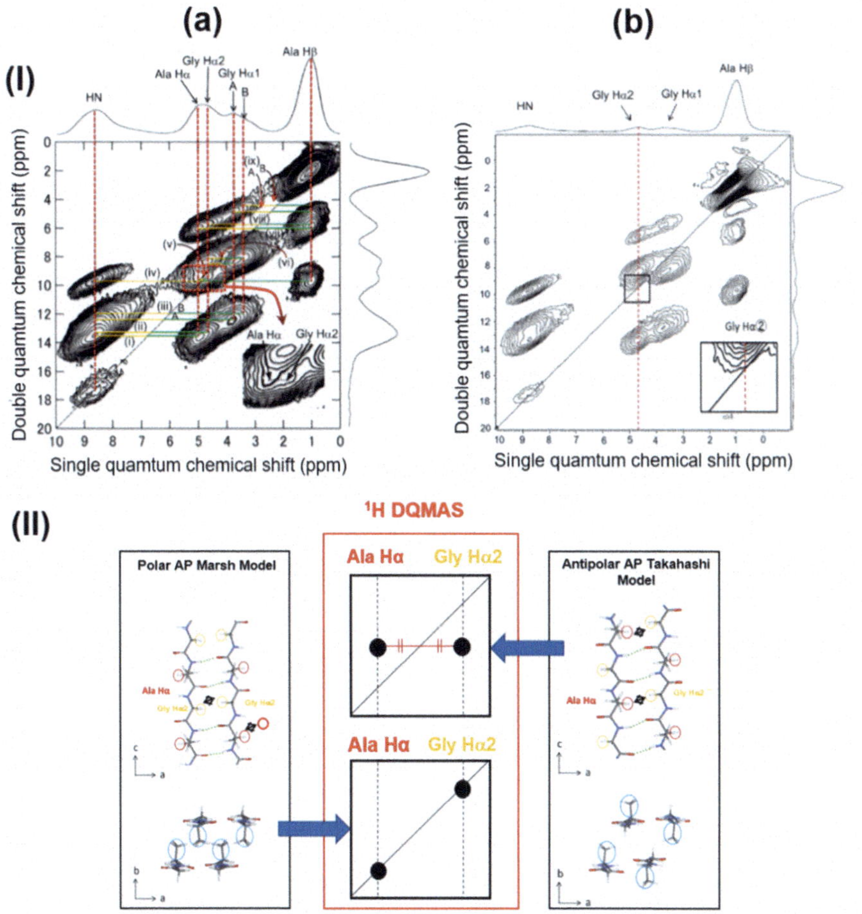

Fig. 1.13 **I** (a) ^{1}H DQMAS spectrum of $(AG)_{15}$ in the Silk II form. (i) AlaHα-AlaHN, (ii) GlyHα2-GlyHN, (iii) GlyHα1-GlyHN, (iv) AlaHβ-AlaHN, (v) AlaHα-GlyHα2, (vi) GlyHα2-GlyHα1, (vii) AlaHα-AlaHβ (viii) GlyHα2-AlaHβ, (ix) GlyHα1-AlaHβ. (b) ^{1}H DQMAS spectrum of deuterium-labeled ($[2\text{-}d_1]AG)_{15}$ in the Silk II form to remove potential overlap in the Hα region (cf. Fig. 1.13I). It is clear that there are no Gly Hα2–Gly Hα2 peaks on the diagonal. Reprinted with permission from Ref. [84]. Modified from Ref. [84]. **II** In the polar Marsh model [24], the Gly $H\alpha_2$ protons in adjacent β-strands are very close to one another (about 2 Å), so a diagonal peak for Gly $H\alpha_2$ should be detected. However, in the antipolar Takahashi model [26], the ^{1}H–^{1}H distances between two Gly $H\alpha_2$ protons in adjacent β-strands are more than 4 Å, so a diagonal peak for Gly $H\alpha_2$ should not be detected. On the other hand, the ^{1}H–^{1}H distances between Gly $H\alpha_2$ and Ala Hα are longer than 4 Å in the polar model, but about 2.4 Å in the antipolar model. Therefore, the cross peak should be observed in the antipolar model, but should not be observed in the polar model. Thus, ^{1}H–^{1}H distance information obtained from the DQMAS ^{1}H NMR method gives a clear conclusion about the packing structure of the AP β-sheet

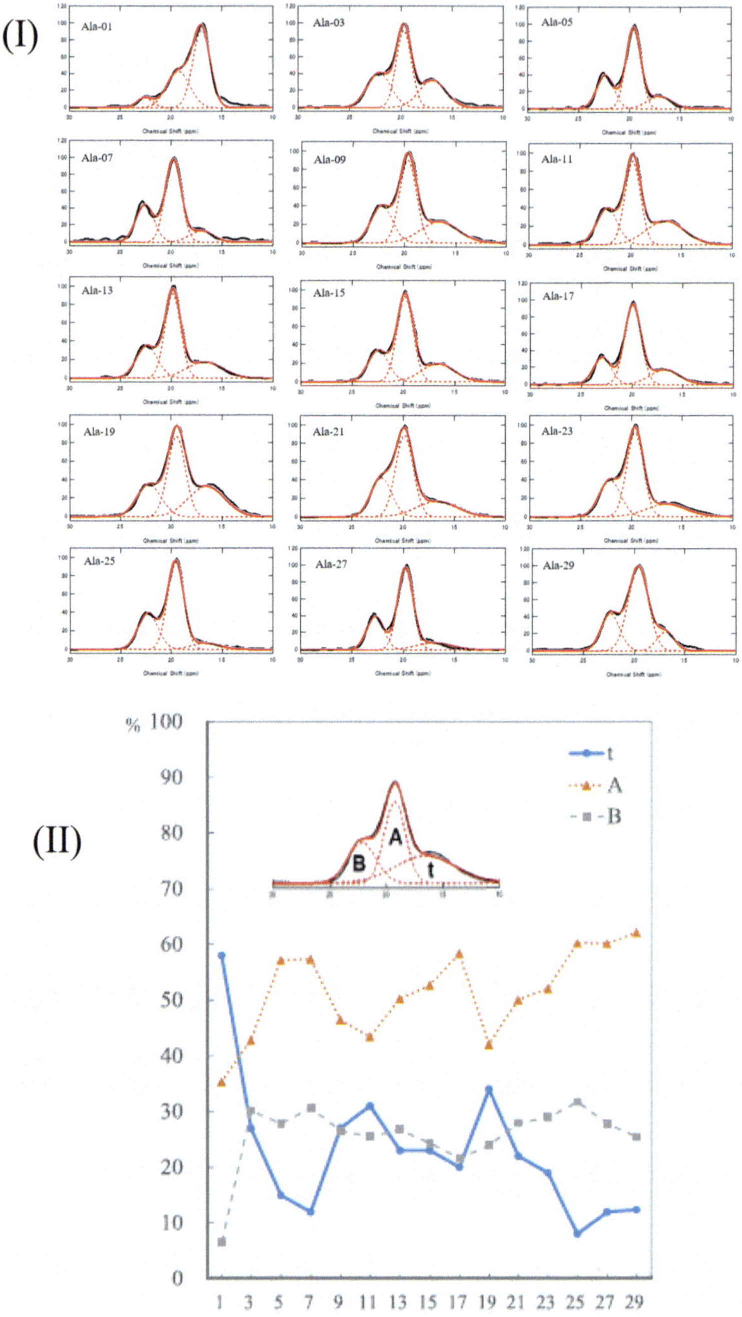

Fig. 1.14 **I** Deconvoluted ^{13}C-labeled Ala Cβ peaks in the ^{13}C CP/MAS NMR spectra of fifteen kinds of [3−^{13}C]Ala-(AG)$_{15}$ peptides with the different ^{13}C labeling positions. **II** Change in the fractions of **t**, **A**, and **B** of ^{13}C-labeled Ala Cβ peaks of 15 [3−^{13}C]Ala-(AG)$_{15}$ peptides plotted as a function of the ^{13}C-labeled site. Reprinted with permission from Ref. [27]

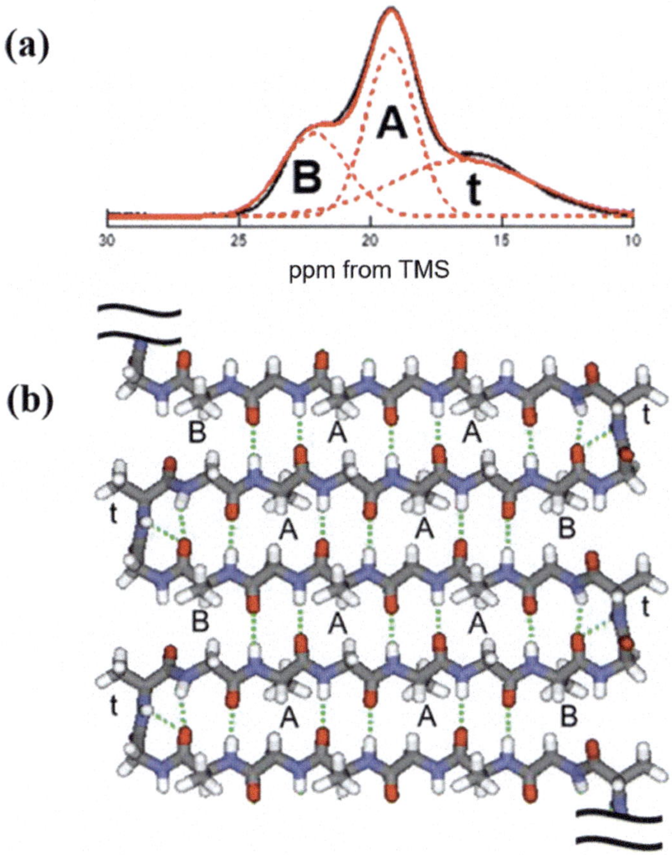

Fig. 1.15 **a** Assignment of Ala Cβ peaks deconvoluted to three peaks, i.e., two β-sheet, **A** and **B**, and one β-turn **t**, of $(AG)_m$ in the Silk II form. **b** Three peaks, **A**, **B**, and **t**, can be assigned to three kinds of Ala Cβ carbons, **A**, **B**, and **t** in the lamellar structure. Reprinted with permission from Ref. [27]

amide groups in the fold [90]. Panitch et al. [91] produced poly(AG) comprising 64 diads $(AG)_{64}$ from recombinant *E. coli* using genetic engineering and recrystallized from dichloroacetic acid solutions in the form of texture-oriented chain-folded lamellae with a lamellar stack periodicity of 32 Å. The proposed structure is shown in Fig. 1.16 [91] which is similar to the lamellar structure proposed by us. However, there are significant differences between the two lamellar models: Their model has polar stacking and γ-turns, but ours has antipolar stacking and β-turns. As mentioned above, our proposed lamellar model was based on the $^{1}H-^{1}H$ distance information on the observed DQMAS ^{1}H NMR spectra and the polar stacking model could not explain the observed distance data. In addition, the structure of SF before spinning, Silk I was type II β-turn, and therefore it seems natural to form the turn part of a lamellar structure of Silk II to be β-turn rather than γ-turn.

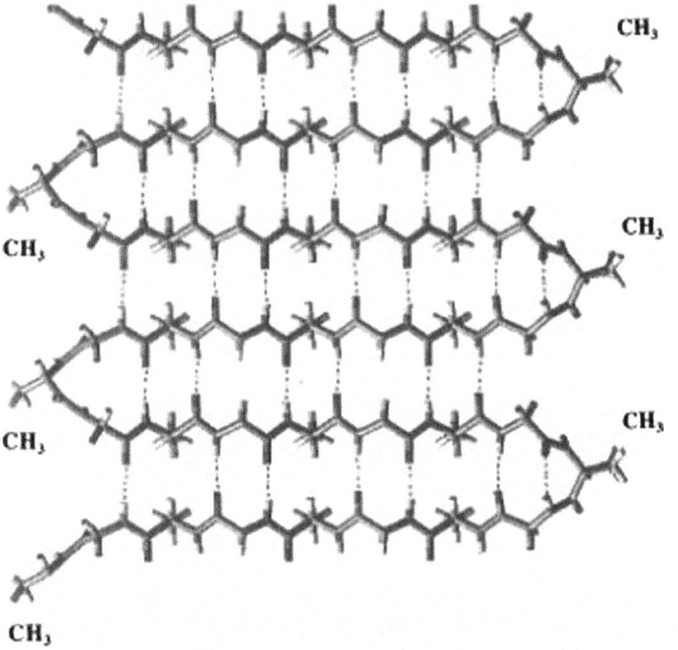

Fig. 1.16 View orthogonal (parallel to *b*-axis) to a single antiparallel β-sheet of poly (AG), folding in phase with the octapeptide periodicity and forming γ-turns. The chains (*c*-direction) run horizontally and the linear hydrogen bonds (shown as dashed lines) are directed vertically (*a*-direction). Reprinted with permission from Ref. [91]

1.8 Conclusion

It is well known that *B. mori* silk fibroin has two crystalline forms, Silk I and Silk II.

Silk I is the solid-state SF structure stored in the middle silk glands after drying without any external forces, and Silk II is the structure after spinning. Here, detailed structures of Silk I and Silk II were determined using mainly solid-state NMR. Silk I was proposed to be a repeated type II β-turn structure with the torsion angles (φ, ψ) = (−62°, 125°) for Ala residue and (φ, ψ) = (77°, 10°) for Gly residue. The intra- and intermolecular hydrogen bonding was formed alternatively along the chain. On the other hand, Silk II was proposed to be a lamella structure, i.e., two kinds of Ala residues in β-sheet structure, plus one distorted β-turn formed by repetitive folding using β-turns every eighth amino acid in an antipolar arrangement.

References

1. Asakura, T., Miller, T.: Biotechnology of Silk. Springer, Berlin (2014)
2. Vepari, C., Kaplan, D.L.: Silk as a biomaterial. Prog. Polym. Sci. **32**, 991–1007 (2007)
3. Koh, L.-D., Cheng, Y., Teng, C.-P., Khin, Y.-W., Loh, X.-J., Tee, S.-Y., Low, M., Ye, E., Yu, H.-D., Zhang, Y.-W., Han, M.-Y.: Structures, mechanical properties and applications of silk fibroin materials. Prog. Polym. Sci. **46**, 86–110 (2015)
4. Thurber, A.E., Omenetto, F.G., Kaplan, D.L.: In vivo bioresponses to silk proteins. Biomater. **71**, 145–157 (2015)
5. Tamara, A.B., DeSimone, E., Scheibel, T.: Biomedical applications of recombinant silk-based materials. Adv. Mater. **30**, 1704636 (2018)
6. Takei, F., Kikuchi, Y., Kikuchi, A., Mizuno, S., Shimura, K.: Further evidence for importance of the subunit combination of silk fibroin in its efficient secretion from the posterior silk gland cells. J. Cell Biol. **105**, 175–180 (1987)
7. Tanaka, K., Mori, K., Mizuno, S.: Immunological identification of the major disulfide-linked light component of silk fibroin. J. Biochem. **114**, 1–4 (1993)
8. Tanaka, K., Kajiyama, N., Ishikura, K., Waga, S., Kikuchi, A., Ohtomo, K., Takagi, T., Mizuno, S.: Determination of the site of disulfide linkage between heavy and light chains of silk fibroin produced by Bombyx mori. Biochim. Biophys. Acta-Protein Struct. Mol. Enzymol. **1432**, 92–103 (1999)
9. Inoue, S., Tanaka, K., Arisaka, F., Kimura, S., Ohtomo, K., Mizuno, S.: Silk fibroin of Bombyx mori is secreted, assembling a high molecular mass elementary unit consisting of H-chain, L-chain, and P25, with a 6:6:1 molar ratio. J. Biol. Chem. **275**, 40517–40528 (2000)
10. Asakura, T., Kuzuhara, A., Tabeta, R., Saito, H.: Conformational characterization of Bombyx mori silk fibroin in the solid state by high-frequency 13C cross polarization-magic angle spinning NMR, X-ray diffraction, and infrared spectroscopy. Macromol. **18**, 1841–1845 (1985)
11. Zhou, C.-Z., Confalonieri, F., Medina, N., Zivanovic, Y., Esnault, C., Yang, T., Jacquet, M., Janin, J., Duguet, M., Perasso, R., Li, Z.-G.: Fine organization of Bombyx mori fibroin heavy chain gene. Nucleic Acids Res. **28**, 2413–2419 (2000)
12. Zhou, C.-Z., Confalonieri, F., Jacquet, M., Perasso, R., Li, Z.-G., Janin, J.: Silk fibroin: Structural implications of a remarkable amino acid sequence. Proteins **44**, 119–122 (2001)
13. Asakura, T., Ogawa, T., Naito, A., Williamson, M.P.: Chain-folded lamellar structure and dynamics of the crystalline fraction of Bombyx mori silk fibroin and of (Ala-Gly-Ser-Gly-Ala-Gly)n model peptides. Int. J. Biol. Macromol. **164**, 3974–3983 (2020)
14. Lotz, B., Cesari, F.C.: The chemical structure and the crystalline structures of Bombyx mori silk fibroin. Biochimie **61**, 205–214 (1979)
15. Fraser, B., MacRae, T.P.: Conformations of Fibrous Proteins and Related Synthetic Polypeptides. Academic Press, New York (1973)
16. Asakura, T., Ashida, J., Yamane, T., Kameda, T., Nakazawa, Y., Ohgo, K., Komatsu, K.: A repeated β-turn structure in Poly(Ala-Gly) as a model for silk I of Bombyx mori silk fibroin studied with two-dimensional spin-diffusion NMR under off magic angle spinning and rotational echo double resonance. J. Mol. Biol. **306**, 291–305 (2001)
17. Asakura, T., Ohgo, K., Komatsu, K., Kanenari, M., Okuyama, K.: Refinement of repeated β-turn structure for silk I conformation of Bombyx mori silk fibroin using 13C solid-state NMR and X-ray diffraction methods. Macromol. **38**, 7397–7403 (2005)
18. Hu, X., Kaplan, D., Cebe, P.: Dynamic protein−water relationships during β-sheet formation. Macromol. **41**, 3939–3948 (2008)
19. Hu, X., Shmelev, K., Sun, L., Gil, E.-S., Park, S.-H., Cebe, P., Kaplan, D.L.: Regulation of silk material structure by temperature-controlled water vapor annealing. Biomacromol **12**, 1686–1696 (2011)
20. Dubey, P., Murab, S., Karmakar, S., Chowdhury, P.K., Ghosh, S.: Modulation of self-assembly process of fibroin: an insight for regulating the conformation of silk biomaterials. Biomacromol **16**, 3936–3944 (2015)

21. Kaewprasit, K., Kobayashi, T., Damrongsakkul, S.: Thai silk fibroin gelation process enhancing by monohydric and polyhydric alcohols. Int. J. Biol. Macromol. **118**, 1726–1735 (2018)
22. Yang, Y., Greco, G., Maniglio, D., Mazzolai, B., Migliaresi, C., Pugno, N., Motta, A.: Spider (Linothele megatheloides) and silkworm (Bombyx mori) silks: comparative physical and biological evaluation. Mater. Sci. Eng. C **107**, 110197 (2020)
23. Park, Y., Jung, Y., Li, T.-D., Lao, J., Tu, R.S., Chen, X.: β-sheet nanocrystals dictate water responsiveness of Bombyx Mori silk. Macromol. Rapid Commun. **41**, 1900612 (2020)
24. Marsh, R.E., Corey, R.B., Pauling, L.: An investigation of the structure of silk fibroin. Biochim. Biophys. Acta **16**, 1–34 (1955)
25. Fossey, S., Némethy, G., Gibson, K.D., Scheraga, H.A.: Conformational energy studies of β-sheets of model silk fibroin peptides. I. Sheets of Poly(Ala-Gly) Chains. Biopolym. **31**, 1529−1541 (1991)
26. Takahashi, Y., Gehoh, M., Yuzuriha, K.: Structure refinement and diffuse streak scattering of silk (Bombyx Mori). Int. J. Biol. Macromol. **24**, 127–138 (1999)
27. Asakura, T., Aoki, A., Komatsu, K., Ito, C., Suzuki, I., Naito, A., Kaji, H.: Lamellar structure in alanine-glycine copolypeptides studied by solid-state NMR spectroscopy: a model for the crystalline domain of Bombyx Mori silk fibroin in silk II form. Biomacromol **21**, 3102–3111 (2020)
28. Asakura, T., Suzuki, H., Watanabe, Y.: Conformational characterization of silk fibroin in intact Bombyx mori and Pilosamia cynthia ricini silkworms by 13C NMR spectroscopy. Macromol. **16**, 1024–1026 (1983)
29. Asakura, T.: Nuclear magnetic resonance studies on the biosynthesis of the silk fibroins in silkworms. JOEL news **23A**, 2–6 (1987)
30. Sezutsu, H., Yukuhiro, K.: The complete nucleotide sequence of the Eri-silkworm(Samia cynthia ricini) fibroin gene. J. Insect Biotechnol. Sericol. **83**, 59–70 (2014)
31. Asakura, T., Murakami, T.: NMR of silk fibroin. 4. Temperature-and urea-induced helix-coil transitions of the-(Ala)n-sequence in Philosamia cynthia ricini silk fibroin protein monitored by 13C NMR spectroscopy. Macromol. **18**, 2614–2619 (1985)
32. Asakura, T., Kashiba, H.,Yoshimizu, H.: NMR of silk fibroin. 8. 13C NMR analysis of the conformation and the conformational transition of Philosamia cynthia ricini silk fibroin protein on the basis of Bixon-Scheraga-Lifson theory. Macromol. **21**, 644–648 (1988)
33. Asakura, T., Yoshimizu, H., Yoshizawa, F.: NMR of silk fibroin. 9. Sequence and conformation analyses of the silk fibroins from Bombyx mori and Philosamia cynthia ricini by 15N NMR spectroscopy. Macromol. **21**, 2038–2041 (1988)
34. Nakazawa, Y., Asakura, T.: Heterogeneous exchange behavior of Samia cynthia ricini silk fibroin during helix-coil transition studied with 13C NMR. Febs Lett. **529**, 188–192 (2002)
35. Nakazawa, Y., Asakura, T.: Structure determination of a peptide model of the repeated helical domain in Samia cynthia ricini silk fibroin before spinning by a combination of advanced solid-state NMR methods. J. Am. Chem. Soc. **125**, 7230–7237 (2003)
36. Suzuki, Y., Kawanishi, S., Yamazaki, T., Aoki, A., Saito, H., Asakura, T.: Structural determination of the tandem repeat motif in Samia cynthia ricini liquid silk by solution NMR. Macromol. **48**, 6574–6579 (2015)
37. Okuyama, K., Takanashi, K., Nakajima, Y., Hasegawa, Y., Hirabayashi, K., Nishi, N.: Analysis of Silk I structure by X-ray and electron diffraction methods. J. Sericultural Sci. Japan **57**, 23–30 (1988)
38. Anderson, J.P.: Morphology and crystal structure of a recombinant silk-like molecule, SLP4. Biopolym. **45**, 307–321 (1998)
39. Asakura, T., Yamane, T., Nakazawa, Y., Kameda, T., Ando, K.: Structure of Bombyx mori silk fibroin before spinning in solid state studied with wide angle x-ray scattering and 13C cross-polarization/magic angle spinning NMR. Biopolym. **58**, 521–525 (2001)
40. Monti, P., Taddei, P., Freddi, G., Ohgo, K., Asakura, T.: Vibrational 13C-cross-polarization/magic angle spinning NMR spectroscopic and thermal characterization of poly(alanine-glycine) as model for silk I Bombyx mori fibroin. Biopolym. **72**, 329–338 (2003)

41. Taddei, P., Asakura, T., Yao, J., Monti, P.: Raman study of poly(alanine-glycine)-based peptides containing tyrosine, valine, and serine as model for the semicrystalline domains of Bombyx mori silk fibroin. Biopolym. **75**, 314–324 (2004)
42. Saito, H., Tabeta, R., Asakura, T., Iwanaga, Y., Shoji, A., Ozaki, T., Ando, I.: High-resolution 13C NMR study of silk fibroin in the solid state by the cross-polarization-magic angle spinning method. Conformational characterization of silk I and silk II type forms of Bombyx mori fibroin by the conformation-dependent. Macromol. **17**, 1405–1412 (1984)
43. Ishida, M., Asakura, T., Yokoi, M., Saito, H.: Solvent-and mechanical-treatment-induced conformational transition of silk fibroins studied by high-resolution solid-state 13C NMR spectroscopy. Macromol. **23**, 88–94 (1990)
44. Robyr, P., Meier, B.H., Ernst, R.R.: Radio-frequency-driven nuclear spin diffusion in solids. Chem. Phys. Lett. **162**, 417–423 (1990)
45. Robyr, P., Tomaselli, M., Straka, J., Grob-Pisano, C., Suter, U.W., Meier, B.H., Ernst, R.R.: RF-driven and proton-driven NMR polarization transfer for investigating local order. Mol. Phys. **84**, 995–1020 (1995)
46. Kümmerlen, J., van Beek, J.D., Vollrath, F., Meier, B.H.: Local structure in spider dragline silk investigated by two-dimensional spin-diffusion nuclear magnetic resonance. Macromol. **29**, 2920–2928 (1996)
47. Levitt, M.H., Raleigh, D.P., Creuzet, F., Griffin, R.G.: Theory and simulations of homonuclear spin pair systems in rotating solids. J. Chem. Phys. **92**, 6347–6364 (1990)
48. van Beek, J.D., Beaulieu, L., Schäfer, H., Demura, M., Asakura, T., Meier, B.H.: Solid-state NMR determination of the secondary structure of Samia cynthia ricini silk. Nature **405**, 1077–1079 (2000)
49. Ashida, J., Ohgo, K., Asakura, T.: Determination of the torsion angles of alanine and glycine residues of Bombyx Mori silk fibroin and the model peptides in the silk I and silk II forms using 2D spin diffusion solid-state NMR under Off magic angle spinning. J. Phys. Chem. B **106**, 9434–9439 (2002)
50. Asakura, T., Ashida, J., Yamane, T.: Structure of Bombyx mori silk Fibroin before spinning in silkworm. In: NMR Spectroscopy of Polymers in Solution and in the Solid State, pp. 71–82 (2003)
51. Asakura, T., Suzuki, Y., Nakazawa, Y., Yazawa, K., Holland, G.P., Yarger, J.L.: Silk structure studied with nuclear magnetic resonance. Prog. Nucl. Magn. Reson. Spectrosc. **69**, 23–68 (2013)
52. Asakura, T., Suzuki, Y., Nakazawa, Y., Holland, G.P., Yarger, J.L.: Elucidating silk structure using solid-state NMR. Soft. Matter. **9**, 11440–11450 (2013)
53. Asakura, T., Okushita, K., Williamson, M.P.: Analysis of the structure of Bombyx mori silk fibroin by NMR. Macromol. **48**, 2345–2357 (2015)
54. Gullion, T., Schaefer, J.: Detection of weak heteronuclear dipolar coupling by rotational-echo double-resonance nuclear magnetic resonance. In: Warren, W.S. (ed.) Advances in Magnetic and Optical Resonance, pp. 57–83. Academic Press, 13 (1989)
55. Gullion, T., Baker, D.B., Conradi, M.S.: New, compensated Carr-Purcell sequences. J. Magn. Reson. **89**, 479–484 (1990)
56. Gullion, T., Schaefer, J.: Elimination of resonance offset effects in rotational-echo, double-resonance NMR. J. Magn. Reson. **92**, 439–442 (1991)
57. Asakura, T., Aoki, A., Demura, M., Joers, J.M., Rosanske, R.C., Gullion, T.: Structure of bombyx mori silk fibroin studied by REDOR NMR spectroscopy. Polym. J. **26**, 1405–1408 (1994)
58. Gullion, T., Kishore, R., Asakura, T.: Determining dihedral angles and local structure in silk peptide by 13C 2H REDOR. J. Am. Chem. Soc. **125**, 7510–7511 (2003)
59. Gullion, T., Yamauchi, K., Okonogi, M., Asakura, T.: 13C–17O REAPDOR NMR as a tool for determining secondary structure in polyamides. Macromol. **40**, 1363–1365 (2007)
60. Lotz, B., Keith, H.D.: Crystal structure of poly(l-Ala-Gly)II. J. Mol. Biol. **61**, 201–215 (1971)
61. Okuyama, K., Somashekar, R., Noguchi, K., Ichimura, S.: Refined molecular and crystal structure of silk I based on Ala-Gly and (Ala–Gly)2-Ser-Gly peptide sequence. Biopolym. **59**, 310–319 (2001)

62. Asakura, T.: NMR of silk fibroin, 6. Structure of bombyx mori silk fibroin in aqueous solution. Makromol. Chemie-Rapid Commun. **7**, 755–759 (1986)
63. Suzuki, Y., Yamazaki, T., Aoki, A., Shindo, H., Asakura, T.: NMR study of the structures of repeated sequences, GAGXGA (X = S, Y, V) in Bombyx mori liquid silk. Biomacromol. **15**, 104–112 (2014)
64. Yamauchi, K., Yamasaki, S., Takahashi, R., Asakura, T.: Microscopic structural analysis of fractured silk fibers from Bombyx mori and Samia cynthia ricini using 13C CP/MAS NMR with a 1mm microcoil MAS NMR probehead. Solid State Nucl. Magn. Reson. **38**, 27–30 (2010)
65. Asakura, T., Suzuki, Y., Yazawa, K., Aoki, A., Nishiyama, Y., Nishimura, K., Suzuki, F., Kaji, H.: Determination of accurate 1H positions of (Ala-Gly)n as a sequential peptide model of bombyx mori silk fibroin before spinning (Silk I). Macromol. **46**, 8046–8050 (2013)
66. Schnell, I., Brown, S.P., Low, H.Y., Ishida, H., Spiess, H.W.: An investigation of hydrogen bonding in benzoxazine dimers by fast magic-angle spinning and double-quantum 1H NMR spectroscopy. J. Am. Chem. Soc. **120**, 11784–11795 (1998)
67. Brown, S.P.: Probing proton–proton proximities in the solid state. Prog. Nucl. Magn. Reson. Spectrosc. **50**, 199–251 (2007)
68. Brinkmann, A., Litvinov, V.M., Kentgens, A.P.M.: Environmentally friendly flame retardants. A detailed solid-state NMR study of melamine orthophosphate. Magn. Reson. Chem. **45**, S231–S246 (2007)
69. Bradley, J.P., Tripon, C., Filip, C., Brown, S.P.: Determining relative proton-proton proximities from the build-up of two-dimensional correlation peaks in 1H double-quantum MAS NMR: insight from multi-spin density-matrix simulations. Phys. Chem. Chem. Phys. **11**, 6941–6952 (2009)
70. Harris, R.K., Hodgkinson, P., Zorin, V., Dumez, J.-N., Elena-Herrmann, B., Emsley, L., Salager, E., Stein, R.S.: Computation and NMR crystallography of terbutaline sulfate. Magn. Reson. Chem. **48**, S103–S112 (2010)
71. Webber, A.L., Elena, B., Griffin, J.M., Yates, J.R., Pham, T.N., Mauri, F., Pickard, C.J., Gil, A.M., Stein, R., Lesage, A., Emsley, L., Brown, S.P.: Complete 1H resonance assignment of β-maltose from 1H–1H DQ-SQ CRAMPS and 1H (DQ-DUMBO)–13CSQ refocused INEPT 2D solid-state NMR spectra and first principles GIPAW calculations. Phys. Chem. Chem. Phys. **12**, 6970–6983 (2010)
72. Brown, S.P.: Applications of high-resolution 1H solid-state NMR. Solid State Nucl. Magn. Reson. **41**, 1–27 (2012)
73. Yazawa, K., Suzuki, F., Nishiyama, Y., Ohata, T., Aoki, A., Nishimura, K., Kaji, H., Shimizu, T., Asakura, T.: Determination of accurate 1H positions of an alanine tripeptide with anti-parallel and parallel β-sheet structures by high resolution 1H solid state NMR and GIPAW chemical shift calculation. Chem. Commun. **48**, 11199–11201 (2012)
74. Asakura, T., Yazawa, K., Horiguchi, K., Suzuki, F., Nishiyama, Y., Nishimura, K., Kaji, H.: Difference in the structures of alanine tri- and tetra-peptides with antiparallel β-sheet assessed by X-ray diffraction, solid-state NMR and chemical shift calculations by GIPAW. Biopolym. **101**, 13–20 (2014)
75. Pickard, C.J., Mauri, F.: All-electron magnetic response with pseudopotentials: NMR chemical shifts. Phys. Rev. B **63**, 245101 (2001)
76. Asakura, T., Demura, M., Date, T., Miyashita, N., Ogawa, K., Williamson, M.P.: NMR study of silk I structure of Bombyx mori silk fibroin with 15N- and 13C-NMR chemical shift contour plots. Biopolym. **41**, 193–203 (1997)
77. Asakura, T., Yao, J., Yamane, T., Umemura, K., Ulrich, A.S.: Heterogeneous structure of silk fibers from bombyx mori resolved by 13C solid-state NMR spectroscopy. J. Am. Chem. Soc. **124**, 8794–8795 (2002)
78. Asakura, T., Yao, J.: 13C CP/MAS NMR study on structural heterogeneity in bombyx mori silk fiber and their generation by stretching. Protein Sci. **11**, 2706–2713 (2009)
79. Strydom, D.J., Haylett, T., Stead, R.H.: The amino-terminal sequence of silk fibroin peptide Cp-a reinvestigation. Biochem. Biophys. Res. Commun. **79**, 932–938 (1977)

80. Saitô, H., Ando, I.: High-resolution solid-state NMR studies of synthetic and biological macromolecules. Ann. Rep. NMR Spectrosc. **21**, 209–290 (1989)
81. Spera, S., Bax, A.: Empirical correlation between protein backbone conformation and Cα and Cβ 13C nuclear magnetic resonance chemical shifts. J. Am. Chem. Soc. **113**, 5490–5492 (1991)
82. Iwadate, M., Asakura, T., Williamson, M.P.: Cα and Cβ carbon-13 chemical shifts in proteins from an empirical database. J. Biomol. NMR **13**, 199–211 (1999)
83. Wishart, D.S.: Interpreting protein chemical shift data. Prog. Nucl. Magn. Reson. Spectrosc. **58**, 62–87 (2011)
84. Asakura, T., Ohata, T., Kametani, S., Okushita, K., Yazawa, K., Nishiyama, Y., Nishimura, K., Aoki, A., Suzuki, F., Kaji, H., Ulrich, A.S., Williamson, M.P.: Intermolecular packing in B. Mori silk fibroin: multinuclear NMR study of the model peptide (Ala-Gly) 15 defines a heterogeneous antiparallel antipola mode of assembly in the silk II form. Macromol. **48**, 28–36 (2015)
85. Asakura, T., Nakazawa, Y., Ohnishi, E., Moro, F.: Evidence from 13C solid-state NMR spectroscopy for a lamella structure in an alanine-glycine copolypeptide: a model for the crystalline domain of bombyx mori silk fiber. Protein Sci. **14**, 2654–2657 (2005)
86. Asakura, T., Sato, H., Moro, F., Nakazawa, Y., Aoki, A.: Lamellar structure in Poly(Ala-Gly) determined by solid-state NMR and statistical mechanical calculations. J. Am. Chem. Soc. **129**, 5703–5709 (2007)
87. McGrath, K.P., Fournier, M.J., Mason, T.L., Tirrell, D.A.: Genetically directed syntheses of new polymeric materials. Expression of artificial genes encoding proteins with repeating-(AlaGly)3ProGluGly-elements. J. Am. Chem. Soc. **114**, 727–733 (1992)
88. Parkhe, A., Fournier, M.J., Mason, T.L., Tirrell, D.A.: Determination of the chain-folding pattern in the crystalline domains of the repetitive polypeptide {(AlaGly)3GluGly(GlyAla)3GluGly}10 by FTIR studies of its blends with a carbon-13 enriched analog. Macromol. **26**, 6691–6693 (1993)
89. Krejchi, M.T., Atkins, E.D.T., Waddon, A.J., Fournier, M.J., Mason, T.L., Tirrell, D.A.: Chemical sequence control of β-sheet assembly in macromolecular crystals of periodic polypeptides. Science **265**, 1427–1432 (1994)
90. Keller, A.: A note on single crystals in polymers: evidence for a folded chain configuration. Philos. Mag.: J. Theor. Exp. Appl. Phys. **2**, 1171–1175 (1957)
91. Panitch, A., Matsuki, K., Cantor, E.J., Cooper, S.J., Atkins, E.D.T., Fournier, M.J., Mason, T.L., Tirrell, D.A.: Poly(LAlanylglycine): multigram-scale biosynthesis, crystallization, and structural analysis of chain-folded lamellae. Macromol. **30**, 42–49 (1997)

Chapter 2
NMR Studies on Polymer Materials

Atsushi Asano

Abstract This chapter presents the NMR investigations of synthetic polymer materials, including polymer blends and alloys, polymer nanocomposites, rubbers and elastomers, and other semicrystalline polymers: polystyrene/poly(vinyl methyl ether), poly(methacrylic acid)/poly(vinyl acetate), and poly-3-hexylthiophene/phenyl-C_{61}-butyric acid methyl ester blends, polyketone/polyamide alloy, a healable polymer blend of poly(octamethylene thiourea)/poly(ether thiourea), poly(vinyl alcohol)/clay, polystyrene-poly(ethylene oxide) block copolymer/clay, nylon-6/clay, and poly(vinyl isobutyl ether)/poly(ε-L-lysine)/clay nanocomposites, and acrylonitrile butadiene rubbers are introduced. High-resolution solid-state ^{13}C NMR spectra are sensitive to the existence of interpolymer interaction, particularly hydrogen bonds making chemical shift value down-field shift. Hetero-correlation NMR spectroscopy shows proof of the existence of the interaction between component polymer and clay. Analysis of the ^{1}H spin–lattice relaxation (T_1^{H}) behavior influenced by the spin diffusion provides the degree of homogeneity, domain size of component polymers, and phase separation process for polymer blends/alloys. The paramagnetic effect on T_1^{H} gives information on the stability of an organic modifier on clay surfaces and of morphology of α and γ crystalline phases of nylon-6 in nylon-6/montmorillonite nanocomposites. Finally, the degradation of rubbers is discussed based on ^{1}H spin–spin relaxation (T_2^{H}) analysis.

Keywords ^{1}H relaxation · Spin diffusion · Polymer blends · Nanocomposites · Rubber · Paramagnetic effect

---

A. Asano (✉)
Department of Applied Chemistry, National Defense Academy, Hashirimizu 1-10-20, Yokosuka 239-8686, Kanagawa, Japan
e-mail: asanoa@nda.ac.jp

The NMR Society of Japan (ed.), *Experimental Approaches of NMR Spectroscopy II*,
https://doi.org/10.1007/978-981-97-6830-1_2

2.1 Introduction

Characterization of morphology, molecular motion, or intermolecular interaction of polymer blends and alloys, nanocomposites, rubbers, and elastomers, and some semicrystalline polymers using the solid-state ^{13}C CPMAS (cross-polarization and magic-angle spinning) NMR is mainly described. The words "blend" and "alloy" are frequently used as the same meaning of "mixed component polymers." On the other hand, polymer "alloy" is used to blend materials having superior mechanical properties after mixing different component polymers in many cases. Hence, it is discriminated here that polymer blends indicate a simple mixing of polymers regardless of having good physical properties and polymer alloys appear some improvement in mechanical/physical properties. The mechanical properties of polymer blends and alloys are significantly influenced by the degree of mixing as well as the characteristic property of a component polymer: this issue is described in Sect. 2.2. Similarly, the dispersion of fillers and the interaction between polymers and fillers for polymer nanocomposites are very important key points to show a great improvement such as gas barrier, impact strength, flame retardants, and thermal properties [1] (Sect. 2.3).

The density and the strength of the cross-links in rubbers and elastomers affect the molecular motion: for example, the existence of the cross-links resists being deformed. The rubbers without the cross-links are deformed easily by the fast MAS, because a strong centrifugal force is produced by the MAS. In addition, the cross-linked rubbers show the MAS rate dependence of ^{1}H spin–lattice relaxation time ($T_1{}^{H}$), indicating that the molecular motion is changed even if the rubbers are not deformed [2]. Thus, it is necessary to be concerned with the effects of MAS on the molecular motion of rubbers and elastomers. On the one hand, many rubbers have a double bond in their unit, and it is easily oxidized and thermally damaged to produce a radical, resulting in the degradation/aging of rubbers [3]. The degradation causes the molecular motion change, and it results in the shortening of ^{1}H spin–spin relaxation time ($T_2{}^{H}$): the investigation of aging is briefly described in Sect. 2.4.

2.2 Polymer Blends and Alloys

2.2.1 Miscibility and Mobile Heterogeneity

Different domain sizes of polymer blends are easily characterized distinctly by $T_1{}^{H}$ and $T_{1\rho}{}^{H}$. Sensitive domain size for $T_{1\rho}{}^{H}$ is one order smaller than $T_1{}^{H}$ [4]. The fast ^{1}H spin diffusion averages various spin temperatures of almost all protons between component polymers in a homogeneous system. Thus, each ^{1}H relaxation time obtained from the well-resolved ^{13}C peaks of the respective component polymer has an identical value. On the other hand, in a heterogeneous blend, the individual ^{1}H relaxation times of component polymers do not fully averaged each other because the large domain size does not allow to be averaged entirely each proton spin temperature.

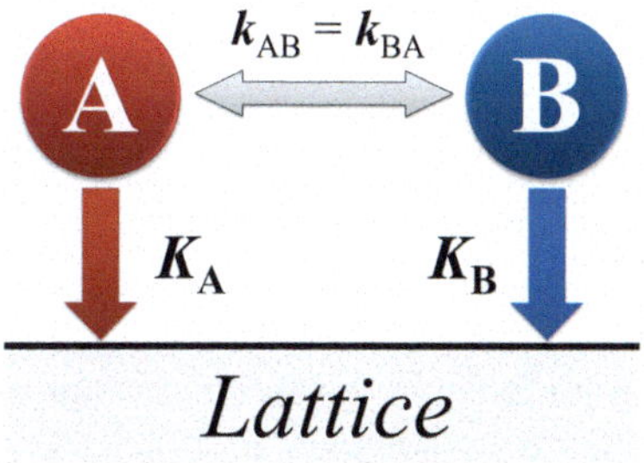

Fig. 2.1 Schematic illustration of the two-spin system. The *arrow* toward the lattice represents the ¹H spin–lattice relaxation rate: $K_A = 1/T_{1A}$ and $K_B = 1/T_{1B}$. The ¹H spin diffusion rate, k_{AB} equals k_{BA}

The simple two-spin system is helpful to evaluate the homogeneity of polymer blends and alloys [4–6]. Figure 2.1 shows the schematic illustration of the two-spin system of polymer blends consisting of polymers A and B. The ¹H spin diffusion rate, k, T_1 rates ($1\ /\ T_1$) of A-spin K_A, and K_B for B-spin are defined as shown in Fig. 2.1. The spin diffusion rates k_{ij} and k_{ji} between spin species i and j are assumed to be the same value k with each other. The proton fraction of the magnetization of polymer A, M_A, is f_A ($= 1—f_B$). The relaxation behavior connected by the ¹H dipolar coupling between M_A and M_B is expressed by the simultaneous differential equations depending on time, t, as follows:

$$\frac{dM_A(t)}{dt} = -(K_A + f_B k) \cdot M_A(t) + f_A k \cdot M_B(t)$$
$$\frac{dM_B(t)}{dt} = -(K_B + f_A k) \cdot M_B(t) + f_B k \cdot M_A(t) \tag{2.1}$$

These simultaneous differential equations are easily solved with the initial conditions of $M_A(0) = M^0{}_A$ and $M_B(0) = M^0{}_B$. The formulae below are the solution.

$$M_A(t) = a_+ e^{r_+ t} + a_- e^{r_- t}$$
$$M_B(t) = b_+ e^{r_+ t} + b_- e^{r_- t} \tag{2.2}$$

where

$$a_\pm = \frac{1}{2} M_A^0 \{1 \pm R^{-1}(K_B + k - K_A)\},$$

$$b_\pm = \frac{1}{2} M_B^0 \{1 \pm R^{-1}(K_A + k - K_B)\},$$

$$r_\pm = \frac{1}{2}\{-(K_A + K_B + k) \pm R\}, \text{ and}$$

$$R = +\sqrt{[(K_A - K_B) + (f_B - f_A)k]^2 + 4f_Af_Bk^2}.$$

In the case that k is much slower than K_A and K_B, Eq. (2.2) becomes $M_A(t) = M_A^0 \cdot e^{-K_At}$ and $M_B(t) = M_B^0 \cdot e^{-K_Bt}$, respectively. This indicates that the magnetizations M_A and M_B have the original relaxation rate K_A and K_B, respectively, even though the mixed sample. Namely, it shows that the blend is a completely immiscible condition on a molecular level. On the other hand, when the ^1H spin diffusion among the component polymers is sufficiently fast: k is much faster than K_A and K_B, and the time-dependent magnetization $M_A(t)$ will be simplified to $M_A(t) = a_+e^{r_+t}$ with the parameters $a_+ = M_A{}^0$, $a_- = 0$, and $r_+ = -(f_AK_A + f_BK_B)$. The similar equation for magnetization $M_B(t)$ is also derived. Thus, the average relaxation rate K_{ave} ($1 / T_{1ave}$) for both magnetizations M_A and M_B is provided as

$$K_{ave} = f_AK_A + f_BK_B = f_AK_A + (1 - f_A)K_B \tag{2.3}$$

The average relaxation time T_{1ave} estimated from Eq. (2.3) would be measured in the system that the ^1H spin diffusion fully and homogeneously occurs. The identical and single-exponential decay with K_{ave} for both polymers A and B is realized when the value of k is 10 – 100 times faster than the fastest K among the components [4, 7]. In a moderate rate case such as k is comparable to K, the observed T_1 relaxations are followed by the curvatures expressed by Eq. (2.2): non-single-exponential decay.

2.2.1.1 PS/PVME Blends

Polystyrene/poly(vinyl methyl ether) (PS/PVME) blends have well been known as a miscible polymer blends. In the early 1970s, DSC and dielectric relaxation measurements had clarified that the PS/PVME blends show a single glass-transition and consist of a single phase [8–10]. The glass-transition temperature, T_g, of PVME, is about 253 K while that of PS is about 373 K. This large contrast of T_g induces a difference in dynamics between PS and PVEM even in a homogeneous state: Kwei et al. called the phenomenon micro-heterogeneity on a segmental scale [9]. The molecular motion of component polymers in the several tens kilohertz range is influenced by blending. The change of molecular motion appears on the NMR spectra and the relaxation curves [11–13].

Figure 2.2 shows the ^{13}C CPMAS NMR spectra of the PS, PVME, and PS/PVME = 5/5 blend at (a) 311 K and (b) 228 K, respectively [7]. The methine (CH) and methoxy (OCH_3) signals of PVME become broad after blending with PS. These linewidth changes indicate the reduction in the molecular motion owing to interacting with rigid PS. The gain of linewidth is due to the reduction in efficiency of ^1H dipolar decoupling (DD). Namely, the molecular motion of pure PVME, which is much faster than the frequency of ^1H DD, was decreased down to the comparable frequency level of ^1H DD (about 55 kHz in this experiment) by blending with PS [11, 12]. This observation is related to the experimental temperature; the T_g of PS/PVME

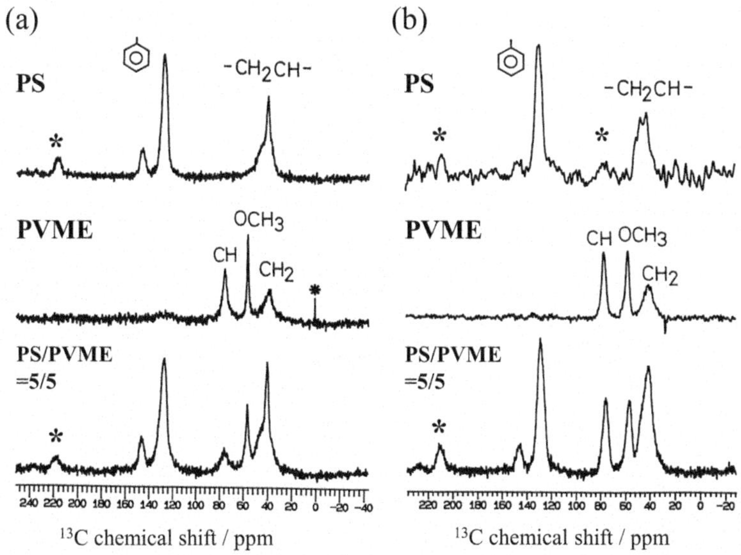

Fig. 2.2 CPMAS ^{13}C NMR spectra for pure PS, pure PVME, and the PS/PVME = 5/5 blend at **a** 311 K and **b** 228 K. An *asterisk symbol* in PVME of **a** comes from a silicon rubber seal. The *other asterisks* (*) are the spinning side band. This figure is redrawn from the original source of Asano et al. [7]

= 5/5 blend (about 259 K) is still much lower than the experimental temperature. In fact, such a line-broadening did not appear at 228 K because the molecular motion is frozen and the ^{1}H DD works effectively. The motion around 50 kHz governs and influences the $T_{1\rho}{}^{H}$ process severely; hence, the observed $T_{1\rho}{}^{H}$ decays are modulated and show non-single-exponential even though the blends are homogeneous [4, 9].

Figure 2.3 shows $T_{1\rho}{}^{H}$ decays obtained from each peak of PS and PVME in the PS/PVME = 5/5 blend [7]. The different and non-single-exponential $T_{1\rho}{}^{H}$ decays for PVME are observed from the two signals, CH and OCH_3 carbons. Usually, the $T_{1\rho}{}^{H}$ decays in solid homopolymers are consistent with each other because of the fast ^{1}H spin diffusion. However, the ^{1}H spin diffusion even in pure PVME was so slow that the $T_{1\rho}{}^{H}$ decays obtained from each signal were not averaged out. This phenomenon is due to the very fast molecular motion of PVME chains at 311 K. Therefore, the observed non-single-exponential $T_{1\rho}{}^{H}$ decays in Fig. 2.3 resulted in a three-spin system consisting of PS (A), side-chain (B), and main-chain (C) of PVME with insufficient ^{1}H spin diffusion rates, not in phase separation: the decays in a case of phase separation are described in the next Sect. 2.2. To estimate the observed $T_{1\rho}{}^{H}$ decays, the simultaneous differential Eq. (2.1) is expanded to adapt the three-spin system as follows:

$$\frac{dM_{\mathrm{A}}(t)}{dt} = -(K_{\mathrm{A}} + f_{\mathrm{B}}k_{\mathrm{AB}} + f_{\mathrm{C}}k_{\mathrm{CA}}) \cdot M_{\mathrm{A}}(t) + f_{\mathrm{A}}k_{\mathrm{AB}} \cdot M_{\mathrm{B}}(t) + f_{\mathrm{A}}k_{\mathrm{CA}} \cdot M_{\mathrm{C}}(t)$$

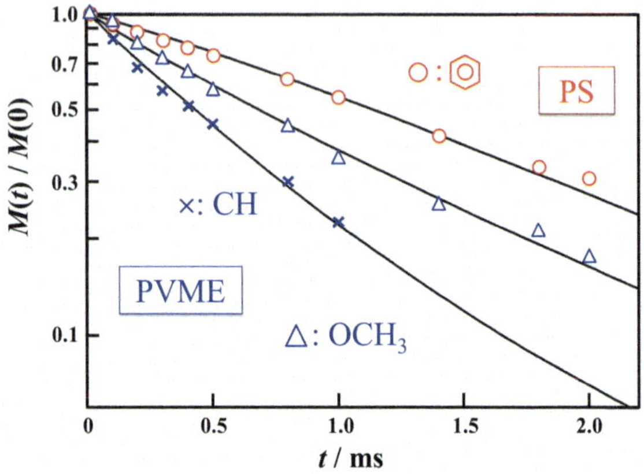

Fig. 2.3 $T_{1\rho}{}^{H}$ decays for the PS/PVME = 5/5 blend observed at 311 K. The *open circles* represent the decay of PS (*red*), the *open triangles* and *crosses* are OCH_3 and CH of PVME (*blue*). The *solid curves* were the "best-fit" obtained from Eq. (2.4). This figure is redrawn from the original source of Asano et al. [7]

$$\frac{dM_B(t)}{dt} = -(K_B + f_A k_{AB} + f_C k_{BC}) \cdot M_B(t) + f_B k_{AB} \cdot M_A(t) + f_B k_{BC} \cdot M_C(t)$$

$$\frac{dM_C(t)}{dt} = -(K_C + f_A k_{CA} + f_B k_{BC}) \cdot M_C(t) + f_C k_{CA} \cdot M_A(t) + f_B k_{BC} \cdot M_B(t)$$

or

$$\frac{d}{dt}\begin{bmatrix} M_A(t) \\ M_B(t) \\ M_C(t) \end{bmatrix} = \begin{bmatrix} -\eta_A & f_A k_{AB} & f_A k_{CA} \\ f_B k_{AB} & -\eta_B & f_B k_{BC} \\ f_C k_{CA} & f_C k_{BC} & -\eta_C \end{bmatrix} \begin{bmatrix} M_A(t) \\ M_B(t) \\ M_C(t) \end{bmatrix} = \mathrm{AM} \qquad (2.4)$$

where η_A, η_B, and η_C are $K_A + f_B k_{AB} + f_C k_{CA}$, $K_B + f_A k_{AB} + f_C k_{BC}$, and $K_C + f_A k_{CA} + f_B k_{BC}$, respectively. The solution of Eq. (2.4) is given by the numeric analysis; the magnetization $M_A(t)$ is given by $M_A(t) = \lambda_{A1}\exp(\varepsilon_1 \cdot t) + \lambda_{A2}\exp(\varepsilon_2 \cdot t) + \lambda_{A3}\exp(\varepsilon_3 \cdot t)$, $\lambda_{A1,2,3}$ and $\varepsilon_{1,2,3}$ are eigenvectors and eigenvalues of matrix **A** for M_A, respectively. The calculated curves, which are obtained from the numeric analysis with the initial condition of $M_A(0)$: $M_B(0)$: $M_C(0) = f_A$:f_B:f_C, are drawn by the solid lines in Fig. 2.3. The solid lines are in excellent agreement with the observed $T_{1\rho}{}^{H}$ decays. The estimated values are $K_A = 0.40 \times 10^3\ s^{-1}$, $K_B = 1.06 \times 10^3\ s^{-1}$, $K_C = 1.85 \times 10^3\ s^{-1}$, $k_{AB} = 1.0 \times 10^3\ s^{-1}$, $k_{CA} = 1.1 \times 10^3\ s^{-1}$, and $k_{BC} = 100\ s^{-1}$.

The value of k_{BC} is one-order smaller than those of k_{AB} and k_{CA}. This indicated that the interpolymer spin diffusion between PS and PVME is much faster and more effective than the *intrapolymer* spin diffusion of PVME. Furthermore, the values of k_{AB} and k_{CA} are comparable to those of K_i values. This means that the interpolymer spin diffusion rate is not enough to average out the respective $T_{1\rho}{}^{H}$ decays in the PS/PVME blends as described above. It has been proved that the PS/PVME blends are homogeneous [8–10]. The disagreement of $T_{1\rho}{}^{H}$ decays among PS and PVME was thus ascribed to the motional effects, namely motional heterogeneity. In fact, the

$T_{1\rho}{}^{H}$ decays observed at below 263 K showed single-exponential decays with the same relaxation rate for both PS and PVME. This observation suggested that the ^{1}H spin diffusion rate largely depends on the molecular motion. Therefore, it is noted to pay attention to the relationship between the experimental temperature and T_g of component polymers.

2.2.1.2 PMAA/PVAc Blends

The interpolymer hydrogen-bonding interaction between the COOH functional group of poly(methacrylic acid) (PMAA) and the CO functional group of poly(vinyl acetate) (PVAc) occurs in the blends [14–16]. The exothermic interaction assists in the Gibbs free energy of mixing to be negative. The hydrogen-bonging interaction causes the lower-field shift and makes the feature of the NMR spectra at the interacted region complex and change. Figure 2.4 shows the observed solid-state ^{13}C NMR spectra of PMAA/PVAc blends expanded at the COOH/CO region [14, 15]. On the left-hand side, decomposed spectra obtained from the least-square fit to the observed ^{13}C NMR spectra by a sum of five Gaussian curves are summarized. The simulated spectra for blends obtained from the simple sum of each observed ^{13}C NMR spectrum of pure PMAA and pure PVAc at the respective molar ratio are also displayed on the right-hand side.

The broad COOH line of pure PMAA is observed at 183 ppm, and the relatively narrow COO peak of PVAc is at 171 ppm. If there is no interpolymer interaction between the side chains of PMAA and PVAc, the observed NMR spectra should be reintroduced by the simple summation as shown on the right-hand side. However, the observed spectra of the PMAA/PVAc blends were obviously different from those obtained from the simple summation. The observed spectra have a much more complicated envelope. The lineshape of the CO region in the blends was divided into apparently five peaks. Two peaks at around 187 and 179 ppm of those five peaks clearly appeared, especially in the spectrum of the PMAA/PVAc = 1/1 blend. A peak at around 175 ppm was also clearly seen in the spectrum of the PMAA/PVAc = 3/1 blend, while the peak for the other blends disappeared behind the broad envelope. These observations clearly suggested the existence of an interpolymer interaction. The composition-dependent ^{13}C spin–lattice relaxation time [14] and two-dimensional (2D) exchange ^{13}C NMR experiments [16] revealed that the peak at 179 ppm is associated with the PMAA-COOH carbon, which the carboxylic hydrogen interacts with PVAc-CO oxygen, and the PVAc-CO carbon interacted with PMAA-COOH has appeared at 175 ppm. The naturally abundant 2D exchange ^{13}C NMR spectrum showed the clear cross-peaks between $\omega_1 = 175$ ppm and $\omega_2 = 179$ ppm, and between $\omega_1 = 179$ ppm and $\omega_2 = 175$ ppm with the same intensity. The distance between the interacted PMAA-COOH carbon and the PVAc-CO carbon was estimated to be 0.37 nm. The length of the hydrogen bond was also estimated from the molecular mechanics calculation to be ca. 0.2 nm [16].

Figure 2.5 shows the observed K values (a: $1/T_1{}^{H}$ and b: $1/T_{1\rho}{}^{H}$) of the PMAA/PVAc blends against the PMAA fraction: the fraction equals the weight ratio of

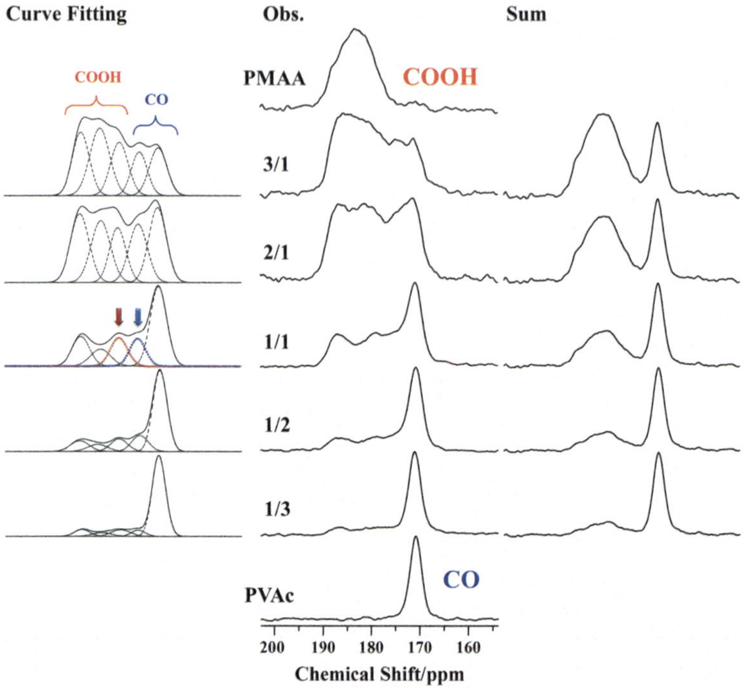

Fig. 2.4 Observed CPMAS ^{13}C NMR spectra (*center*) and the decomposed simulated spectra using five Gaussian curves (the *left-hand side*) of the expanded CO region in the PMAA/PVAc blends: a *broken line* depicts each decomposed peak and a *solid line* is the sum of those *broken lines*. The ^{13}C NMR spectra obtained by the sum of those of pure PMAA and pure PVAc at the respective molar unit ratio are drawn on the *right-hand side*. This figure is redrawn from the original source of Asano et al. [14, 15]

PMAA or unit ^{1}H molar ratio because the unit weight and number of protons in a unit structure of PMAA is the same as those of PVAc [14]. This plot indicated that both original *K* values of pure PMAA and pure PVAc are averaged depending on the compositions in the blend. All the obtained *K* values, $T_1{}^{H}$ rates in Fig. 2.5a, of PMAA in the blends are in excellent agreement with those of PVAc. This result suggested that the complete averaging of the *K* values by the ^{1}H spin diffusion occurs, and then, PMAA and PVAc are in close proximity to each other. Namely, the PMAA/PVAc blend is homogeneous on a scale of 20–50 nm for all compositions.

Similarly, Fig. 2.5b shows that the observed *K* values ($T_{1\rho}{}^{H}$ rates) of PMAA in the PMAA-rich/PVAc blends, which are the PMAA/PVAc = 3/1, 2/1 and 1/1 blends, are fully consistent with those of PVAc. For the PMAA-poor (PMAA/PVAc-rich = 1/2 and 1/3) blends, both rates of PMAA and PVAc changed together and became close, although the $T_{1\rho}{}^{H}$ rates of PMAA were different from those of PVAc. This plot indicated that the partial averaging of $T_{1\rho}{}^{H}$ rates by the ^{1}H spin diffusion occurs in the PMAA/PVAc-rich blends. These results showed that the PMAA-rich/PVAc blends are homogeneous on a scale of 2–5 nm as well as 20–50 nm, in contrast, the

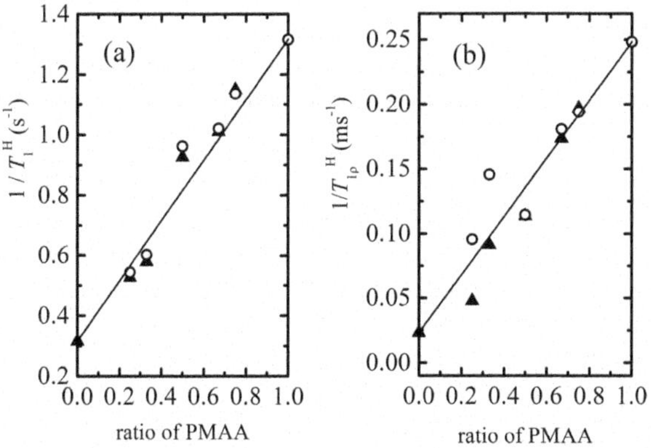

Fig. 2.5 Observed $1/T_1{}^H$ **a** and $1/T_{1\rho}{}^H$ **b** values, namely the ^{1}H relaxation rates K, against the ^{1}H molar fraction f_{PMAA} of PMAA/PVAc blends: *circle*; CH_2 carbon of PMAA, *triangle*; OCH carbon of PVAc, and each *solid straight line* represents the calculated one from Eq. (2.3). Every *error bar* for each data point is covered with the *symbols* itself. In this case, the PMAA weight ratio of the blends is equivalent to that of ^{1}H molar fraction because both unit weight and number of protons in a unit structure of PMAA [–(CH_2CCH_3(COOH))–] are the same as those of PVAc [–($CH_2CH(OCOCH_3)$)–]. This figure is newly drawn from the original source of Asano et al. [14, 15]

PMAA/PVAc-rich blends are inhomogeneous on a scale of 2–5 nm. The PMAA/PVAc-rich blends are probably homogeneous on a 5–10 nm scale because the $T_{1\rho}{}^H$ rates are affected significantly by the ^{1}H spin diffusion.

The solid straight lines in Fig. 2.5a and b were obtained using Eq. (2.3) with K values of pure PMAA and pure PVAc. The observed $T_1{}^H$ rates were in good agreement with the calculated line. Similarly, the observed $T_{1\rho}{}^H$ rates for PMAA-rich/PVAc blends are consistent with the calculated line. These agreements indicated that the molecular motions of both polymers in the blends did not change drastically with each other compared with those before mixing, and the very fast ^{1}H spin diffusion occurs in the blends.

2.2.1.3 PK/PA Alloys

Polyketone/polyamide (PK/PA) alloys showed superior impact resistance in the wet condition, particularly for the range of PK/PA = 1/9 to 7/3 [17]. The impact energy of 25 kJ m^{-2} under the dry condition became greater than 160 kJ m^{-2} under the over 1.5% humid condition. Interestingly, no improvement was seen for either pure PA or pure PK. The three-dimensional (3D) morphology of the PK/PA alloys detected by the transmission electron microscope (TEM) analysis showed a characteristic structure of lamellar-like structure with co-continuous spherical domains. It was clarified by the electron energy loss spectroscopy (EELS) measurement that the lamellar-like structure mainly originated in PK and the spherical domain came from

PA [17–21]. Both 3D-TEM and EELS images also showed a characteristic spherical amorphous domain filled with lamellar structure. Furthermore, small angle X-ray scattering (SAXS) studies revealed that the long period of lamellae for PK-rich alloys is estimated to be ca. 14 nm but that for PA-rich alloys decreases from 14 to 9 nm gradually with the PA content [17, 19–21]. The long periods of both pure PK and pure PA were ca. 14 nm and ca. 8 nm, respectively. The lamellar network for the PA-rich alloys consisted of a mixture network of both PK and PA. Thus, the existence of both lamellae and co-continuous spherical amorphous regions is strongly related to the improvement in impact resistance.

Because the solid-state NMR is sensitive to the conformational change in polymer chains, the NMR spectra give information on crystalline and amorphous phases regardless of wet conditions. Furthermore, the ^{1}H spin diffusion analysis indirectly obtained from the $T_1{}^{H}$ recovery curves gives information on the domain size of both polymers.

Figure 2.6 shows the ^{15}N CPMAS NMR spectra of (a) the PK/PA = 6/4 alloy and (b) pure PA under wet conditions [17, 18]. The CP efficiency of the amorphous phase of PA was depressed because the amorphous phase was predominantly water-absorbed. Thus, Fig. 2.6 is enhanced mainly by the crystalline phase of PA. The geometry of the hydrogen bond is different between the α and γ crystalline phases of PA. Such a difference influences the ^{15}N chemical shift values. The signal at 84 ppm is ascribed to the α crystalline phase of PA and that at 89 ppm to the γ crystalline phase. Figure 2.6 clearly indicates that the PA in the PK/PA = 6/4 alloy changes remarkably from their γ crystalline phase to α form. This observation suggested that PK induces the growth of the α crystalline phase of PA in the blend rather than that of γ crystallites. The relative fraction of the α crystalline phase of PA changed from 40 to 70% in the PK/PA = 6/4 alloy. The change in the crystalline form of PA is presumably related to the complicated morphology detected in the TEM image and to the high-impact properties under the wet condition.

Figure 2.7 shows the $T_1{}^{H}$ curves of both PK and PA observed in the PK/PA = 6/4 alloy under the dry condition [17]. The $T_1{}^{H}$ curves of both pure PK and pure PA (dashed lines) show single-exponential decays with the K_{PK} ($= 1/T_1{}^{H}$) of 0.485 s^{-1} ($T_1{}^{H} = 2.06 \pm 0.03$ s) and K_{PA} of 0.943 s^{-1} ($T_1{}^{H} = 1.06 \pm 0.01$ s), respectively. For the PK/PA = 6/4 alloy, both $T_1{}^{H}$ curves of PK and PA were close together, indicating that the insufficient but moderately fast ^{1}H spin diffusion occurs between PK and PA. The observed curves suggested that both PK and PA are not completely phase separated but are located in a region where both polymer protons diffuse each other to some extent. The solid lines show the least-squared fit curves using Eq. (2.2). Both PK and PA are semicrystalline polymers, so there are four different phases in the alloy at least. However, the amorphous and crystalline phases in PK showed identical $T_1{}^{H}$ value, and the same phenomenon was also detected for PA. These phenomena indicated that the fast ^{1}H spin diffusion occurs between the crystalline and non-crystalline phases for both PK and PA. This fact allowed us to analyze the observed $T_1{}^{H}$ curves by the two-spin system using Eq. (2.2).

To obtain the information on domain size, the observed data points were fitted by Eq. (2.2) with an f_{PA} of 0.548. The solid non-single exponential lines were in

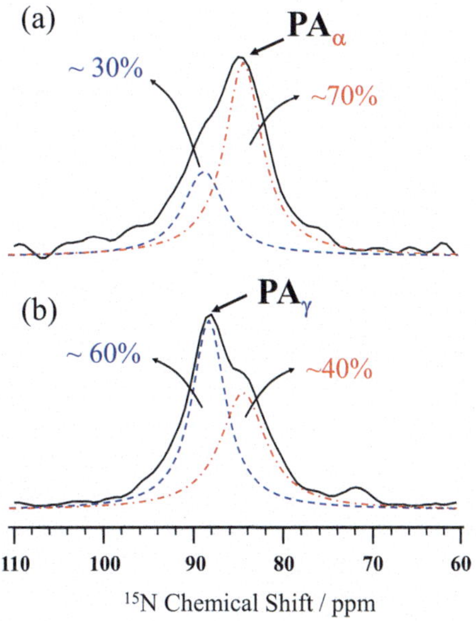

Fig. 2.6 Observed ^{15}N CPMAS NMR spectra of **a** PK/PA = 6/4 alloy and **b** pure PA. This figure is redrawn from the original source of Asano et al. [17, 18]

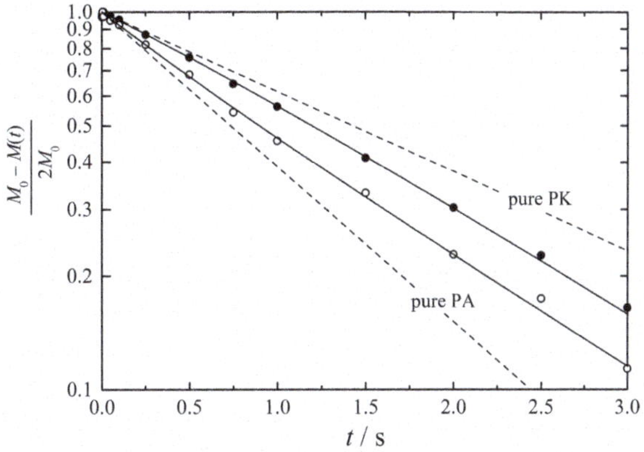

Fig. 2.7 Semilogarithmic plots of normalized $T_1{}^{H}$ decay curves for PK (*filled circle*) and PA (*empty circle*) in the PK/PA = 6/4 alloy under the dry condition. *Dashed straight lines* represent $T_1{}^{H}$ decay curves for pure PA and pure PK. The *solid lines* were obtained with Eq. (2.2). This figure is redrawn from the original source of Asano et al. [17]

good agreement with the observed $T_1{}^{H}$ decay curves of both PK and PA in the PK/PA = 6/4 alloy. The K values obtained were $0.53 \pm 0.02\ s^{-1}$ for PK and $0.81 \pm 0.02\ s^{-1}$ for PA, respectively, and the ^{1}H spin diffusion rate, k, of $0.8 \pm 0.3\ s^{-1}$. This analysis suggested that PA is dispersed and located in the PK matrix. The k value is comparable to the initial relaxation rate of both polymers, indicating that ^{1}H spin diffusion is slow, but that the two polymers are not completely phase separated on a scale of several tens to several hundreds of nanometers. The estimated average domain size was approximately 150 nm. The maximum diffusive path length with the assumption of the 3D diffusion model and using the effective diffusion coefficient D_{eff} of ca. 700 $nm^2 s^{-1}$ was obtained to be 75 nm, and the domain size is estimated to be twice as large as the diffusive path length because the spin diffusion occurs both sides of PK and PA. Here, the value of D_{eff} was calculated from both $D_{PK} = 840$ nm^2s^{-1} and $D_{PA} = 600$ nm^2s^{-1}: these D values were obtained from each ^1H spin–spin relaxation rate of PK and PA [17, 22]. The value of 150 nm was comparable to the average overall size of the elliptical domains estimated from the 3D TEM images. Furthermore, the k value under the wet condition was consistent with that under the dry condition. Thus, the domain size under the wet condition did not change as compared to that under the dry condition.

The creation of the co-continuous spherical amorphous region and the presence of PK crystalline lamellae only in the PK/PA alloys were detected by 3D TEM images. The NMR studies revealed the conversion to the α crystalline phase from the γ crystalline phase for PA in the PK/PA alloys and the selective water absorption in the amorphous region of PA. Furthermore, SAXS analyses clarified that the long period of PK lamellae becomes wider by approximately 1 nm under wet condition [17, 19–21]. These various morphological changes play an important role in exhibiting high-impact resistance and toughness in the PK/PA alloys under wet condition. The combination of both the rigidity of PK lamellae and the elasticity of the water-absorbing amorphous phase of PA is needed to obtain high-impact properties, modulus, and toughness.

2.2.1.4 Poly(Octamethylene Thiourea)/Poly(Ether Thiourea) Blends

Poly(ether thiourea) ($PTUEG_3$) self-heals at 12°C and has a mechanically robust property in dry condition. On the other hand, Poly(octamethylene thiourea) ($PTUC_8$) does not show the self-healing property. The $PTUC_8/PTUEG_3$ = 70/30 and 50/50 blends self-heal in temperature ranges above 32 and 20°C, respectively, and the $PTUC_8/PTUEG_3$ = 80/20 blend also self-heals in a temperature range above 32°C even though the $PTUEG_3$ component is only 20 mol% [23]. However, the $PTUC_8/PTUEG_3$ = 90/10 blend does not show any healing properties at all. A single glass transition temperature (T_g) was detected between the T_g values for those pure homopolymers for all blends, indicating that every blend is homogeneous on a range of 10 nm scale: T_g values of pure $PTUEG_3$ and pure $PTUC_8$ are 27°C and 50°C, respectively.

Figure 2.8 shows the $T_{1\rho}{}^{H}$ decays for OCH_2 groups of $PTUEG_3$ (*filled circle*) and CH_2 groups of $PTUC_8$ (*empty circle*) obtained from $PTUC_8/PTUEG_3$ = 80/20 (*left-hand side*) and 90/10 (*right-hand side*). Healing behaviors for the homopolymers and blends are also inserted into the *right-hand side* figure.

From the healing properties, the fractured parts of the $PTUC_8/PTUEG_3$ = 80/20 and 50/50 blends show a gradual increase in stress at the break because of self-healing onset at temperatures lower than 32°C (see the inserted picture in the *right-hand side* figure). On the other hand, the 90/10 blend does not show such a healing property. The $T_{1\rho}{}^{H}$ decay of $PTUEG_3$ in the 90/10 blend is in good agreement with that of $PTUC_8$. For the 80/20 blend, the $T_{1\rho}{}^{H}$ decay of $PTUEG_3$ is slightly different from that of $PTUC_8$. These observations indicate that the $PTUC_8/PTUEG_3$ = 90/10 blend

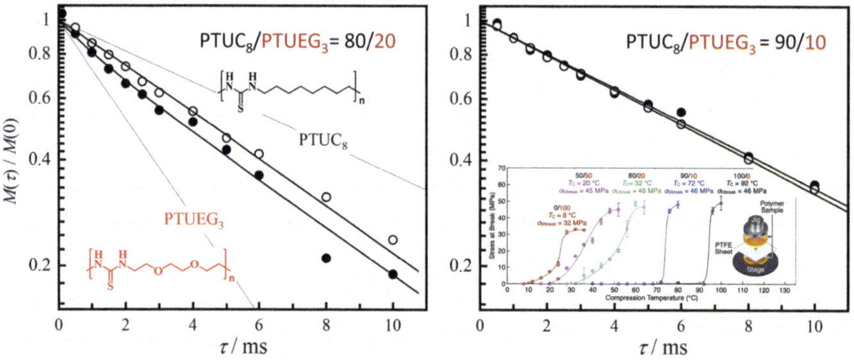

Fig. 2.8 Observed $T_{1\rho}{}^{H}$ decay curves for $PTUC_8$ (*empty circle*) and $PTUEG_3$ (*filled circle*) in the PTUC8/PTUEG3 = 80/20 (*left-hand side*) and 90/10 (*right-hand side*) blends. The *thin straight lines* in the *left* figure represent the homopolymer's $T_{1\rho}{}^{H}$ decays. The *solid lines* in the *left* figure were obtained with Eq. (2.2) and those in the *right* figure were least-square fitted by the conventional single-exponential decay function. Healing behaviors are also inserted in the *right* figure: $PTUC_8$ (*black*), polymer blends $PTUC_8/PTUEG_3$ (90/10 in mol%; *blue*, 80/20 in mol%; *green*, and 50/50 in mol%; *pink*), $PTUEG_3$ (*red*). T_C and $\sigma_{0/\mathrm{break}}$ denote the critical temperature for the onset of healing and pristine stress at break, respectively. The source of the inserted illustration in the *right* figure is Fujisawa et al. [23], and the current $T_{1\rho}{}^{H}$ decay curves are newly redrawn and re-analyzed using the original data by A. Asano who is one of the authors of ref [23]

is homogeneous on a scale of 2–5 nm, while the 80/20 blend is inhomogeneous on the scale. Using Eq. (2.2) with parameters of a ^{1}H fraction $f_{\mathrm{PTUEG3}} = 0.16$, the observed data points for the 80/20 blend are least-square fitted and the ^{1}H spin diffusion rate, k of about 0.6 ms^{-1} is obtained (*solid lines* in the *left* figure): K values are 0.23 $\pm$0.02 ms^{-1} for $PTUEG_3$ and 0.143 $\pm$ 0.005 ms^{-1} for $PTUC_8$. The estimated average domain size for a 3D diffusion model was approximately 6 nm (2.9 nm $\times$ 2) with k and $D = 800$ nm^2 s^{-1}; a typical diffusion coefficient D for a polymer is employed. The estimated domain size suggests that the $PTUC_8/PTUEG_3 = 80/20$ blend has a nanometer order microdomain (nanophase separation), which is not detectable from the DSC measurement. Consequently, nanophase separation is certainly essential for the polymer blend to exert its excellent self-healable property [23].

2.2.2 *Phase Separation*

Homogeneous single phase of a polymer blend for a lower critical solution temperature (LCST) phase diagram is thermodynamically unstable: the two types of the phase separation undergo, and they are referred to the spinodal decomposition and the nucleation growth. The phase is easily collapsed when a blend is heat-treated above the LCST line, which is the binodal point. The homogeneous and transparent phase becomes heterogeneous and opaque in general. The concentration change during the spinodal decomposition at the initial stage can also be investigated by analyzing the ^{1}H spin–lattice relaxation [4, 7, 15].

2.2.2.1 PS/PVME Blends

The homogeneous phase of the PS/PVME blends is phase separated after heat treatment above 373–393 K (binodal point) because the PS/PVME blends exhibit the LCST phase diagram. Figure 2.9 shows the observed $T_{1\rho}{}^{H}$ decay curves of the PS/PVME = 5/5 blend, which was quickly cooled in liquid nitrogen to quench the phase separation progress after heating at 413 K for 2 min [7]. The $T_{1\rho}{}^{H}$ decay curves were measured at 263 K to freeze the molecular motion of PVME and let the ^{1}H spin diffusion work effectively. Before the heat treatment, both observed $T_{1\rho}{}^{H}$ decays obtained from PS and PVME signals showed the same single-exponential curve. With increasing the heat treatment time, they deviated from the single exponential to show a double-exponential behavior.

A model based on a simple assumption for phase separation provides the double-exponential behavior [4, 7, 15]. The homogeneous polymer blend *x*/*y*, in which *x* composition ratio is represented by $\rho^{0}{}_{x}$, is phase separated into the *x*-rich and the *y*-rich domains after the heat treatment. This model represents that the fast ^{1}H spin diffusion occurs within each rich domain in the phase-separated blend; polymers *x* and *y* are homogeneously mixed in the respective *x*-rich and *y*-rich domains from the ^{1}H spin diffusion point of view. There is no effective ^{1}H spin diffusion between the *x*-rich and the *y*-rich domains because the sizes of both *x*-rich and *y*-rich domains become beyond 5 nm for $T_{1\rho}{}^{H}$ or 50 nm for $T_{1}{}^{H}$. Each domain has an individual ^{1}H spin–lattice relaxation time. In a word, the T_1 value of polymer *y* in the *x*-rich domain has the same value as that of polymer *x*, T_{1x}. Similarly, both polymers *x* and *y* in the

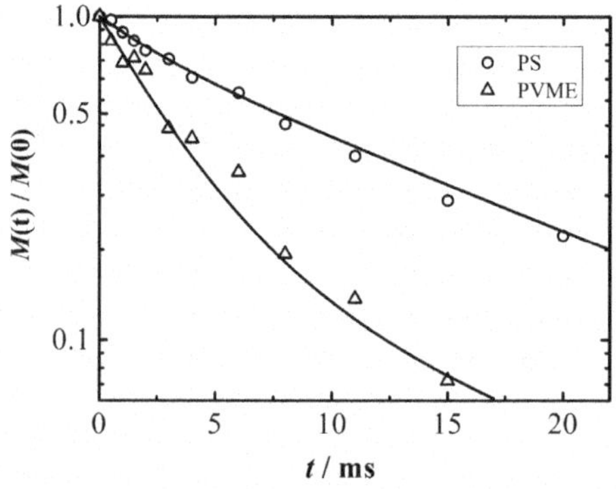

Fig. 2.9 Observed $T_{1\rho}{}^{H}$ decay curves for PS (*circle*) and PVME (*triangle*) in the phase-separated PS/PVME = 5/5 blend after 2 min of heating at 413 K. These decay curves were measured at 263 K. The *solid lines* are the "best-fit" curves from Eq. (2.5) with the parameters of $\xi_x = 0.81 \pm 0.05$, $\xi_y = 0.20 \pm 0.03$, $T_{1x} = 15.3 \pm 0.6$ ms and $T_{1y} = 3.1 \pm 0.2$ ms. This figure is redrawn from the original source of Asano et al. [4, 7]

y-rich domain have T_{1y}. According to this assumption, it is obvious that the relaxation curves observed from polymers x and y are expected to be double-exponential decays as follows:

$$M_x(t) = \xi_x exp(-t/T_{1x}) + (1 - \xi_x) exp(-t/T_{1y})$$

$$M_y(t) = \xi_y exp(-t/T_{1x}) + (1 - \xi_y) exp(-t/T_{1y}) \tag{2.5}$$

where

$$\xi_x = \frac{\phi_x \cdot r}{\rho_x^0}, \xi_y = \frac{(1 - \phi_x) \cdot r}{1 - \rho_x^0}, \text{ and } \rho_x^0 = \phi_x \cdot r + \psi_x(1 - r).$$

Here, $M_x(t)$ and $M_y(t)$ represent the normalized magnetization. The parameters ξ_x and ξ_y are ^{1}H molar fractions of polymers x and y in the x/y blend, respectively. The parameters φ_x and ψ_x are the concentration of polymer x in the x-rich and y-rich domains, respectively, and the r is the fraction of the x-rich domain. The stoichiometric values φ_x and ψ_x of the two separated phases and the r value are easily deduced from the fractions ξ_x and ξ_y. The φ_x or ψ_x are any of ^{1}H molar/weight/volume fractions, depending on the determination of the $\rho^0{}_x$ value.

The solid lines in Fig. 2.9 were the "best-fit" curves obtained from Eq. (2.5). The adjusted values for ξ_X and ξ_Y were 0.81 ± 0.05 and 0.20 ± 0.03, respectively, and T_{1X} and T_{1Y} were 15.3 ± 0.6 ms and 3.1 ± 0.2 ms. Here, X and Y represent PS and PVME, respectively. By using values of ξ_x and ξ_y, the parameters of φ_X, ψ_X, and r were calculated to be 0.81 ± 0.06, 0.20 ± 0.03, and 0.51 ± 0.04, respectively.

The $T_{1\rho}{}^H$ decays after heating at 413 K for 20 s, 40 s, 1 min, and 1.5 min were similarly examined. Thus, the gradual composition change for both PS-rich and PVME-rich phases and constancy of the fraction of the PS-rich domain r were found. For example, the values of ξ_x, ξ_y, T_{1X}, T_{1Y} at 1 min heat treatment were 0.62 ± 0.06 and 0.40 ± 0.05, 17.8 ± 0.2 ms, and 6.7 ± 0.2 ms, respectively [7]. These values produced the φ_X, ψ_X, and r to be 0.62 ± 0.05, 0.39 ± 0.05, and 0.51 ± 0.05, respectively. The r value was not changed from 1 to 2 min heating. The PS composition in the PS-rich phase increased from 0.62 to 0.81, and that in the PVME-rich phase decreased from 0.39 to 0.20. This concentration analysis through $T_{1\rho}{}^H$ curves during phase separation revealed that the spinodal decomposition undergoes for the PS/PVME blends with heat treatment at 413 K. The rate of the concentration change during the spinodal decomposition was verified by comparing to the results obtained from the other spectroscopy [24–26].

The appreciable change on the $T_{1\rho}{}^H$ curves was not detected after further heating longer than 30 min. No change in $T_{1\rho}{}^H$ curves indicated the end of the initial stage of the spinodal decomposition. After finishing the initial stage, the morphological change occurs on a larger scale that is not reflected in T_1. The growth of domain size at the late stage of the phase separation of the PS/PVME blend was revealed from the

2D exchange ^{129}Xe NMR [27]. The 2D exchange ^{129}Xe NMR showed that the cross-peaks, which are observed after heating 30 min, do not appear after 1200 min. This observation indicated that the estimated average domain size was beyond 5000 nm after 1200 min at least.

2.2.2.2 P3HT/PCBM Blends

For rigid and immobile polymers, the CRAMPS spectrum with moderate MAS is also useful to detect high-resolution ^{1}H NMR spectra. The domain size distribution for poly-3-hexylthiophene (P3HT) and phenyl-C_{61}-butyric acid methyl ester (PCBM) blend, which is one of the candidates for an organic photovoltaic device, was analyzed via the direct ^{1}H spin diffusion measurement and found that the size depends on spin-coating speed and thermal annealing significantly [28]. The observed ^{1}H spin diffusion behavior was simulated by the linear combination of theoretical spin diffusion curves obtained from a rods-in-a-matrix or a lamellar model assumption with several domain sizes. The simulation gave the information on the domain size distribution: the as-cast P3HT/PCBM film has mainly less than 10 nm domains, but the size distribution shifts toward greater than 100 nm upon annealing at 413 K for 30 min [28].

The P3HT/PCBM blends showed the phase separation upon heat treating at 423 K. Figure 2.10 shows the ^{1}H CRAMPS NMR spectra of (a) P3HT/PCBM blend, (b) P3HT, and (c) PCBM [29]. Despite the existence of overlap of both ^{1}H signals of P3HT and PCBM, the sinusoidal magnetization profile produced by the MREV8 preparation step gave the negative signal for PCBM and the positive signal for P3HT (dashed line in Fig. 10d). Figure 10e–h displays the ^{1}H signal intensity change by the ^{1}H spin diffusion during mixing time. The positive and negative magnetizations effectively diffuse each other when both polymers are miscible: the measured ^{1}H spin diffusion curves are inconsistent with each other. When the phase is heterogeneous, the identical ^{1}H spin diffusion curves deviate and show characteristic curvature depending on the morphology. The ^{1}H spin diffusion behaviors were satisfactorily explained by a two-phase model, in which the P3HT-rich phase contains the grown-up PCBM crystals (composition of 11% by mass in the phase) and the PCBM-rich phase with 6% P3HT [29].

2.3 Polymer Nanocomposites

Organic polymer/inorganic clay nanocomposites have the potential to show a great improvement in gas barrier, mechanical or thermal properties, particularly in the exfoliate and the intercalated states [1]. Paramagnetic Fe^{3+}-ions which are naturally contained in montmorillonite (mmt) clay influenced the $T_1{}^H$ curves of a polymer in a polymer/mmt nanocomposite, and the analysis of the $T_1{}^H$ curves gives morphological information [30, 31]. The interaction between inorganic fillers and polymers can be

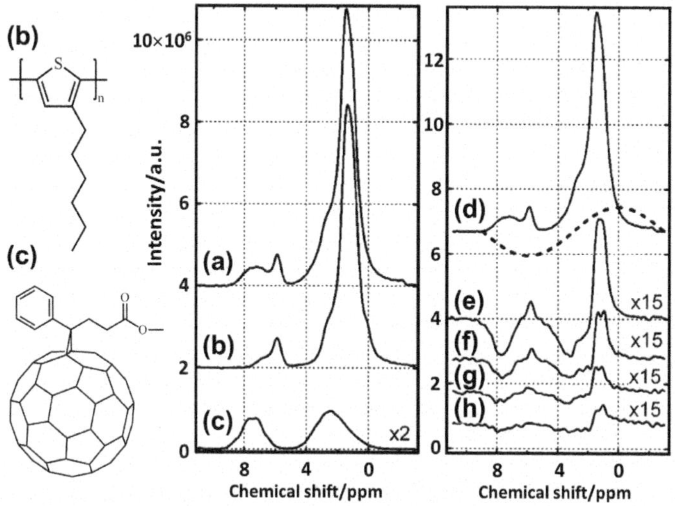

Fig. 2.10 Stack plot of ^{1}H CRAMPS NMR spectra: **a** P3HT/PCBM blend, **b** P3HT, and **c** PCBM: the chemical structures of **b** and **c** are depicted on the *left-hand side*. Spectrum **d** is the same as **a** and the sinusoidal excitation profile (*dashed line*) by the MREV8 preparation is also depicted. Spectrum **e** was observed after a mixing time of 2 ms subsequent to the MREV8 preparation, **f** 60 ms, **g** 180 ms, and **h** 550 ms. The original source of this figure comes from Nieuwndaal et al. [29]

detected and analyzed by NMR techniques. In this section, the paramagnetic effect on T_1^{H} phenomenon is shown in the first. Next, the investigation of the interaction between inorganic fillers and polymers is described. Finally, the morphological study for polymer nanocomposites is presented.

2.3.1 Paramagnetic Effect on Relaxation

2.3.1.1 PVA/Montmorillonite Clay Nanocomposites

Figure 2.11 shows the T_1^{H} values obtained from the poly(vinyl alcohol) (PVA)/ mmt nanocomposites [32]. The T_1^{H} recovery curves showed a single-exponential function. Because PVA is severely moisture-absorbed and the T_1^{H} is significantly sensitive to humidity, the nanocomposites and pure PVA were dried under vacuum at 353 K for 3 days after preparation. In addition, the sample after putting in a sample rotor was further dried under the same condition for a day before measuring T_1^{H} to eliminate the effect of humidity: the T_1^{H} value under the humid condition becomes very short as compared to that under the dry condition. The original T_1^{H} value of PVA, which is represented by a cross, decreased steeply with the amount of mmt. The original T_1^{H} value of 15.5 ± 0.2 s for PVA became half at only 2 wt% (7.9 ±

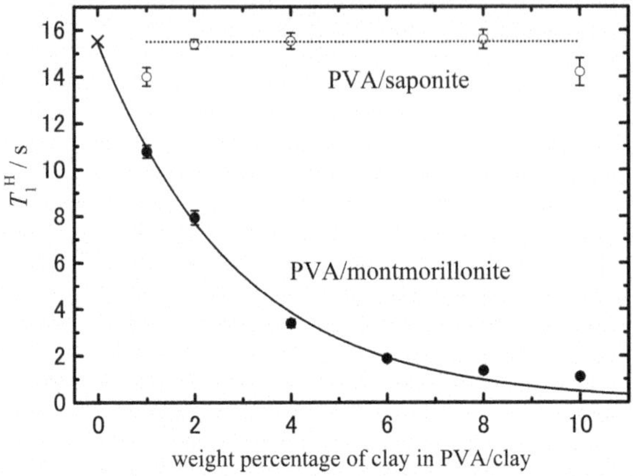

Fig. 2.11 Observed $T_1{}^H$ values of the PVA/mmt and PVA/saponite nanocomposites against the weight percentage of clay at 11.74 T: pure PVA is represented by *cross*, PVA/mmt is *solid circles* and PVA/saponite *open circles*. The *solid line* is the "best-fit" curve to the observed data points using the simple-exponential decay function. The *botted-line* is drawn to guide the reader's eyes and represents the $T_1{}^H$ value of 15.5 s. This figure is redrawn from the original source of Asano et al. [32]

0.3 s) in the PVA/mmt nanocomposites. At beyond 6 wt%, the value became less than 2 s and saturated till 10 wt%. This phenomenon indicated that the mmt clay is well dispersed and exfoliated in the PVA matrix up to 10 wt%. The $T_1{}^H$ value of the PVA neighboring Fe^{3+} ions becomes extremely short due to the paramagnetic effect. The effect of Fe^{3+} ions on $T_1{}^H$ of PVA is, however, limited in a very short range. The very short $T_1{}^H$ relaxation averages out the original $T_1{}^H$ values of the bulk PVA by the fast ^1H spin diffusion occurring among PVA protons. Therefore, the good dispersion is important to show the single-exponential decay and causes the extreme depression of $T_1{}^H$ with the increment of mmt. At an excess amount of 10 wt% of mmt, such as 20 or 30 wt%, the $T_1{}^H$ curves showed a non-single exponential, suggesting the poor dispersion of mmt in the PVA matrix.

In contrast, the $T_1{}^H$ values of PVA in the PVA/saponite clay nanocomposites were not affected by the coexistence of the saponite clay. The saponite is artificially produced clay without any paramagnetic sources. Therefore, this simple experiment suggested again that the effect of a paramagnetic source on $T_1{}^H$ is quite remarkable.

2.3.1.2 Nylon-6/mmt Nanocomposites

The influence of Fe^{3+} in the mmt clay layers on the spins of a polymer is within about 0.4 nm of the clay surface [33]. The paramagnetic contribution to $T_1{}^H$ can be estimated from the approximate "sum-of-rates" equation as follows [30, 34]:

$$\frac{1}{T_1^{\mathrm{para}}} = \frac{1}{T_1^{\mathrm{H}}(\mathrm{nanocomposite})} - \frac{1}{T_1^{\mathrm{H}}(\mathrm{polymer})} \tag{2.6}$$

where $T_1{}^{\mathrm{para}}$ is the paramagnetic contribution.

For nylon-6 (N6)/mmt nanocomposites, it was revealed that the $T_1{}^{\mathrm{para}}$ value changes by depending on the degree of mmt dispersion and exfoliation. The contribution ($T_1{}^{\mathrm{para}}$) hardly changes with magnetic field strength: the $T_1{}^{\mathrm{H}}$ of N6 depends on the magnetic field, of course, 1.63 s for 7.05 T and 0.52 s for 2.35 T. The $1/T_1{}^{\mathrm{para}}$ rate for well dispersed mmt in the N6/mmt nanocomposite was 1.05 s^{-1} (7.05 T) to 1.11 s^{-1} (2.35 T) and 0.45 to 0.46 s^{-1} for a bad mmt dispersed N6/mmt nanocomposite [30, 34]. The difference in clay dispersion is also related to the quality of the physical properties of a nanocomposite. For example, the tensile yield strength of 64 MPa for N6 became 85 MPa for the well dispersed and 69 MPa for the badly dispersed N6/mmt nanocomposites.

2.3.2 Interaction Between Polymers and Fillers

2.3.2.1 Nylon-6/Montmorillonite Clay Nanocomposites

Figure 2.12 shows the ^{1}H MAS NMR spectra of N6 and N6/mmt nanocomposites [30, 34, 35]. Due to the strong ^{1}H homonuclear dipole interaction, the ^{1}H MAS NMR spectrum of N6 shows a very broad and featureless peak as shown in Fig. 2.12a. For N6/mmt nanocomposites, particularly the spectrum of Fig. 2.12c, there is a clear extremely narrow line sited on top of the broad ^{1}H signal of N6; for the spectrum in Fig. 2.12b, the narrow line is very small and less visible. Because a mmt contains paramagnetic Fe^{3+}-ions inside layers as described above, the NMR signals of the organic modifiers (OM) located on the clay surface, which is treated for clay layers to be exfoliated in and to interact with polymer chains in a nanocomposite, are significantly broaden and hard to be detected. Thus, the visible but very tiny peak sited on top of the N6 ^{1}H signal shown in Fig. 2.12b was not a surprising feature; Fig. 2.12b indicated that the OM (b) is still located on the surface of clay layers.

In contrast, the strong narrow peak measured in Fig. 2.12c suggested that the OM (c) is apart from the clay surface at least 2 nm: the paramagnetic effect is negligible beyond the distance [33]. The dissimilarity between Fig. 2.12b and c was related to the processing condition. For an in situ polymerization from ε-caprolactam, the sharp peak was invisible (Fig. 2.12b), but for mechanical mixing, the sharp peak was visible (Fig. 2.12c). This observation indicated that the role of OM is different in the way of making nanocomposites even though both OM is mixed for clays to be compatible with polymer chains. The mmt layers in the N6/mmt nanocomposite shown in Fig. 2.12c were well dispersed, and the $1/T_1{}^{\mathrm{para}}$ rate was 1.05 s^{-1} at 7.05 T.

The chemical shifts for the sharp peaks were 0.9, 1.3, and 2.2 ppm, and they were associated with methyl, methylene, and protons on carbons attached to amine nitrogen, respectively. The peak integral (area) of 2.2 ppm was 10% and the sum

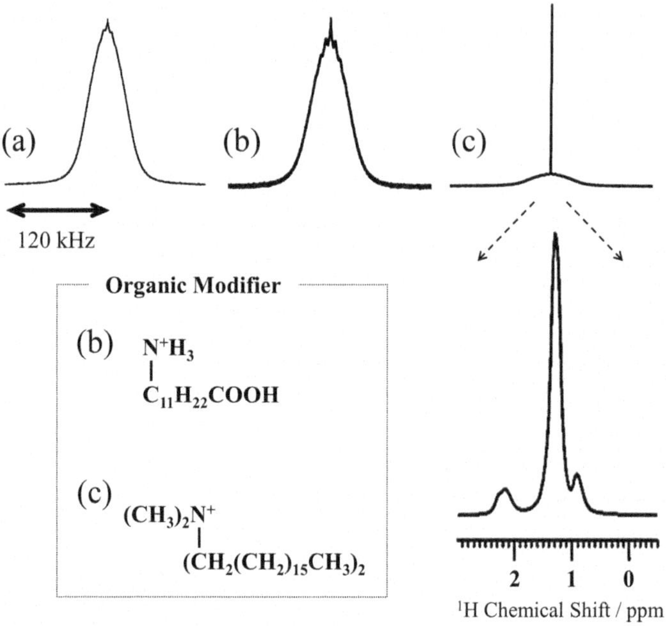

Fig. 2.12 ^{1}H MAS NMR signals of N6/mmt nanocomposites at MAS = 5 kHz: **a** N6, **b** in situ polymerized N6/mmt from a ε-caprolactam solution in the presence of treated mmt, and **c** mechanical blending N6/mmt. Expanded *narrow lines* that sit on top of the broad N6 ^{1}H signal **c** are enlarged under the spectrum. The organic modifiers treated on mmt for **b** and **c** are also illustrated on the *left-hand lower side*. This figure is newly drawn using the original data of VanderHart et al. [30, 34, 35]

of other peaks 90%. The peak area ratio revealed that the OM (c) was degraded by removing CH_3 functional group and changed to $CH_3N(CH_2(CH_2)_{15}CH_3)_2$ [30, 34]. Therefore, the remained CH_3^+-cation on the mmt clay layers became a contact point between the anionic environment of clay surface and N6 as $CH_3N^+H_2–(CH_2)_5CO–(NH–(CH_2)_5–CO)_n–$. The removal of a small cation from the OM (c) plays a very important role for clay layers to exfoliate in hydrophobic polymer chains, and then the decomposed OM (c) is dispersed in the N6 matrix.

For the in situ polymerized N6/mmt, in contrast, the role of the OM (b) with carboxyl group (COOH) was quite different from that of mechanical mixed N6/mmt. The ring-open polymerization of ε-caprolactam starts with the reaction between the COOH group of the OM (b) and the amine group of ε-caprolactam. The N6 chains were created on the clay surface and aligned with the direction of the clay layer, resulting in parallel orientation. This suggested the large amounts of the γ crystalline phase of N6 and the no dispersion of the OM (b) in the in situ polymerized N6/mmt nanocomposites.

2.3.2.2 PS-PEO Block Copolymer/Hectorite Clay Nanocomposites

The simplest idea for proving the existence of the interaction between inorganic fillers and polymers in the nanocomposites is to observe the hetero- or homonuclear correlation between their spins. Figure 2.13 shows the ^{29}Si-^{1}H HETCOR spectra for PS-poly(ethylene oxide) (PEO) block copolymer and hectorite clay (HCT) nanocomposites: (a) PS30EO-HCT and (b) PS4EO-HCT [36]. The strong cross-peak appeared on the PEO ^{1}H resonance position. This observation clearly indicated that the PEO chains are in the vicinity of HCT layers for PS4EO-HCT nanocomposite. On the other hand, the absence of the cross-peak for PS30EO-HCT nanocomposite showed that the PEO chains are far from the HCT layers. These two observations suggested that PEO blocks of PS4EO do intercalate significantly into the silicate galleries, while the HCT clays are mostly aggregated in the PS30EO-HCT nanocomposite. This different result detected in the PS30EO nanocomposite was ascribed to the small fraction of PEO segments: a hydrophilic group of PEO was a key component to intercalating into the clay layers [36].

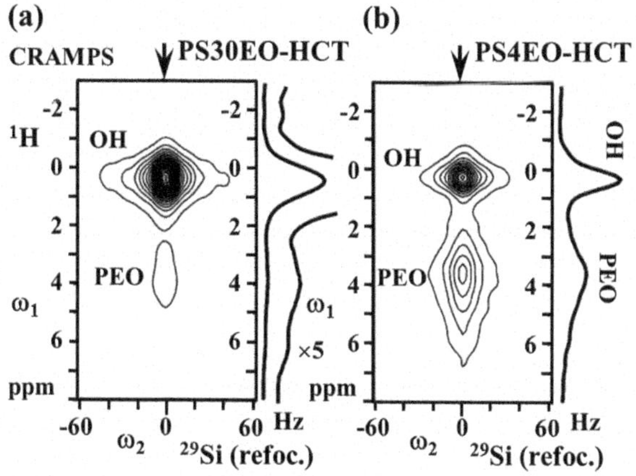

Fig. 2.13 ^{29}Si-^{1}H HETCOR spectra for **a** PS30EO-HCT and **b** PS4EO-HCT with MAS = 5 kHz and frequency-switched Lee-Goldburg (FSLG) ^{1}H decoupling: the numbers 30 and 4 indicate the number-average molecular weight (M_n) of PS, corresponding to M_n = 30,000 and 4000, respectively. The *arrows* represent the center of the Si signal. The LGCP contact time for ^{1}H-^{29}Si was 1.5 ms. The OH signal of ω_1 dimension comes from the clay surface. The source of this figure is Hou et al. [36]

2.3.3 Morphology

2.3.3.1 PVIBE/ε-PL/Saponite Clay Blends

The melting point (T_m) of the crystalline phase depends on the lamellar thickness referred to as the Gibbs–Thomson effect [37]. The T_m values of poly(ε-l-lysine) (ε-PL) for poly(vinyl isobutyl ether) (PVIBE)/ε-PL blends decreased with an increase in the amounts of PVIBE, while the T_m values of PVIBE were not affected by the amounts of ε-PL [38]. The decrease in T_m value was attributed to the decrease in crystal domain size of ε-PL, whose size was estimated from the T_1^H analysis.

Semicrystalline polymers PVIBE and ε-PL have small crystal (CR) domains less than ca. 50 nm, so that the T_1^H of CR is consistent with that of non-crystalline (NC) phase: the fast ^{1}H spin diffusion occurs between CR and NC phases [38, 39]. The blends of PVIBE and ε-PL were inhomogeneous, but the T_1^H curves could be reproduced by Eq. (2.2), suggesting that the moderate ^{1}H spin diffusion occurs between PVIBE and ε-PL. For the PVIBE/ε-PL = 10/1 and PVIBE/ε-PL/saponite clay = 10/1/0.03 blends, however, the T_1^H curves of CR and NC for ε-PL did not agree with each other. Therefore, the T_1^H curves for those blends were fitted by the three-spin system of Eq. (2.4). The T_1^H analyses gave the repeat length of both PVIBE and ε-PL domains. The domain size of ε-PL was estimated from the repeat length using the proton fraction in the blends. The CR size was finally calculated by using the crystallinity that was estimated from the ^{13}C CPMAS NMR spectra editing using $T_{1\rho}^H$ difference between CR and NC. The determined CR size of ε-PL showed the PVIBE content dependence for the PVIBE/ε-PL blends. The dependency is plotted in Fig. 2.14, and was fitted by the Gibbs–Thomson effect successfully on the assumption that the obtained CR size is comparable to the lamellar thickness [38].

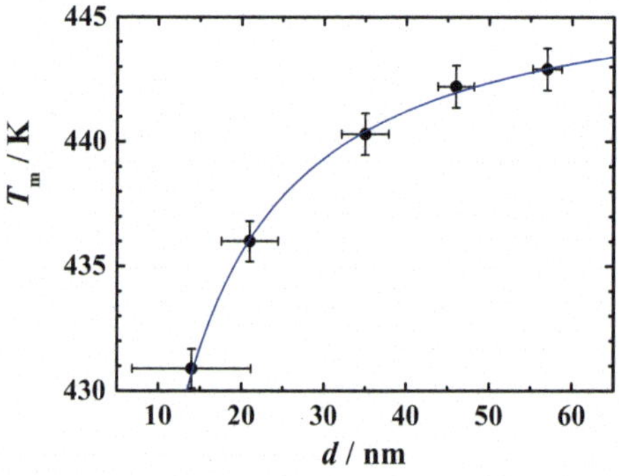

Fig. 2.14 Relationship between the melting temperature, T_m, and the CR thickness of ε-PL, *d*, for the PVIBE/ε-PL blends. The *solid line* represents the Gibb–Thomson relation expressed by 446.8 (K)–220 (K nm) /*d* (nm). This figure is reprinted with permission from the e-Journal of Soft Materials 2007, 3, 1–8 by Asano et al. [38]. Copyright 2007 The Society of Rubber Science and Technology, JAPAN

For the PVIBE/ε-PL/saponite clay = 10/3/0.09 to 10/5/0.15 blends, the CR size of ε-PL was constant at ca. 33 nm. The value was in good agreement with that of the PVIBE/ε-PL = 10/3 blend (35 nm). Furthermore, the T_m values for those of PVIBE/ε-PL/saponite clay = 10/3/0.09–10/5/0.15 blends were constant of 439 K and the value was also consistent with that of the PVIBE/ε-PL = 10/3 blend (440 K) within an experimental uncertainty. These results showed that clays influence the growth of CR. The constancy of T_m and CR size proved that the decrease in T_m is caused by the CR size [39].

2.3.3.2 Nylon-6/Montmorillonite Clay Nanocomposites

The thermal stability of the α crystallite of N6 is better than that of γ crystallite. The N6 chain near the surface of the clay layers aligns parallel orientation; thus, the N6 chains create the γ crystalline phase in the N6/clay nanocomposites with very slow cooling after the melting state of crystallites. With fast or natural cooling, both α and γ crystallites exist in the N6/clay nanocomposites. The N6 located between the exfoliated mmt clay layers and the γ crystalline phase is placed closer to the clay surfaces, indicating that the $T_1{}^H$ of the γ crystalline phase becomes shorter because of the paramagnetic effects. So, only if the α crystalline phase were mostly well isolated from mmt clay surfaces would the $T_1{}^H$ of these latter crystallites be notably longer than the average [30].

Figure 2.15 shows the $T_1{}^H$ curves observed from the N6/mmt nanocomposites with and without annealing [30]. The $T_1{}^H$ behaviors of α and γ crystalline phases (*open circles* and *triangles*) without annealing are in excellent agreement with each other, even though the observed nanocomposites have substantial amounts of both crystallites. This observation indicated that the average distance from the clay surfaces was the same for both crystallites, indicating that not only γ crystallites located closer to the clay surfaces but also α crystallite were in close proximity. In contrast, the $T_1{}^H$ behavior of γ crystalline phase with annealing was different from those of α crystalline and non-crystalline phases (NC). The amounts of α crystallites increased by ca. 30% after annealing. The initial slope of the γ crystalline phase (*solid triangles*) was steeper than those of α crystalline (*solid circles*) and NC phases (*open squares*); the protons in the γ crystalline phase are more rapidly decaying and influenced by the paramagnetic Fe^{3+} relative to the protons of the α crystalline phase. This behavior strongly suggested that the γ crystalline phase is closer in proximity to the clay surfaces than the α crystalline phase.

To explain the observed $T_1{}^H$ behaviors for both α and γ crystalline phases, the simplified spin diffusion model was proposed: the inset shows the schematic illustration [30, 31]. Basics of this model are as follows: (1) the protons of N6 decay initially with the original $T_1{}^H$, (2) the very short $T_1{}^H$ directly affected by the paramagnetic Fe^{3+} for the γ crystalline phase, which is directly contacting with the clay existed in the range of 0.4 nm, influences on the neighboring protons in the γ crystalline phase, and finally (3) ^1H spin diffusion (SD) from the γ crystalline phase to the α crystalline phase produces the relatively longer initial slope. The utilized parameters

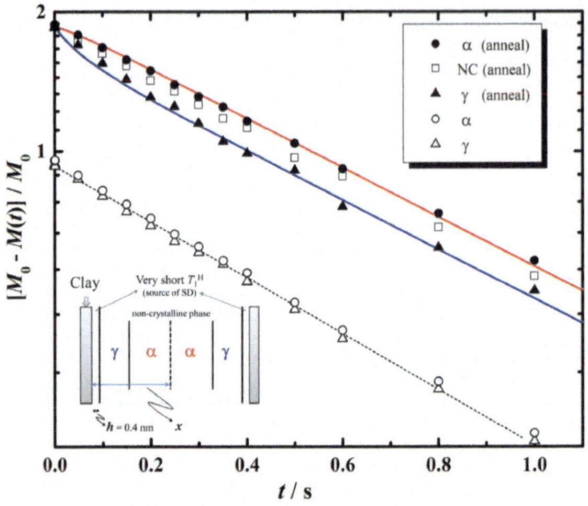

Fig. 2.15 Observed $T_1{}^H$ behaviors for the N6/mmt nanocomposites: *open triangle* and *circle symbols* are without annealing and those *solid symbols* with annealing at 487 K for 18 h. *Open squares* are the $T_1{}^H$ behaviors observed from the non-crystalline phase (NC) of the annealed sample. The *solid curves* are obtained from the simplified spin diffusion (SD) computational model with two parameters of the clay layers' spacing (2*x*) and the very short $T_1{}^H$ value affected by the paramagnetic Fe^{3+}: N6 chains neighboring clay surface have such a short $T_1{}^H$ and the range (*h*) is assumed to be ca. 0.4 nm. This figure is newly drawn using the original data of VanderHart et al. [30]

were the very short $T_1{}^H$ value and the clay–clay spacing: the SD constant and the γ crystallite fraction are 0.7 nm^{2}ms^{-1} and 0.25, respectively. The simulated curves were in excellent agreement with the observed data points. The estimated clay–clay spacing (2*x*) of ca. 46 nm was comparable to that estimated from the TEM image [30, 33].

2.4 Rubbers and Elastomers

Rubbers and elastomers are one of the indispensable materials for agricultural, medical, electrical appliances, and various engineering fields. The rubbery properties of them, such as damping or sealing properties, are closely related to their molecular motion. For example, the relationship between rigidity and elasticity would be interpreted by the storage and loss moduli; the wet gripping performance or the rolling resistance for a tire is concerned with these parameters. Furthermore, the degradation of rubbers causes a significant change in molecular motion: the reduction in molecular motion occurs and rubbers become rigid and fragile. Thus, understanding molecular motion is very important to investigate the intrinsic rubbers' character. To detect the change in the molecular motion, the ^{1}H spin–spin relaxation process of overall magnetizations detected by a low-resolution and low-field NMR is useful

[40]. Here, the reduction in molecular motion caused by the thermal degradation is addressed briefly by describing an aging investigation for the seal ability of rubbers [41].

Figure 2.16 shows (a) the relationship between the compression set and the acetone soluble fraction for acrylonitrile butadiene rubbers (NBR), which were used as sealing materials, and (b) the correlation between the ^{1}H spin–spin relaxation time (T_2^{H}) and the compression set: the compression set (CS) is determined from the remaining permanent deformation after compression [41]. A larger CS value indicates that a sample has incurred severe damage much more and become rigid and non-flexible: the sample having a CS value of 80% is regarded to be degraded and out of the limitation for use. In particular, elasticity and flexibility are essential for sealing materials. The used samples were kept at 333, 353, and 473 K up to 12,800 h in the air to make reach a higher CS value. The two NBRs had different shapes and different acrylonitrile contents.

The symbols less than 10% of acetone soluble fraction and beyond 90% CS represent the aging period of 12,800 h at 80 and 100 °C. On the other hand, at 60 °C aging (*open* and *solid circles*), NBRs did not reach such a high CS and lower acetone soluble fraction region. Similar results were also observed for the correlation between the toluene swelling ratio and the CS. The acetone soluble fraction or the toluene swelling ratio relates to the amount of included plasticizer or of cross-link density, respectively. The lower value of the acetone soluble fraction shows a decrease in plasticizer by volatilization or bleed-out accompanied by degradation. Similarly, the lower value of the toluene swelling ratio indicates the increase in the

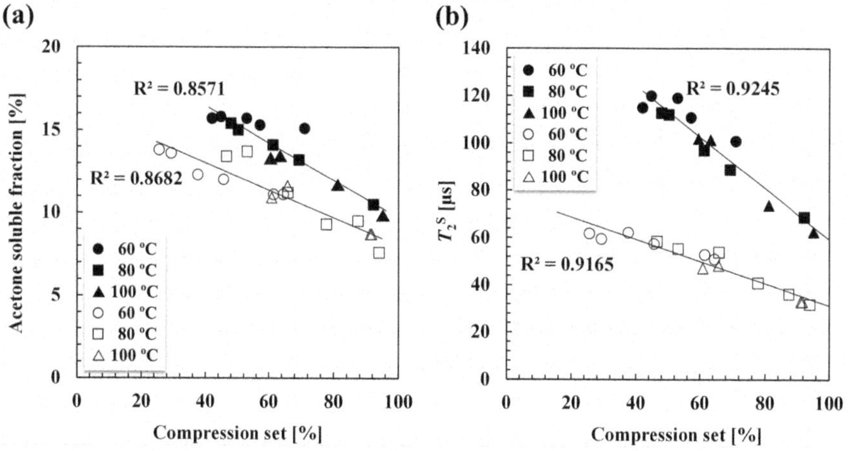

Fig. 2.16 Relationship between the acetone soluble fraction and the compression set **a** and between the short T_2^{H} component (T_2^{S}) and the compression set **b**. *Open symbols* represent plate shape and 37% acrylonitrile content NBR, and *solid symbols* are O-ring and 46% acrylonitrile content NBR. The aging was executed under the air condition and up to 12,800 h at a constant temperature. The temperature was kept within ± 1 K for every aging period. The source of this figure is Numata et al. [40, 41] and the current figure is arranged using the original source

cross-link density occurring during degradation. Thus, the good correlation between the acetone soluble fraction and the CS as shown in Fig. 2.16a indicates that the molecular interaction strengthened by the decrease in plasticizer concentration with aging reduces the molecular mobility.

The reduction in molecular motion causes the shortening of T_2^H. The observed T_2^H decays were fitted by the conventional double-exponential function, and the analysis gave the short T_2^H (T_2^S) and long T_2^H (T_2^L) values and their fractions. Among the parameters, T_2^S values reflected the reduction in molecular motion caused by the high cohesion much better [41]. Figure 2.16b shows the good proportionality between the reduction in T_2^S values and the increase in CS for both NBRs. This observation strongly supported that the reduction in T_2^S represents the decrease in sealability of rubber materials in the thermal degradation process. The degradation by compression in the long term, called aging, causes the rigidness and brittleness of rubbers with changes in chemical structure and polymer segmental motion. It is very important to know the aging process and its degree to ensure the lifetime of rubber products.

2.5 Conclusion

In this chapter, I described the characterization of polymer blends/alloys and nanocomposites via the various high-resolution solid-state ^{13}C NMR techniques, and of rubbers via the low-resolution ^{1}H NMR method. To know morphology and/or interaction between components is significant to understand the physical and chemical properties. Information of molecular motion would directly connect to the physical properties, such as flexibility, rigidity, entanglement of polymer chains, resistance to impact or friction. The analyses of NMR relaxation behaviors via the spin diffusion process or utilization of paramagnetic effect provided specific characterization on a molecular level such as domain size of crystalline phase or components, phase separation process, change in molecular motion of polymer chain, and the degree of dispersion of fillers. The correlation NMR spectroscopy directly proved the evidence of interaction between clay surfaces and polymer chains. Consequently, advanced solid-state NMR techniques would certainly be useful and helpful for the characterization of synthetic or natural polymers, amorphous or semicrystalline polymers, ordered and oriented polymers, and various polymer nanocomposites with organic or inorganic fillers.

References

1. Pinnavaia, T.J., Beall, G.W. (eds.): Polymer-Clay Nanocomposites. Wiley, Chichester (2000)
2. Asano, A., Hori, S., Kitamura, M., Nakazawa, C.T., Kurotsu, T.: Influence of magic angle spinning on T_1^H of SBR studied by solid-state ^{1}H NMR. Polym. J. **44**, 706–712 (2012)

3. bin Samusuri, A.: Degradation of natural rubber and synthetic elastomers. In: Richardson, T. (ed.) Shreir's Corrosion, **3**. pp. 2407–2438. Elsevier, Durham (2010)
4. Asano, A., Takegoshi, K.: Polymer blends and miscibility. In: Ando, I., Asakura, T. (eds.) Solid State NMR of Polymers, pp. 351–414. Elsevier, Amsterdam (1998)
5. Solomon, I.: Relaxation process in a system of two spins. Phys. Rev. **99**, 559–565 (1955)
6. Stejskal, E.O., Schaefer, J., Sefcik, M.D., Mckay, R.A.: Magic-angle carbon-13 nuclear magnetic resonance study of the compatibility of solid polymeric blends. Macromolecules **14**, 275–279 (1981)
7. Asano, A., Takegoshi, K., Hikichi, K.: ^{13}C c.p./m.a.s. n.m.r. study on the miscibility and phase separation of a polystyrene/poly(vinyl methyl ether) blend. Polymer **35**, 5630–5636 (1994)
8. Bank, M., Leffingwell, J., Thies, C.: The Influence of solvent upon the compatibility of polystyrene and poly(vinyl methyl ether). Macromolecules **4**, 43–46 (1971)
9. Kwei, T.K., Nishi, T., Roberts, R.F.: A study of compatible polymer mixtures. Macromolecules **7**, 667–674 (1974)
10. Nishi, T., Kwei, T.K.: Cloud point curves for poly(vinyl methyl ether) and monodisperse polystyrene mixtures. Polymer **16**, 285–290 (1975)
11. Takegoshi, K., Hikichi, K.: Effects of blending on local chain dynamics and glass transition: polystyrene/poly(vinyl methyl ether) blends as studied by high-resolution solid-state 13C nuclear magnetic resonance spectroscopy. J. Chem. Phys. **94**, 3200–3206 (1991)
12. Menestrel, C.L., Kenwright, A.M., Sergot, P., Lauprêtre, F., Monnerie, L.: Carbon-13 NMR investigation of local dynamics in compatible polymer blends. Macromolecules **25**, 3020–3026 (1992)
13. Asano, A., Takegoshi, K.: Free volume study of amorphous polymers detected by solid state ^{13}C NMR linewidth experiments. J. Chem. Phys. **115**, 8665–8669 (2001)
14. Asano, A., Eguchi, M., Shimizu, M., Kurotsu, T.: Miscibility and molecular motion of PMAA/PVAc blends investigated by high-resolution solid state CPMAS ^{13}C NMR. Macromolecules **35**, 8819–8824 (2002)
15. Asano, A.: Polymer blends. In: Webb, G.W. (ed.) Modern Magnetic Resonance, pp. 627–631. Springer, Dordrecht (2006)
16. Asano, A.: Hydrogen-bond Interaction of PMAA/PVAc blends: a natural abundant two-dimensional ^{13}C exchange NMR investigation. Polym. J. **36**, 23–27 (2004)
17. Asano, A., Nishioka, M., Takahashi, Y., Kato, A., Hikasa, S., Iwabuki, H., Nagata, K., Sato, H., Hasegawa, T., Sawabe, H., Arao, M., Suda, T., Isoda, A., Mukai, M., Ishikawa, D., Izumi, T.: High impact properties of polyketone/polyamide-6 alloys induced by characteristic morphology and water absorption. Macromolecules **42**, 9506–9514 (2009)
18. Nishioka, M., Takahashi, Y., Kato, A., Hikasa, S., Iwabuki, H., Nagata, K., Asano, A.: Influence of water absorption on high-impact properties of polyketone/polyamide polymer alloy. Kobunshi Ronbunshu **66**, 570–576 (2009)
19. Takahashi, Y., Nishioka, M., Kato, A., Hikasa, S., Iwabuki, H., Nagata, K., Asano, A.: Nanostructure of polyketon/polyamide polymer alloy. Kobunshi Ronbunshu **66**, 577–584 (2009)
20. Kato, A., Nishioka, M., Tkahashi, Y., Suda, T., Sawabe, H., Isoda, A., Drozdova, O., Izumi, T., Nagata, K., Hikasa, S., Iwabuki, H., Asano, A.: Phase separation and mechanical properties of polyketone/polyamide polymer alloys. J. Appl. Polym. Sci. **116**, 3056–3069 (2010)
21. Kato, A., Isono, Y., Nagata, K., Asano, A., Ikeda, Y.: Study on polymeric nanocomposites by 3D-TEM and related techniques. In: Kumar, C.S.S.R. (ed.) Transmission Electron Microscopy Characterization of Nanomaterials, pp. 139–193. Springer, Berlin (2014)
22. Assink, A.: Nuclear spin diffusion between polyurethane microphases. Macromolecules **11**, 1233–1237 (1978)
23. Fujisawa, Y., Nan, Y., Asano, A., Yanagisawa, Y., Yano, K., Itoh, Y., Aida, T.: Blending for making nonhealable polymers to be healable: nanophase separation observed by CP/MAS ^{13}C NMR Analysis. Angew. Chem. Int. Ed. **62**, e202214444 (2023)
24. Gelles, R., Frank, C.W.: Phase separation in polystyrene/poly(vinyl methyl ether) blends as studied by excimer fluorescence. Macromolecules **15**, 1486–1491 (1982)

25. Gelles, R., Frank, C.W.: Effect of molecular weight on polymer blend phase separation kinetics. Macromolecules **16**, 1448–1456 (1983)
26. Nishi, T., Wang, T.T., Kwei, T.K.: Thermally induced phase separation behavior of compatible polymer mixtures. Macromolecules **8**, 227–234 (1975)
27. Miyoshi, T., Takegoshi, K., Terao, T.: ^{129}Xe n.m.r, study of free volume and phase separation of the polystyrene/poly(vinyl methyl ether) blend. Polymer **38**, 5475–5480 (1997)
28. Nieuwendaal, R.C., Ro, H.W., Germack, D.S., Kline, R.J., Toney, M.F., Chan, C.K., Agrawal, A., Gundlach, D., VanderHart, D.L., Delongchamp, D.M.: Measuring domain sizes and compositional heterogeneities in P3HT-PCBM bulk heterojunction thin films with ^{1}H spin diffusion NMR spectroscopy. Adv. Func. Mater. **22**, 1255–1266 (2012)
29. Nieuwendaal, R.C., Snyder, C.R., Kline, R.J., Lin, E.K., VanderHart, D.L., DeLongchamp, D.M.: Measuring the extent of phase separation in poly-3-hexylthiophene/phenyl-C61-butyric acid methyl ester photovoltaic blends with ^{1}H spin diffusion NMR spectroscopy. Chem. Mater. **22**, 2930–2936 (2010)
30. VanderHart, D.L., Asano, A., Gilman, J.W.: Solid state NMR investigation of paramagnetic nylon-6 clay nanocomposites. 2. Measurement of clay dispersion, crystal stratification, and stability of organic modifiers. Chem. Mater. **13**, 3796–3809 (2001)
31. Bourbigot, S., VanderHart, D.L., Gilman, J.W., Awad, W.H., Davis, R.D., Morgan, A.B., Wilkie, C.A.: Investigation of nanodispersion in polystyrene-montmorillonite nanocomposites by solid-state NMR. J. Polym. Sci.: Part B: Polym. Phys. **41**, 3188–3213 (2003)
32. Asano, A., Shimizu, M., Kurotsu, T.: Effect of paramagnetic Fe^{3+} on $T_1{}^{H}$ in PVA/montmorillonite-clay nanocomposites. Chem. Lett. **33**, 816–817 (2004)
33. VanderHart, D.L., Asano, A., Gilman, J.W.: Solid state NMR investigation of paramagnetic nylon-6 clay nanocomposites. 1. Crystallinity, morphology, and the direct influence of Fe^{3+} on nuclear spins. Chem. Mater. **13**, 3781–3795 (2001)
34. VanderHart, D.L., Asano, A., Gilman, J.W.: NMR Measurements related to clay-dispersion quality and organic-modifier stability in nylon-6/clay nanocomposites. Macromolecules **34**, 3819–3822 (2001)
35. Asano, A.: Polymer blends and composites. In: Webb, G.W. (ed.) Modern Magnetic Resonance, 2nd edn., pp. 793–807. Springer, Switzerland (2018)
36. Hou, S.-S., Bonagamba, T.J., Beyer, F.L., Madison, P.H., Schmidt-Rohr, K.: Clay intercalation of poly(styrene-ethylene oxide) block copolymers studied by two-dimensional solid-state NMR. Macromolecules **36**, 2769–2776 (2003)
37. Strobl, G.R.: The Physics of Polymers. Springer, Berlin (1997)
38. Asano, A., Murata, Y., Kurotsu, T.: Crystallinity and miscibility of poly(vinyl isobutyl ether)/poly(ε-l-lysine) blends by solid state ^{13}C NMR study. E-J. Soft Mater. **3**, 1–8 (2007)
39. Asano, A., Tanaka, C., Kurotsu, T.: Melting point and domain Size of PVIBE/ε-PL/saponite clay investigated by solid-state ^{13}C NMR. Macromolecules **41**, 9469–9473 (2008)
40. Asano, A.: NMR relaxation studies of elastomers. In: Webb, G.W. (ed.) Annual Reports on NMR Spectroscopy, pp. 1–72. Academic Press (Elsevier), Oxford (2015)
41. Numata, K., Kurokawa, H., Kawaguchi, S., Sekine, S., Nakazawa, Y., Asano, A.: Evaluation of sealability for aged rubber seals by spin-spin relaxation time. Polym. Test. **49**, 147–155 (2016)

Chapter 3
Solid-State ^{2}H NMR Studies of Molecular Motion in Functional Materials

Motohiro Mizuno

Abstract In this chapter, the analysis of molecular motion in solid materials, which is connected with the function of those materials, using ^{2}H NMR spectroscopy is described. The line shape of a ^{2}H NMR spectrum, which is dominated by the quadrupole interaction, is very sensitive to molecular motion. The mode and rate of molecular motion in the dynamic range of 10^4~10^7 Hz can be obtained by the line shape analysis of a ^{2}H NMR broad line spectrum of a powder sample using spectral simulation. Molecular motion in the order of 10^3 and 10^8 Hz can be analyzed by a line shape and line width of a ^{2}H NMR Quadrupolar Carr-Purcell-Meiboom-Gill (QCPMG) spectrum. The analysis of molecular motion in paramagnetic materials is possible by simulation of the ^{2}H NMR spectrum, including paramagnetic effects. The methods used to simulate these spectra are explained. In addition, the application of these methods to porous coordination polymer (PCP)/metal–organic framework (MOF), proton conducting material, and spin-crossover material is introduced.

Keywords Molecular motion · Paramagnetic effect · MOF/PCP · Proton conduction · Spin-crossover

3.1 Introduction

The function of solid materials is significantly affected by the structure and dynamics of local space, which is formed by molecules and ions in solid materials. Therefore, to develop innovative functional solid materials, the importance of material design based on detailed local structure and dynamics analysis is growing. Solid-state NMR spectroscopy is a very powerful tool for investigating molecular dynamics and local

M. Mizuno (✉)
Nanomaterials Research Institute, Kanazawa University, Kakuma-Machi, Kanazawa 920-1192, Ishikawa, Japan
e-mail: mizuno@se.kanazawa-u.ac.jp

Department of Chemistry, Graduate School of Natural Science and Technology, Kanazawa University, Kakuma-Machi, Kanazawa 920-1192, Ishikawa, Japan

The NMR Society of Japan (ed.), *Experimental Approaches of NMR Spectroscopy II*,
https://doi.org/10.1007/978-981-97-6830-1_3

structure in solid materials. Solid-state ^{2}H NMR spectroscopy is especially effective for the analysis of molecular motion in solid materials [1–3]. The nuclear spin quantum number of deutron is $I = 1$ and the line shape of a ^{2}H NMR spectrum of solid materials is mainly dominated by the quadrupole interaction. The change of the quadrupole interaction due to molecular motion creates the characteristic line shape of a ^{2}H NMR spectrum. The mode and rate of molecular motion can be revealed by the line shape analysis of a ^{2}H NMR broad line spectrum and a Quadrupolar Carr-Purcell-Meiboom-Gill (QCPMG) spectrum of a powder sample using spectral simulation [1–10].

For paramagnetic materials, because of the fast relaxation due to the interaction between the unpaired electron spin and resonant nuclei, the use of spin–lattice relaxation time (T_1) and a two-dimensional exchange spectrum to study molecular dynamics is difficult. So, line shape analysis of one-dimensional NMR spectra becomes very important for studying molecular dynamics in paramagnetic materials. For the analysis of molecular motion in paramagnetic materials, the simulation of a ^{2}H NMR spectrum, including not only the quadrupole interaction but also paramagnetic effects, is necessary [1, 11–16].

In this chapter, the methods used to simulate a ^{2}H NMR spectrum are explained. The application of these methods to the analysis of molecular motion in functional materials is then introduced.

3.2 Measurements of Solid-State ^{2}H NMR Spectrum

To measure a solid-state ^{2}H NMR spectrum, echo methods are usually used in order to avoid a rapid decrease of the NMR signal due to the quadrupole interaction. Figure 3.1a and b shows a quadrupolar echo sequence for the measurement of a ^{2}H NMR broad line spectrum and the pulse sequence for a ^{2}H NMR QCPMG spectrum, respectively [1–3, 10]. For the QCPMG measurement, echo signals are connected to form a single echo-train FID signal for Fourier transfer (Fig. 3.1c). Figure 3.2 shows a typical line shape of a ^{2}H NMR broad line and QCPMG spectra of a powder sample without molecular motion. The broad line spectrum becomes a symmetric line shape with two horns (the Pake pattern) in the absence of molecular motion. The width of the two horns depends on the quadrupole coupling constant (e^2qQ/h) and on the asymmetry parameter (η).

The QCPMG spectrum consists of sharp peaks separated by $1/\tau_a$ where τ_a is the detection period of each echo signal (Fig. 3.1b) and is similar to the spinning side-band pattern of the ^{2}H NMR MAS spectrum. For paramagnetic materials, the interactions between deuteron and unpaired electrons cannot be negligible. Since the quadrupolar echo sequence only refocuses the dephasing of the quadrupole interaction, distortion of the line shape is caused by a paramagnetic shift [1, 11–16]. Hence the pulse sequences of Fig. 3.3 are used for the measurement of paramagnetic materials. Figure 3.3a is a shift-compensated quadrupolar echo sequence with the insertion of π-pulses in the quadrupolar echo sequence midway between

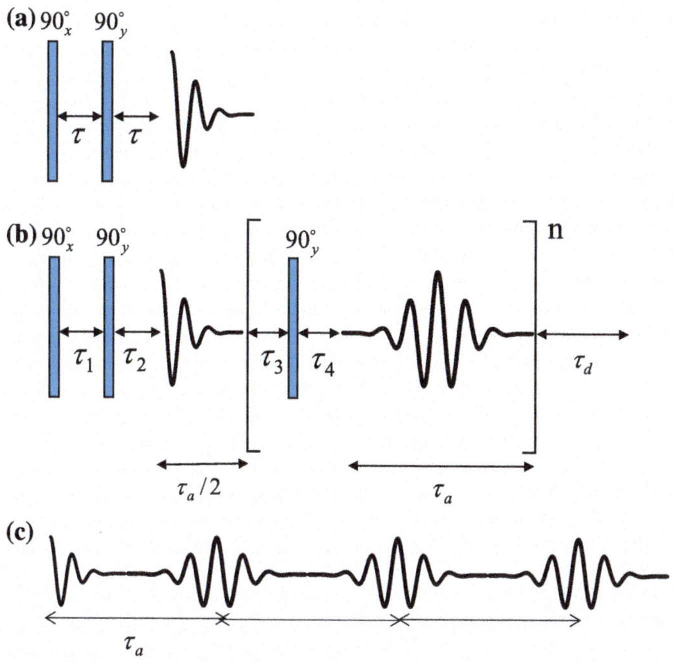

Fig. 3.1 Pulse sequences for the ^{2}H NMR broad line spectrum and the QCPMG spectrum. **a** Quadrupolar echo sequence, **b** QCPMG sequence, **c** NMR signal obtained by the QCPMG sequence

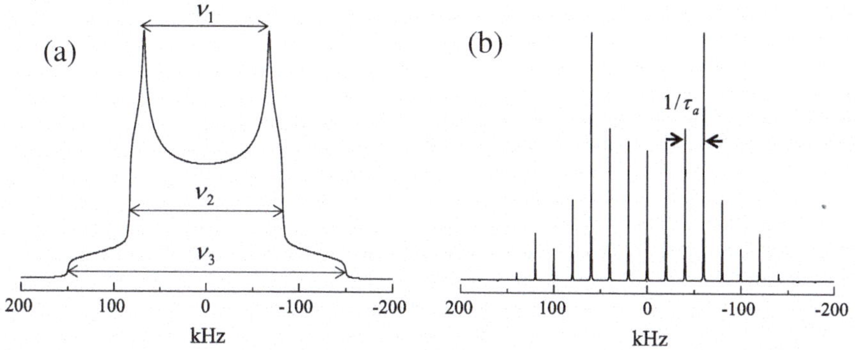

Fig. 3.2 Typical line shape of a ^{2}H NMR broad line spectrum and a QCPMG spectrum without molecular motion simulated by $e^2qQ/h = 200$ kHz and $\eta = 0.1$. **a** Broad line spectrum, **b** QCPMG spectrum. $\nu_1 = \frac{3e^2qQ}{4h}(1-\eta)$, $\nu_2 = \frac{3e^2qQ}{4h}(1+\eta)$, $\nu_3 = \frac{3e^2qQ}{2h}$

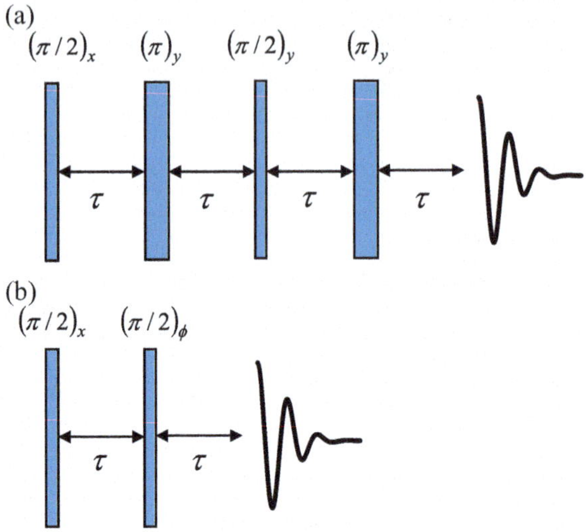

Fig. 3.3 Pulse sequences of a ^{2}H NMR broad line spectrum for paramagnetic materials. **a** Shift-compensated quadrupolar echo sequence, **b** exorcycled quadrupolar echo sequence

the π/2-pulses to refocus the paramagnetic shift [11]. Figure 3.3b is an exorcycled quadrupolar echo sequence with the selection of the coherence transfer pathway [12]. The signal intensity of the exorcycled quadrupolar echo sequence becomes half of that of the shift-compensated quadrupolar echo sequence. However, the exorcycled quadrupolar echo sequence that uses only two π/2-pulses is effective for materials with a very short spin–lattice relaxation time due to paramagnetic relaxation.

The ^{2}H NMR spectra in this chapter were obtained by applying a resonance frequency of 45 MHz and a π/2-pulse of 2~3 μs.

3.3 Simulation Methods of Solid-State ^{2}H NMR Spectrum

The calculation of a spectrum is performed by considering the ^{2}H NMR frequency and the spin–spin relaxation rate at each deuteron site and the jump frequency of deuterons between sites. For diamagnetic materials, the ^{2}H NMR frequency is dominated by the quadrupole interaction, and the frequency at site i is written as [1, 2, 4, 17]

$$\omega_i = \pm\omega_{Qi} \tag{3.1}$$

$$\omega_{Q_i} = \sqrt{\frac{3}{2}} \sum_{n,m=-2}^{2} D_{0n}^{(2)*}(\psi, \theta, \phi) D_{nm}^{(2)*}\left(\alpha_{Q_i}, \beta_{Q_i}, \gamma_{Q_i}\right) \rho_{2m}^{Q} \tag{3.2}$$

$$\rho_{20}^{Q} = \sqrt{\frac{3}{8}} \frac{e^2 qQ}{\hbar}, \quad \rho_{2\pm2}^{Q} = \frac{\eta}{4} \frac{e^2 qQ}{\hbar} \tag{3.3}$$

where $D_{nm}^{(2)*}(\Omega)$ is the second-order Wigner rotation matrix. $(\alpha_{Q_i}, \beta_{Q_i}, \gamma_{Q_i})$ and (ψ, θ, ϕ) are the Euler angles for transformation from the principal axes of the quadrupole tensor to the molecular axes and from the molecular axes to the laboratory axes, respectively. e^2qQ/h and η are the quadrupole coupling constant and asymmetry parameter, respectively. The echo signal $G(t, \theta, \phi)$ is written as

$$G(t,\theta,\phi) = \mathbf{W} \cdot \hat{\mathbf{B}}^3_{\pi/2}\exp\left[\hat{A}t\right]\exp\left[\hat{A}\left(\tau + t_{\pi/2}\right)\right]\exp\left[\hat{A}^*\left(\tau + t_{\pi/2}\right)\right] \cdot 1 \quad (3.4)$$

where $\hat{A}$ is the matrix with the elements $i\omega_i - k_{ii} - R_2$ on the diagonal and k_{ij} off the diagonal. k_{ij} is the jumping rate between sites i and j. R_2 is the spin–spin relaxation rate [4, 5]. For the N-site jump, $\hat{A}$ becomes an N × N dimensional matrix and the relation $k_{ii} = (N-1)k_{ij}$ is established. **W** and **1** are a vector of site populations and a vector written as **1** = (1,1, ⋯, 1), respectively. The matrix for finite width of π/2-pulse $\hat{\mathbf{B}}_{\pi/2}$ is written as [5]

$$\hat{\mathbf{B}}_{\pi/2} = \omega_1 \sin\left(t_{\pi/2}\sqrt{\omega_1^2 + \Omega_i^2/4}\right)/\sqrt{\omega_1^2 + \Omega_i^2/4} \quad (3.5)$$

where $t_{\pi/2}$ and $\omega_1 (=\gamma B_1)$ represent the width and the strength of the π/2-pulse, respectively. Ω_i is the imaginary part of the ith eigenvalue of the matrix $\hat{A}$. For the second π/2-pulse, the effect of pulse width becomes $\hat{\mathbf{B}}^2_{\pi/2}$ because of the time evolution of multi-quantum coherence arising from the quadrupole interaction under the RF field during $t_{\pi/2}$. Therefore $\hat{\mathbf{B}}^3_{\pi/2}$ is multiplied in Eq. (3.5) [5, 18]. The signal of the powder sample was calculated by

$$G(t) = \int_0^{2\pi}\int_0^{\pi} G(t,\theta,\phi)\sin\theta d\theta d\phi \quad (3.6)$$

The broad line spectrum was obtained by the Fourier transformation of $G(t)$. The QCPMG spectrum can be obtained by the repeated calculation of time evolution due to $\hat{A}^*$ and $\hat{A}$ after calculation of Eq. (3.4).

Figure 3.4 shows the simulation of a broad line spectrum in the presence of a fast n-site jump of the X–^{2}H axis around the n-hold symmetry axis. The principal z-axis of the quadrupole tensor was assumed to lie on the X–^{2}H axis. θ is an angle between the X–^{2}H axis and the n-hold symmetry axis. k_{ij} was set as 1×10^9 Hz. When the jumping rate is fast enough, the line shape of the spectrum becomes the same and η becomes 0 for the rotation of $n \geq 3$. e^2qQ/h decreases as θ approaches 54.7° and becomes 0 at $\theta = 54.7°$. The line shape of the spectrum for the rotation of $n = 2$, which corresponds to the 180° flip motion, is different from that of $n \geq 3$. η increases as θ approaches 54.7° and becomes 1 at $\theta = 54.7°$ for $n \geq 3$. The line shape of the ^{2}H NMR spectrum with a large η is observed when a fast 180° flip of the water molecule occurs. Figure 3.5 shows the simulation of a broad line spectrum in the presence of a 3-site jump and a 6-site jump of the X–^{2}H axis around the n-hold axis with an intermediate rate. In the dynamic range of $k = 10^4$~10^7 Hz, the line

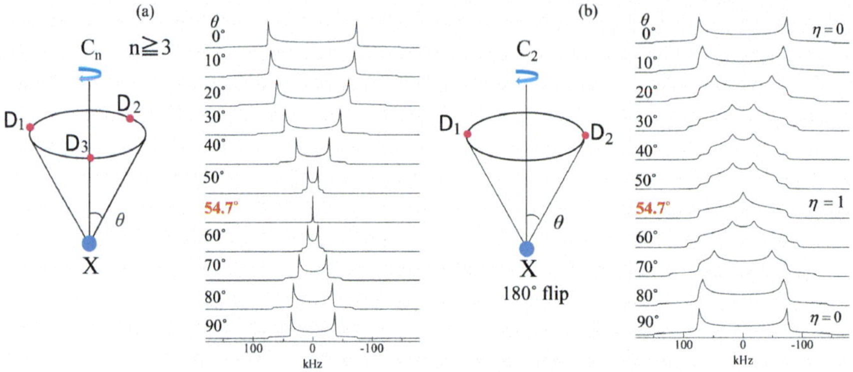

Fig. 3.4 Simulation of a ^{2}H NMR broad line spectrum in the presence of a fast n-site jump of the X$-^2$H axis around the n-hold axis. **a** $n \geq 3$, **b** $n = 2$. $e^2qQ/h =$ 200 kHz, $\eta = 0$, rotational rate $k = 1 \times 10^9$ Hz were used for the simulation

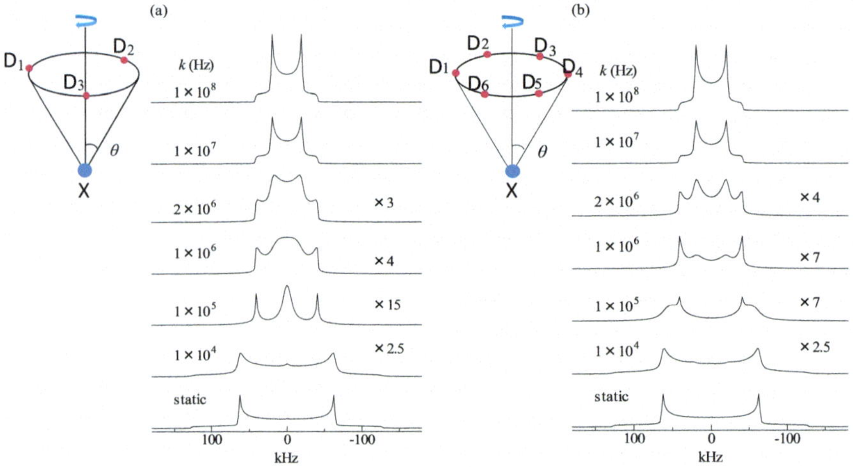

Fig. 3.5 Simulation of a ^{2}H NMR broad line spectrum in the presence of an n-site jump of the X$-^2$H axis around the n-hold axis with an intermediate rotational rate. **a** $n = 3$, **b** $n = 6$. $e^2qQ/h =$ 170 kHz, $\eta = 0$, $\theta = 70°$ were used for the simulation

shape of the ^{2}H NMR spectrum depends on the mode and rate of reorientational motion. Figure 3.6 shows a simulation of the QCPMG spectrum in the presence of a 3-site jump of the X$-^2$H axis around the n-hold axis with an intermediate rate. The pronounced decrease in the intensity of the QCPMG spectrum takes place in the dynamic range of $k = 10^4$~10^6 Hz. The line broadening of each peak appears in the QCPMG spectrum in the presence of molecular rotation with an order of 10^3 and 10^8 Hz. Thus, by using a QCPMG spectrum, it is possible to obtain information of molecular motion in the order of 10^3 and 10^8 Hz, which is difficult to analyze using a broad line spectrum.

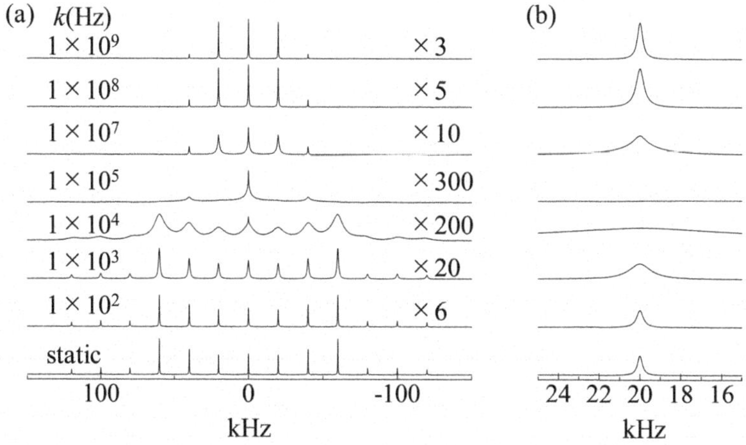

Fig. 3.6 Simulation of a ^{2}H NMR QCPMG spectrum in the presence of a 3-site jump of the X−^{2}H axis around a 3-hold axis with an intermediate rotational rate. **a** Whole spectrum, **b** Expansion around 20 kHz. $e^2qQ/h = 170$ kHz, $\eta = 0$, $\theta = 70°$ were used for the simulation

For paramagnetic materials, not only the quadrupole interaction but also the paramagnetic interaction affects the ^{2}H NMR frequency, and the frequency at site i is written as [1, 13–16]

$$\omega_i = \pm\omega_{Qi} - \sum_j \omega_{P_{ij}} + \omega_{CS_i} \tag{3.7}$$

$$\omega_{P_{ij}} = \sum_{n=-2}^{2} D_{0n}^{(2)*}(\psi, \theta, \phi) D_{n0}^{(2)*}\left(\alpha_{P_{ij}}, \beta_{P_{ij}}, \gamma_{P_{ij}}\right) \rho_{20}^{P_{ij}} \tag{3.8}$$

$$\rho_{20}^{P_{ij}} = \left(\frac{\mu_0}{4\pi}\right) \frac{2\gamma_D B_0 g_j^2 \mu_B^2}{3k_B T r_{ij}^3} S(S+1) \tag{3.9}$$

where $\omega_{P_{ij}}$ is the contribution of the dipolar interaction between the ^{2}H nucleus and the jth paramagnetic ion and ω_{CS_i} is the contribution of the isotropic paramagnetic shift (Fermi's contact shift). $\left(\alpha_{P_{ij}}, \beta_{P_{ij}}, \gamma_{P_{ij}}\right)$ are the Euler angles for transformation from the principal axes of the paramagnetic dipolar tensor to the molecular axes. r_{ij} is the distance between the ^{2}H nucleus and the jth paramagnetic ion. T, γ_D, B_0, μ_B, k_B, S and g_j are temperature, gyromagnetic ratio of the ^{2}H nucleus, strength of the static magnetic field, Bohr magneton, Boltzmann constant, electron spin quantum number and g-factor of the jth paramagnetic ion, respectively.

Equation (3.9) can be rewritten as Eq. (3.10) using magnetic susceptibility χ_M

$$\rho_{20}^{P_{ij}} = \left(\frac{\mu_0}{4\pi}\right) \frac{2\gamma_D B_0 \chi_M}{N_A r_{ij}^3} \tag{3.10}$$

where N_A is Avogadro's constant. The anisotropic g-factor can be represented by

$$g_j^2 = g_{iso_j}^2 + \sum_{n,m=-2}^{2} D_{0n}^{(2)*}(\psi, \theta, \phi) D_{nm}^{(2)*}\left(\alpha_{g_j}, \beta_{g_j}, \gamma_{g_j}\right) \rho_{2m}^{g_j} \tag{3.11}$$

$$\rho_{20}^{g_j} = \sqrt{\frac{2}{3}}\left(g_{zz_j}^2 - \frac{1}{2}\left(g_{xx_j}^2 + g_{yy_j}^2\right)\right), \quad \rho_{2\pm2}^{g_j} = \frac{1}{2}\left(g_{yy_j}^2 - g_{xx_j}^2\right) \tag{3.12}$$

$$g_{iso_j}^2 = \frac{g_{xx_j}^2 + g_{yy_j}^2 + g_{zz_j}^2}{3} \tag{3.13}$$

where $\left(\alpha_{g_j}, \beta_{g_j}, \gamma_{g_j}\right)$ are the Euler angles for transformation from the principal axes of the g-tensor to the molecular axes. Figure 3.7 shows the temperature dependence of the ^{2}H NMR broad line spectrum for the $[Ni(D_2O)_6][SiF_6]$ crystal [15]. The spectrum was observed by a shift-compensated quadrupolar echo sequence. The spin quantum number of the Ni^{2+} ion is $S = 1$. Consequently, the line shape of the ^{2}H NMR spectrum becomes asymmetric due to the paramagnetic interaction. For $[Ni(D_2O)_6]^{2+}$, there are two molecular motions, a 180° flip of D_2O and the reorientation of $[Ni(D_2O)_6]^{2+}$ about the 3-hold axis (Fig. 3.8). In this temperature range, the 180° flip of D_2O is in the fast motion limit and temperature variation of the spectral line shape is caused by reorientation of $[Ni(D_2O)_6]^{2+}$. The principal axes system of the quadrupole tensor (x_p, y_p, z_p), which is averaged by a fast 180° flip of D_2O was assigned as follows: the z_p-axis is perpendicular to the water molecular plane, the y_p-axis lies in the water molecular plane and the x_p-axis is parallel to the bisector of D-O-D plane. The rate of reorientation k was obtained at each temperature after fitting (Fig. 3.7).

In the case of a strong paramagnetic interaction, fast spin–spin relaxation due to the paramagnetic interaction affects the line shape of the spectrum. The anisotropic spin–spin relaxation rate due to the paramagnetic dipolar interaction can be written as [16]

$$\begin{aligned} lR_{2i}^P = {} & \frac{4}{3}\left(\frac{\mu_0}{4\pi}\right)^2 \gamma_D^2 \mu_B^2 S(S+1) \\ & \times \sum_j g_j^2 \left\{ \frac{1}{6}\left|F_{20}^{ij}\right|^2 \left(4\tau_e + \frac{\tau_e}{1 + (\omega_D - \omega_e)^2 \tau_e^2}\right) \right. \\ & + \frac{1}{2}\left|F_{21}^{ij}\right|^2 \left(\frac{2\tau_e}{1 + \omega_e^2 \tau_e^2} + \frac{\tau_e}{1 + \omega_D^2 \tau_e^2}\right) \\ & \left. + \left|F_{22}^{ij}\right|^2 \left(\frac{\tau_e}{1 + (\omega_D + \omega_e)^2 \tau_e^2}\right) \right\} \end{aligned} \tag{3.14}$$

$$F_{2m}^{ij} = \sum_{n=-2}^{2} D_{mn}^{(2)*}(\psi, \theta, \phi) D_{n0}^{(2)*}\left(\alpha_{P_{ij}}, \beta_{P_{ij}}, \gamma_{P_{ij}}\right) \rho_{20}^{D_{ij}} \tag{3.15}$$

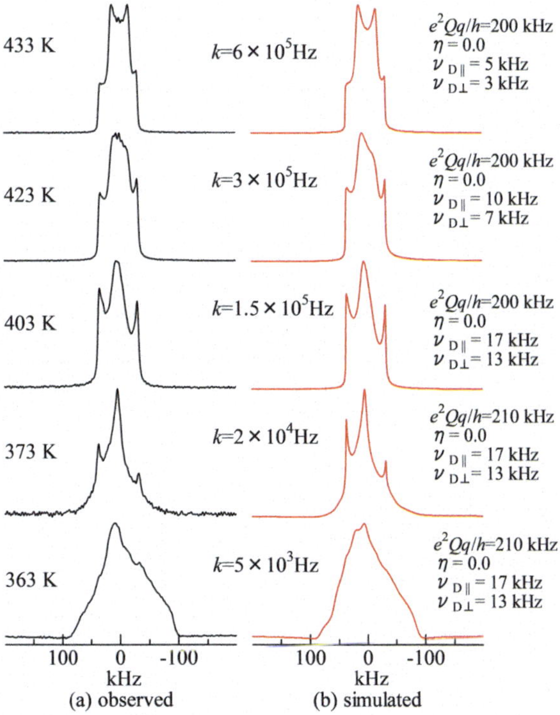

Fig. 3.7 Temperature dependence of a ^{2}H NMR broad line spectrum for the $[Ni(D_2O)_6][SiF_6]$ crystal [15]

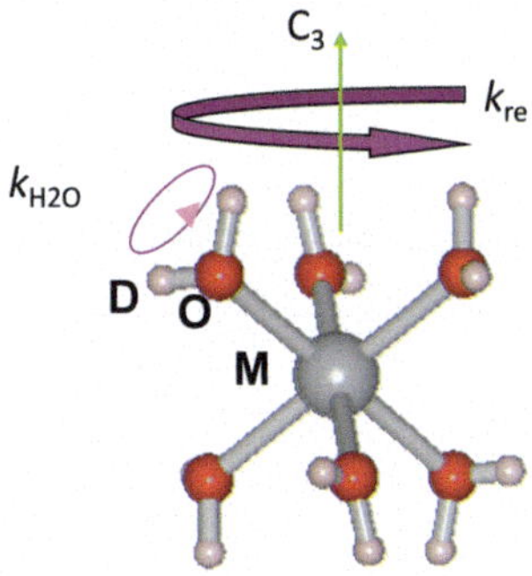

Fig. 3.8 Reorientation of $[M(H_2O)_6]^{2+}$ ($M = Ni^{2+}$, Mn^{2+}) and 180° flip of H_2O

$$\rho_{20}^{D_{ij}} = \sqrt{\frac{3}{2}} r_{ij}^{-3} \tag{3.16}$$

where ω_D and ω_e are angular NMR and ESR frequencies, respectively. τ_e is the correlation time of the electron spin.

Figure 3.9 shows the temperature dependence of the ^{2}H NMR spectrum for the $[Mn(D_2O)_6][SiF_6]$ crystal [16]. The spin quantum number of the Mn^{2+} ion is $S = 5/2$. Thus, the Mn^{2+} ion applies strong paramagnetic effects to the ^{2}H NMR spectrum. The contribution of the paramagnetic dipolar interaction between the ^{2}H nuclei and the nearest neighboring Mn^{2+} ion exceeds 100 kHz and even contribution from the second nearest neighbor Mn^{2+} ion causes the frequency to become *ca.* 30 kHz. The spin–lattice relaxation time (T_1) was ~70 μs. The exorcycled quadrupolar echo sequence was used to measure the spectrum of $[Mn(D_2O)_6][SiF_6]$. For the simulation in Fig. 3.9, the contribution of Mn^{2+} in 3^3 hexagonal unit cells of the crystal (81 Mn^{2+}) is applied. The anisotropic spin–spin relaxation rate was calculated using a crystal structure and $\tau_e = 3.0 \times 10^{-10}$ s obtained by a ^{2}H NMR T_1 of $[Mn(D_2O)_6][SiF_6]$. For $[Mn(D_2O)_6][SiF_6]$, both the 180° flip of D_2O and reorientation of $[Mn(D_2O)_6]^{2+}$ affect the change in line shape of the ^{2}H NMR spectrum in Fig. 3.9. Thus, a 6-site jump of deuteron around the C_3 axis was assumed. The principal z-axis of the quadrupole tensor was assumed to lie on the O-D axis. From the simulation analysis, the rotational rates $k_{180°}$ and k_{re} were obtained for the 180° flip of D_2O and for the reorientation of $[Mn(D_2O)_6]^{2+}$, respectively as shown in Fig. 3.9 [16].

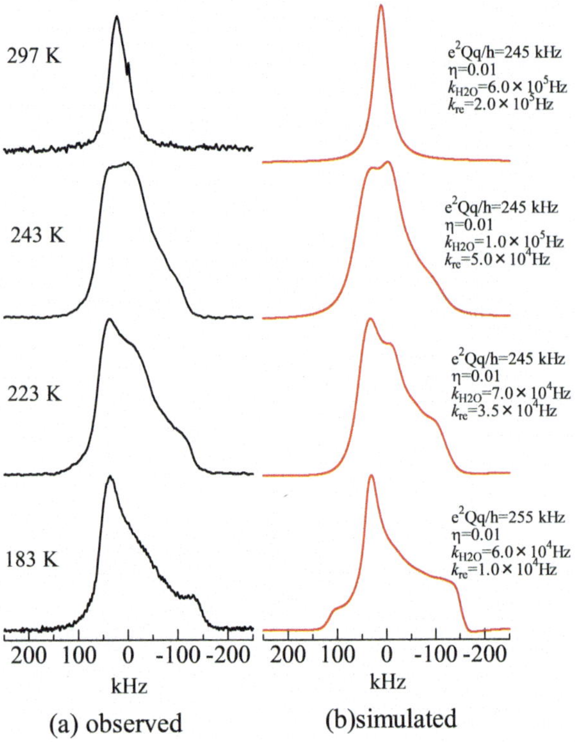

Fig. 3.9 Temperature dependence of a ^{2}H NMR broad line spectrum for the $[Mn(D_2O)_6][SiF_6]$ crystal [16]

3.4 Analysis of MOF/PCP

Metal–organic frameworks (MOFs) or porous coordination polymers (PCPs) have attracted attention since they can be used to make tunable pore sizes and topologies, which lead to versatile architectures and applications, and have been actively studied due to their applications in gas storage, separation, catalysis, and chemical sensing. The dynamics of nano-spaces (rotation of bridging ligand, adsorbed molecules, etc.) in MOFs (PCPs) have been investigated by solid-state ^{2}H NMR spectroscopy [19–24]. For $[Zn_2(1,4\text{-ndc})_2(dabco)]_n$ (1,4-ndc = 1,4-naphthalenedicarboxylate, dabco = 1,4-diazabicyclo[2,2,2]octane, Fig. 3.10) [20, 25], rotational motion of a naphthalene ligand before and after benzene adsorption was investigated. In this framework, paddle-wheel Zn^{2+} dimers are linked by 1,4-naphthalenedicarboxylate ions, resulting in regular square grid-type layers parallel to the ab plane, and these layers are pillared by 1,4-diazabicyclo[2,2,2]octane molecules (Fig. 3.10a). Rectangular channels of 5.7 × 5.7 Å^2 run along the c direction. At 296 K, the naphthalene ring of the dicarboxylate ion is disordered over four positions which are defined by orientation angles of 0, 70, 180, and 250° about the C1–C4 axis (Fig. 3.10b). The ^{2}H NMR spectrum of the naphthalene ring was selectively observed by using $[Zn_2([D_6]1,4\text{-ndc})_2(dabco)]_n$ where only 6 hydrogens of the naphthalene ring were replaced by deuterons. Figure 3.11 shows ^{2}H NMR spectra of $[Zn_2([D_6]1,4\text{-ndc})_2(dabco)]_n$ [20]. Figure 3.11a is the temperature dependence of the ^{2}H NMR spectrum before the adsorption of benzene. Figure 3.11b is the ^{2}H NMR spectrum after the adsorption of benzene at 293 K. The rapid rotation of the naphthalene ring around the C1–C4 axis takes place before the adsorption of benzene, even below room temperature, which can be ascertained from the line shape of the ^{2}H NMR spectrum in Fig. 3.11a. In contrast, the naphthalene ring is static at room temperature after the adsorption of benzene. The broad component with peaks at about ± 60 kHz in Fig. 3.11a corresponds to D2 of naphthalene (Fig. 3.10c). Since the C-D2 bonding direction corresponding to the principal axis direction of quadrupole interaction is almost consistent with the direction of the rotational axis, rotation of the naphthalene ring around the C1–C4 axis does not affect the spectrum of D2. Inner components with peaks at ± 15 kHz and ± 5 kHz correspond to D1 and D3. The broken lines in Fig. 3.11a represent the spectral simulation assuming rotation of the naphthalene ring among four positions in a disordered structure. $e^2qQ/h = 180$ kHz and $\eta = 0.02$ were used for the simulation. Euler angles were $(\alpha_{Q1}, \alpha_{Q2}, \alpha_{Q3}, \alpha_{Q4}) = (0°, 70°, 180°, 250°)$ for four sites, $\beta_Q = 56°, 60°$ for D1, D3 and $\beta_Q = 0°$ for D2 in Fig. 3.10c. Rotation rates ranging from 3.0×10^6 Hz at 203 K to 5×10^7 Hz at 223 K were determined from the fitting. The activation energy and the pre-exponential factor of the rotation of the naphthalene ring were obtained as 53 kJmol^{-1} and 9.5×10^{19} Hz, respectively, from an Arrhenius plot of rotation rate.

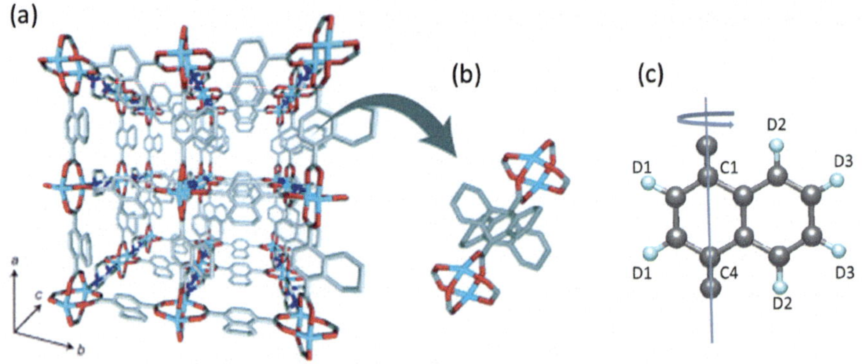

Fig. 3.10 Crystal structure of $[Zn_2(1,4\text{-ndc})_2(\text{dabco})]_n$ at 296 K. **a** The pillared layer structure viewed in the c direction. **b** Four-site disorder of the naphthalene ring. Hydrogen atoms are omitted; Zn blue, C gray, O red, N dark blue. Reproduced from ref. [20] by permission of John Wiley & Sons Ltd. **c** Position of deuterons in the naphthalene ring

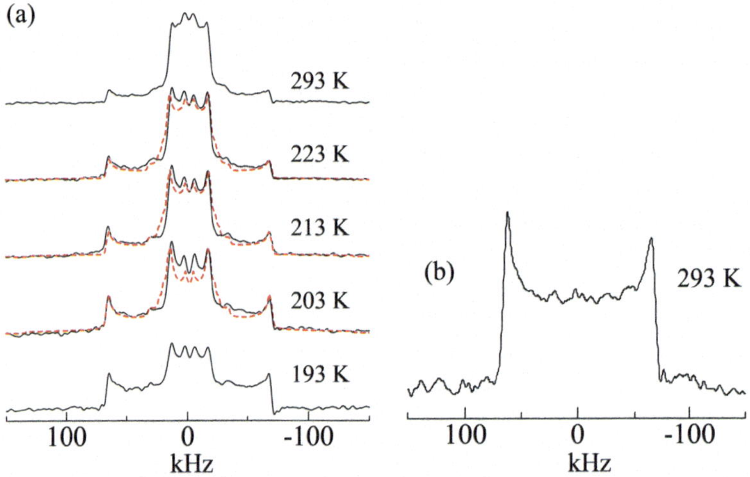

Fig. 3.11 ^{2}H NMR broad line spectra of $[Zn_2([D_6]1,4\text{-ndc})_2(\text{dabco})]_n$. **a** Temperature dependence of a ^{2}H NMR spectrum before benzene adsorption. **b** ^{2}H NMR spectrum after benzene adsorption at 293 K. Broken lines are simulation spectra assuming the rotation of the naphthalene ring about the C1-C4 axis [20]

3.5 Analysis of Solid Proton-Conducting Material

Anhydrous proton-conducting solids operating above 100 °C are required for fuel cell applications. Heterocyclic organic molecules (HOMs) such as imidazole (Im) and triazole (Tz) have attracted considerable attention for this purpose because they are non-volatile molecules with high boiling points. In such materials, proton transport occurs between HOMs through a hydrogen-bonded chain. When proton conductivity is dominated by continuous proton transport in the hydrogen bond network between HOMs (the Grotthuss mechanism), the reorientational motion of HOMs is considered to play an important role [26]. For poly(vinylphosphonic acid) (PVPA)−imidazole (Im) composite material (PVPA/xIm, where x represents the number of moles of Im per mole of polymer repeat unit, Fig. 3.12), proton conductivity increased with increasing x, and a sudden increase in proton conductivity was observed in the range from $x = 1$ to 2 [27]. In the PVPA/xIm composite, Im molecules are intercalated into host polymer PVPA and provide a migration path for excess protons caused by the dissociation of the acid functions. The reorientational motion of Im molecules in the PVPA/xIm composite was investigated using ^{2}H NMR. ^{2}H NMR spectrum of the imidazole was selectively observed by using imidazole-d$_3$ (Im-d$_3$), where only hydrogen bonded to carbon was replaced by deuterium. In polymer materials, dynamic inhomogeneity is closely related to these physical properties. In order to investigate the flexible region and the rigid region in PVPA/xIm, a combination of ^{2}H NMR broad line spectra and QCPMG spectra is very effective.

The ^{2}H NMR broad line and QCPMG spectra of PVPA/2Im-d$_3$ above 30 °C are shown in Fig. 3.13 [28]. For the broad line spectrum, the spectral intensity around ± 60 kHz decreased and that of the central portion increased with increasing temperature. These changes in the line shape of the ^{2}H NMR broad line spectrum are observed when the rate of isotropic rotation of molecules gradually increases. For the simulation of thc ^{2}H NMR spectrum due to the isotropic rotation of molecules, the multisite jump on the cone at the magic angle (Fig. 3.14a) and the tetrahedral jump (Fig. 3.14b) are conventional models [28, 29]. In polymers, there is often a large distribution of

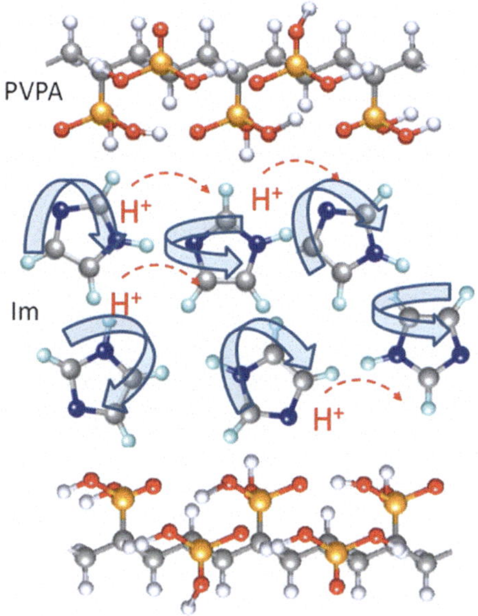

Fig. 3.12 Poly(vinylphosphonic acid) (PVPA)−imidazole (Im) composite material (PVPA/xIm)

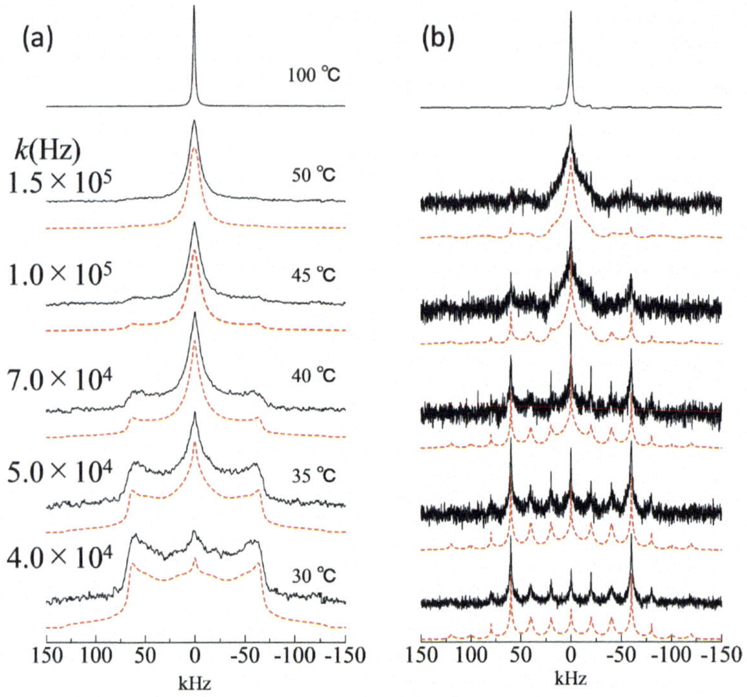

Fig 3.13 Temperature dependence of a ^{2}H NMR spectrum for PVPA/2Im [28]

Fig. 3.14 Models of isotropic rotation. **a** 8-site jump on a cone at the magic-angle model, **b** Tetrahedral 4-site jump model

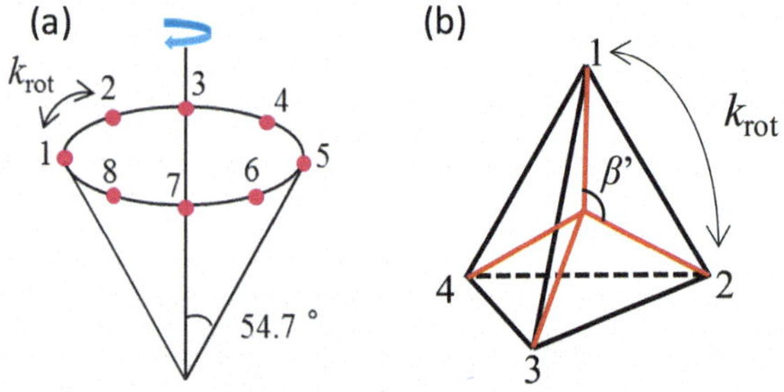

correlation time of molecular motion [30]. Simulation of the ^{2}H NMR broad line and QCPMG spectra at 35 °C was performed using an eight-site jump on a cone at the magic angle (Fig. 3.14a) with a Gaussian distribution of the jumping rate k_{rot}. The observed and simulated spectra are shown in Fig. 3.3. The signal intensity was reduced by the molecular motion with an intermediate jumping rate. The reduction factor of signal intensity due to the eight-site jump on a cone at the magic angle was estimated. The distribution of the jumping rate used for the spectral simulation, the estimated reduction factor, and the resulting signal intensity are shown in Fig. 3.15e. The line width of the central portion of the observed broad line spectrum is broader

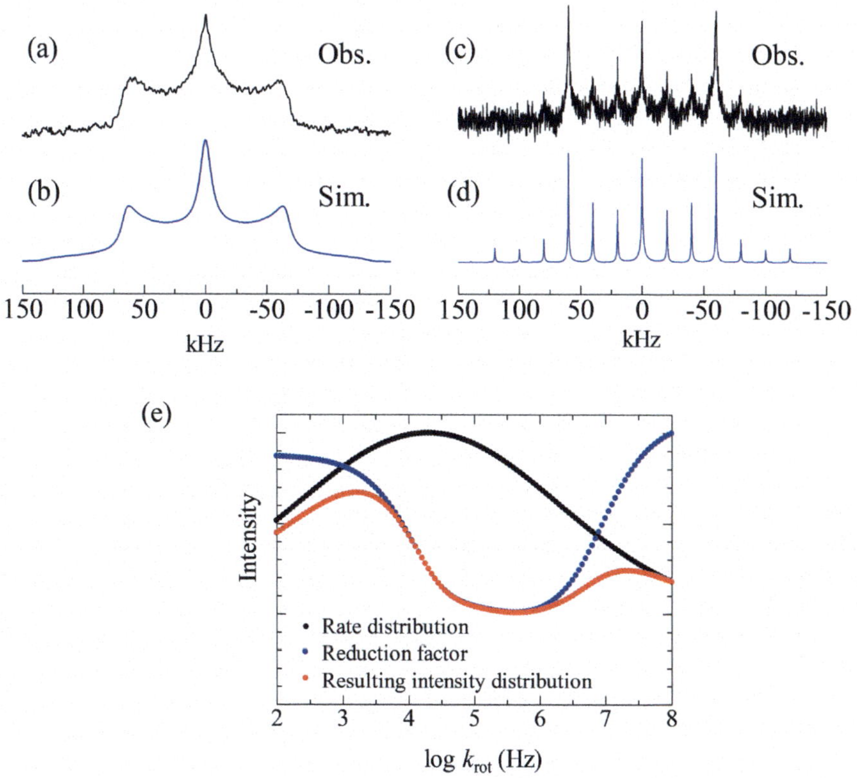

Fig. 3.15 A ^{2}H NMR broad line and a QCPMG spectrum at 35 °C for PVPA/2Im. **a** and **c** are observed broad line and QCPMG spectra, respectively. **b** and **d** are a simulation using the cone model at the magic angle (Fig. 3.14a) and a Gaussian distribution of the jumping rate. **e** is the distribution of the jumping rate, reduction factor, and resulting signal intensity [28]

than the simulation (Fig. 3.15a and b). The line broadening of the central component of the broad line spectrum indicates that the motion of Im molecules is not perfectly isotropic. Rather, there is pseudo-isotropic rotation with small anisotropy. Each peak of the observed QCPMG spectrum was broad, and the intensity of peaks within ± 50 kHz was weak relative to the simulated one. These differences between observed and simulated QCPMG spectra are caused by the rotational vibration of Im. Thus, the ^{2}H NMR spectrum could be analyzed by three kinds of Im molecules, namely Im undergoing pseudo-isotropic rotation with small anisotropy, rotational vibration, and fast isotropic rotation. The Im molecules undergoing rotational vibration interact strongly with phosphonic acid, and their mobility is highly restricted in the tight segment. The Im molecules undergoing pseudo-isotropic rotation do not have a strong interaction with phosphonic acid and are relatively movable in the flexible segment. The Im molecules undergoing fast isotropic rotation are predicted to be in a flexible space around the end of PVPA chains. In order to investigate the rate and anisotropy of pseudo-isotropic rotation, and the ratio of three kinds of motions of the Im molecules in PVPA/xIm, simulations of the ^{2}H NMR broad line and QCPMG spectra were performed. The line shape of the ^{2}H NMR spectrum due to the isotropic rotation and the pseudo-isotropic rotation of the Im molecules was simulated by the tetrahedral jump model (Fig. 3.14b). The four deuteron sites were determined by Euler angles (α, β) of Eq. (3.2) as (0°, 0°), (0°, β'), (120°, β'), and (240°, β'). Here, the anisotropy parameter A for molecular rotation is defined as [28]

$$A = \frac{1}{8}\left(2 + 3(3\cos\cos^2\beta' - 1)\right) \tag{3.17}$$

For the isotropic rotation, β' is 109.4° and A becomes 0. In the present work, β' changed in the range of 109.4°–180°, and the range of A was 0–1. As the A value increased, the line width of the central portion of the ^{2}H NMR spectrum increased (see Fig. 3.16). The observed and simulated ^{2}H NMR broad line and QCPMG spectra of $x = 2$ at 35 °C are shown in Fig. 3.17. Simulation spectra were obtained by the superposition of three components: pseudo-isotropic rotation with small anisotropy, rotational vibration, and fast isotropic rotation of the Im molecules. The rate of

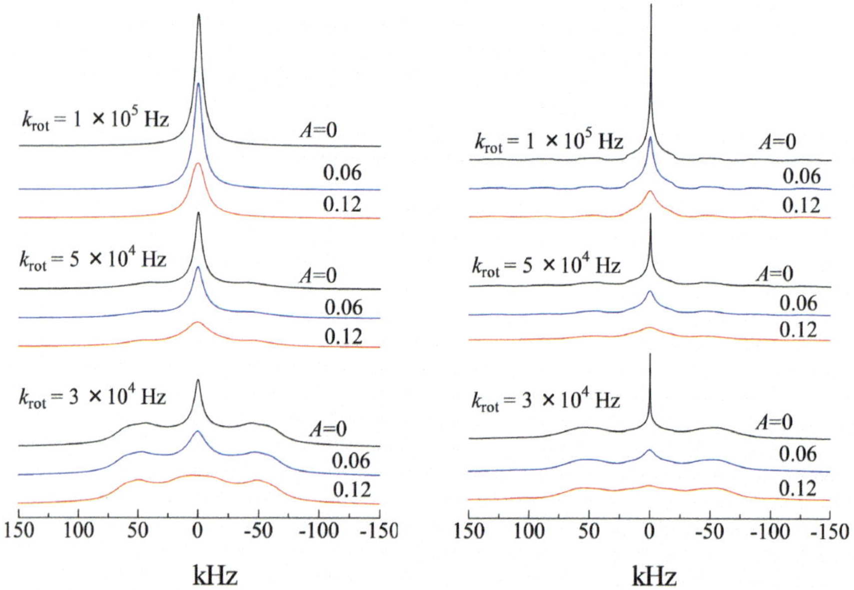

Fig. 3.16 Anisotropy parameter A dependence of the ^{2}H NMR spectrum for several motional rates. **a** Broad line spectra. **b** QCPMG spectra. k_{rot} is the jumping rate among four tetrahedral sites. $e^2qQ/h = 176$ kHz and $\eta = 0$ were used for simulation. Spin–spin relaxation time T_2 values of 0.5 and 2.2 ms were used to simulate broad line and QCPMG spectra, respectively. To obtain a simulation of the broad line spectrum, a Lorentzian broadening factor of 3 kHz, which is the same as that used for the simulation of the rigid spectrum, was added for the fourier transformation of the calculated time-dependent NMR signal [28]

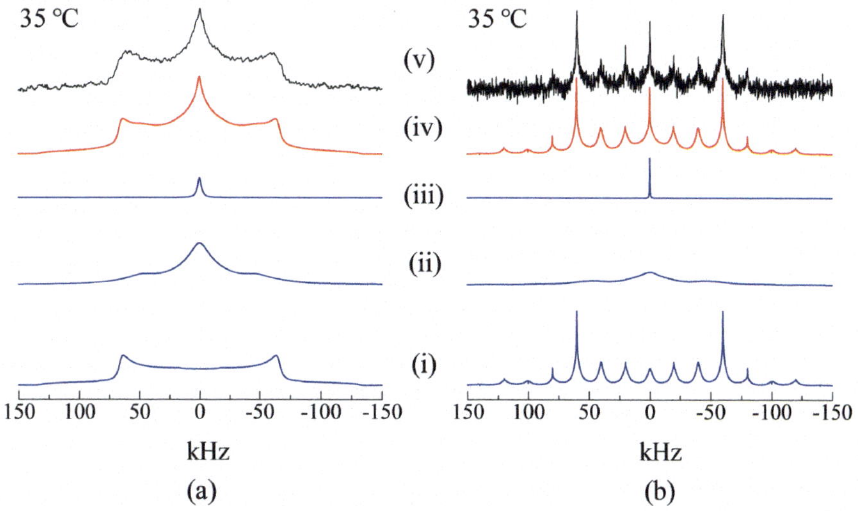

Fig. 3.17 Observed and simulated ^{2}H NMR spectra for $x = 2$. **a** and **b** are broad line spectra and QCPMG spectra at 35 °C, respectively. (i) Simulation spectra of the rotational vibration of Im with $k_{lib} = 1 \times 10^4$ Hz and $\theta_{lib} = \pm 30°$; (ii) simulation spectra of the pseudo-isotropic rotation of Im with $k_{rot} = 5 \times 10^4$ Hz and $A = 0.12$; (iii) simulation spectra of the isotropic rotation of Im with $k_{iso} = 1 \times 10^8$ Hz; (iv) superposition of (i), (ii), and (iii) components; (v) observed spectra. $e^2qQ/h = 176$ kHz and $\eta = 0$ were used for the simulation. The ratio of Im molecules undergoing pseudo-isotropic rotation, rotational vibration, and isotropic rotation was 96.2, 3.7, and 0.09, respectively [28]

rotational vibration (1.0×10^4 Hz), estimated by extrapolation from the rates at low temperatures, was used for the simulation. A rate of 1.0×10^8 Hz was used for fast isotropic rotation. The rate k_{rot} and anisotropy parameter A of the pseudo-isotropic rotation and the ratio of each component which reproduced the ^{2}H NMR broad line and QCPMG spectra simultaneously were determined. There was an extreme decrease in the intensity of the ^{2}H NMR spectrum in the presence of molecular rotational motion at a rate of 10^4–10^5 Hz. Therefore, although the abundance of Im undergoing pseudo-isotropic rotation exceeds 90%, the contribution of this component to the spectrum is strongly suppressed. The broken lines in Fig. 3.13 show the results of the spectral simulation. k_{rot} and A for the pseudo-isotropic rotation and the ratio of each component were estimated from a spectral simulation at each temperature. In the temperature range between 30 and 50 °C, the spectrum could be reproduced by $A = 0.12$ for $x = 2$. A was 0.14 in the temperature range between 50 and 80 °C for $x = 1$. Thus, the rotation of Im molecules was more suppressed with $x = 1$ than with $x = 2$. The efficient proton conduction accompanied by the pseudo-isotropic rotation of Im molecules exists for PVPA/2Im. The enhanced pseudo-isotropic rotation of Im molecules is correlated with an increase in the proton conductivity of PVPA/xIm.

In the ^{2}H NMR analysis of materials where multiple constituents with different kinetic states coexist, a non-negative least squares (NNLS) method is valuable for separating the components of the spectra [31, 32]. When the intensity of the observed spectrum at frequency ν_i is d_i ($i = 1 \sim n$), the intensity of the jth simulated spectrum at frequency ν_i is c_{ij}, and the weight of the jth simulated spectrum is x_j ($j = 1 \sim N$), d_i is expressed as $d_i = \sum_{j=1}^{N} c_{ij}x_j$, where n is the number of data points in the observed and simulated spectra, and N is the total number of simulated spectra. It is also expressed as $\mathbf{C}\vec{x} = \vec{d}$ using $n \times N$ matrix $\boldsymbol{C}$ with elements c_{ij}, vector $\vec{x}$ with components x_j, and vector $\vec{d}$ with components d_i. The ratio of each spectral component ($\vec{x}$) is obtained by solving the following NNLS problem:

$$\min_{x \geq 0} \left\| \mathbf{C}\vec{x} - \vec{d} \right\|^2 \tag{3.18}$$

The ^{2}H NMR broad line spectra of the proton-conducting composite, polyacrylic acid (PAA)/alginic acid (AA) /1,2,4-triazole (Tz-d$_2$) ((1−x)PAA/xAA/yTz), where x is the molar ratio of AA in polymers (PAA and AA), and y is the number of Tz moles per mole of the polymer repeating unit), are shown in Fig. 3.18 [32]. A Pake doublet line shape is observed for all samples at 30 °C with no effect on the molecular motion. As the temperature increases, a sharp component appears around the center of the spectrum; only sharp components are observed at relatively high temperatures. These changes in the spectral line shape indicate the isotropic or pseudo-isotropic rotation of Tz in the composites. The line width of the central peak can be effectively reproduced using a four-site tetrahedral jump model with a larger β' value (109.4°) than those for regular tetrahedral structures. This result indicates the residual anisotropy in the rotation of Tz, corresponding to pseudo-isotropic rotation. The distributions of the k_{rot} values of the pseudo-isotropic rotation of Tz in the composites were investigated through the simulation analysis of the ^{2}H NMR spectra based on NNLS analysis [32]. A series of ^{2}H NMR spectra (N = 161) was calculated with equally spaced

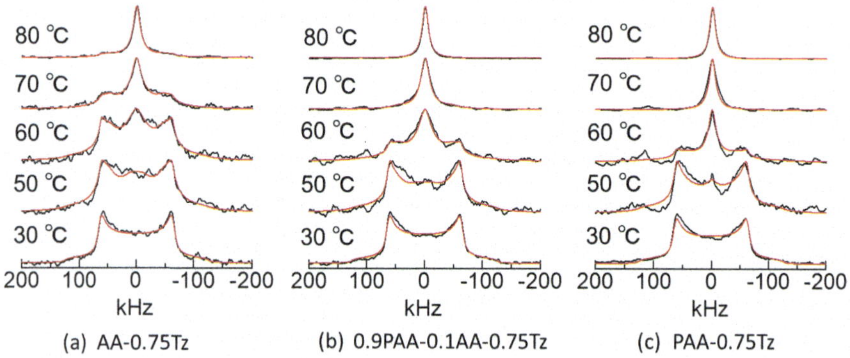

Fig. 3.18 Temperature variations in the ^{2}H NMR broad line spectra. The black and red lines indicate the observed spectra and the superposition of the simulated spectra of the isotropic rotation of Tz, respectively. For the simulation, e^2Qq/h = 166 kHz and η = 0 were employed. β' was 117° for AA–0.75Tz, 0.9PAA–0.1AA–0.75Tz and 113° for PAA–0.75Tz. Reproduced from ref. [32] by permission of Elsevier

jumping rates (10^2 Hz $\leq k_{rot} \leq 10^{10}$ Hz) on a logarithmic scale to create matrix $\boldsymbol{A}$. The ratios ($\overrightarrow{x}$) of each of the simulated spectra were determined by Eq. (3.18). The solution to Eq. (3.18) was obtained by a home-built program using Basic Linear Algebra Subprograms (BLAS) and Linear Algebra PACKage (LAPACK). It can also be computed using various off-the-shelf solvers, such as MATLAB (lsqnonneg). The distributions of k_{rot} obtained through the NNLS analysis are illustrated in Fig. 3.19. The spectra of each component and their superpositions are presented in Figs. 3.18 and 3.20. For PAA–0.75Tz and 0.9PAA–0.1AA–0.75Tz, the most frequent k_{rot} value of the distribution (2.2×10^5 Hz) is obtained at 80 °C. Contrarily, the most frequent

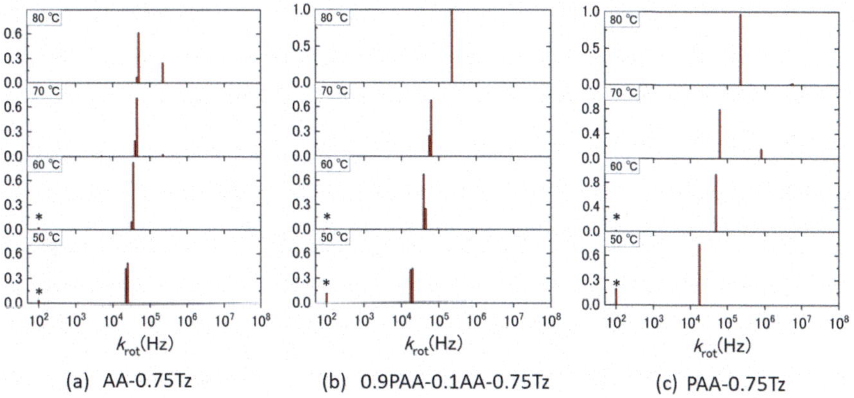

Fig 3.19 Distributions of the jumping rates of the isotropic rotation (k_{rot}) of Tz obtained based on the simulation analysis of the ^{2}H NMR broad line spectra. Reproduced from ref. [32] by permission of Elsevier

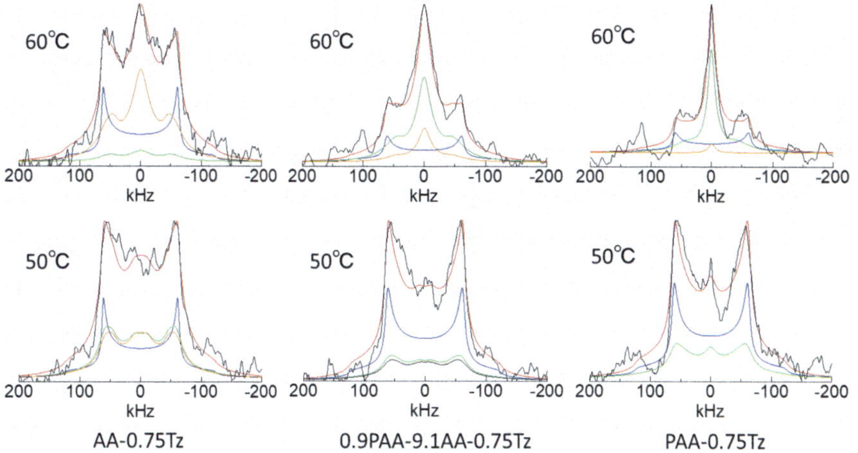

Fig. 3.20 ^{2}H NMR spectral components (blue, green, and orange lines) correspond to the distribution of k_{rot} (Fig. 3.19) and their superpositions (red lines) [32]

k_{rot} value of AA–0.75Tz at 80 °C is 5.0×10^4 Hz, which is lower than those of PAA–0.75Tz and 0.9PAA–0.1AA–0.75Tz. At relatively low temperatures, some minor components (denoted by asterisks in Fig. 3.19) exhibit k_{rot} values that are over two orders of magnitude lower than that of the main component in all composites. Thus, differences in the motility due to the differences in the temperature and composition of the composite can be studied in detail.

3.6 Analysis of Spin-Crossover Material

The spin-crossover phenomenon is well known in Fe^{2+} coordination materials, whose electron configurations can move between paramagnetic high-spin (HS) and diamagnetic low-spin (LS) states under extrinsic stimuli, affording changes in magnetic, optical, dielectric, and structural properties. When the cooperativity of the spin crossover is large enough, these materials present a first-order spin transition with hysteretic behavior.

$\{Fe(pz)[Pt(CN)_4]\}$ (pz = pyrazine) undergoes a first-order spin transition with a hysteresis of ~25 K width (transition temperatures: T_c = 285 and 309 K on cooling and heating, respectively) as shown in Fig. 3.21 [33]. The crystal structure of $\{Fe(pz)[Pt(CN)_4]\}$ can be described as 2D $\{Fe[Pt(CN)_4]\}_\infty$ layers connected along the (001) direction by pyrazine ligands, which occupy the apical positions of the Fe octahedrons (Fig. 3.22). The pyrazine rings are rotationally disordered in both spin states, and this is reflected in the appearance of electronic density corresponding to the carbon atoms of the pyrazine in two positions of the ring at 90° relative to each other.

The rotation of the pyrazine ligands in the LS and HS states was investigated by solid-state ^2H NMR spectroscopy [33]. The ^2H NMR spectrum of pyrazine

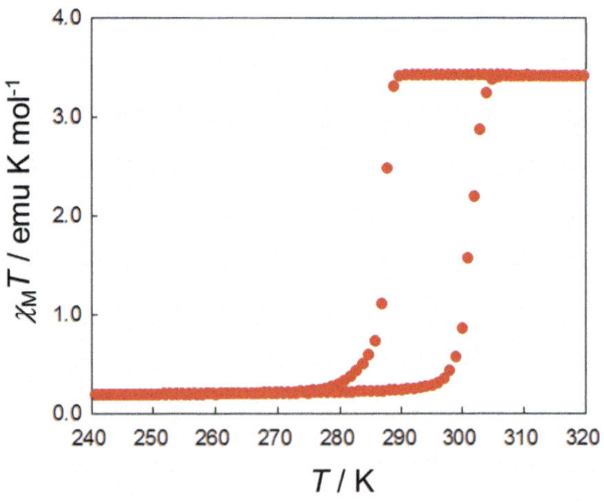

Fig 3.21 Temperature dependence of magnetic susceptibility for $\{Fe(pz)[Pt(CN)_4]\}$ [33]

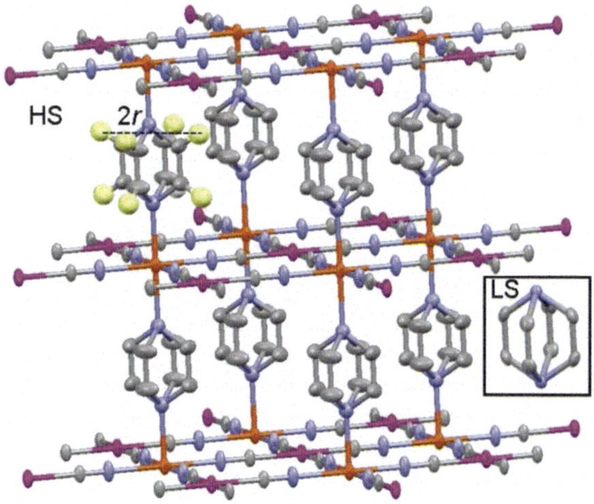

Fig 3.22 Crystal structure of {Fe(pz)[Pt(CN)$_4$]} in the HS state. Fe (orange), Pt (pink), N (blue), C (gray), H (yellow). The H atoms are omitted for clarity, except in one pz ring. Inset: detail of the thermal ellipsoids of the pz carbon atoms in the LS state. Reprinted with permission from ref. [33]. Copyright 2011 American Chemical Society

was selectively observed by using {Fe([D$_4$]pz)[Pt(CN)$_4$]} where only 4 hydrogens of the pyrazine ring were replaced by deuterons. For the measurement of the ^{2}H NMR spectrum in the HS state, a shift-compensated quadrupolar echo sequence was used. Figure 3.23 shows the temperature dependence of the ^{2}H NMR spectrum for {Fe([D$_4$]pz)[Pt(CN)$_4$]}. In the diamagnetic LS state, the line shape of the ^{2}H NMR spectrum is dominated by only the quadrupole interaction and becomes symmetric. In the paramagnetic HS state, however, not only the quadrupole interaction but also the paramagnetic interaction (dipolar interaction between ^{2}H and paramagnetic Fe^{2+}) affects the line shape of the ^{2}H NMR spectrum. The dipolar interaction between ^{2}H and paramagnetic ion causes the asymmetric ^{2}H NMR line shape, since this interaction is anisotropic and the directions of its principal axes are different from those of the quadrupole interaction (Fig. 3.24). The spectra for the paramagnetic HS state were simulated as a four-site flip of pyrazine rings along the C_4 axis, considering the contribution of the paramagnetic shift of three kinds of nearest-neighbor Fe^{2+} (Fig. 3.23). $e^2qQ/h = 165$ kHz and $\eta = 0$ were used for the simulation. The distance and angle between deuterium and Fe^{2+} for the simulation were extrapolated from the crystal structure of {Fe(pz)[Pt(CN)$_4$]}. The distances between deuterium and the first-, second-, and third-nearest Fe^{2+} are 3.17, 5.39, and 5.78 Å, respectively. Euler angles were (α_{Q1}, α_{Q2}, α_{Q3}, α_{Q4}) = (0°, 90°, 180°, 270°), $\beta_Q = 60°$ for the quadrupole interaction, and $\beta_{Pi1} = 42.9°$, $\beta_{Pi2} = 23.7°$, $\beta_{Pi3} = 66.3°$ for dipolar interactions between deuterium and three nearest-neighbor Fe^{2+}. The dipolar interactions between deuterium and Fe^{2+} were calculated using observed magnetic susceptibility values and Eq. (3.10). The anisotropic spin–spin relaxation was not considered for the simulation. It was determined that the rotation rate, k, of pyrazine was > 10^8 Hz at 300 and 290 K in the hysteresis loop. For comparison, the simpler 2-site jump model was also simulated with flip angles of both 90° and 180°, but these simulations failed to reproduce the measured spectra (Fig. 3.25). The contribution of the paramagnetic shift from nearest-neighbor iron ions to the ^{2}H NMR spectrum was

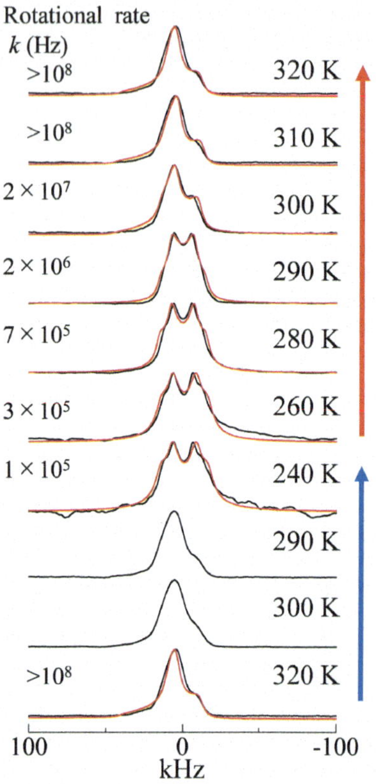

Fig 3.23 Temperature dependence of a ^2H NMR broad line spectrum for {Fe([D_4]pz)[Pt$(CN)_4$]} [33]

confirmed (Fig. 3.26). In the cooling process from 300 K, the rotation of pyrazine was remarkably suppressed by the spin transition below 280 K. The symmetric spectra for the diamagnetic LS state were also analyzed as a 4-site flip motion and gave a rotational rate of 3×10^5 Hz at 260 K. In the heating process from 260 to 300 K, the rotational rate gradually increased with heating (7×10^5 Hz at 280 K and 2×10^6 Hz at 290 K), and then the shape returned to the initial distorted line shape at 300 K. The activation energy for the rotation of pyrazine in the LS state was estimated to be 33 kJ mol^{-1} by an Arrhenius plot of k, with a pre-exponential factor of 1.1×10^{12} Hz.

Thus, as can be assessed from a ^2H NMR investigation, the pyrazine ring rotates almost freely in the HS state, while rotation is suppressed in the LS state. The connection of the mobility with the switch of the spin state is expected, taking into account that spin-crossover is an entropy-driven phenomenon: The LS state is the ground state of the system at very low temperatures, while the HS state becomes the ground state at high temperatures. The difference in energy is thermally compensated by the higher entropy of the HS state, which mainly stems from electronic and vibrational contributions. In the case of {Fe(pz)[Pt$(CN)_4$]}, the change in rotation

Fig 3.24 Principal axis systems of quadrupole interaction and paramagnetic dipolar interaction

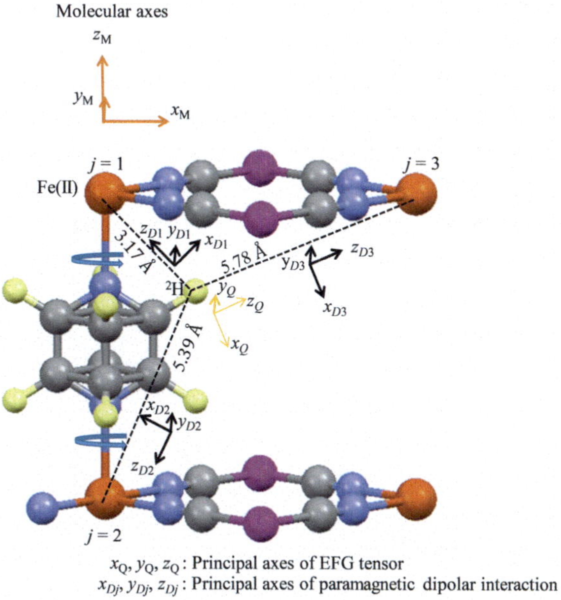

Fig 3.25 Spectral simulation of two-site jump and four-site jump models of the pz ring. Reprinted with permission from ref. [33]. Copyright 2011 American Chemical Society

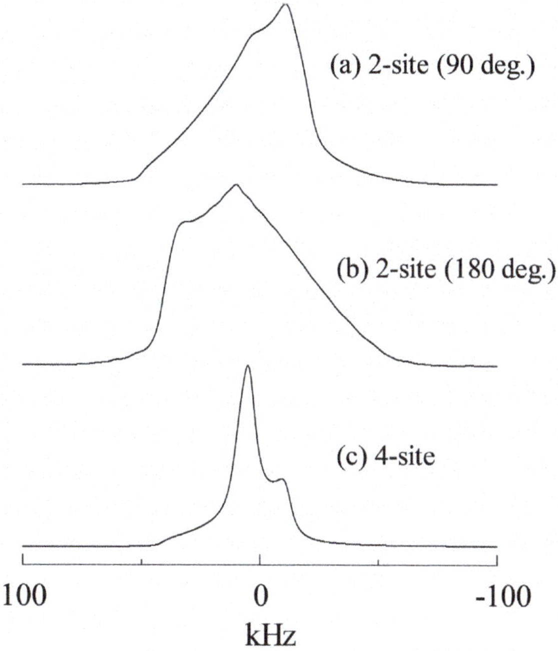

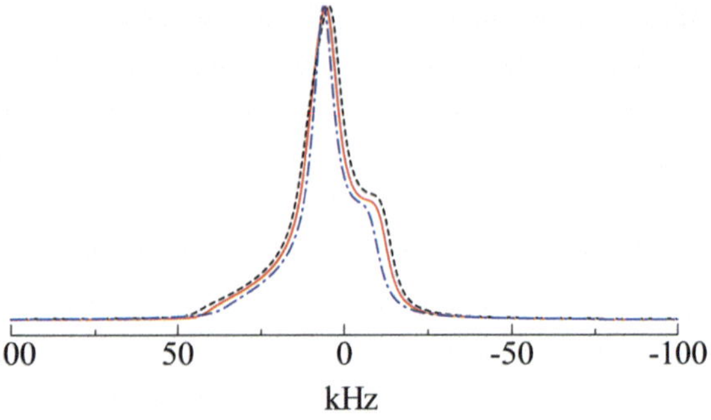

Fig 3.26 Contribution of paramagnetic shifts to the ^{2}H NMR spectrum. Blue dashed line, black broken line, and red solid line show the simulation spectrum including one, two, and three nearest-neighbor iron ions, respectively

velocity of the pyrazine rings may contribute considerably to this variation in entropy, thereby influencing the spin transition process.

3.7 Summary

Line shape analysis of a ^{2}H NMR spectrum using spectral simulation is a powerful method for studying molecular motions in solid materials. The dynamics in the nano-space of porous materials, which is correlated with their adsorption and reaction properties, can be analyzed by the ^{2}H NMR spectral simulation method. The mode and rate of molecular motion over a wide dynamic range can be investigated using a broad line spectrum and a QCPMG spectrum. A combination of ^{2}H NMR broad line spectra and QCPMG spectra is very effective for the analysis of dynamic inhomogeneity, which is closely related to the functions of polymers. The simulation of a ^{2}H NMR spectrum, including the quadrupole interaction and paramagnetic effects, is indispensable for the analysis of molecular motion in paramagnetic materials. Although the point particle model, which ignores the delocalization of the unpaired electron and which was used to calculate the paramagnetic dipolar interaction in this chapter, seems rather simple, the ^{2}H NMR spectra of the paramagnetic metal complex were well reproduced by the simulation. Thus, the application of this simulation method to the analysis of various paramagnetic materials is expected.

References

1. Vold, R.R.: Deuterium NMR studies of dynamics in solids and liquid crystals. In: Tycko, R. (ed.) Nuclear Magnetic Resonance Probes of Molecular Dynamics, pp. 27–112. Kluwer Academic Publishers, Norwell, MA (1994)
2. Vold, R.R., Vold, R.L.: Deuterium relaxation in molecular solids. In: Warren, W.S. (ed.) Advances in Magnetic and Optical Resonance, vol. 16, pp. 85–171. Academic Press Inc., San Diego (1991)

3. Schmidt-Rohr, K., Spiess, H.W.: Multidimensional Solid-State NMR and Polymers. Academic Press, London (1994)
4. Greenfield, M.S., Ronemus, A.D., Vold, R.L., Vold, R.R., Ellis, P.D., Raidy, T.E.: Deuterium quadrupole-echo NMR spectroscopy. III. Practical aspects of lineshape calculations for multiaxis rotational processes. J. Magn. Reson. **72**, 89–107 (1987)
5. Barbara, T.M., Greenfield, M.S., Vold, R.L., Vold, R.R.: Deuterium quadrupole echo NMR spectroscopy. I. Effects of chemical exchange during single and composite pulses. J. Magn. Reson. **69**, 311–330 (1986)
6. Long, J.R., Ebelhauser, R., Griffin, R.G.: ^{2}H NMR line shapes and spin–lattice relaxation in $Ba(ClO_3)_2{\cdot}^2H_2O$. J. Phys. Chem. A **101**, 988–994 (1997)
7. Wittebort, R.J., Olejniczak, E.T., Griffin, R.G.: Analysis of deuterium nuclear magnetic resonance line shapes in anisotropic media. J. Chem. Phys. **86**, 5411–5420 (1987)
8. Hiyama, Y., Silverton, J.V., Torchia, D.A., Gerig, J.T., Hammond, S.J.: Molecular structure and dynamics of crystalline p-fluoro-D,L-phenylalanine. A combined X-ray/NMR investigation. J. Am. Chem. Soc. **108**, 2715–2723 (1986)
9. Araya, T., Niwa, A., Mizuno, M., Endo, K.: Dynamics of $[Zn(D_2O)_6]^{2+}$ in $[Zn(D_2O)_6][SiF_6]$ crystal as studied by 1D, 2D spectra and spin-lattice relaxation time of ^{2}H NMR. Chem. Phys. **344**, 291–298 (2008)
10. Larsen, F.H., Jakobsen, H.J., Ellis, P.D., Nielsen, N.C.: Molecular dynamics from ^{2}H Quadrupolar Carr–Purcell–Meiboom–Gill solid-state NMR spectroscopy. Chem. Phys. Lett. **292**, 467–473 (1998)
11. Shiminovitch, D.J., Rance, M., Jeffrey, K.R., Brown, M.F.: The quadrupolar spectrum of a spin $I = 1$ in a lipid bilayer in the presence of paramagnetic ions. J. Magn. Reson. **58**, 62–75 (1984)
12. Antonijevic, S., Wimperis, S.: Refocussing of chemical and paramagnetic shift anisotropies in ^{2}H NMR using the quadrupolar-echo experiment. J. Magn. Reson. **164**, 343–350 (2003)
13. Lin, T.-H., DiNatale, J.A., Vold, R.R.: Determination of molecular reorientation rates and electron-nuclear coupling in paramagnetic materials by deuterium solid echo NMR spectroscopy. J. Am. Chem. Soc. **116**, 2133–2134 (1994)
14. Mizuno, M., Hamada, Y., Kitahara, T., Suhara, M.: Deuterium NMR study of unstable phenomena and water molecular dynamics in samarium nitrate hexahydrate crystal. J. Phys. Chem. A **103**, 4981–4985 (1999)
15. Mizuno, M., Iijima, T., Suhara, M.: Dynamical structure of paramagnetic $[M(H_2O)_6][SiF_6]$ ($M=Fe^{2+}$, N^{2+}) crystal studied by means of ^{2}H nuclear magnetic resonance. J. Phys. Condens. Matter **12**, 7261–7273 (2000)
16. Mizuno, M., Itakura, N., Endo, K.: Effects of strong paramagnetic interactions on solid-state deuterium NMR spectra. Chem. Phys. Lett. **416**, 358–363 (2005)
17. Rose, M.E.: Elementary Theory of Angular Momentum. Wiley, New York (1957)
18. Bloom, M., Davis, J.H., Valic, M.I.: Spectral distortion effects due to finite pulse widths in deuterium nuclear magnetic resonance spectroscopy. Can. J. Phys. **58**, 1510–1517 (1980)
19. Tanaka, D., Masaoka, S., Horike, S., Furukawa, S., Mizuno, M. Endo, K., Kitagawa, S.: Porous coordination polymer with π lewis acidic pore surface, $\{[Cu_3(CN)_3\{hat\text{-}(CN)_3(OEt)_3\}]{\cdot}3THF\}_n$. Angew. Chem. Int. Ed. **45**, 4628–4631 (2006)
20. Horike, S., Matsuda, R., Tanaka, D., Matsubara, S., Mizuno, M., Endo, K., Kitagawa, S.: Dynamic motion of building blocks in porous coordination polymers. Angew. Chem. Int. Ed. **45**, 7226–7230 (2006)
21. Takamizawa, S., Nataka, E., Akatsuka, T., Miyake, R., Kakizaki, Y., Takeuchi, H., Maruta, G., Takeda, S.: Crystal transformation and host molecular motions in CO_2 adsorption process of a metal benzoate pyrazine (M^{II} = Rh, Cu). J. Am. Chem. Soc. **132**, 3783–3792 (2010)
22. Hijikata, Y., Horike, S., Tanaka, D., Groll, J., Mizuno, M., Kim, J., Takatade, M., Kitagawa, S.: Differences of Crystal Structure and Dynamics between Soft Porous Nanocrystal and Bulk Crystal. Chem. Commun. **47**, 7632–7634 (2011)
23. Zhu, K., Vukotic, V.N., O'Keefe, C.A., Schurko, R.W., Loeb, S.J.: Metal-organic frameworks with mechanically interlocked pillars: controlling ring dynamics in the solid-state via a reversible phase change. J. Am. Chem. Soc. **136**, 7403–7409 (2014)

24. Inukai, M., Fukushima, T., Hijikata, Y., Ogiwara, N., Horike, S., Kitagawa, S.: Control of molecular rotor rotational frequencies in porous coordination polymers using a solid-solution approach. J. Am. Chem. Soc. **137**, 12183–12186 (2015)
25. Chun, H., Dybtsev, D.N., Kim, H., Kim, K.: Synthesis, X-ray crystal structures, and gas sorption properties of pillared square grid nets based on paddle-wheel motifs: implications for hydrogen storage in porous materials. Chem. Eur. J. **11**, 3521–3529 (2005)
26. Bureekaew, S., Horike, S., Higuchi, M., Mizuno, M., Kawamura, T., Tanaka, D., Yanai, N., Kitagawa, S.: One-dimensional imidazole aggregate in aluminum porous coordination polymers with high proton conductivity. Nature Mater. **8**, 831–836 (2009)
27. Sevil, F., Bozkurt, A.: Proton conducting polymer electrolytes on the basis of poly(vinylphosphonic acid) and imidazole. J. Phys. Chem. Solids **65**, 1659–1662 (2004)
28. Mizuno, M., Iwasaki, A., Umiyama, T., Ohashi, R., Ida, T.: Local structure and dynamics of imidazole molecules in proton-conducting poly(vinylphosphonic acid)-imidazole composite material. Macromol. **47**, 7469–7476 (2014)
29. Lee, Y.J., Murakhtina, T., Sebastiani, D., Spiess, H.W.: ^{2}H solid-state NMR of mobile protons: it is not always the simple way. J. Am. Chem. Soc. **129**, 12406–12407 (2007)
30. Hansen, M.R., Graf, R., Spiess, H.W.: Solid-state NMR in macromolecular systems: insights on how molecular entities move. Acc. Chem. Res. **46**, 1996–2007 (2013)
31. Bro, R., Jong, S.D.: A fast non-negativity-constrained least squares algorithm. J. Chemom. **11**, 393–401 (1997)
32. Watanabe, R., Kurihara, T., Shigeta, Y., Amemori, S., Yamaguchi, N., Ishii, F., Ida, T., Mizuno, M.: Structural and physical properties of proton-conducting polyacrylic acid/alginic acid/1,2,4-triazole composite membranes. Solid State Ion. **399**, 116299 (2023)
33. Rodríguez-Velamazán, J.A., González, M.A., Real, J.A., Castro, M., Muñoz, M.C., Gaspar, A.B., Ohtani, R., Ohba, M., Yoneda, K., Hijikata, Y., Yanai, N., Mizuno, M., Ando, H., Kitagawa, S.: A Switchable molecular rotator: neutron spectroscopy study on a polymeric spin-crossover compound. J. Am. Chem. Soc. **134**, 5083–5089 (2012)

Chapter 4
NMR Spectral Observations of the Gases in Polymer Materials

Hiroaki Yoshimizu

Abstract The gas transport properties of polymeric materials, including sorption, diffusion, and permeation, are closely related to the structural characteristics of the polymers. As a result, the analysis of polymers exposed to certain gases by NMR spectroscopy can determine detailed correlations between gas sorption and diffusion in the polymer and high-order structures. One such method involving the analysis of ^{129}Xe NMR chemical shifts is presented herein. This ^{129}Xe NMR spectroscopy was also applied to the first discovery of continuous nanopores within fluid materials. This chapter also discusses the estimation of gas diffusion properties from the peak widths in the NMR spectra of polymers, as well as from data generated using pulsed field gradient NMR.

Key words ^{129}Xe NMR · PFG NMR · Polymer · Gas sorption · Gas diffusion

4.1 Introduction

The fuel sources used by humans have progressed from coal (a solid fuel) to petroleum (a liquid) to natural gas. At present, another gas, hydrogen, is one of the most important gaseous fuels associated with the pursuit of sustainable energy, so new technologies related to the controlled transport of gases are required. These technologies include the development of novel materials for applications such as transfer tubes, cylinders, gaskets, and permselective membranes. Advanced techniques for the characterization of such materials are also necessary. In particular, the development of polymeric materials acting as high-performance gas permeable membranes or barriers necessitates an understanding of the gas transport properties in such materials, including sorption, diffusion, and permeation. Conversely, because gas transport properties in a polymer are affected by both static and dynamic structural characteristics, it is possible to analyze the structures of macromolecules by assessing gas transport properties. Our own research group has been examining the

H. Yoshimizu (✉)
Graduate school of engineering, Nagoya institute of technology, Nagoya, Aichi, Japan
e-mail: yoshimizu.hiroaki@nitech.ac.jp

The NMR Society of Japan (ed.), *Experimental Approaches of NMR Spectroscopy II*,
https://doi.org/10.1007/978-981-97-6830-1_4

applications of NMR techniques for the assessment of various structural properties of polymeric materials, focusing on the correlations of structure with gas transport characteristics. The present chapter discusses the manner in which the fine structure of polymeric solids can be characterized based on NMR spectroscopy, using gaseous probe molecules. One advantage of this NMR technique is that both gas transport properties and high-order structural information for the polymer can be obtained. For additional information, the reader is referred to our publications regarding the structural analyses of polymeric materials by means of ^{129}Xe NMR spectroscopy [1–4].

4.2 Basics of the NMR Analysis of Gas-Polymer Systems

This chapter does not address the newest and/or advanced NMR techniques but rather presents data primarily recorded using the single-pulse method in conjunction with standard commercial NMR spectrometers with typical solution NMR probes. The exception is that the analysis of relatively high-pressure gases requires special NMR tubes with thick, hardened walls. Fortunately, commercial NMR tubes that can withstand internal pressures up to approximately 1 MPa are readily available, and the development of improved tubes capable of safely containing pressures up to 10 MPa is anticipated in the near future.

4.2.1 Apparatus for the Preparation of NMR Sample Tubes Containing High-Pressure Gas Samples

Herein, the sample tubes and gaseous introduction apparatus are described. Our group typically uses commercially available NMR tubes with thick glass walls and a PTFE plug valve that can be easily and repeatedly connected to a narrow-bore stainless steel gas line. This valve is readily attached and removed so that the sample gas (up to a pressure of approximately 1 MPa) can be easily introduced solely by operating the valve. Inserting and removing a polymer sample is also simple. Fig. 4.1 shows a schematic diagram of an apparatus for introducing gas samples to NMR tubes of our own design. Using the integrated pressure gauge and transducer, the operator can introduce gas into the NMR tube to the desired pressure. The stainless steel gas line purposely has a length of several meters so that the sample tube can be situated within the field of the superconducting NMR magnet. This apparatus also allows for real-time monitoring of the pressure in the NMR tube throughout the trial, as well as variable temperature measurements. Following each experiment, the gas used for the measurements can be easily retrieved from the NMR tube by cooling an attached 50-ml stainless steel gas storage cylinder with liquid nitrogen.

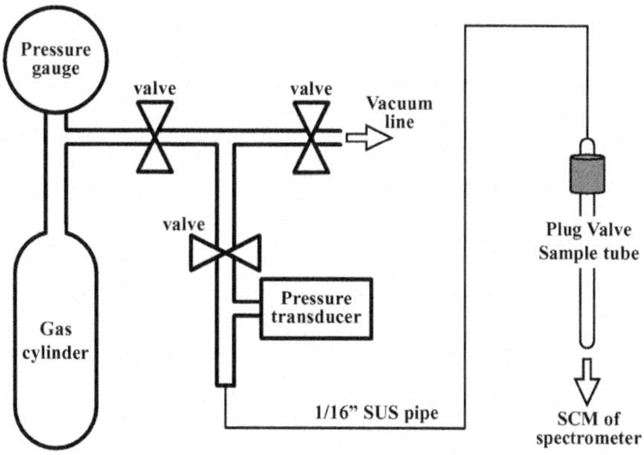

Fig. 4.1 A schematic diagram of the laboratory-made apparatus employed for introducing gas into an NMR sample tube having thick walls and incorporating a plug valve

4.2.2 Gas Sorption Properties of Polymers

The gas transport properties of rubbery polymers can generally be described based on Fick's theory [5, 6] of gas-liquid systems, but this theory is not applicable in the case of glassy polymers. A glassy polymer is in a state of thermodynamic non-equilibrium and thus includes excess free volume (unrelaxed volume). While the gas sorption isotherms of rubbery polymers will increase in direct proportion to pressure, those of glassy polymers are typically concave along the pressure axis. This effect has been successfully explained by the dual-mode sorption model [7], in which the amount of gas adsorbed consists of the quantity predicted by Henry's law (C_D: where the subscript indicates "dissolved"), and that predicted by the Langmuir mode (C_H: where the subscript indicates "hole"). This dual-sorption is represented by the equation

$$C = C_D + C_H = k_D p + \frac{C'_H bp}{1 + bp}, \tag{4.1}$$

where C is the equilibrium sorption amount at a pressure, p, defined as the molar concentration of gas per unit volume of polymer (usually expressed in units of cm^3 (STP)/$cm_{polym}{}^3$), corresponding to the density of the gas in the polymer. The hole saturation constant in the Langmuir sorption mode, C_H' generally corresponds to the unrelaxed volume in the glassy polymer.

As an example, Fig. 4.2 shows the isotherms for the sorption of Xe in a polyphenylene oxide (PPO) sample at 25 °C, as obtained using the conventional gravimetric method (◇) and a quantitative analysis of NMR signal intensity (○). The data obtained from both methods are in good agreement, and give almost the same line upon nonlinear least-squares regression with Eq. (4.1), as indicated by the solid curve in Fig. 4.2. This result suggests that NMR spectroscopy is a viable means of measuring gas sorption. There are two ways of the conversion from the NMR signal intensity to the absolute sorption amounts. One is using the sorption data obtained elsewhere,

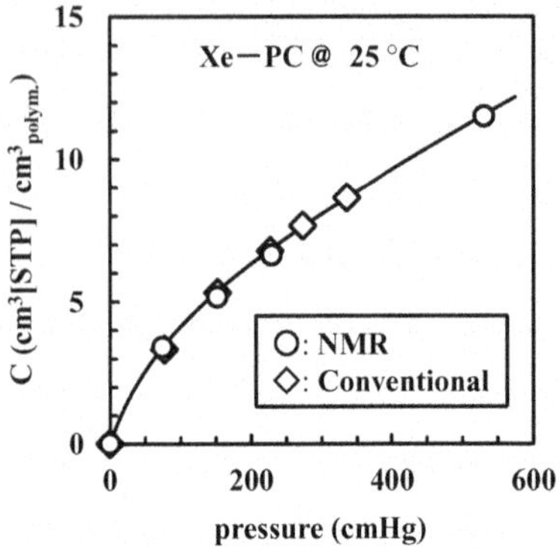

Fig. 4.2 Xe sorption isotherms of poly 2,6-dimethyl phenylene oxide (PPO) at 25 °C as obtained by NMR and conventional (gravimetric) methods

another is using the intensity of the NMR peak could be assigned to that of a free gas observed simultaneously. However, the assignment of free gas peak needs attention. As a further example, the 100 MHz ^{13}C NMR spectra of the $^{13}CO_2$ in NMR sample tubes packed with the drawn syndiotactic polystyrene (SPS) films at approximately 0.8 MPa and 25 °C are presented in Fig. 4.3. Each of these spectra was observed at the same temperature and pressure, but the direction of the film sample in the NMR tube was varied, as shown in the figure. ^{13}C-enriched CO_2 was used in this experiment, so the sharp peaks observed in each spectrum are readily assigned to the free gas in the tube. These peaks are split into doublets or triplets, with the exception of the spectrum obtained when small pieces of the film are inserted into the tube at random. These results are primarily attributed to the anisotropic volume susceptibilities of the drawn SPS film samples. The CO_2 molecules near the top and bottom edges, and middle planes of the regularly stacked SPS films experience different magnetic fields to some extent. However, this susceptibility effect is sterically averaged when the small pieces of SPS films were randomly inserted into an NMR tube. This same effect will occur for any nuclei, so it must be taken into account during experimental analyses.

4.3 Characterization of the High-Order Structure of a Glassy Polymer Based on ^{129}Xe NMR Chemical Shifts

^{129}Xe is an isotope with a natural abundance of approximately 26.4% and spin number $I = 1/2$. ^{129}Xe NMR allows for relatively high sensitivity, so Xe gas can be easily observed by NMR at relatively low pressures. The gyromagnetic ratio of ^{129}Xe is about 1.1 times that of ^{13}C, such that the resonant frequency for ^{129}Xe is close to that for ^{13}C, so ^{129}Xe spectroscopy can be performed using commercial spectrometers with standard multinuclear NMR probes. Furthermore, ^{129}Xe NMR is highly sensitive to the sorption environment due to the very high polarizability of the Xe atom.

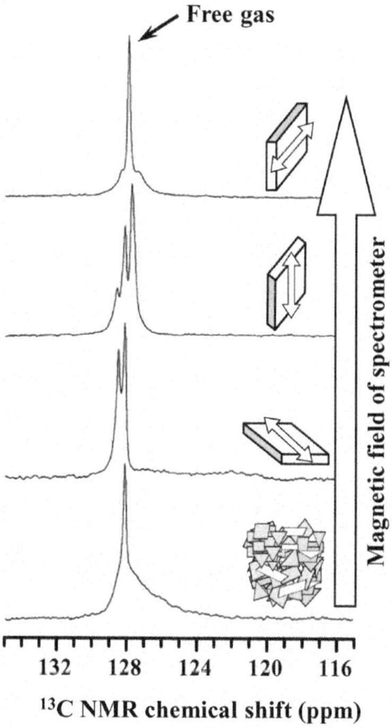

Fig. 4.3 The 100 MHz ^{13}C NMR spectra of $^{13}CO_2$ in NMR sample tubes containing mono-axially drawn syndiotactic polystyrene (SPS) films at approximately 0.8 MPa and 25 °C. The inserted illustrations indicate the direction of the film relative to the magnetic field of the spectrometer. ^{13}C-enriched carbon dioxide (approximately 98%) was employed in these trials

Therefore, the ^{129}Xe NMR chemical shifts are strongly correlated with the size and nature of micropores in the polymeric material, because interactions with the host system can modify the electron density of the Xe atom. This unique property has been demonstrated in studies utilizing inorganic cage compounds such as zeolites [8–12]. Recently, it was reported that various gas-polymer systems can be successfully characterized by ^{129}Xe NMR [13–22]. A review of this prior work is provided in references [1–4], but additional interpretations are presented herein.

The unrelaxed volume in a glassy polymer can play an important role in gas sorption by providing a source of microvoids, meaning the interspaces between polymer chains and ^{129}Xe NMR porosimetry can be applied for the determination of microvoid size in some glassy polymers. The chemical shift of Xe in a polymer is, in general, explained by Xe-Xe interactions as well as interactions between Xe and other atoms. These interactions increase when atomic distances are short, generating chemical shifts to lower-field values, the magnitudes of which depend on the size of the empty holes and the contribution of Xe-Xe interactions. The latter interactions will be proportional to the Xe density, meaning the extent of Xe sorption in the polymer. In the case of glassy polymers, a nonlinear relationship is observed between pressure and chemical shift, which is explained by assuming that the contributions of the Xe-Xe interactions at the Henry and Langmuir sites are different. Specifically, the effect of the sorption amount at Langmuir sites on the ^{129}Xe NMR chemical shift is larger than that of sorption at Henry sites. This assumption is based on the dual-mode sorption model. That is, only a limited number of Langmuir sorption sites are formed, so that the extent of gas sorption at these sites is less than that at Henry sites at high pressures, and gas molecules are preferentially adsorbed at the Langmuir sites. In addition, the disruption of Xe-Xe interactions by the molecular motions of polymer chains occurs much more frequently at Henry sites.

The above considerations are summarized by the model schematic shown in Fig. 4.4. Here, the straight bold line indicates the center line of a polymer chain, and the gray part along this line indicates the effective volume of the polymer chain including local molecular motions. This gray region, therefore, represents the volume occupied by the polymer and the free volume due to local molecular motions, which will be reduced with decreasing temperature. Because the larger segmental motions are frozen in the glassy state, the region that remains empty (the domain indicated by the slanted triangle) can be imagined as existing at the center of a triangle surrounded by three polymer chains. This domain can be defined as a microvoid, explaining why microvoids tend to expand with decreasing temperature. Previous studies of variable temperature measurements using PPO [21] and polycarbonate (PC) [23] have confirmed that the microvoid size as determined by analysis of ^{129}Xe NMR chemical shifts increases with decreasing temperature. It is also evident that the temperature-extrapolated microvoids become smaller than an Xe atom as the temperature approaches the glass transition temperature of each polymer.

Since the Xe atoms in polymeric materials are in constant diffusive motion, the inter-atomic distance that affects the ^{129}Xe NMR chemical shift can be related to the collision frequency of these atoms. An Xe atom contacting the gray region in Fig. 4.4 will collide with a polymer chain, and such collisions will occur repeatedly and continuously, such that the Xe atoms tortuously diffuse through the polymer. Xe atoms in this region will also collide with polymer chains much more frequently than with other Xe atoms. The scenario within a microvoid will be reversed, because of a lack of contact with polymer chains. These Xe atoms will collide with other Xe atoms more frequently. Experimental data are in good agreement with these predictions. The ^{129}Xe NMR chemical shift of the Xe at Henry sites is predicted to be lower than that of Xe in Langmuir sites, indicating that the distance between Xe atoms and polymer chains at Henry sites is very short, such that they are almost in contact with one another. The chemical shift of the Xe at Langmuir sites is estimated to appear at

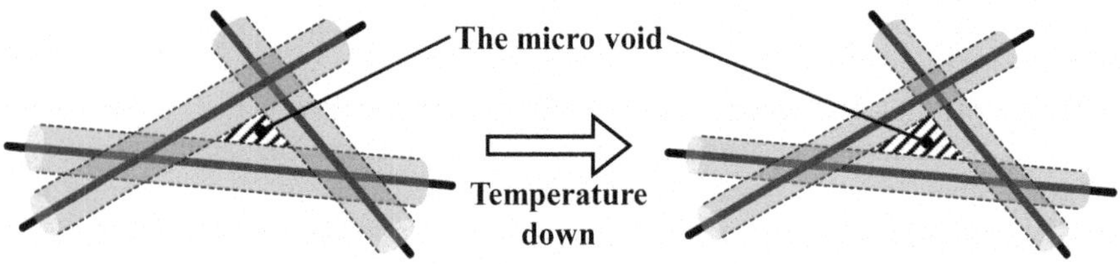

Fig. 4.4. Schematic representation of the microvoid model (*slanted region*). The *solid line* indicates the center of the polymer chain. The *gray* regions along the *center line* indicate the occupied volume, with the free volume dependent on the molecular motions of the polymer chain. Details are provided in the text

high-field values, while Xe-Xe interactions will lead to significant low-field shifts. Thus, information regarding the microvoid structures in a glassy polymer can be obtained from the effect of the sorption amount on the ^{129}Xe NMR chemical shift.

4.4 Characterization of the High-Order Structure of Rubbery Polymers from ^{129}Xe NMR Chemical Shifts

This section describes the analysis of rubbery polymers by ^{129}Xe NMR. The micro-Brownian segmental motions of polymer chains are active in the rubbery state, meaning that the gray region in Fig. 4.4 expands and thus can be drawn without the slanted triangle (meaning the microvoid region). As discussed above, the gas sorption isotherms of glassy and rubbery polymers are different from one another, so it is necessary to use different analysis procedures and interpretations for each case.

As examples, the ^{129}Xe NMR spectra of Xe in conjunction with a rubbery polymer (acrylonitrile-butadiene copolymer rubber with 29 mol% acrylonitrile, NBR29) and with a glassy polymer (polyimide, PI) at 25 °C are shown in Fig. 4.5. The sharp peaks observed at approximately 0 ppm are assigned to free Xe gas, and the associated chemical shifts are consistent with those of pure Xe gas at the same pressure and temperature. In contrast, the peaks assigned to Xe in these polymers are observed at approximately 205 ppm (in PI) and 228 ppm (in NBR29). Interestingly, the peaks obtained from the rubbery polymer are always sharper than those from the glassy material. This occurs because the gas diffusivity in a rubbery polymer is higher than that in a glassy polymer. This relationship between the peak width and diffusivity is discussed in more detail further on. The ^{129}Xe NMR chemical shifts obtained from the rubbery polymer undergo a slight low-field shift that is linearly correlated with increasing pressure. This linear shift can be explained by Henry's law because the sorption amount will be proportional to the pressure. The variation in the chemical shift of the Xe-rubbery polymer system is also considerably smaller than that in the Xe-glassy polymer system; in the range of Xe pressure from 0.1 to 0.8 MPa, a low-field shift of 2.9 ppm was observed for the Xe-NBR29 system, while the Xe-PI system showed a shift of 25 ppm. These data suggest that collisions between the Xe atoms tend to be disturbed by active micro-Brownian segmental motions.

The value of the ^{129}Xe NMR chemical shift in a rubbery polymer extrapolated to a pressure value of nil depends on the degree of interaction between the Xe atoms and the polymer. This interaction becomes strong when the inter-atomic distance is short and, as a result, a low-field shift is induced. In Fig. 4.6, the extrapolated ^{129}Xe NMR chemical shifts determined using liquid alkanes [24] and rubbery polymers are plotted against the host material densities. All these data are well represented by a single straight line, indicating that the ^{129}Xe NMR chemical shift in a liquid or a rubbery polymer is useful for the estimation of density. It has been reported that, in the case of a liquid crystalline polyester with alkyl side chains, gas sorption only occurs within the aggregated side chain layer, allowing the density of this layer

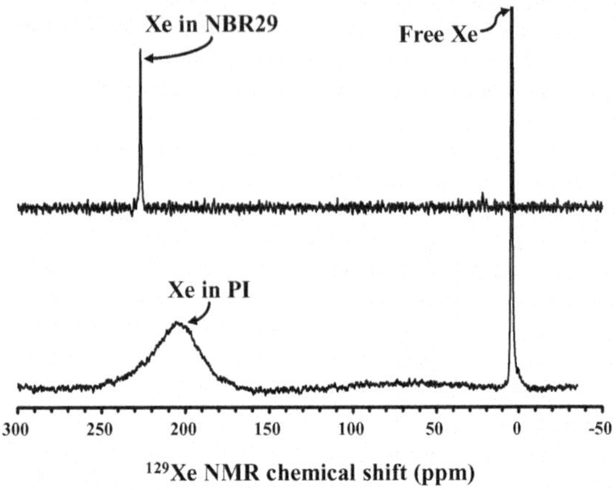

Fig. 4.5 The 193.6 MHz ^{129}Xe NMR spectra of free (gaseous) Xe and Xe in acrylonitrile-butadiene copolymer rubber (with 29 mol% acrylonitrile; NBR29) and polyimide (PI) at 0.7 (NBR29) and 0.8 (PI) MPa and 25 °C

to be successfully estimated [20] even when conventional density measurements cannot be applied because of the coexistence of the polyester main chain. The density values estimated from the ^{129}Xe NMR chemical shift, in this case, agreed with those predicted from X-ray structural analysis and gas solubility data. However, it should be noted that the ^{129}Xe NMR peak generated by the PI appears at a higher-field position compared with that of the NBR29 (Fig. 4.5), although the density of PI (1.42 g/cm^3) is higher than that of NBR29 (0.947 g/cm^3). This result demonstrates that the density of a glassy polymer cannot always be estimated from the ^{129}Xe NMR chemical shift. That is, the interpretation of the Xe data from a glassy polymer must take into consideration the effect of microvoids.

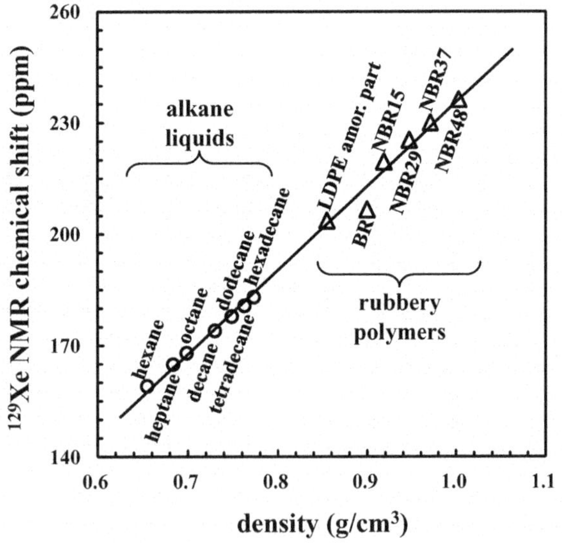

Fig. 4.6 ^{129}Xe NMR chemical shifts as a function of the density of various liquid alkanes and rubbery polymers at 25 °C

4.4.1 *Analysis of Gas Diffusion Characteristics in Polymers Based on NMR Peak Width*

During the technique described herein, small samples of polymeric solids are transferred into a sample tube, and NMR spectra are acquired from the gas-polymer system without the use of deuterated solvents. Therefore, there is the possibility of significant sample inhomogeneity because of the anisotropic nature of the sample. Since the NMR peak width will be affected by inhomogeneity and spin-spin relaxation, quantitative analysis of gas sorption in the polymer based on peak width is not possible. However, variations in the peak width with changes in temperature, pressure (which in turn affects the sorption amount), and polymer can be assessed. In this section, analyses based on the widths (as reflected in the full-width at half-maximum, FWHM) of NMR peaks generated by gases in polymers are described. As shown in Fig. 4.5, the FWHM of the Xe peaks obtained from a rubbery polymer is narrower than those obtained from a glassy polymer. This same effect is also observed during ^{1}H NMR analyses using CH_4 [25]. The effects of pressure on the FWHM values of Xe peaks in conjunction with polyvinyl chloride (PVC), PPO, and NBR29 are summarized in Fig. 4.7. While the NBR29 spectra exhibit almost no FWHM changes with pressure, both the PVC and PPO undergo significant peak sharpening as the pressure is increased. The gas diffusion coefficients in these polymers as estimated from gas permeation, and sorption experiments are in the order PVC < PPO < NBR29. The FWHM values are also decreased by increasing the sample temperature, so it can be concluded that the FWHM of the peaks resulting from a gas in a polymer represents a measure of gas diffusivity.

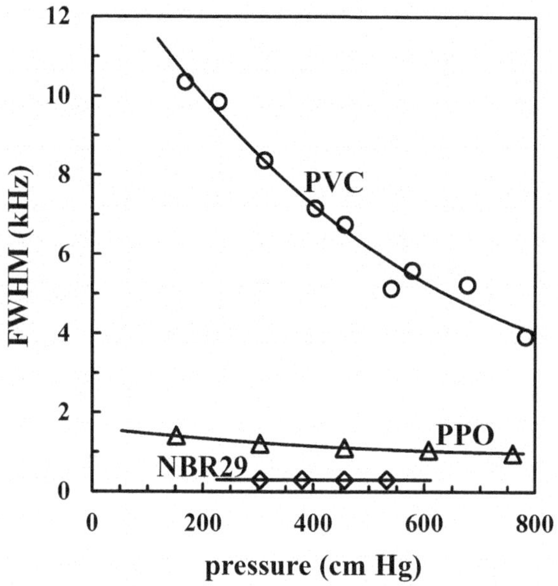

Fig. 4.7 ^{129}Xe NMR peak widths (FWHMs) obtained from analysis of poly(vinyl chloride) (PVC), PPO, and NBR29 at 25 °C as functions of Xe pressure. The *solid curves* indicate the results of curve-fitting based on the dual-sorption model. Details are provided in the text

The larger variation in the FWHM of peaks generated by glassy polymers can also be interpreted based on the dual-mode sorption model. The Xe atoms at Henry and Langmuir sorption sites are known to undergo rapid exchange, meaning that the experimental FWHM values should be considered as the weighted averages of the fractions of the sorption amounts at both sites. Therefore, the experimentally observed FWHM ($FWHM_{obs.}$) values can be described by the equation

$$FWHM_{obs.} = \phi_D FWHM_D \phi_H FWHM_H, \tag{4.2}$$

where

$$\phi_D + \phi_H = C_D/C + C_H/C = 1, \tag{4.3}$$

and $FWHM_D$ and $FWHM_H$ are the virtual FWHM values of the peaks originating from the Xe at the Henry and Langmuir sites, respectively. The solid curves in Fig. 4.7 represent the fitting of the Xe sorption data using Eq's. (4.2) and (4.3). These results confirm that the value of $FWHM_D$ is much smaller than that of $FWHM_H$ for the Xe-PVC and PPO systems. In addition, similar results have been obtained for other Xe-glassy polymer systems, demonstrating that the diffusivity of a gas at Henry sites is much greater than that at Langmuir sites. The low diffusivity in glassy polymers can also be explained based on the effect of Langmuir sites, and the remarkably low diffusivity exhibited by PVC can likely be attributed to the presence of only a small number of isolated microvoids. The estimated $FWHM_H$ value for PPO also changes with temperature and with the sample preparation procedure (such as slow cooling and melt quenching), all of which can affect the gas diffusivity. In contrast, the $FWHM_D$ value shows only minimal variation. Together, these results indicate that the NMR peak width of a gas in a polymer can be applied to the characterization of the diffusivity of the gas in that material.

In an attempt to perform quantitative analysis based on peak widths, the spin-spin relaxation time (T_2) of Xe in polymers was determined using the solid echo method. In some experiments, CH_4 was also used instead of Xe to confirm the universality of this method, since the sorption properties of CH_4 are similar to those of Xe but the diffusivity of CH_4 is higher than that of Xe due to molecular size effects. The experimental T_2 values were readily determined from single exponential decay data and were found to correspond to the reciprocals of the FWHM values, demonstrating that FWHM data may be used to study diffusivity regardless of the nuclei being assessed. Fig. 4.8 plots FWHM values against decay times obtained from solid echo experiments for CH_4 and Xe-PPO systems. In the case of PPO, both the peak intensity and FWHM are reduced during the T_2 process. These findings indicate the distributions of T_2. If only one component of T_2 exists in the system, the peak intensity decays exponentially, but FWHM is not changed. The reciprocal value of FWHM is apparently related to T_2, the change of FWHM during the T_2 measurement, therefore, means the existence of the multi-component T_2. In other words, the diffusivity of the CH_4 and Xe in PPO over very short timescales is distributed to some extent, because the value of FWHM is influenced by diffusivity as mentioned above. The degree of reduction of the FWHM is, therefore, likely to be correlated with the distribution of diffusivity. Both optical and X-ray diffraction analyses demonstrated that the PPO was completely amorphous. However, from the viewpoint of gas diffusivity, the amorphous structure of the PPO may have been slightly inhomogeneous, and it may be possible to detect this inhomogeneity on a very small timescale. The diffusing gas molecules can be expected to move through different regions of the samples in which the diffusivity varies, such that the observed diffusion behavior is actually an averaged value when the measurement is performed over a relatively long time span, or if the diffusivity of the gas is especially rapid. Since the observation time span used to assess the distribution of diffusivities is equal to an apparent T_2, this distribution will become narrower with increasing T_2 values. Based on this discussion, the observed order of the reduction in FWHM values (CH_4 at 80 °C < CH_4 at 25 °C < Xe at 25 °C) is considered to be reasonable. It should also be noted that the fine high-order structure of PPO in the glassy state is inhomogeneous and thus will exhibit different gas diffusivities on the scale of several tens of nm based on the mean pathway length calculated using the diffusion coefficient and the apparent T_2 at various experimental conditions. The glassy polymer PC, whose gas diffusion properties are lower than those of PPO, showed no change in FWHM values during solid echo measurements, indicating that the fine glassy state structure of PC is more homogeneous than that of PPO. T_2 observations of gases in glassy polymers are therefore a useful means of characterizing the glassy state.

The FWHM values of the Xe peaks obtained from rubbery polymers are only minimally affected by pressure or by the ^{129}Xe chemical shift, although the chemical shift is correlated with the density of the polymer, as discussed earlier. In general, the diffusion coefficient of a gas in a polymer has been understood based on the free volume theory [26]. Thus, in the case of rubbery polymers, the diffusion coefficient can be predicted from the ^{129}Xe NMR chemical shift, while the FWHM values of the Xe peaks obtained from rubbery polymers also qualitatively correspond to

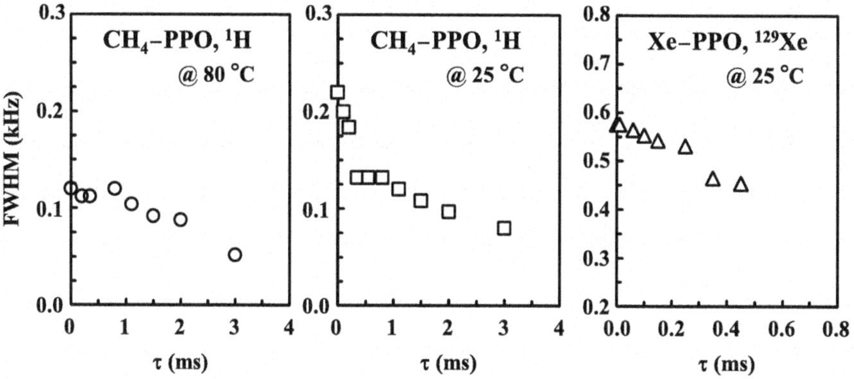

Fig. 4.8 ^{1}H and ^{129}Xe NMR peak widths (FWHMs) obtained from CH_4 and Xe in PPO as functions of the relaxation decay time, τ, to determine spin-spin relaxation times by the solid echo method. Experimental conditions are indicated

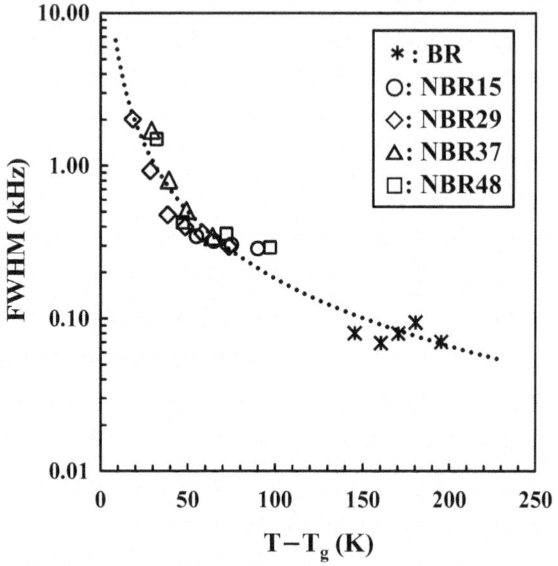

Fig. 4.9. Plot of FWHMs as a function of the difference between observation (T) and glass transition (T_g) temperatures

the diffusivity. Despite this, the accuracy with which diffusion coefficients can be determined from FWHM data is less than that is possible using ^{129}Xe NMR chemical shifts because the FWHM is also affected by structural factors such as inhomogeneity due to the volume susceptibility of the polymer pieces. However, the FWHM values of Xe peaks obtained from rubbery polymers exhibit an interesting trend with changes in temperature and type of polymer, indicating that the FWHM roughly correlates with the diffusivity of the Xe. Figure 4.9 shows a semi-logarithmic plot of FWHM as a function of the difference between the experimental (T) and glass transition (T_g) temperatures. Although both density and T_g varied among the rubbery polymers employed to obtain these data, the FWHM values can be nearly normalized by the temperature difference (T - T_g). This plot is similar to the temperature dependence of the shift factor, a_T, which is derived from the well-known WLF equation [27]. Thus, the FWHM values in rubbery polymers, and by extension the Xe diffusivity, evidently are also affected by the polymer chain mobility.

4.5 Characterization of Oriented Structures by Pulsed Field Gradient NMR

It is known that self-diffusion coefficients can be determined by pulsed field gradient (PFG) NMR spectroscopy. Since this is a one-dimensional technique, the structural anisotropy of the polymeric material can be estimated from the anisotropic diffusion properties of gas and/or solvent molecules. As an example, the long-channel cavities in hexagonal columns composed of an oriented polyester having dodecyl side chains have been successfully characterized using PFG NMR in conjunction with ethane as the probe molecule [28]. However, in many gas-glassy polymer systems, the gas diffusivity is too slow to determine the self-diffusion coefficient by the PFG method. In this section, a study [29] in which the gas diffusion anisotropy of molecularly oriented samples was determined is detailed. Figure 4.10 summarizes the gas diffusion paths in randomly and mono-axially oriented samples. Here, a single crystallite having anisotropic diffusivity is indicated by a striped square. When the crystallites are oriented randomly, the gas molecules must move tortuously, so diffusivity is low. In contrast, the diffusion pathway is relatively straight for long-range motion in the mono-axially oriented domains.

The form I crystal of poly 4-methyl-1-pentene (PMP) has four tetragonal 7/2 helical chains [30] and presents empty channel-like inter-chain spaces with mean diameters of approximately 0.4 nm along the helical axis of the PMP chains. It has been reported that only small gas molecules can penetrate these spaces [16, 31]. Analyses using infrared dichroism and X-ray diffraction have confirmed that mono-axially oriented crystalline regions can be obtained by mono-axial drawing of a PMP film, with rod-like helical chains in these crystallites positioned parallel to the drawing direction. The relationship between the direction in which the film is positioned inside the NMR sample tube and molecular orientation is presented in Fig. 4.11, in which the drawing direction is indicated by stripes. Comparing the self-diffusion coefficients determined using two specially prepared sample tubes, information can be obtained concerning the anisotropy of the gas diffusivity. As an example, the ^{1}H PFG NMR results obtained using CH_4 in conjunction with a PMP film drawn perpendicular to the PFG direction (see the left side of Fig. 4.11) are shown in Fig. 4.12. The molecular mobility within the PMP film sample is too slow, such that peaks resulting from protons in the polymer are absent from the spectra. Therefore, all peaks in these spectra are assignable to CH_4. An intense peak appears in the low field at a PFG strength (g) of 0 and can be assigned to free gaseous CH_4. This peak disappears upon applying the smallest g value of 170 G/cm, indicating rapid diffusion. In contrast, a relatively small peak appears in the high field, and the intensity of this peak decreases gradually with increasing g, meaning that this

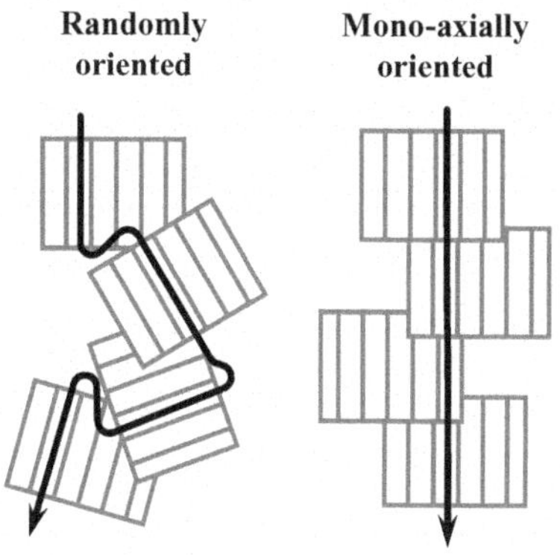

Fig. 4.10 Schematic representing the model for the gas diffusion pathways (*arrows*) at a cross-section in a polymer whose crystallites are arranged either randomly or mono-axially. The direction of gas diffusion is from *top to bottom*. The striped *square* indicates a single crystallite with anisotropic gas diffusion properties

peak can be assigned to CH_4 in the PMP film, having relatively low diffusivity. These findings demonstrate another application of PFG NMR as a technique for peak assignments.

In the case of undrawn PMP film, in which chains are randomly oriented in the crystallites, the same self-diffusion coefficient was obtained regardless of the direction in which the film was set in the NMR tube. However, the self-diffusion coefficient obtained when using a drawn PMP film set parallel to the PFG direction

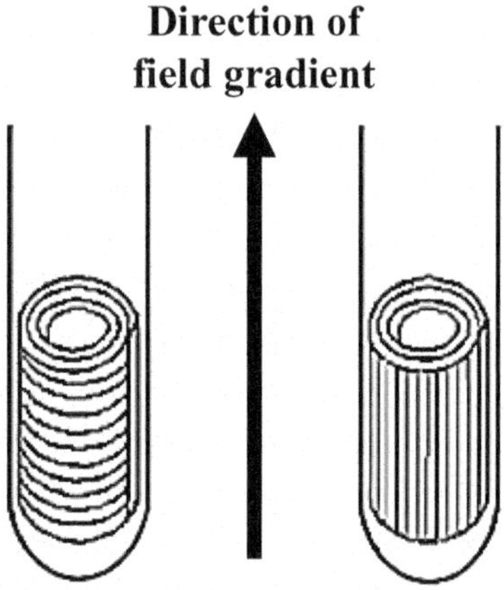

Fig. 4.11 Schematic illustration of the directions of mono-axially drawn poly 4-methyl-1-pentene (PMP) films in NMR sample tubes. The drawing direction is indicated by *stripes*

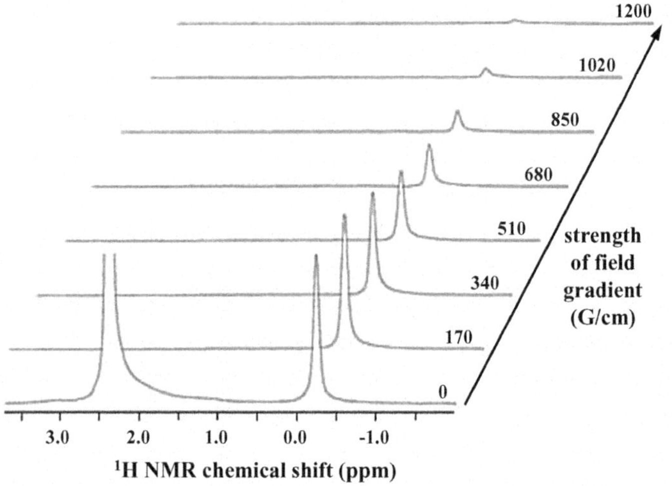

Fig. 4.12 The 400 MHz ^{1}H NMR spectra of the CH_4 in an NMR tube packed with a drawn PMP film set perpendicular to the drawing axis (see the *left side* of Fig. 4.11), acquired at various PFG strengths (g) at approximately 0.4 MPa and 25 °C. The PFG length (δ) and interval (Δ; the so-called diffusion and/or observation time) were 1.0 and 10 ms, respectively. A stimulated-echo (STE) pulse sequence was used

(see the right side of Fig. 4.11) was approximately three times that obtained using a different orientation. This result indicates that the CH_4 in a mono-axially oriented PMP film diffuses preferentially along the helical chains in the crystallites, as shown in Fig. 4.10. Similar studies using liquid crystalline polyesters [32] and polypeptides [33] have been reported elsewhere.

4.6 Summary

This chapter discussed the NMR spectroscopy data that can be obtained from numerous gas-polymer systems, in order to demonstrate that this technique allows the characterization of the fine structures and gas transport properties of polymeric materials. ^{129}Xe NMR spectroscopy is one of the most powerful of these methods. In particular, the ^{129}Xe NMR chemical shifts of polymeric materials can provide information regarding the density of rubbery polymers or the presence of microvoids in glassy materials, on both a qualitative and quantitative basis. In addition, the spectral intensity can be converted to units of gas sorption, and the peak width (equivalent to the spin-spin relaxation time) presents a measure of diffusivity. These features are independent of the nuclei being assessed, indicating that various gases can be applied. Because peak widths are typically in the range of several tens to tens of thousands of Hz, gas diffusivity can be observed on very short timescales. Furthermore, the PFG NMR analysis of gases in polymeric materials is useful for the determination of anisotropic gas diffusion properties and molecular orientations. Recently, by using ^{129}Xe NMR, it was confirmed that a thermotropic columnar liquid-crystalline material composed of a shape-persistent macrocycle formed a continuous nano-void inside the columnar assembly [34]. The chemical shifts and intensities of the Xe entrapped in this liquid-crystalline material depended on the pressure of Xe and phase-transition temperature. These results indicate that the encapsulation of Xe within the discrete and oriented nanospaces of nanoporous liquid crystals is different from the homogeneous dissolution of the solute in an ordinary solution. Relatively fast movement of the Xe within the channel is confirmed by the PFG NMR method, in addition, the diameter of the nanospace was estimated to be ~4.9 Å from the chemical shift values. The ^{129}Xe NMR spectroscopy assisted in the first discovery of the continuous nanopores within fluid materials. In conclusion, the NMR spectral observation of gases in polymeric substances has significant potential as a means of studying both gas transport properties and fine high-order structures.

References

1. Yoshimizu, H.: NMR characterizations of polymers using gases as probes. J. Jpn. Soc. Color Material **88**, 265–270 (2015). https://doi.org/10.4011/shikizai.88.265

2. Yoshimizu, H.: Application of NMR spectroscopy to the characterization of gas separation membranes. Sen'I Gakkaishi J. Soc. Fiber Sci. Tech. Jpn. **68**, pp. 323–328 (2012). https://doi.org/10.2115/fiber.68.P_323
3. Yoshimizu, H.: Dependence of the amount of Xe sorption on the ^{129}Xe NMR chemical shift in glassy polymers. Cheng, H.N.: Asakura, T.: English, A.D. (eds.) ACS Symposium Series **1077**; Am. Chem. Soc., Chapter 31, 509–523 (2011). https://doi.org/10.1021/bk-2011-1077.ch031
4. Yoshimizu, H.: Suzuki, T.: Tsujita, Y.: Characterization of polymers in the solid state by ^{129}Xe NMR spectroscopy. Sen'I Gakkaishi. J. Soc. Fiber Sci. Tech. Jpn. **60**, pp. 498–503 (2004). https://doi.org/10.2115/fiber.60.P_498
5. Kesting, R.E.: Fritzsche, A.K. (eds.) Polymeric gas separation membranes. Wiley Interscience, New York (1993)
6. Crank, J.: Park, G.S. (eds.) Diffusion in polymers. Academic Press, London and New York (1968)
7. Micaels, A.S.: Vieth, W.R.: Barrie, J.A.: Solution of gases in polyethylene terephthalate. J. Appl. Phys. **34**, 1–12 (1963). https://doi.org/10.1063/1.1729066
8. Demarquay, J.: Fraissard, J.: ^{129}Xe NMR of xenon adsorbed on zeolites: Relationship between the chemical shift and the void space. Chem. Phys. Lett. **136**, 314–318 (1987). https://doi.org/10.1016/0009-2614(87)80258-0
9. Ripmeester, J.A.: Ratcliffe, C.I.: Tse, J.S.: The nuclear magnetic resonance of ^{129}Xe trapped in clathrates and some other solids. J. Chem. Soc. Faraday Trans. 1: Phys. Chem. Condensed Phases **84**, 3731-3745 (1988). https://doi.org/10.1039/F19888403731
10. Fraissard, J.: Ito, T.: ^{129}Xe n.m.r. study of adsorbed xenon: A new method for studying zeolites and metal-zeolites. Zeolites **8**, 350–361 (1988). https://doi.org/10.1016/S0144-2449(88)80171-4
11. Ito, T.: Springuel-Huet, M.A.: Fraissard, J.: ^{129}Xe n.m.r. study of xenon adsorbed on Na-Omega and Na and K ferrierite: Characterization of void spaces, defects, and intergrowth of ferrierite-mordenite structures. Zeolites **9**, 68–73 (1989). https://doi.org/10.1016/0144-2449(89)90011-0
12. Ripmeester, J.A.: Ratcliffe, C.I.: Application of xenon-129 NMR to the study of microporous solids. J. Phys. Chem. **94**, 7652–7656 (1990). https://doi.org/10.1021/j100382a061
13. Suzuki, T.: Miyauchi, M.: Takekawa, M.: Yoshimizu, H.: Tsujita, Y.: Kinoshita, T.: Characterization of the microvoids in poly(2,6-dimethyl-1,4-phenylene oxide) by means of ^{129}Xe NMR spectroscopy. Macromolecules **34**, 3805-3807 (2001). https://doi.org/10.1021/ma001656f
14. Suzuki, T.: Miyauchi, M.: Yoshimizu, H.: Tsujita, Y.: Characterization of microvoids in glassy polymers by means of ^{129}Xe NMR spectroscopy. Polymer J. **33**, 934–938 (2001). https://doi.org/10.1295/polymj.33.934
15. Suzuki, T.: Yoshimizu, H.: Tsujita, Y.: Characterization of microvoids in PPO/PS polymer blend by means of ^{129}Xe NMR spectroscopy. Desalination **148**, 359–361 (2002). https://doi.org/10.1016/S0011-9164(02)00730-0
16. Suzuki, T.: Tanaka, T.: Nakajima, M.: Yoshimizu, H.: Tsujita, Y.: Characterization of the cavity in poly(4-methyl-1-pentene) crystal by gas permeation and ^{129}Xe NMR measurements. Polym J. **34**, 891–896 (2002). https://doi.org/10.1295/polymj.34.891
17. Suzuki, T.: Yoshimizu, H.: Tsujita, Y.: Analysis of gas transport properties of PPO/PS blends by ^{129}Xe NMR spectroscopy. Polymer **44**, 2975–2982 (2003). https://doi.org/10.1016/S0032-3861(03)00191-5
18. Sivakumar, M.: Suzuki, T.: Yamamoto, Y.: Mahesh, K.P.O.: Yoshimizu, H.: Tsujita, Y.: Structure and properties of the mesophase of syndiotactic polystyrene membrane: VII. Isothermal sorption behavior of xenon. J. Membr. Sci. **238**, 75–81 (2004). https://doi.org/10.1016/j.memsci.2004.03.022
19. Tsukahara, T.: Tsujita, Y.: Yoshimizu, H.: Kinoshita, T.: Structure, gas sorption, and gas diffusion of a low density liquid crystalline polyester. Pinnau, I.: Freeman, B.D. (eds.) ACS Symposium Series. Am. Chem. Soc., Chapter 8, **876** 129–138 (2004). https://doi.org/10.1021/bk-2004-0876.ch008

20. Yoshimizu, H.: Tsukahara, M.: Suzuki, T.: Toida, J.: Ando, A.: Watanabe, J.: Tsujita, Y.: Layered structure and Xe sorption and diffusion properties of low-density liquid-crystalline polyesters with n-alkyl side chains. J. Mol. Struct. **739**, 19–26 (2005). https://doi.org/10.1016/j.molstruc.2004.07.044
21. Yoshimizu, H.: Ohta, S.: Asano, T.: Suzuki, T.: Tsujita, Y.: Temperature dependence of the mean size of microvoids of polyphenyleneoxide studied by analyses of Xe sorption and ^{129}Xe NMR chemical shift. Polymer J. **44**, 821–826 (2012). https://doi.org/10.1038/pj.2012.123
22. Yoshimizu, H.: Murakami, T.: Suzuki, T.: Tsujita, Y.: Characterization of the microvoids of a tetramethyl polycarbonate/polystyrene blend system using Xe sorption measurements and ^{129}Xe NMR spectroscopy. Polymer J. **44**, 827–831 (2012). https://doi.org/10.1038/pj.2012.128
23. Higuchi, T.: Yoshimizu, H.: Characterization of a glass state of polycarbonate by Xe sorption properties and Xe-129 NMR. Polym. Prep. Jpn. **61**, 4348–4349 (2012)
24. Stengle, T.R.: Williamson, K.L.: Nuclear magnetic resonance of xenon absorbed in solid polymers: a probe of the amorphous state. Macromolecules **20**, 1428–1430 (1987). https://doi.org/10.1021/ma00172a050
25. Yoshimizu, H.: Okazawa, M.: Jinno, S.: Asano, T.: Suzuki, T.: Characterizations of gas diffusion properties of polyphenyleneoxide in several glassy states by means of NMR spectroscopy. Kobunshi Ronbunshu **69**, 424–434 (2012). https://doi.org/10.1295/koron.69.424
26. Bondi, A.: Physical properties of molecular crystals, liquids, and glasses. Wiley, New York (1968)
27. Williams, M.L.: Landel, R.F.: Ferry, J.D.: The temperature dependence of relaxation mechanisms in amorphous polymers and other glass-forming liquids. J. Am. Chem. Soc. **77**, 3701–3707 (1955). https://doi.org/10.1021/ja01619a008
28. Matsui, M.: Yamane, Y.: Kuroki, S.: Ando, I.: Fu, K.: Watanabe, J.: Long channel cavity and gas diffusion in liquid-crystalline polyester with n-alkyl side chains. Ind. Eng. Chem. Res. **44**, 8694–8698 (2005). https://doi.org/10.1021/ie048785v
29. Yoshimizu, H.: Okumura, Y.: Gas transport properties of the crystalline phase controlled its orientation of poly(4-methyl-1-pentene). Kobunshi Ronbunshu **71**, 601–607 (2014). https://doi.org/10.1295/koron.71.601
30. Kusanagi, H.: Takase, M.: Chatani, Y.: Tadokoro, H.: Crystal structure of isotactic poly(4-methyl-1-pentene). J. Polym. Sci. Polym. Phys. Edit. **16**, 131–142 (1978). https://doi.org/10.1002/pol.1978.180160112
31. Yoshimizu, H.: Fukatsu, H.: Suzuki, T.: Tsujita, Y.: Kinoshita, T.: CO_2 permeation and diffusion properties of semicrystalline poly(4-methyl pentene-1) membranes. Polym J. **30**, 981–984 (1998). 0.1295/polymj.30.981
32. Yoshimizu, H.: Karakasa, T.: Magnetic orientation and gas diffusion properties of layered structures of liquid crystalline copolyesters with alkyl side chains of different length. Kobunshi Ronbunshu **71**, 325–333 (2014). https://doi.org/10.1295/koron.71.325
33. Iwamoto, J.: Yoshimizu, H.: Structural analysis and gas diffusion properties of the solid oriented PBLG. Prep. 53rd Ann. Meeting NMR Soc. Jpn., 348–349 (2014).
34. Kawano, S.: Yoshimizu, H.: Tanaka, K.: Continuous Nanospace in Nanoporous Liquid Crystal Investigated by ^{129}Xe NMR Spectroscopy. *Angew. Chem. Int. Ed.*, e202316523 (2023). https://doi.org/10.1002/anie.202316523

Chapter 5
Solid-State NMR of Amorphous Materials—Application to Organic Light-Emitting Diodes

Katsuaki Suzuki and Hironori Kaji

Abstract In this chapter, we introduce some examples of the structure analysis by solid-state and liquid NMR for materials in organic light-emitting diodes (OLEDs). The isotropic chemical shift, chemical shift anisotropy, and quadrupolar parameters are closely related to the intra- and intermolecular structures of molecules. In this chapter, these parameters are obtained typically by the 1D ^{13}C, ^{15}N cross-polarization/magic angle spinning (CP/MAS) NMR measurements, 2D ^{27}Al multiple quantum magic angle spinning (MQMAS), and ^{15}N isotropic/anisotropic separation measurements. Also, we show the results of dynamic nuclear polarization enhanced solid-state NMR (DNP-ssNMR) experiments, which enable us to measure samples with very limited sample amounts and/or low natural abundance nuclei. Quantum chemical calculations, especially NMR chemical shift calculations by gauge-including atomic orbital (GIAO) and gauge-including projector-augmented wave (GIPAW) methods are also effectively employed. The combination of experimental NMR results and quantum chemical calculations provides fruitful structural information, which is deeply related to the origin of charge carrier transporting and light-emitting properties of OLEDs.

Keywords Organic light-emitting diodes · Dynamic nuclear polarization · Thermally activated delayed fluorescence

5.1 Introduction

This chapter introduces nuclear magnetic resonance (NMR) investigations of organic light-emitting diodes (OLEDs). Since Prof. Tang and co-authors published a paper on the first pioneering OLED [1] using organic thin films as the emissive layer, OLEDs have been an active research area of next-generation solid-state lighting and display technology. The OLED devices are generally fabricated on glass or plastic substrates

K. Suzuki · H. Kaji (✉)
Institute for Chemical Research, Kyoto University, Uji, Kyoto 611-0011, Japan
e-mail: kaji@scl.kyoto-u.ac.jp

The NMR Society of Japan (ed.), *Experimental Approaches of NMR Spectroscopy II*,
https://doi.org/10.1007/978-981-97-6830-1_5

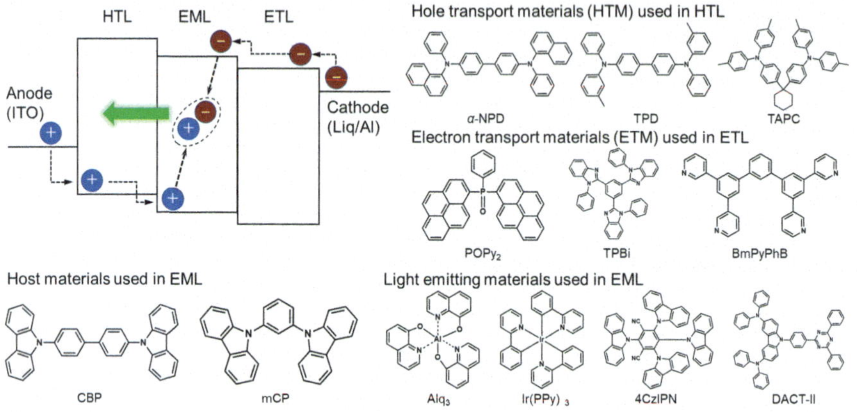

Fig. 5.1 A schematic representation of OLED device structure and representative materials used in OLEDs

with multilayer-stacked structures, as shown in Fig. 5.1. A typical OLED device consists of an anode, a hole-transport layer (HTL), a light-emitting layer (EML), an electron-transport layer (ETL), and a cathode. In some devices, a hole-blocking layer (HBL), an electron-blocking layer (EBL), and a singlet/triplet exciton-blocking layer (EBL) are introduced to achieve excellent performances. The light-emitting mechanism illustrated in Fig. 5.1 is summarized by the following processes.

1. *Hole and Electron Injection From the Anode and the Cathode*

When an electric field is applied to the device, holes and electrons are injected from the anode and the cathode, respectively. Holes can be effectively injected into HTL by reducing the energy barrier between the anode and HTL. Similarly, electrons can be efficiently injected into ETL by reducing the energy barrier between the cathode and ETL.

2. *Hole and Electron Transport Through the HTL and ETL*

Once holes and electrons have been injected, molecules in the HTL and ETL become radical cations and anions, respectively, which are referred to as (positive and negative) polarons. The formed polarons transport through the HTL and ETL by the so-called charge hopping mechanism. Materials with excellent hole and electron transport properties are used as the hole and electron transport materials (HTM and ETM), respectively.

3. *Charge Recombination, Exciton Formation, and Emission*

Next, both holes and electrons are injected into EML. EML typically consists of a light-emitting dopant material and a host material having both the hole and electron transport properties. The electrons and holes are recombined to form singlet and triplet excitons with a 1:3 ratio. Triplet excitons are typically converted as heat rather than as light for normal fluorescent emitters. Meanwhile, phosphorescent emitters containing heavy metal atoms can directly convert triplet excitons into light due to the heavy atom effect. Newly emerging thermally activated delayed fluorescence (TADF) emitters can also convert triplets to light by reverse intersystem crossing which converts triplet excitons into singlets.

To develop materials used for OLEDs, we should have a deep understanding of fundamental relationships between molecular structures and properties such as hole/electron mobilities and light-emission of materials. In the following sections, we describe NMR approaches to characterize OLED-related materials.

5.2 Analysis of Charge Transport Materials

5.2.1 Conformation Analysis of Hole Transport Materials

In OLEDs, organic materials are in an amorphous state. Therefore, clarification of conformations of organic molecules in such amorphous states is essential to understanding their charge (hole/electron) transporting properties. The detailed molecular structures of these materials are often studied by X-ray diffraction analysis of single crystals or density functional theory (DFT) calculations. However, the molecular structures in single crystals often differ from those of the DFTs and would differ from those in amorphous states. Here, we show an example of solid-state NMR analysis to reveal the conformation of an HTM in the amorphous state [2]. As an HTM, *N*, *N*'-diphenyl-*N*, *N*'-di(*m*-tolyl)benzidine (TPD; Fig. 5.1) [3–7] is used because it has been widely used in HTL in OLEDs.

Figure 5.2 shows a DFT-optimized structure of a TPD molecule. The three torsion angles, denoted as α, β, and γ, in Fig. 5.2 determine the structure around the central nitrogen atom. The three carbons directly bonded to nitrogen and the nitrogen itself are in the same plane, and three phenyl rings around the nitrogen form a propeller-type structure. Due to the presence of a tolyl ring, TPD has forty conformers; the energy differences between each conformer are within 0.8 kJ mol^{-1}, indicating that these conformers are equally probable in an amorphous state.

Figure 5.3a shows the experimental cross-polarization/magic angle spinning (CP/MAS) ^{15}N NMR spectrum of ^{15}N enriched amorphous TPD (^{15}N-aTPD). The observed resonance line was relatively broad, reflecting structural disorder in the amorphous state. In the chemical shift anisotropy (CSA) spectrum of ^{15}N-aTPD in Fig. 5.3b, clear shoulders were not observed, and precise principal values of σ_{11}, σ_{22}, and σ_{33} could not be determined. Thus, the CSA span defined by σ_{11}–σ_{33} is used for the analysis. Here, gauge-including atomic orbital (GIAO) calculations [8] for the NMR chemical shift parameters are carried out to correlate the experimental CSA span and the molecular torsion angles (Figs. 5.4 and 5.5). In these calculations, the torsion angles γ are assumed to be the same for β due to the structural similarity and all the other parameters are from the DFT-optimized structure shown in Fig. 5.2. Figure 5.4 shows the DFT-calculated CSA spans plotted as a function of the torsion

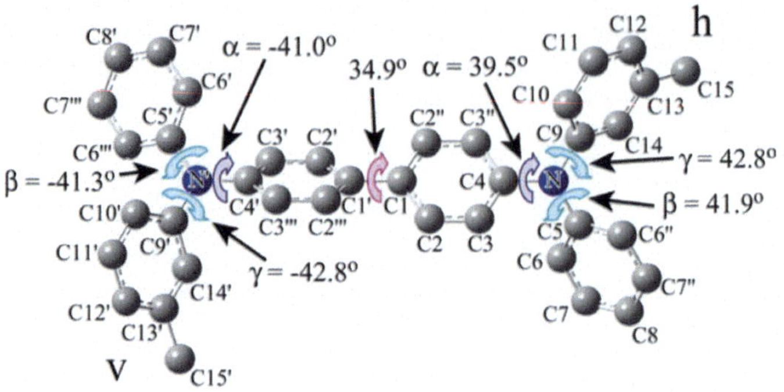

Fig. 5.2 DFT-optimized structure of TPD. The DFT-optimized torsion angles are shown along with the notation of the relevant torsion angles, α, β, and γ. Reprinted from Ref. [2]. With permission from Elsevier

angles, α, β. The CSA spans are strongly dependent on the torsion angles and thus can be used for the torsion angle determination. The two local minima of the CSA span are observed. One is near $(\alpha, \beta) = (40°, 40°)$, and the other is near $(\alpha, \beta) = (140°, 140°)$; the two conformers are geometrically equivalent.

Figure 5.5a and b depicts more detailed α-and β-dependences of the CSA spans (filled red squares). Two lines in the figures correspond to two nitrogens in one TPD

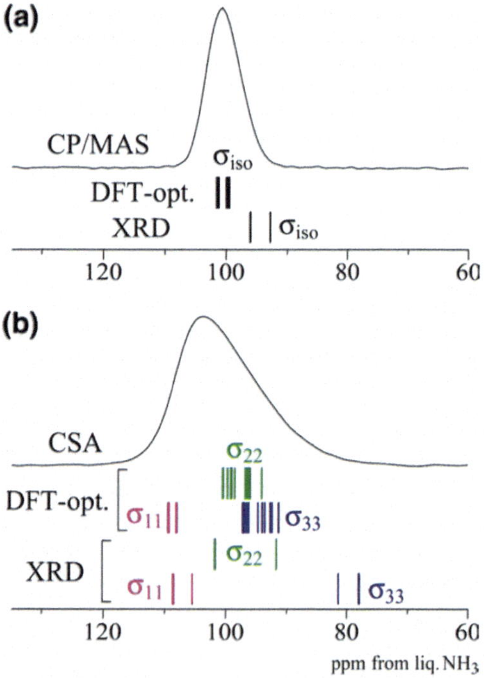

Fig. 5.3 **a** Experimental ^{15}N CP/MAS NMR spectrum of ^{15}N-aTPD. **b** Experimental ^{15}N CSA NMR spectrum of ^{15}N-aTPD. The isotropic chemical shift (σ_{iso}) and σ_{11}, σ_{22}, σ_{33} for DFT-optimized TPD and TPD based on XRD results [9] are also shown by the vertical bars. Reprinted from Ref. [2]. With permission from Elsevier

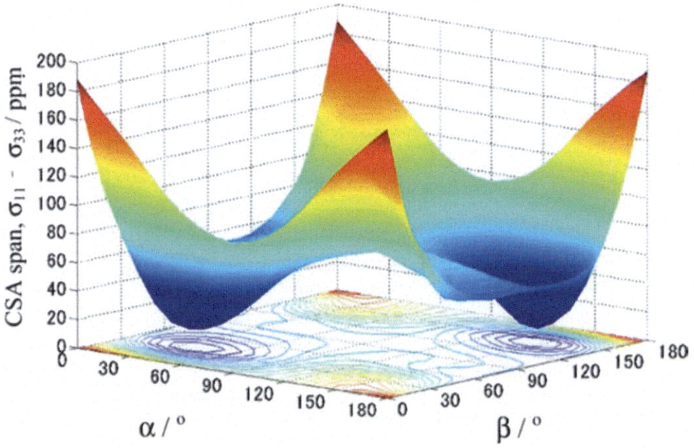

Fig. 5.4 DFT-calculated CSA span (σ_{11}–σ_{33}) values of TPD as a function of torsion angles α and β under the assumption of $\beta = \gamma$. The calculations were carried out with 30° intervals, and a two-dimensional interpolation was applied for the graphical presentation. Only the values for the right nitrogen of TPD are shown. Reprinted from Ref. [2]. With permission from Elsevier

molecule; the CSA spans of the two nitrogens are almost identical to each other. The experimentally determined CSA span of 15 ppm from Fig. 5.3b corresponds to the narrowest CSA span among the DFT-calculated values. The small anisotropy of the nitrogens in a TPD reflects the nearly spherical symmetry of electronic shielding. The torsion angles determined by the experimental CSA span are very close to those obtained from the DFT-optimized structure. The calculated σ_{iso}, σ_{11}, σ_{22}, and σ_{33} values for the DFT-optimized conformers are consistent with the observed line-shape, as shown by the vertical bars in Fig. 5.3b, although the σ_{22} values are slightly different. A significant inconsistency is observed for the CSA values calculated from the structure obtained from powder X-ray diffraction (XRD) results [9], especially for σ_{33} values, which may originate from the structure modification owing to the crystalline packing.

5.2.2 *Orientation Distribution Analysis of Electron Transport Materials by DNP-ssNMR*

Molecular orientation in thin films has a strong impact on the performances of OLEDs, particularly on charge mobility, light emission, and light out-coupling efficiency [10, 11]. The molecular orientation of amorphous materials in OLEDs has been found to be non-random, which is typically determined by angular-dependent photoluminescence measurements and variable angle spectroscopic ellipsometry. These methods provide a so-called *order parameter*, which only shows the averaged molecular orientation in a thin film; the *"distribution"* of molecular orientations has

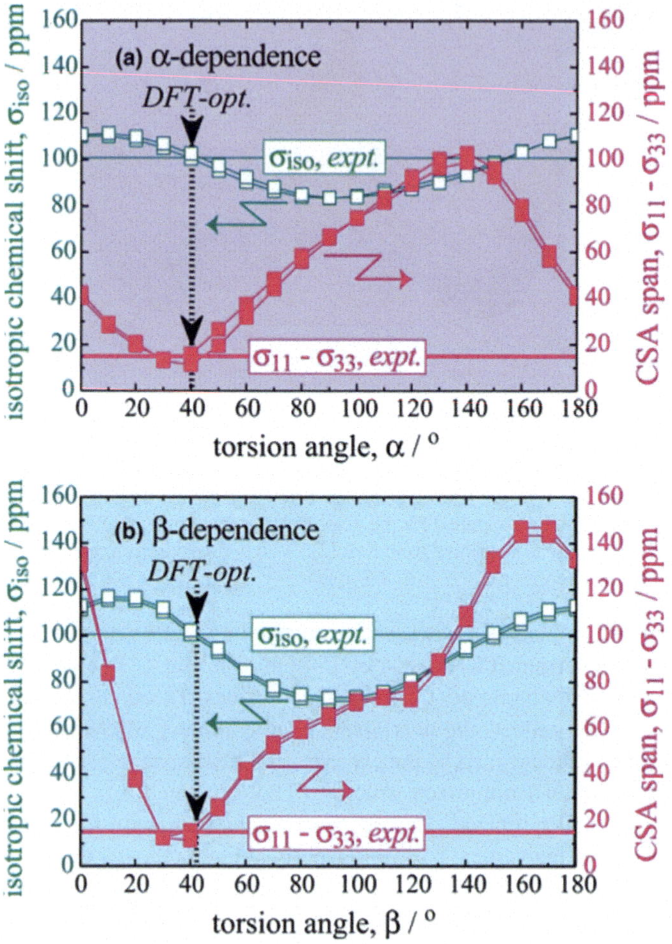

Fig. 5.5 Torsion angle dependence of DFT-calculated σ_{iso} (open squares) and σ_{11}–σ_{33} (filled squares) values of TPD; dependence of **a** α and **b** β under the assumption of β=γ. The values for the two nitrogens in TPD are shown as duplicated lines. Reprinted from Ref. [2]. With permission from Elsevier

not yet been revealed with either experiments or calculations so far. Meanwhile, solid-state NMR is a promising candidate tool for analyzing molecular orientation including the distribution, providing deeper insight into OLED performances at the molecular level.

For the analysis of thin film samples by solid-state NMR, the most crucial problem is the inherent low sensitivity of NMR because the sample amount in a thin film is very limited (the thickness of each layer in OLEDs is typically 10–40 nm). To overcome the problem, we used dynamic nuclear polarization (DNP) enhanced solid-state NMR (DNP-ssNMR) [12–15], which largely enhances the sensitivity by transferring high electron polarization under microwave irradiation to nuclei. Although most recent DNP-ssNMR measurements have been performed under magic angle spinning conditions [7], we used the "*static*" DNP-ssNMR technique for the analysis of molecular orientation in amorphous thin films [16]. The sample used in this work is an electron transport material, phenyldi(pyren-1-yl)phosphine oxide ($POPy_2$; Fig. 5.6a) [17]. $POPy_2$ shows different electron transport behaviors depending on film preparation methods (vacuum deposition and drop-casting deposition). Thus, orientation analysis is important to examine the difference at the molecular level. For a polarizing agent, we used a bisnitroxide radical, bis-TEMPO-bisketal (bTbK; Fig. 5.6a), which was thermally stable at the sublimation temperature of 178 °C. An amorphous $POPy_2$ film with 0.25 wt.% of bTbK was fabricated on glass (SiO_2) substrates. The thicknesses of $POPy_2$ were 1.5 μm (about 52 μg) for a vacuum-deposited sample and 2.8 μm (96 μg) for a drop-cast sample (Fig. 5.6b). These layer thicknesses were thicker than those of devices (typically 100–200 nm), but thinner than typical sample thicknesses required for time-of-flight mobility measurements (several μm).

The $POPy_2$ films on the substrates were set perpendicular to the external magnetic field, $\boldsymbol{B}_0$ (Fig. 5.6b), and then, static ^{31}P CP DNP-ssNMR experiments were conducted to obtain chemical shift anisotropy (CSA) spectra and to reveal the P=O orientations. The ^{31}P CSA spectra of $POPy_2$ for vacuum-deposited and solution-processed films with and without microwave irradiation are shown in Fig. 5.7. The sensitivity enhancement factor ($\varepsilon_{on/off}$) for the vacuum-deposited films (12 sheets) and drop-cast films (15 sheets) were calculated to be 3.0 (Fig. 5.7a) and 2.0 (Fig. 5.7b), respectively, while the $\varepsilon_{on/off}$ of a single sheet vacuum-deposited film was 6.0 (Fig. 5.7c). Even a single sheet sample provided sufficient S/N signal for

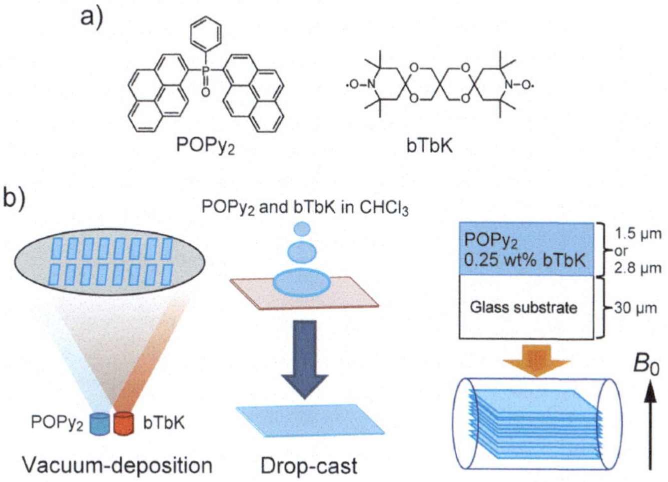

Fig. 5.6 **a** Molecular structures of $POPy_2$ and bTbK. **b** Schematic representation of sample preparations. The samples on glass (SiO_2) substrates are inserted into a 5 mmφ quartz tube. Reprinted from Ref. [16]. With permission from John Wiley & Sons Inc.

further analysis within 16 h of measurement. The lower $\varepsilon_{on/off}$ values for the multi-sheet samples are due to microwave absorption by the stacked SiO_2 substrates and the resulting temperature increase.

The CSA spectra of both the vacuum-deposited and drop-cast $POPy_2$ films showed axially symmetric patterns with the chemical shift principal values of $\sigma_\perp$ of 114.3 ppm (vacuum-deposited) and 110.1 ppm (drop-cast), and $\sigma_{//}$ of −93.4 ppm (vacuum-deposited) and −88.4 ppm (drop-cast). A clear difference was observed in the signal intensity at around −100 ppm, indicating the difference in molecular orientations. On the basis of GIAO calculations, the principal axes were assigned as shown in Fig. 5.8a. The $\sigma_{//}$ corresponded to the P=O direction and the $\sigma_\perp$ corresponded to the direction perpendicular to the P=O axis, as expected from the molecular symmetry.

The molecular orientation distributions were quantified by fitting the experimental CSA spectra using the Legendre moment expansion approach [18, 19] expressed by the following equations:

$$S(\sigma) = \int_0^\pi f(\theta - \sigma(\theta))p(\theta)\sin\theta d\theta \tag{5.1}$$

$$\sigma(\theta) = \sigma_{//}\cos^2\theta + \sigma_\perp\sin^2\theta \tag{5.2}$$

where θ is the angle between the P=O axis and $\boldsymbol{B_0}$, $\sigma(\theta)$ denotes the angular dependent chemical shift, and $p(\theta)$ is the orientation distribution function. To model the experimental CAS spectra, we applied a Gaussian-type line broadening, $f(\theta - \sigma(\theta))$. On the basis of axially symmetric CSA spectra of $POPy_2$, the $p(\theta)$ can be defined as an expansion of Legendre polynomials with the following equations:

$$p(\theta) = \sum_n \frac{2n+1}{2}\langle P_n\rangle P_n(\cos\theta),\ n = 0, 2, 4, 6, \cdots \tag{5.3}$$

$$\int_0^\pi p(\theta)\sin\theta d\theta = 2\langle P_0\rangle = 1 \tag{5.4}$$

$$\langle P_n\rangle = \frac{\int_0^\pi p(\theta)P_n(\cos\theta)\sin\theta d\theta}{\int_0^\pi p(\theta)\sin\theta d\theta} \tag{5.5}$$

The experimental data were fitted by least squares minimization with the fitting parameters of $\sigma_{//}$, $\sigma_\perp$, nth order parameters of $\langle P_n\rangle$ (n = 2, 4, 6), and a Gaussian line broadening factor. Higher order parameters (n $\geq$ 8) were ignored because the contribution of the 6th order parameter was already negligible (see Fig. 5.8d and e).

As shown in Fig. 5.8b and c, the experimental vacuum-deposited and drop-cast ^{31}P CSA spectra (the black curves) were well matched with the best-fit simulated spectra (the red curves). The $p(\theta)$ with best-fit parameters are shown as thick red curves in Fig. 5.8d and e with the respective contributions of $P_n(\cos\theta)$. For the vacuum-deposited films, the probability of finding a P=O orientation at $\theta = 0$ was

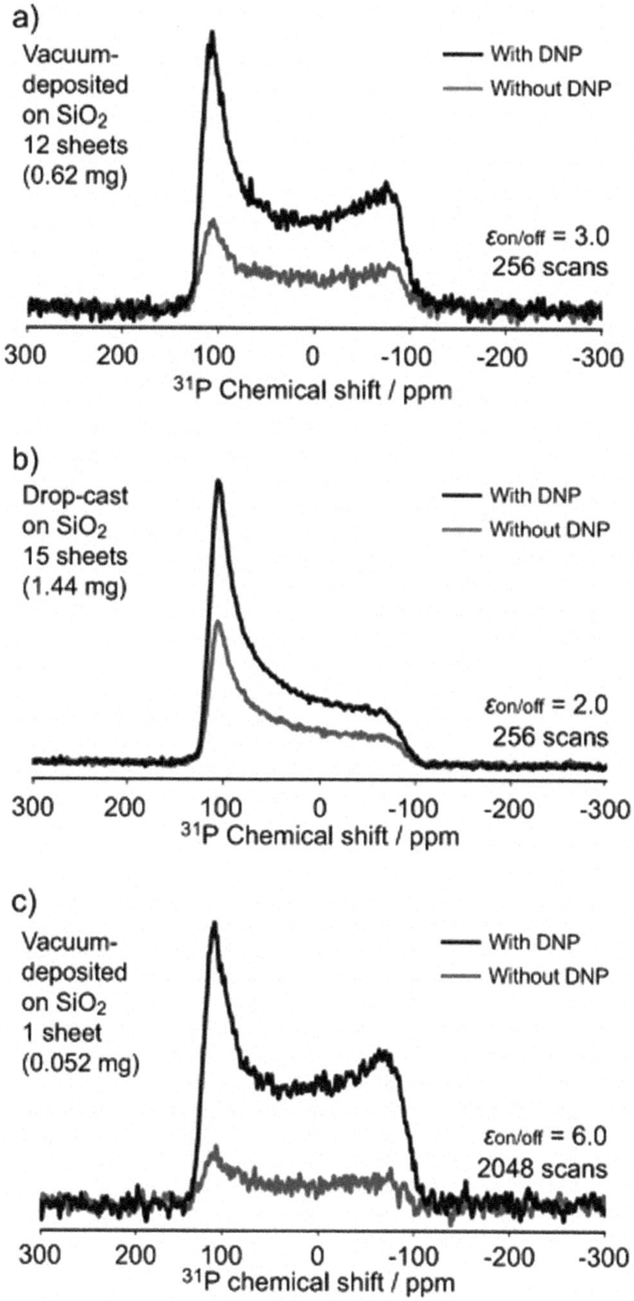

Fig. 5.7 ^{31}P CSA spectra of POPy$_2$ films with (black) and without (grey) DNP enhancement: **a** vacuum-deposited (12 sheets), **b** drop-cast (15 sheets), and **c** vacuum-deposited (1 sheet). Reprinted from Ref. [16]. With permission from John Wiley & Sons Inc.

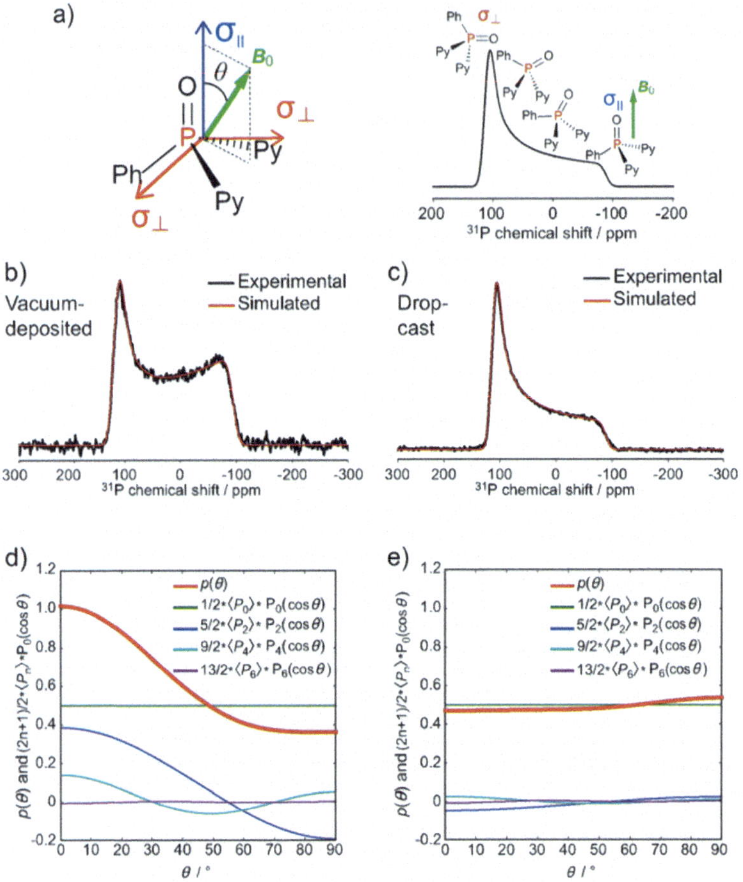

Fig. 5.8 **a** Schematic representation of chemical shift tensor orientation for $POPy_2$. The relationship between the P=O orientation and the chemical shift is also shown in the calculated CSA spectrum with completely random orientations. **b** and **c** Experimental ^{31}P CSA spectra of $POPy_2$ films for **b** the vacuum-deposited and **c** the drop-cast films with DNP enhancement (black lines). Simulated best-fit spectra are shown by the red lines. **d** and **e** Orientation distributions $p(\theta)$ (thick red lines) and contributions of respective terms (thin lines) of Legendre polynomials $P_n(\cos\theta)$ for **d** the vacuum-deposited and **e** the drop-cast films. Owing to the symmetry, values of $p(\theta)$ with $\theta >$ 90° are folded in between 0 and 90°. Reprinted from Ref. [16]. With permission from John Wiley & Sons Inc.

highest and decreased with increasing θ, which originates from the large contribution of the second-order parameter $\langle P_2 \rangle$. For the drop-cast film, $p(\theta)$ is almost constant against θ, indicating the random distribution of the P=O axis. The difference in molecular orientations results in a difference in charge transport behavior between vacuum-deposited and drop-casted films.

5.3 NMR Analysis of Light-Emitting Materials

5.3.1 *Structure Analysis of Tris(8-Hydroxyquinoline) Aluminum(III) (Alq_3) by Solid-State NMR*

Tris(8-hydroxyquinoline)aluminum(III) (Alq_3; Fig. 5.1) is one of the most well-known compounds in OLEDs, which exhibits fluorescence as well as electron transporting properties. The Alq_3 has two isomeric forms: facial and meridional. The facial isomer has C_3 symmetry with three equivalent ligands, whereas the meridional isomer has lower C_1 symmetry (Fig. 5.9). In addition, four crystalline polymorphs α-, β-, γ-, and δ-Alq_3 have been found in addition to the vacuum-deposited amorphous (amo)-Alq_3 [20–25]. α-, β-, and amo-Alq_3 show green emission, while γ-, and δ-Alq_3 show blue emission as shown in Fig. 5.10; the photoluminescence quantum yield of the blue-emitting Alq_3 is twofold greater than that of green-emitting Alq_3. The origin of the differences in emission colors has been controversial, with one suggesting that the origin is intermolecular packing (*intermolecular* interactions) and the other from the effect of isomeric states (*intramolecular* structure). In the following, we investigated intramolecular (isomeric state) and the intermolecular packing structures of α-, γ-, δ-, and amo-Alq_3 to reveal the origin of different emission colors.

To clarify the above, solid-state CP/MAS ^{13}C NMR measurements have been conducted for α-, γ-, δ-, and amo-Alq_3 [23]. Figure 5.11b–e shows the experimental CP/MAS ^{13}C NMR spectra of α-, γ-, δ- and amo-Alq_3. The solution spectrum (Fig. 5.11a) and the DFT-calculated ^{13}C isotropic chemical shifts of Alq_3 in the two isomeric forms (Fig. 5.11f and g) are also shown. The DFT-calculated resonance lines reflect the symmetries of the meridional and facial Alq_3 molecules; the calculated resonance lines of meridional Alq_3 (Fig. 5.11f) split into three for each carbon species, whereas those of the facial Alq_3 (Fig. 5.11g) show only one for each.

All the resonance lines in the CP/MAS ^{13}C NMR spectrum of α-Alq_3 are broadly distributed. Since the WAXD diffraction pattern of α-Alq_3 is sharp, the broadening

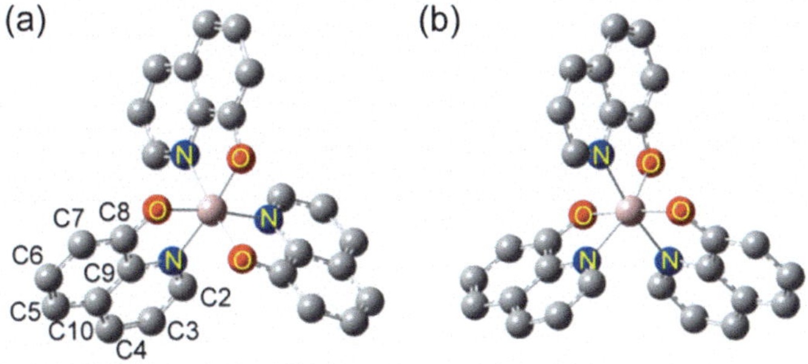

Fig. 5.9 Molecular structures of Alq_3 in **a** facial and **b** meridional isometric states. Hydrogen atoms are omitted for clarity. Reprinted with permission from [26]. Copyright 2013 American Chemical Society

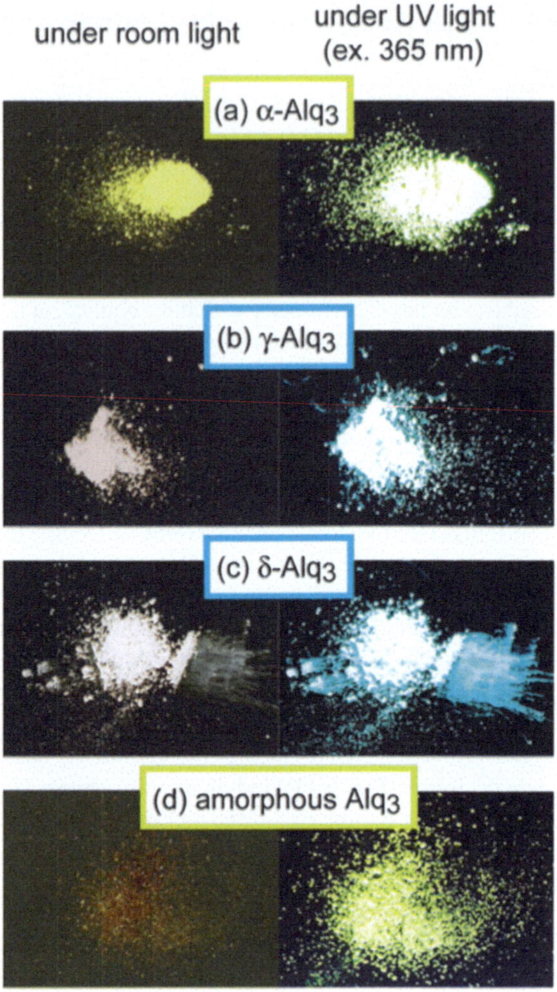

Fig. 5.10 Photographs of **a** α-, **b** γ-, **c** δ-, and **d** amo-Alq_3 under room light (left column) and UV irradiation at an excitation wavelength of 365 nm (right column). The γ- and δ-Alq_3, whitish under room light, show blue luminescence under UV irradiation. Reprinted with permission from [23]. Copyright 2006 American Chemical Society

of NMR signals originated from the local structural disorder for α-Alq_3. The CP/MAS ^{13}C NMR spectrum of amo-Alq_3 is quite similar to that of α-Alq_3 but clearly different from the sharp spectral patterns of γ- and δ-Alq_3. Because the isomeric state of α-Alq_3 is determined to be meridional [20], we concluded that amo-Alq_3 is mainly composed of the meridional state.

In contrast, sharp single resonance lines were observed for each carbon of γ- and δ-Alq_3 (Fig. 5.11c and d), indicating the well-ordered structures of the two polymorphs. For the γ-Alq_3 spectrum (Fig. 5.11c), a single resonance line is observed for each carbon, indicating the equivalence of the three ligands; the C2 carbon exhibits a single Lorentzian line shape with a line width of 51.0 Hz (the line widths of the other carbons were between 24.8 and 44.2 Hz). We could not observe any splitting due to intermolecular packing. These results clearly indicate that γ-Alq_3 is in the

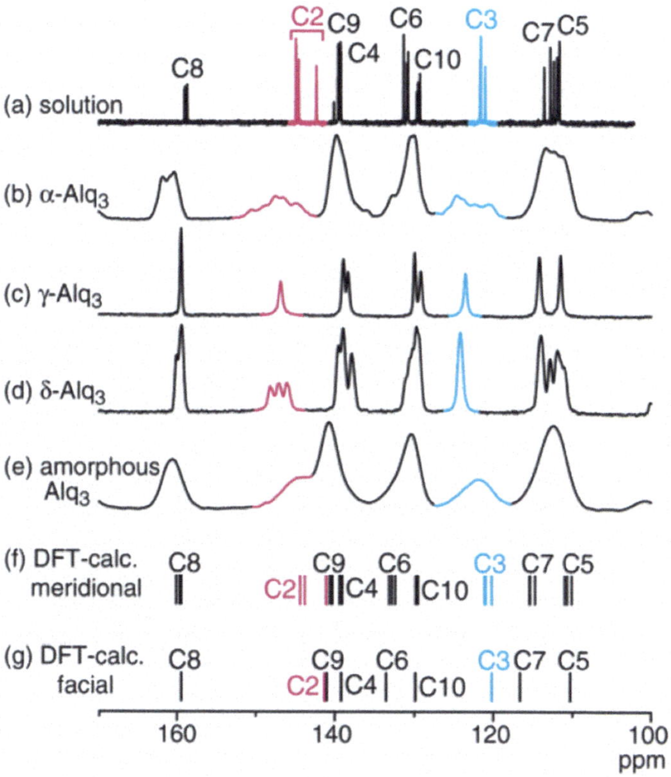

Fig. 5.11 **a** Experimental solution ^{13}C NMR spectrum of Alq_3. Experimental CP/MAS ^{13}C NMR spectra of **b** α-, **c** γ-, **d** δ-, and **e** amo-Alq_3. In **f** and **g**, DFT-calculated isotropic resonance lines are shown as vertical bars for meridional and facial Alq_3, respectively. Reprinted with permission from [23]. Copyright 2006 American Chemical Society

facial isomeric state. For the δ-Alq_3 (Fig. 5.11d), the splitting of resonance lines was observed for each carbon species. Particularly, the C2 resonance line exhibits three equally split resonance lines. The δ-Alq_3 spectrum could not be explained by DFT calculations of meridional and facial isolated molecules (Fig. 5.11f and g, respectively). The origin of the line split in the δ-Alq_3 spectrum is intermolecular packing as shown later by gauge-including projector-augmented wave (GIPAW) calculations.

Isomeric states of α-, γ-, δ-, and amo-Alq_3 were further investigated by 2D ^{1}H decoupled ^{27}Al multiple quantum magic angle spinning (MQMAS) [28]. The MQMAS experiment gives a correlation between the usual MAS dimension (quadrupolar anisotropy (QA) axis) and an isotropic dimension (ISO axis). In addition to the vertical ISO axis and the horizontal QA axis, two tilted axes, the chemical shift (CS) axis and the quadrupolar induced shift (QIS) axis, are often shown in MQMAS spectra for the eye guide, as shown in Figs. 5.12 and 5.13. If the sample has a single set of quadrupolar coupling constant (χ), asymmetry parameter (η), and isotropic chemical shift (δ_{cs}), the slice along the ISO axis provides a sharp resonance and that along the QA axis shows a powder pattern identical to a typical 1D MAS NMR spectrum. If the quadrupolar parameter χ and/or η is distributed, the 2D spectra spread along the QIS axis, whereas the distribution of δ_{cs} provides 2D spectra spread along the CS axis.

The left parts of Fig. 5.12a and b show ^{1}H decoupled ^{27}Al MQMAS spectra of γ- and δ-Alq_3, respectively. The gravity centers of MQ signals for both Alq_3s appear at (QA, ISO) = (~20 ppm, 38 ppm). The slice spectra along the ISO dimension exhibit sharp resonances, which is typical for a well-defined crystal sample, indicating the ordered structure of γ- and δ-Alq_3. The quadrupolar and isotropic chemical parameters (χ, η, δ_{cs}) were found to be (4.90 MHz, 0.0, 31.3 ppm) for γ-Alq_3 and (4.84 MHz, 0.22, 31.0 ppm) for γ- and δ-Alq_3 from spectral simulations (right of Fig. 5.12a and b). The asymmetry parameter η of 0 also confirms the facial isomeric state of γ-Alq_3. The origin of the small but non-zero η of 0.22 for δ-Alq_3 is discussed later.

The ^{1}H decoupled ^{27}Al MQMAS spectra of α- and amo-Alq_3 are shown on the left side of Fig. 5.12a and b, respectively. Although 1D CP/MAS spectra of α- and amo-Alq_3 are very similar to each other (Fig. 5.11b and e), the 2D MQMAS spectra

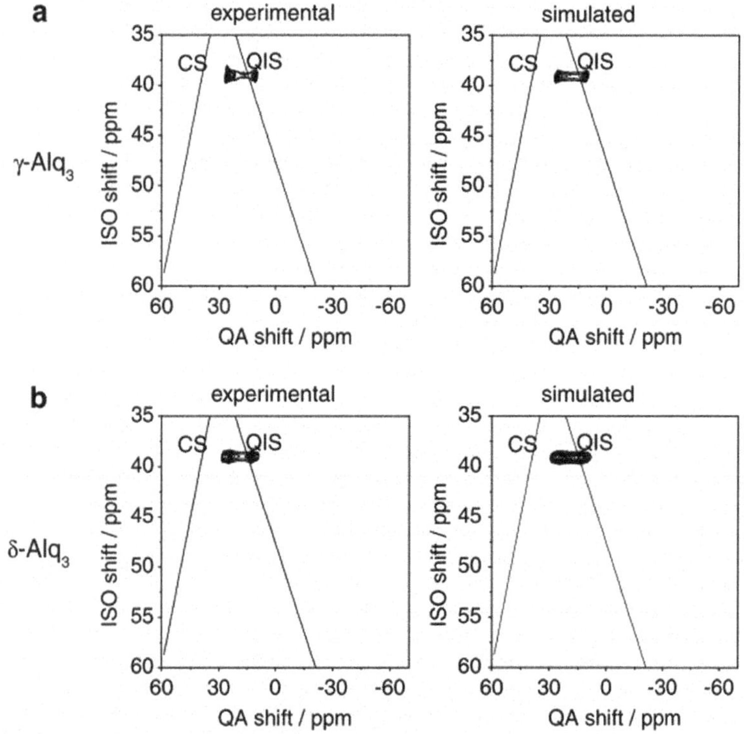

Fig. 5.12 Experimental (left) and simulated (right) ^{1}H decoupled ^{27}Al MQMAS spectra of **a** γ- and **b** δ-Alq_3 under the static magnetic field of 9.4 T. The samples were spun with a spinning frequency of 15.015 kHz. The CS and QIS axes are eye guides. Reprinted from Ref. [28]. With permission from Elsevier

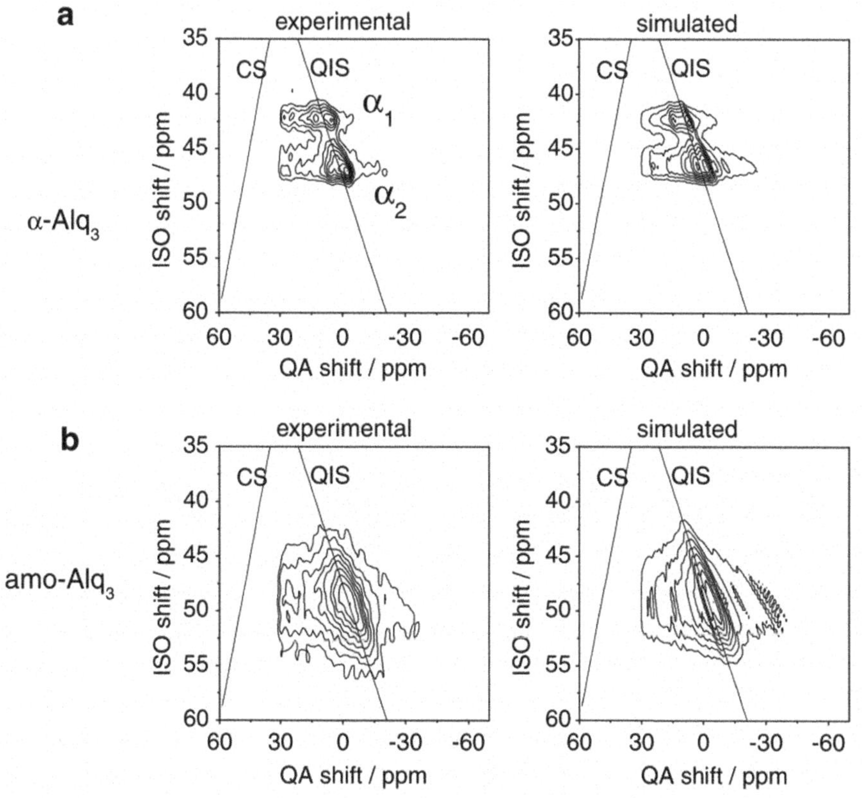

Fig. 5.13 Experimental (left) and simulated (right) ^{1}H decoupled ^{27}Al MQMAS spectra of **a** α- and **b** amo-Alq$_3$ under the static magnetic field of 9.4 T. The samples were spun with a spinning frequency of 15.015 kHz. The CS and QIS axes are eye guides. Reprinted from Ref. [28]. With permission from Elsevier

of α- and amo-Alq$_3$ are clearly different. In α-Alq$_3$, two distinctly different sites, referred to as α_1- and α_2-components, are clearly observed. The powder patterns of the α_1- and α_2-components in the α-Alq$_3$ are extended not along the CS axis but along the QIS axis. The amo-Alq$_3$ has only a single component but is widely distributed along the CS axis. In both the α- and amo-Alq$_3$s, the broadenings are caused by the distributions of quadrupolar parameter χ and/or η, and not by the distribution of δ_{cs} or rapid relaxation because rapid relaxation simply broadens the spectra along the QA and ISO axis. These experimental 2D spectra can be well modeled by assuming the following asymmetric Gaussian distributions of the χ values.

$$P(\chi) = \frac{1}{\sigma^+ + \sigma^-}\sqrt{\frac{2}{\pi}}\exp\frac{(\chi_0 - \chi)^2}{2\sigma(\chi)^2} \tag{5.6}$$

$$\sigma(\chi)^2 = \begin{cases} \sigma^+ \text{for } \chi < \chi_0 \\ \sigma^- \text{for } \chi \geq \chi_0 \end{cases} \tag{5.7}$$

The best-fit spectrum for α-Alq_3 (right of Fig. 5.13a) was obtained when the parameters (χ_0, σ^+, σ^-, η) are (5.5 MHz, 0.2 MHz, 0.25 MHz, 0.7) for the α_1-component and (6.4 MHz, 0.15 MHz, 0.35 MHz, 0.84) for the α_2-component. The best-fit 2D pattern for amo-Alq_3 was obtained with (χ_0, σ^+, σ^-, η) = (7.0 MHz, 0.4 MHz, 0.8 MHz, 0.9), as shown in Fig. 5.13b (right). In contrast, the calculation assuming the distribution of η does not agree with the experimental results. The resonance lines for the α- and amo-Alq_3 appear at the ISO shift of > 41 ppm, while those for the γ- and δ-Alq_3 appear at 38 ppm. Therefore, the resonance lines for the meridional and facial isomers can be clearly separated in the MQMAS spectra, which overlap with each other in the 1D experiments [25].

To clarify the origin of the non-splitting and three-line-splitting in the CP/MAS spectra of γ- and δ-Alq_3, and to reveal their intermolecular packing, we have carried out GIPAW calculations [27] for these Alq_3s [26]. Figure 5.14 shows the GIPAW-calculated ^{13}C isotropic chemical shifts of facial γ-Alq_3 together with the experimental CP/MAS ^{13}C NMR spectrum. The GIPAW calculations were performed for three different crystal structures. ***F*-γ-M-orig** is the original crystal structure of γ-Alq_3 determined by XRD [29]. The crystal structure of ***f*-γ-M-Hopt** is obtained by optimization of ***f*-γ-M-orig** only for H atoms with all the other C, N, O, and Al atoms being fixed. The structure of ***f*-γ-M-ALLopt** is obtained by all-atom optimization. The calculated ^{13}C isotropic chemical shifts of ***f*-γ-M-orig** (Fig. 5.14b) show relatively good agreement for the experimental chemical shifts for C8, C3, and C5 carbons, but insufficient agreement for C2 and C10 carbons. The GIPAW calculation was also performed for the H-optimized γ-Alq_3 (***f*-γ-M-Hopt**) because it is well-known that the atomic coordinates of protons by XRD analysis are not reliable. However, the result shown in Fig. 5.14c becomes worse compared to that of ***f*-γ-M-orig**. In contrast, the calculated chemical shifts of ***f*-γ-M-ALLopt** (Fig. 5.14d) show excellent agreement with the experiment; the deviation of chemical shift values from the experiment was within 1.6 ppm for all carbons for ***f*-γ-M-ALLopt**. This result clearly shows that the crystal structure determined by XRD analysis can be refined by the combination of CP/MAS ^{13}C NMR experiments and GIPAW calculations. Here, we concluded that ***f*-γ-M-ALLopt** is the most reliable structure for γ-Alq_3. We also performed GIPAW chemical shift calculations for an isolated molecule extracted from the crystal structure of ***f*-γ-M-ALLopt** without changing the intramolecular structures (Fig. 5.14e). The results show a large discrepancy with the experiment, indicating that the isotropic chemical shifts include information on intermolecular packing which should be taken into account for the quantitative agreement.

The top of Fig. 5.15a shows the experimental CP/MAS ^{13}C NMR spectrum of δ-Alq_3, representing all carbon species of δ-Alq_3. In the spectrum, the resonance lines of the C8, C2, and C3 carbons are well separated from the others. However, significant overlaps of signals are observed for the C4 and C9, the C6 and C10, and the C7 and C5 carbons. To distinguish the tertiary (C4 and C6) and quaternary (C9 and C10) carbons, we also performed CP/MAS ^{13}C NMR with a short contact time of 50 μs, denoted as

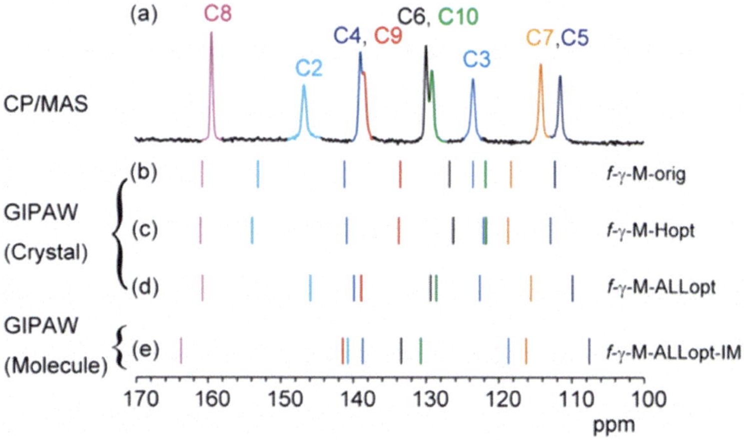

Fig. 5.14 **a** Experimental CP/MAS ^{13}C NMR spectrum of γ-Alq_3. **b–e** Isotropic resonance lines of γ-Alq_3 calculated using the GIPAW method. The calculations were carried out for the crystal structures **b** proposed by Muccini et al. [29] without further structural optimization (***f*-γ-M-orig**), **c** after the optimization of the atomic coordinates for protons alone (***f*-γ-M-Hopt**), and **d** after the optimization of the atomic coordinates for all atoms (***f*-γ-M-ALLopt**). Calculations of **e** were carried out for a single molecule extracted from the crystal structure of ***f*-γ-M-ALLopt**. Reprinted with permission from [26]. Copyright 2013 American Chemical Society

short-CP/MAS (middle of Fig. 5.15a). In the short-CP/MAS spectrum, quaternary C8, C9, and C10 resonance lines disappear, and tertiary carbons are observed. The difference spectrum between the CP/MAS and short-CP/MAS spectra shown at the bottom of Fig. 5.15a provides the spectrum for quaternary carbons (this can also be obtained by dipolar dephasing method). From these spectra, the isotropic chemical shifts of the C4 and C9 carbons are distinguished. The separation of the C6 and C10 carbons is also successful, however, the tertiary carbons of the C7 and C5 cannot be separated.

For GIPAW calculations of δ-Alq_3, we carried out all-atom optimization for the crystal structures reported by Cölle et al. [22] and Rajeswaran et al. [27], which are abbreviated as ***f*-δ-C-ALLopt** and ***f*-δ-R-ALLopt**, respectively. The GIPAW-calculated chemical shifts for ***f*-δ-C-ALLopt** and ***f*-δ-R-ALLopt** are shown in Fig. 5.15b and c, respectively. The calculated chemical shifts for ***f*-δ-C-ALLopt** and ***f*-δ-R-ALLopt** are almost indistinguishable. We should note that the crystal structures proposed by Cölle and Rajeswaran were different and their original or proton-optimized structures provided different GIPAW-calculated chemical shifts. However, the different crystal structures converge on indistinguishable structures after all atomic coordinate optimization. The GIPAW results for ***f*-δ-C-ALLopt** (Fig. 5.15b) and ***f*-δ-R-ALLopt** (Fig. 5.15c) excellently describe the resonance line splitting for all carbon species. Especially, the three equally spaced C2 resonance lines and mostly single C3 resonance lines are excellently modeled. The deviation of chemical shift values from the experiment was within 1.54 ppm. These

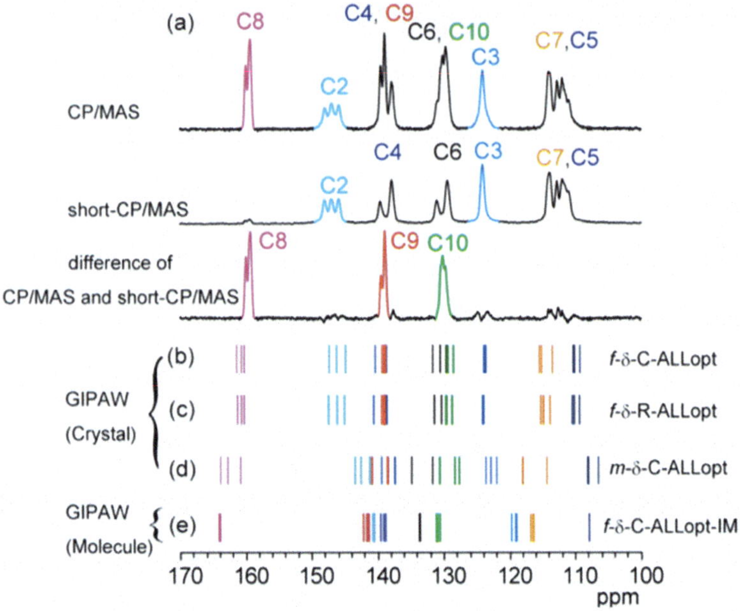

Fig. 5.15 **a** Experimental CP/MAS (contact time: 8 ms), short-CP/MAS (contact time: 50 μs), and the difference spectra of δ-Alq_3. **b–d** Isotropic resonance lines of δ-Alq_3 calculated using the GIPAW method. The calculations were carried out for the crystal structures proposed by **b** Cölle et al. [22] after the all atoms optimization (***f*-δ-C-ALLopt**) and by **c** Rajeswaran et al. [27] after the all atoms optimization (***f*-δ-R-ALLopt**). Calculation of **d** was carried out for the crystal structure reported by Cölle et al. [22], consisting entirely of meridional isomers, after the all atoms optimization (***f*-δ-C-ALLopt**). **e** The GIPAW calculation was also carried out for a single molecule extracted from the crystal structure of ***f*-δ-C-ALLopt**. Reprinted with permission from [26]. Copyright 2013 American Chemical Society

results confirm that the GIPAW-optimization provides the most reliable crystalline structure. Cölle et al. [22] also provided a crystal structure based on meridional isomers. Figure 5.15d represents the GIPAW result for the all-atom-optimized crystal structure (***m*-δ-C-ALLopt**) structure. The result is significantly different from the experimental chemical shifts, showing that the δ-Alq_3 is not meridional isomers but composed of facial isomers. Also, the calculated chemical shifts of an isolated molecule (***f*-δ-C-ALLopt-IM**) (Fig. 5.15e) extracted from the ***f*-δ-C-ALLopt** differ from those in the crystalline ***f*-δ-C-ALLopt**; the three-line-splitting of C2 carbon is found to originate from intermolecular packing.

The structure analysis of γ- and δ-Alq_3 was further conducted by ^{1}H dipolar decoupled ^{27}Al MAS NMR and GIPAW calculations [30]. Figure 5.16 shows the experimental ^{27}Al MAS spectrum of γ-Alq_3. The quadrupolar parameters, η_q and C_q, determined from the experimental spectrum are also shown in the figure. The η_q reflects the spectral line shape and the C_q determines the spectral width. The experimental spectrum is axially symmetric ($\eta_q = 0$). This result indicates the axially symmetric facial structure of γ-Alq_3. Figure 5.16b–e displays the GIPAW-calculated

^{27}Al MAS spectra result for the three structures of ***f*-γ-M-orig** (Fig. 5.16b), ***f*-γ-M-Hopt** (Fig. 5.16c), and ***f*-γ-M-ALLopt** (Fig. 5.16d) along with an isolated molecule ***f*-γ-M-ALLopt-IM** (Fig. 5.16e) extracted from ***f*-γ-M-Allopt**. All of the spectra in Fig. 5.16b–d are axially symmetric with $\eta_q = 0$, reflecting C_3 symmetry of facial γ-Alq_3, irrespective of the extent of geometry optimization employed. In contrast, the spectral widths significantly differ from each other; those for ***f*-γ-M-orig** and ***f*-γ-M-Hopt** are significantly narrower compared with the experimental spectra, whereas that of ***f*-γ-M-ALLopt** is comparable with the experiment. This result clearly shows that ***f*-γ-M-ALLopt** is the most appropriate structure, confirming the validity of ^{13}C CP/MAS NMR and GIPAW results above (Fig. 5.14). Different from the ^{13}C case, the GIPAW-calculated ^{27}Al MAS spectrum for ***f*-γ-M-ALLopt-IM** is similar to that of ***f*-γ-M-ALLopt**, indicating that the ^{27}Al MAS spectrum does not significantly depend on intermolecular interactions, probably the Al atom is far from neighboring molecules.

A corresponding ^{27}Al NMR study was carried out for δ-Alq_3 (Fig. 5.17). The η_q and C_q values determined from the experimental spectrum (Fig. 5.17a) are 0.23 and 4.8 MHz, respectively. As is in the case of γ-Alq_3, GIPAW results for the original and proton-optimized structures composed of facial isomers (Fig. 5.17b–d) do not match the experimental spectrum. Among the GIPAW-calculated spectra in this study, the spectrum for ***f*-δ-R-ALLopt** (Fig. 5.17f) agrees well with the experiment ($\eta_q = 0.27$ and $C_q = 4.5$ MHz) and that for ***f*-δ-C-ALLopt** (Fig. 5.17g) shows the best agreement ($\eta_q = 0.23$ and $C_q = 4.6$ MHz). The results show that all-atoms-optimized crystals composed of facial isomers provide more refined structures compared with the XRD-determined structures. The calculated spectra for meridional δ-Alq_3 (Fig. 5.17e

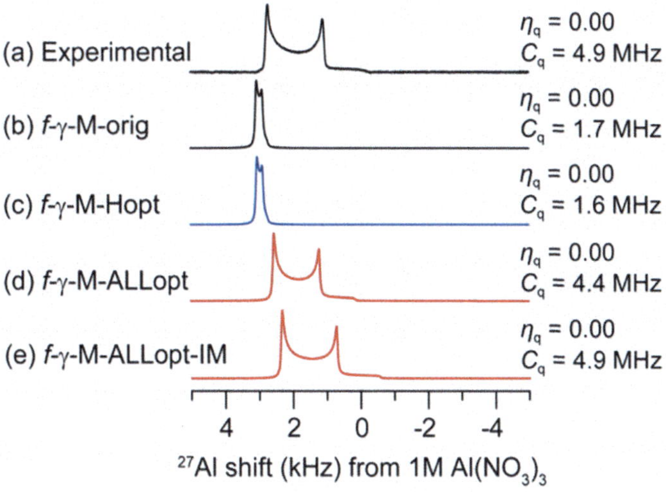

Fig. 5.16 **a** Experimental and **b–e** GIPAW-calculated dipolar decoupled ^{27}Al MAS spectra of γ-Alq_3. For all spectra, $\eta_q = 0$. Reprinted from Ref. [30]. With permission from Elsevier

and h) differ significantly from the experiment, regardless of whether they are optimized or not. Thus, we can conclude that δ-Alq_3 is composed of facial isomers, though the experimental η_q is not zero. Comparison with isolated molecules extracted from all-atoms-optimized crystals (***f*-δ-R-ALLopt-IM**, ***f*-δ-C-ALLopt-IM**, ***m*-δ-C-ALLopt-IM**) shown in Fig. 5.17i, j, and k again confirm that ^{27}Al spectra of Alq_3 do not depend on intermolecular interactions. This comparison also shows that the non-zero η_q value of δ-Alq_3 does not originate from the intermolecular interaction but due to structural deviation from C_3 symmetry.

The intra- and intermolecular structures of γ-Alq_3 and δ-Alq_3 determined in the above studies suggest that the origin of the blue emitting Alq_3 originated from the facial isomeric states and not the intermolecular interactions. Since ^{13}C and ^{27}Al NMR spectra contain precise information on inter- and intramolecular structures, the crystal structures determined by XRD analysis can further be refined through the combined use of NMR experiments and GIPAW calculations. The method shown here can reveal the coexistence of different isomeric states and different polymorphs, which are difficult to distinguish by XRD experiments alone.

5.3.2 Torsion Angle Analysis of a TADF Emitter by DNP-ssNMR

Thermally activated delayed fluorescence (TADF) materials have attracted considerable attention as emitters of OLEDs because TADF emitters can up-convert originally non-emissive triplet excitons to emissive singlet via reverse intersystem crossing (RISC), which enables 100% conversion of electrically generated excitons into light (Fig. 5.18a) [10, 31–33]. To promote RISC at room temperature, the energy gap (ΔE_{ST}) between the lowest excited singlet (S_1) and the lowest triplet states (T_1) should be small in TADF emitters, which can be realized by spatially separating the HOMO and LUMO in a molecule. Thus, TADF molecules have been designed to be composed of donor (D) and acceptor (A) units connected by a single bond. The torsion angle between the D and A units (D–A torsion angle) determines the HOMO–LUMO overlap, which controls the ΔE_{ST} [34, 35]. Simultaneously, the D–A torsion angle also controls the oscillator strength (f) between S_1 and the ground (S_0) states as well as spin–orbit coupling (SOC) between S_1 and T_1, both of which have large effects on the light-emitting performance of the molecule, through the rate constants of radiative decay and RISC [36, 37].

In view of this importance, we experimentally evaluated the torsion angle of a highly efficient TADF emitter, DACT-II, by DNP-ssNMR [38]. DACT-II has a diphenylamino carbazole donor and a triphenyltriazine acceptor unit, which are connected by a single bond (Fig. 5.18b). Due to the adequate D–A torsion angle and resulting suitable HOMO–LUMO separation, DACT-II is a highly efficient TADF emitter that exhibits 100% photoluminescence quantum yield and the smallest ΔE_{ST} (9.0 meV) as far as we know [10]. Among various solid-state NMR observables,

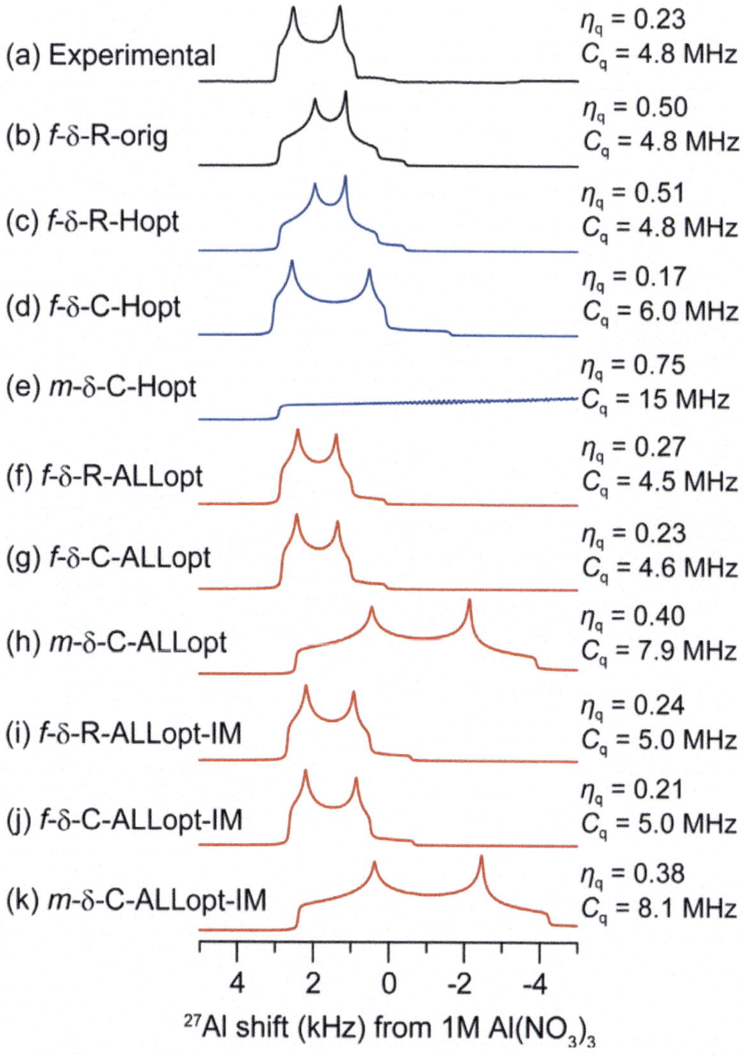

Fig. 5.17 **a** Experimental and **b–k** GIPAW-calculated dipolar decoupled ^{27}Al MAS spectra of δ-Alq_3. Reprinted from Ref. [30]. With permission from Elsevier

principal values of CSA spectra are sensitive to the electronic structure of molecules and thus can be used for torsion angle determination. Here, we examine the D–A torsion angle (β in Fig. 5.18b) of DACT-II in an amorphous state by measuring CSA spectra of ^{15}N in the carbazole group (N_{CZ}).

We carried out GIAO chemical shift calculations for DACT-II. Figure 5.18d shows the β dependence of CSA principal values of ^{15}N atoms in N_{Cz}. The principal values varied markedly with β. Although δ_{CN} and δ_{iso} were insensitive to β, both $\delta_{\perp}$ and $\delta_{\perp\perp}$ appreciably changed from 55 to 100 ppm and from 180 to 120 ppm, respectively.

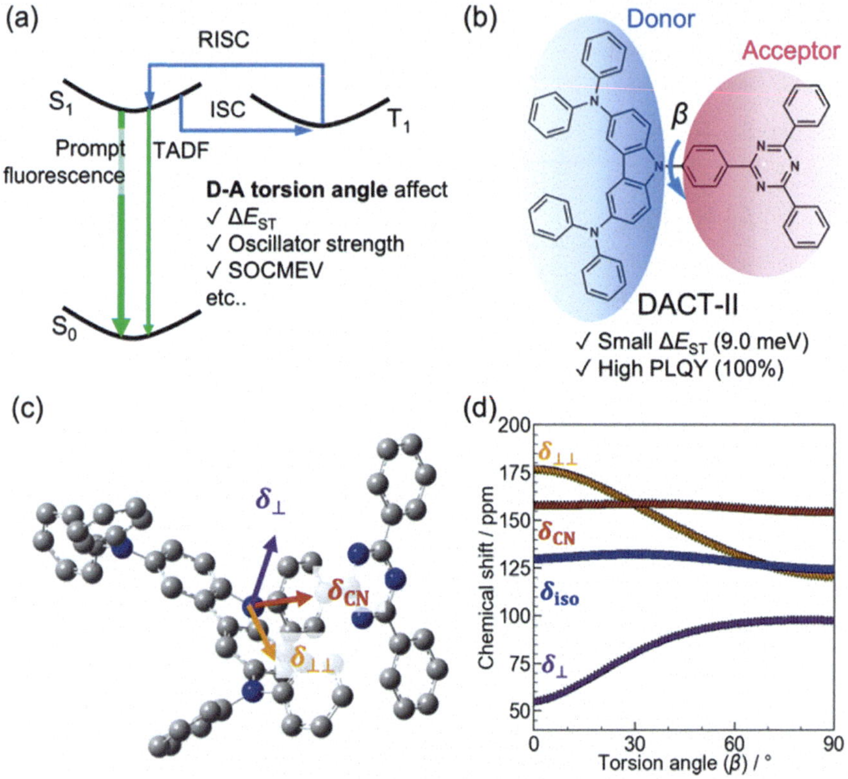

Fig. 5.18 **a** Schematic of TADF mechanism. **b** Molecular structure of DACT-II. **c** CSA principal axes of the nitrogen at the carbazole. **d** DFT-calculated chemical shift principal values (δ_{CN}, $\delta_{\perp}$, and $\delta_{\perp\perp}$) and isotropic chemical shift (δ_{iso}) as a function of D–A torsion angle, β. Reprinted with permission from [38]. Copyright 2023 American Chemical Society

Here, we denote the principal axes for N_{CZ} as follows: δ_{CN} is parallel to the C–N bond connecting carbazole and triphenyltriazine, $\delta_{\perp}$ is perpendicular to the plane of the carbazole, and $\delta_{\perp\perp}$ is perpendicular to both δ_{CN} and $\delta_{\perp}$. In contrast, the principal values of the ^{15}N atoms in diphenylamine (N_{DP}) and triazine (N_{TRZ}), which are far from the D–A bond, were almost constant against β.

The analysis of the ^{13}C spectrum was difficult for DACT-II, because of the severe overlap of the resonance lines (Fig. 5.19b). Therefore, ^{15}N of N_{CZ} is the best probe for determining β. However, ^{15}N NMR measurements of natural-abundance samples are generally time-consuming and low signal-to-noise ratio (SNR) due to the low natural abundance (0.37%) and the low gyromagnetic ratio ($\gamma_N/\gamma_H = 0.10$) of ^{15}N nuclei. To improve the SNR of ^{15}N NMR spectra, we used DNP–ssNMR. The ^{15}N CP/MAS spectrum with a good SNR ratio was obtained within an experimental time of less than 1 h by microwave irradiation, while we could not observe any signal without it (Fig. 5.19a). The three well-separated ^{15}N signals were observed at 249.9,

130.0, and 96.7 ppm, which can be assigned as N_{TRZ}, N_{CZ}, and N_{DP}, respectively. The DNP enhancement factor was determined to be 30 from the ^{13}C CP/MAS spectra with and without microwave irradiation (Fig. 5.19b).

To obtain ^{15}N CSA principal values, we applied the extended CSA amplification (xCSA) method developed by Hung and Gan [39, 40], which is useful for determining CSA principal values with small anisotropy. The xCSA pulse sequence used in this study and the obtained amplified spinning sideband pattern are shown in Fig. 5.20a and b, respectively. The sideband intensities fitted by Herzfeld–Berger analysis [41, 42] well agree with the experimental results (Fig. 5.20c). The determined CSA principal values were $\delta_{CN} = 157$ ppm, $\delta_{\perp\perp} = 139$ ppm, and $\delta_{\perp} = 94$ ppm.

The torsion angle β in the amorphous state was determined by evaluating root mean squared error (RMSE) values between the experimental and GIAO-calculated CSA principal values for N_{CZ}. The GIAO calculation gave the smallest RMSE of

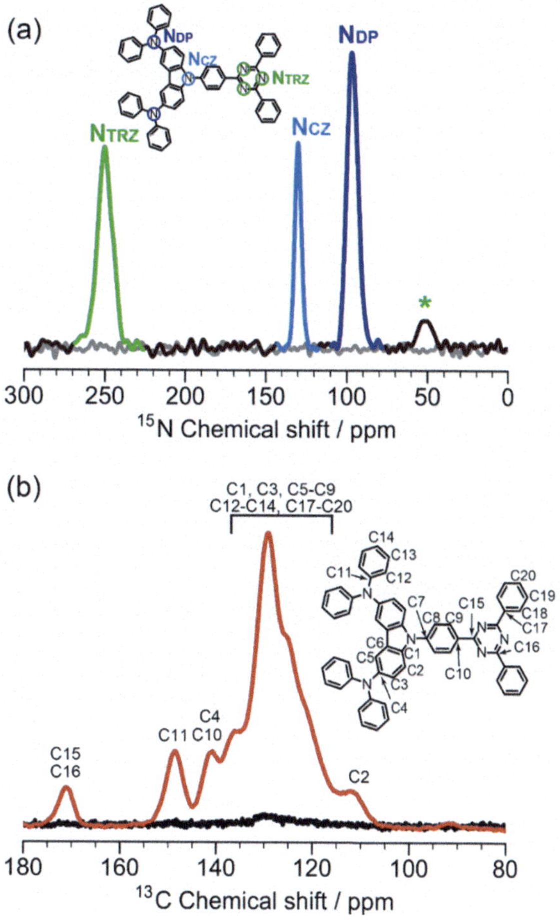

Fig. 5.19 **a** ^{15}N CP/MAS spectra recorded by using a Hahn-echo pulse sequence with (color) and without (gray) DNP enhancement. The spinning sideband of the N_{TRZ} signal (50 ppm) is marked with an asterisk. **b** ^{13}C CP/MAS spectra of DACT-II with (red) and without (black) microwave irradiation. All spectra were taken under 8 kHz MAS frequency. Reprinted with permission from [38]. Copyright 2023 American Chemical Society

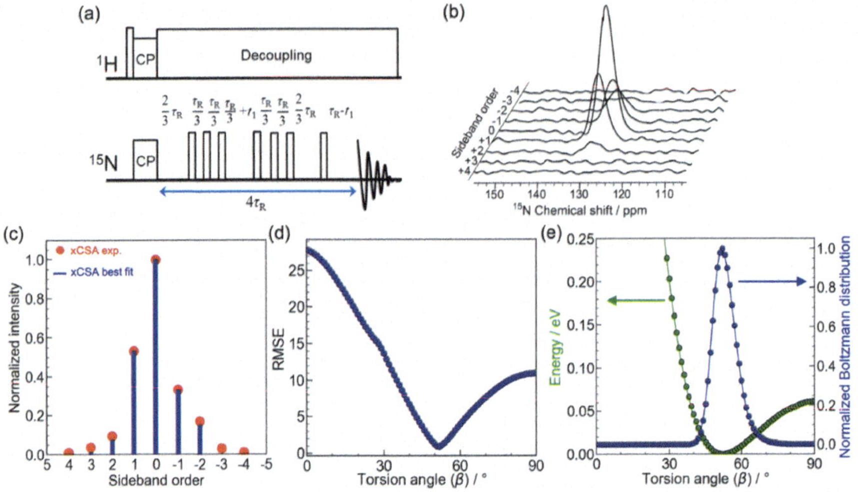

Fig. 5.20 **a** Pulse sequence used for xCSA measurements. **b** Stacked plot of xCSA spectra for N_{CZ} of DACT-II. **c** Experimental sideband intensities of xCSA measurements and the best-fit result. **d** RMSE values as a function of β. **e** DFT-calculated total energies (green) of DACT-II and corresponding Boltzmann distribution at 100 K (blue) as a function of β. Reprinted with permission from [38]. Copyright 2023 American Chemical Society

1.0 at $\beta = 52°$. From this result, we conclude that DACT-II molecules in the amorphous solid have a torsion angle of 52° on average (Fig. 5.20d). When we assume that the torsion angle distribution obeys a Boltzmann distribution (Fig. 5.20e), the RMSE value is 1.2. The method used here can be used as a routine to determine torsion angles of TADF emitters because it does not require isotopic labeling and is not time-consuming. This method can be applied to other functional organic optoelectronic materials, in which molecular conformation has a considerable impact on photoelectric conversion, carrier transport, and resulting device performance.

5.4 Material Degradation Analysis of Liquid OLEDs by Solution NMR

To realize truly flexible displays, liquid OLEDs, which are composed of liquid semiconductors sandwiched between two substrates with electrodes, have been examined [43, 44]. Thus, unlike standard OLEDs, both cracks of the emitting layer and detachments between the liquid emitting layer and electrodes do not occur when the displays are bent. However, the device performance of liquid OLEDs is still low compared to standard OLEDs, and the device lifetime is quite short. Thus, the analysis of the degradation mechanisms in liquid OLEDs is important. To reveal the degradation mechanisms of liquid OLEDs, it is important to identify the degraded molecules produced during the device operation, which will provide guidelines for material designs with high durability. In general, solution NMR is useful to identify the chemical structures of organic molecules. Therefore, in this study, we have investigated the material degradation in liquid OLEDs by solution NMR.

In this study, we have investigated the degradation of 9-{2-[2-(2-methoxyethoxy)ethoxy]ethyl}-9H-carbazole (TEGCz) in liquid OLEDs. The chemical structure is shown in Fig. 5.21a. Tributylmethylphosphonium bis(trifluoromethanesulfonyl)imide (TMP-TFSI) is used as an electrolyte in the device. Figures 5.21b–e show ^{1}H NMR spectra of the aromatic region for materials in the liquid OLEDs. Figure 5.21b shows the NMR spectrum before driving, and Fig. 5.21c, d, and e shows the NMR spectra after driving for 11, 30, and 50 s, respectively. The resonance lines at 8.43 ppm (D1′), 8.18 ppm (D1), and 7.86 ppm (D3′), which were not observed before driving, were clearly found after driving, as shown in Fig. 5.21c–e. The intensity of these new resonance lines increased with the increase in drive time. Therefore, these new peaks can reasonably be assigned to ^{1}H formed by the degradation of TEGCz in the liquid OLEDs.

To identify the structure and number of compounds produced by the degradation, a diffusion-ordered NMR spectroscopy (DOSY) experiment [45–47] was conducted for the materials in the active layer of the liquid OLED after driving for 50 s. 2D DOSY experiment can correlate isotropic chemical shifts and self-diffusion coefficients and thus can separate the spectra of different compounds in a mixture using the difference in the diffusion coefficients for respective chemical species. Figure 5.22 represents a DOSY contour plot of the materials in the device after driving for 50 s. The resonance lines for TEGCz are observed at 8.08, 7.49, 7.45, and 7.21 ppm as black lines at the same diffusion coefficient of $D \sim 10^{-9.0}$ m^2 s^{-1}. In contrast, the resonance lines at 8.43, 8.18, 7.86, 7.6, 7.53, and 7.26 ppm, shown by red lines, appeared at a different diffusion coefficient at around $D \sim 10^{-9.3}$ m^2 s^{-1}. Since the diffusion coefficient depends primarily on the molecular weights, the above results clearly show that the molecular weight of the degraded molecule is larger than TEGCz, contrary to our expectations. Since it is well known that the para position of carbazole is most reactive, we assumed that 9,9′-{2-[2-(2-methoxyethoxy)ethoxy]ethyl}-3,3′-bis(9H-carbazole) $(TEGCz)_2$ was formed during device operation. The resonance lines of $(TEGCz)_2$ in Fig. 5.21f agree well with newly appeared peaks in the device after driving. Therefore, we found from this solution NMR experiments that the dimerization of TEGCz occurred during the driving of the solution OLEDs, which was the origin of the device degradation.

5.5 Summary

In this chapter, we have presented NMR approaches for investigating organic functional materials used in OLEDs. The isotropic chemical shift, chemical shift anisotropy, and quadrupolar parameters obtained from solid-state NMR measurements provide important information on the intra- and intermolecular structures of

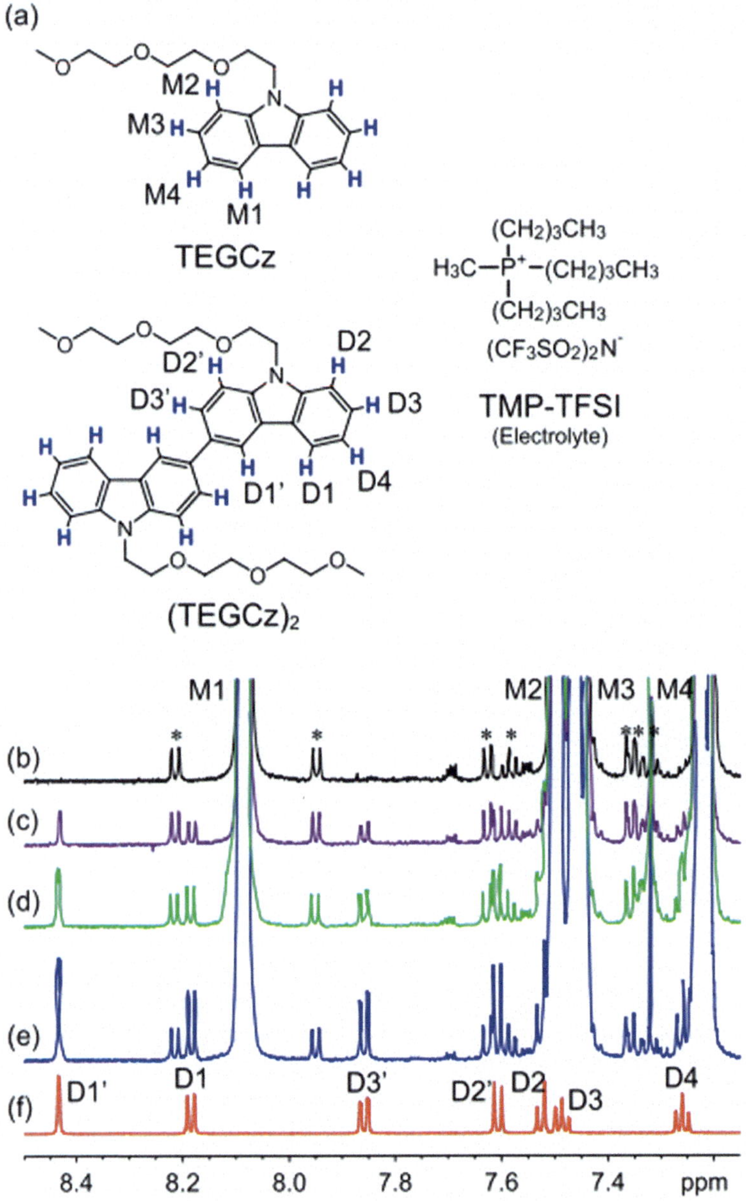

Fig. 5.21 **a** Chemical structures of TEGCz, $(TEGCz)_2$, and TMP-TFSI. Hydrogen atoms in the ethylene glycol chain are omitted for clarity. The assignments of protons are also shown. ^{1}H NMR spectra of TEGCz with 0.25 wt.% TMP-TFSI; **b** before, after driving for **c** 11 s, **d** 30 s, and **e** 50 s. **f** ^{1}H NMR spectrum of $(TEGCz)_2$ with 0.25 wt.% TMP-TFSI before driving is shown for comparison. $*$ marks in the figure represent carbon satellites for the peak of TEGCz

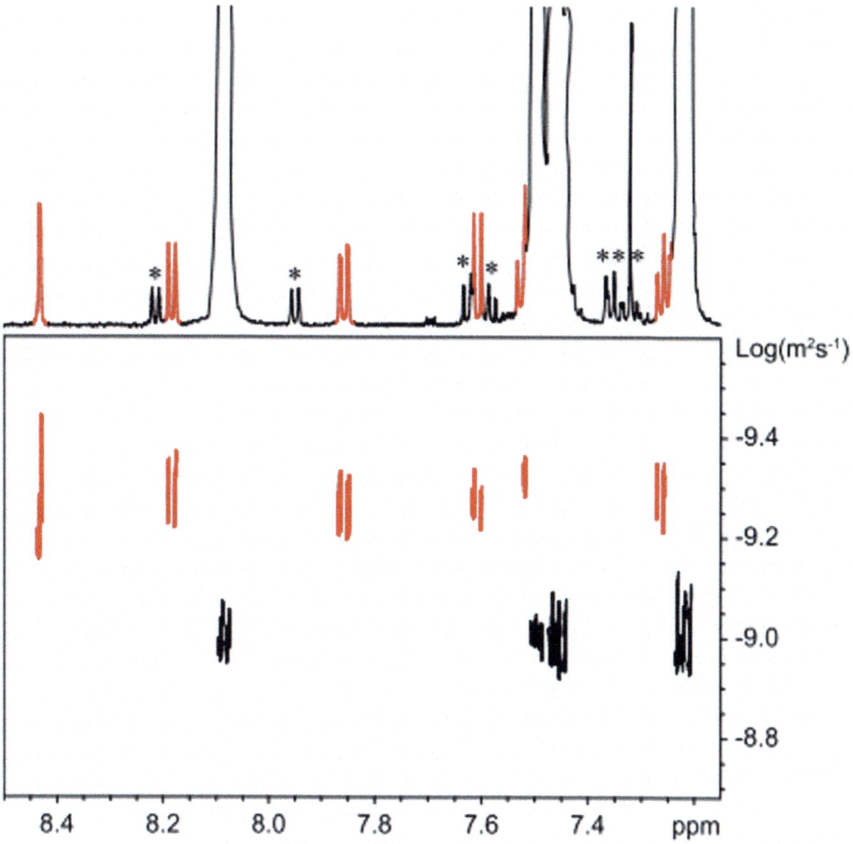

Fig. 5.22 DOSY contour plot of TEGCz with 0.25 wt.% TMP-TFSI after driving for 50 s. For TEGCz, contour lines are plotted in black between 0.25 and 59.9% of the maximum intensity. For the degraded material, contour lines are plotted in red between 12.9 and 65.1% of the maximum intensity of the degraded material

target compounds. In addition, NMR studies combined with quantum chemical calculations are effective in analyzing detailed structural information, which provides a deep understanding of charge transport and light-emitting properties of OLED materials. Liquid NMR analysis also provides important information, especially on device degradations. DNP-ssNMR allows measurements of very limited sample amounts, and of nuclei with low natural abundance and/or low gyromagnetic ratio, which will extend the range of applicability of NMR spectroscopy towards even more challenging systems in the future. We are currently attempting *operando* measurements under light irradiation and under device operation using DNP-ssNMR.

References

1. Tang, C.W., VanSlyke, S.A.: Organic electroluminescent diodes. Appl. Phys. Lett. **51**, 913–915 (1987)
2. Kaji, H., Yamada, T., Tsukamoto, N., Horii, F.: A combined experimental and theoretical study of the conformation of *N*, N′-diphenyl-N, N′-d*i*(m-tolyl)benzidine using solid-state 15N NMR and DFT calculations. Chem. Phys. Lett. **401**, 246–253 (2005)
3. Adachi, C., Tokito, S., Tsutsui, T., Saito, S.: Electroluminescence in organic films with three-layer structure. Jpn. J. Appl. Phys. **27**, L269 (1988)
4. Tamoto, N., Adachi, C., Nagai, K.: Electroluminescence of 1,3,4-oxadiazole and triphenylamine-containing molecules as an emitter in organic multilayer light emitting diodes. Chem. Mater. **9**, 1077–1085 (1997)
5. Strohriegl, P., Grazulevicius, J.V.: Charge-transporting molecular glasses. Adv. Mater. **14**, 1439–1452 (2002)
6. Yamada, T., Sato, T., Tanaka, K., Kaji, H.: Percolation paths for charge transports in *N*, N′-diphenyl-N, N′-d*i*(m-tolyl)benzidine (TPD). Org. Electron. **11**, 255–265 (2010)
7. Chapran, M., Sahalianov, I., Karaush-Karmazin, N.N., Wiosna-Salyga, G., Glowacki, I., Luszczynska, B., Pander, P., Baryshnikov, G.V.: Electronic structure of exciplexes and the role of local triplet states on efficiency of thermally activated delayed fluorescence. ACS Appl. Electron. Mater. **5**, 1489–1501 (2023)
8. Wolinski, K., Hinton, J.F., Pulay, P.: Efficient implementation of the gauge-independent atomic orbital method for NMR chemical shift calculations. J. Am. Chem. Soc. **112**, 8251–8260 (1990)
9. Kennedy, A.R., Smith, W.E., Tackley, D.R., David, W.I.F., Shankland, K., Brown, B., Teat, S.J.: Tetraaryl biphenyl diamine hole transport materials: a structural study utilizing both single crystal and high resolution powder diffraction. J. Mater. Chem. **12**, 168–172 (2002)
10. Kaji, H., Suzuki, H., Fukushima, T., Shizu, K., Suzuki, K., Kubo, S., Komino, T., Oiwa, H., Suzuki, F., Wakamiya, A., Murata, Y., Adachi, C.: Purely organic electroluminescent material realizing 100% conversion from electricity to light. Nat. Commun. **6**, 8476 (2015)
11. Shibata, M., Sakai, Y., Yokoyama, D.: Advantages and disadvantages of vacuum-deposited and spin-coated amorphous organic semiconductor films for organic light-emitting diodes. J. Mater. Chem. C **3**, 11178–11191 (2015)
12. Hall, D.A., Maus, D.C., Gerfen, G.J., Inati, S.J., Becerra, L.R., Dahlquist, F.W., Griffin, R.G.: Polarization-enhanced NMR spectroscopy of biomolecules in frozen solution. Science **276**, 930–932 (1997)
13. Lilly Thankamony, A.S., Wittmann, J.J., Kaushik, M., Corzilius, B.: Dynamic nuclear polarization for sensitivity enhancement in modern solid-state NMR. Prog. Nucl. Magn. Reson. Spectrosc. **102–103**, 120–195 (2017)
14. Plainchont, B., Berruyer, P., Dumez, J.-N., Jannin, S., Giraudeau, P.: Dynamic nuclear polarization opens new perspectives for NMR spectroscopy in analytical chemistry. Anal. Chem. **90**, 3639–3650 (2018)
15. Biedenbänder, T., Aladin, V., Saeidpour, S., Corzilius, B.: Dynamic nuclear polarization for sensitivity enhancement in biomolecular solid-state NMR. Chem. Rev. **122**, 9738–9794 (2022)
16. Suzuki, K., Kubo, S., Aussenac, F., Engelke, F., Fukushima, T., Kaji, H.: Analysis of molecular orientation in organic semiconducting thin films using static dynamic nuclear polarization enhanced solid-state NMR spectroscopy. Angew. Chem. Int. Ed. **56**, 14842–14846 (2017)
17. Matsushima, T., Adachi, C.: Extremely low voltage organic light-emitting diodes with p-doped alpha-sexithiophene hole transport and n-doped phenyldipyrenylphosphine oxide electron transport layers. Appl. Phys. Lett. **89** (2006)
18. Schmidt-Rohr, K., Spiess, H.W.: Multidimensional Solid-State NMR and Polymers. Academic Press (1994)
19. Gabriëlse, W., Angad Gaur, H., Veeman, W.S.: Molecular orientation in differently processed poly(ethylene terephthalate) yarns as studied by ^{13}C 2D CP-MAS NMR. Macromolecules **29**, 4125–4133 (1996)

20. Brinkmann, M., Gadret, G., Muccini, M., Taliani, C., Masciocchi, N., Sironi, A.: Correlation between molecular packing and optical properties in different crystalline polymorphs and amorphous thin films of *mer*-Tris(8-hydroxyquinoline)aluminum(III). J. Am. Chem. Soc. **122**, 5147–5157 (2000)
21. Braun, M., Gmeiner, J., Tzolov, M., Coelle, M., Meyer, F.D., Milius, W., Hillebrecht, H., Wendland, O., von Schütz, J.U., Brütting, W.: A new crystalline phase of the electroluminescent material tris(8-hydroxyquinoline) aluminum exhibiting blue shifted fluorescence. J. Chem. Phys. **114**, 9625–9632 (2001)
22. Cölle, M., Dinnebier, R.E., Brütting, W.: The structure of the blue luminescent δ-phase of tris(8-hydroxyquinoline) Aluminium(III) (Alq_3). Chem. Commun., 2908–2909 (2002)
23. Kaji, H., Kusaka, Y., Onoyama, G., Horii, F.: CP/MAS ^{13}C NMR characterization of the isomeric states and intermolecular packing in tris(8-hydroxyquinoline) Aluminum(III) (Alq_3). J. Am. Chem. Soc. **128**, 4292–4297 (2006)
24. Utz, M., Nandagopal, M., Mathai, M., Papadimitrakopoulos, F.: Characterization of isomers in aluminum tris(quinoline-8-olate) by one-dimensional ^{27}Al nuclear magnetic resonance under magic-angle spinning. Appl. Phys. Lett. **83**, 4023–4025 (2003)
25. Kaji, H., Kusaka, Y., Onoyama, G., Horii, F.: Relationships between light-emitting properties and different isomers in polymorphs of tris(8-hydroxyquinoline) Aluminum(III) (Alq_3) analyzed by solid-state ^{27}Al NMR and density functional theory (DFT) calculations. Jpn. J. Appl. Phys. **44**, 3706 (2005)
26. Suzuki, F., Fukushima, T., Fukuchi, M., Kaji, H.: Refined structure determination of blue-emitting tris(8-hydroxyquinoline) Aluminum(III) (Alq_3) by the combined use of cross-polarization/magic-angle spinning ^{13}C solid-state NMR and first-principles calculation. J. Phys. Chem. C **117**, 18809–18817 (2013)
27. Rajeswaran, M., Blanton, T.N., Klubek, K.P.: Refinement of the crystal structure of the δ-modification of tris(8-hydroxyquinoline) aluminum(III), δ-$Al(C_9H_6NO)_3$, the blue luminescent Alq_3. Z. Kristallogr.—New Cryst. Struct. **218**, 439–440 (2003)
28. Nishiyama, Y., Fukushima, T., Takami, K., Kusaka, Y., Yamazaki, T., Kaji, H.: Characterization of local structures in amorphous and crystalline tris(8-hydroxyquinoline) aluminum(III) (Alq_3) by solid-state ^{27}Al MQMAS NMR spectroscopy. Chem. Phys. Lett. **471**, 80–84 (2009)
29. Muccini, M., Loi, M.A., Kenevey, K., Zamboni, R., Masciocchi, N., Sironi, A.: Blue luminescence of facial tris(quinolin-8-olato)aluminum(III) in solution, crystals, and thin films. Adv. Mater. **16**, 861–864 (2004)
30. Suzuki, F., Nishiyama, Y., Kaji, H.: Clarification of isomeric structures and the effect of intermolecular interactions in blue-emitting aluminum complex Alq_3 using first-principles ^{27}Al NMR calculations. Chem. Phys. Lett. **605–606**, 1–4 (2014)
31. Uoyama, H., Goushi, K., Shizu, K., Nomura, H., Adachi, C.: Highly efficient organic light-emitting diodes from delayed fluorescence. Nature **492**, 234–238 (2012)
32. Hirata, S., Sakai, Y., Masui, K., Tanaka, H., Lee, S.Y., Nomura, H., Nakamura, N., Yasumatsu, M., Nakanotani, H., Zhang, Q., Shizu, K., Miyazaki, H., Adachi, C.: Highly efficient blue electroluminescence based on thermally activated delayed fluorescence. Nat. Mater. **14**, 330–336 (2015)
33. Wu, T.-L., Huang, M.-J., Lin, C.-C., Huang, P.-Y., Chou, T.-Y., Chen-Cheng, R.-W., Lin, H.-W., Liu, R.-S., Cheng, C.-H.: Diboron compound-based organic light-emitting diodes with high efficiency and reduced efficiency roll-off. Nat. Photon. **12**, 235–240 (2018)
34. Penfold, T.J.: On predicting the excited-state properties of thermally activated delayed fluorescence emitters. J. Phys. Chem. C **119**, 13535–13544 (2015)
35. Shizu, K., Tanaka, H., Uejima, M., Sato, T., Tanaka, K., Kaji, H., Adachi, C.: Strategy for designing electron donors for thermally activated delayed fluorescence emitters. J. Phys. Chem. C **119**, 1291–1297 (2015)
36. Wada, Y., Wakisaka, Y., Kaji, H.: Efficient direct reverse intersystem crossing between charge transfer-type singlet and triplet states in a purely organic molecule. ChemPhysChem **22**, 625–632 (2021)

37. Shi, Y., Ma, H., Sun, Z., Zhao, W., Sun, G., Peng, Q.: Optimal dihedral angle in twisted donor–acceptor organic emitters for maximized thermally activated delayed fluorescence. Angew. Chem. Int. Ed. **61**, e202213463 (2022)
38. Suzuki, K., Kaji, H.: Torsion angle analysis of a thermally activated delayed fluorescence emitter in an amorphous state using dynamic nuclear polarization enhanced solid-state NMR. J. Am. Chem. Soc. **145**, 16324–16329 (2023)
39. Hung, I., Gan, Z.: An efficient amplification pulse sequence for measuring chemical shift anisotropy under fast magic-angle spinning. J. Magn. Reson. **213**, 196–199 (2011)
40. Hung, I., Ge, Y., Liu, X., Liu, M., Li, C., Gan, Z.: Measuring $^{13}C/^{15}N$ chemical shift anisotropy in [$^{13}C,^{15}N$] uniformly enriched proteins using CSA amplification. Solid State Nucl. Magn. Reson. **72**, 96–103 (2015)
41. Herzfeld, J., Berger, A.E.: Sideband intensities in NMR spectra of samples spinning at the magic angle. J. Chem. Phys. **73**, 6021–6030 (1980)
42. De Groot, H.J.M., Smith, S.O., Kolbert, A.C., Courtin, J.M.L., Winkel, C., Lugtenburgt, J., Herzfeld, J., Griffin, R.G.: Iterative fitting of magic-angle-spinning NMR spectra. J. Magn. Reson. **1969**(91), 30–38 (1991)
43. Xu, D., Adachi, C.: Organic light-emitting diode with liquid emitting layer. Appl. Phys. Lett. **95** (2009)
44. Hirata, S., Kubota, K., Jung, H.H., Hirata, O., Goushi, K., Yahiro, M., Adachi, C.: Improvement of electroluminescence performance of organic light-emitting diodes with a liquid-emitting layer by introduction of electrolyte and a hole-blocking layer. Adv. Mater. **23**, 889–893 (2011)
45. Morris, K.F., Johnson, C.S., Jr.: Resolution of discrete and continuous molecular size distributions by means of diffusion-ordered 2D NMR spectroscopy. J. Am. Chem. Soc. **115**, 4291–4299 (1993)
46. Jerschow, A., Müller, N.: Diffusion-separated nuclear magnetic resonance spectroscopy of polymer mixtures. Macromolecules **31**, 6573–6578 (1998)
47. Johnson, C.S.: Diffusion ordered nuclear magnetic resonance spectroscopy: Principles and applications. Prog. Nucl. Magn. Reson. Spectrosc. **34**, 203–256 (1999)

Chapter 6
NMR Studies on Natural Product—Stereochemical Determination and Conformational Analysis in Solution and in Membrane

Nobuaki Matsumori and Michio Murata

Abstract In this chapter, we overview two topics on NMR methodologies for small molecules, mostly natural products; one is about the solution NMR-based methods used for stereochemical determination of natural products in Sect. 6.1, and the other is on the solid-state and solution NMR techniques for investigating natural product-membrane interactions in Sect. 6.2. Since important two methods for stereochemical analysis of natural products, namely the J-based configuration analysis (JBCA) and universal NMR database (UDB) methods, were reported in the 1990s, both methods have been widely used in the field of natural products. The newly coming residual dipolar coupling (RDC) method is not the major method in the field of natural products yet, but will surely be an important tool for the stereochemical correlation between distant stereogenic centers, which could provide invaluable information as to the whole shape of natural products in solution. In the latter part of this chapter, we discuss the application of solid-state and other NMR techniques to membrane interaction analysis of natural products. In particular, we describe four examples of natural products that interact with biological membranes such as saponins, amphidinols, erythromycin A, and theonellamide A. As shown in these NMR studies, natural products often reveal very high affinities for phospholipids and sterols in bilayer membranes. Solid-state NMR, therefore, provides a very promising approach toward the structure studies of membrane-active complexes formed by natural products that have a high affinity to lipids. In addition, solution NMR techniques can be applied to elucidate the structural features of membrane-bound small molecules such as antibiotic erythromycin A and membrane-disrupting cyclic peptide theonellamide A. Standard 2D ^{1}H–^{1}H experiments such as COSY (correlation spectroscopy),

N. Matsumori
Department of Chemistry, Graduate School of Science, Kyushu University, Fukuoka 819-0395, Japan

M. Murata (✉)
Department of Chemistry, Graduate School of Science, Osaka University, Osaka 560-0043, Japan
e-mail: murata@chem.sci.osaka-u.ac.jp

Institute for Protein Research, Osaka University, Osaka 565-0871, Japan

The NMR Society of Japan (ed.), *Experimental Approaches of NMR Spectroscopy II*,
https://doi.org/10.1007/978-981-97-6830-1_6

NOESY (nuclear Overhauser effect spectroscopy), and DOSY (diffusion-ordered spectroscopy) are often helpful in elucidating the membrane interaction between natural products and lipids.

Keywords Natural products · Stereochemical determination · Bicelles · Solid-state NMR of membranes

6.1 Stereochemical Determination of Natural Products

In natural products chemistry, NMR spectroscopy is undoubtedly the most useful and important tool because NMR parameters provide fundamental information about not only the planar structures of natural products, but also their configuration and conformation [1]. In most cases, the molecular skeletons of natural products can be characterized by through-bond ^{1}H–^{1}H and ^{1}H–^{13}C correlation experiments such as COSY and HMBC. In addition, the evaluation of ^{1}H–^{1}H spin coupling constants ($^{3}J_{HH}$), chemical shifts, and nuclear Overhauser effects (NOEs) allows the determination of the relative stereochemistry of cyclic compounds with rigid three to six-membered rings. In contrast, the relative configuration assignment of multiple-substituted acyclic chains and macrocycles as exemplified in Fig. 6.1 is more difficult, because the geometrical flexibility of such systems makes NOE intensities less reliable for stereochemical assignments. For example, assuming that a molecule has major and minor conformations and the interproton distance of interest is small in the minor conformation, the corresponding NOEs would be relatively strongly observed, even though the distance is out of NOE detection in the major conformation. This would cause misinterpretation of NOEs and mislead the stereochemical assignment of flexible natural products.

To address these issues in the relative configuration assignment of organic molecules, NMR-based approaches, *J*-based configuration analysis (JBCA) [2, 3], universal NMR database (UDB) [3–5], and theoretical calculation of NMR parameters [3, 6], were reported after the 1990s. More recently, residual dipolar couplings (RDCs) were utilized as a tool to determine the relative configuration of natural products [7, 8]. The modified Mosher's method involving the derivatization of chiral secondary alcohols and primary amines with α-methoxy-α-trifluoromethyl-α-phenylacetic acid (MTPA) and comparison of chemical shifts has been widely used to determine absolute stereochemistry [9]. Although many other protocols such as circular dichroism measurement are also currently available [10], relative stereochemistry is mostly determined by the following methods: JBCA, UDB, and theoretical methods.

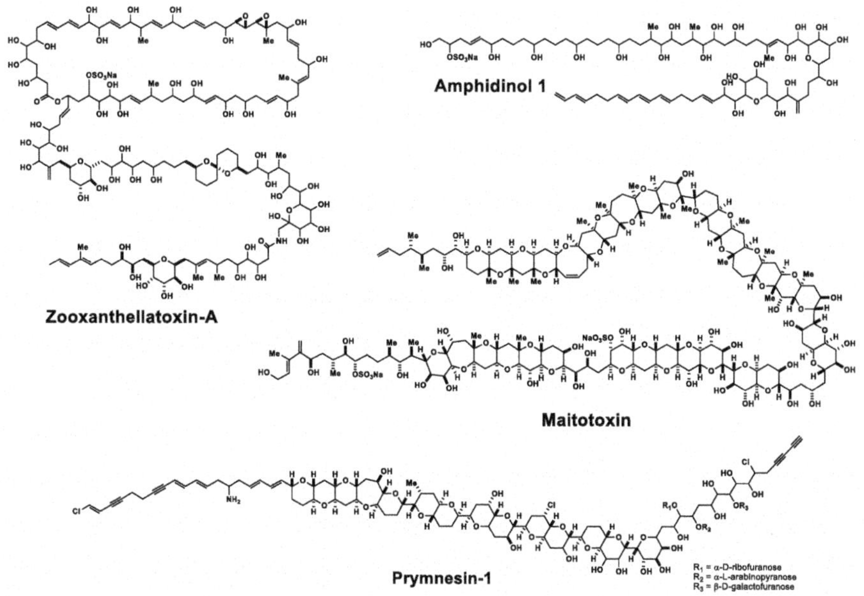

Fig. 6.1 Representative marine natural products bearing multiple-substituted acyclic chains and macrocycles

6.1.1 JBCA Method

Until the 1990s, long-range C–H coupling constants ($^{2,3}J_{C,H}$) were seldom used for the conformational or configurational analysis of natural products, because, unlike biomolecules such as proteins and nucleic acids, it is difficult to achieve the isotopic enrichment of natural products. In the 1990s, more sensitive methods such as hetero-half-filtered TOCSY (HETLOC) became available by detecting proton magnetization [11, 12]; this paved the way for the application of $^{2,3}J_{C,H}$ values in the conformational and configurational analysis of isotopically unenriched natural samples. Under these circumstances, the JBCA method was developed by Murata and co-workers in 1999 [2]. This method utilizes the empirical and qualitative Karplus-type dihedral angle dependence of not only $^{3}J_{H,H}$, but also $^{2,3}J_{C,H}$ values (Fig. 6.2) to determine the relative stereochemistry of two adjacent or alternate stereocenters, commonly and frequently observed in the polysubstituted acyclic chains of natural products as shown in Fig. 6.1. In this section, the basic concept of the JBCA method is briefly described.

As shown in Fig. 14.2, $^{3}J_{C,H}$ is known to follow a Karplus-type equation similar to $^{3}J_{H,H}$ [13]. In addition, $^{2}J_{C,H}$ also provides dihedral angle information when the α-carbon has an electronegative substituent [14] (Fig. 6.2). In the JBCA method, the magnitude of coupling constants is qualitatively classified as small and large under the assumption that C–C bonds can be represented by three staggered rotamers (one *anti* and two *gauche*). This assumption is generally reasonable for acyclic carbon

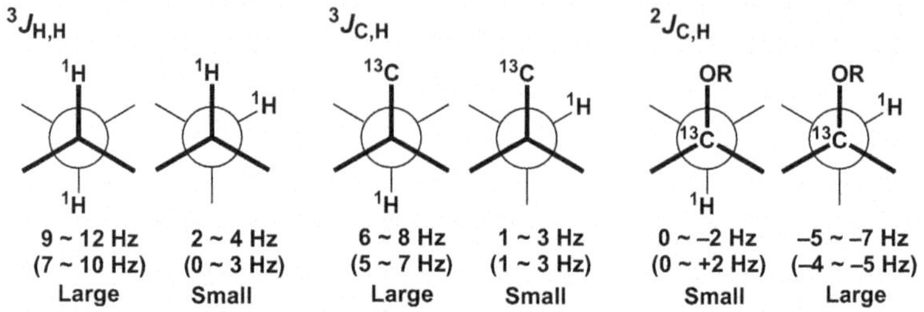

Fig. 6.2 Empirical and qualitative Karplus-type dihedral angle dependence of coupling constants and their classification into large and small according to the magnitude of the absolute values. The parentheses indicate the values for 1,2-dioxygenated systems. (Reproduced with permission from Ref. [2]. Copyright 1999 American Chemical Society)

chains with less bulky substituents such as hydroxy and methyl groups, which hardly cause significant deviations from the staggered positions.

To assign the relative stereochemistry (*threo* and *erythro*) of a 1,2-dimethine system, six staggered rotamers possible in the *threo* and *erythro* diastereomers should be distinguished from one another (Fig. 6.3). Among these six rotamers, four rotamers, A-1, A-2, B-1, and B-2, can be specified using the qualitative magnitudes of $^3J_{H,H}$, $^2J_{C,H}$, and $^3J_{C,H}$ values. However, two rotamers A-3 and B-3 with an H/H-*anti* orientation are still indistinguishable only from coupling constants. To identify these two *anti*-conformers, NOE information is obtained. In the B-3 rotamer (Fig. 6.3), weak or no NOE (or ROE) should be observed between H-1 and H-4, whereas in the A-3 rotamer, H-1 and H-4 should be closer to the NOE distance. Based on these criteria, all six rotamers are distinguished, and their relative configuration (*threo* or *erythro*) can be determined accordingly.

The question is how many coupling constants are necessary to assign a specific rotamer out of the six rotamers. Because each rotamer shown in Fig. 6.3 can be defined by two *anti*-arrangements, the coupling constants that specify two antiperiplanar relationships are enough to assign a single rotamer out of the six rotamers. For example, A-2 shown in Fig. 6.3 can be defined by the *anti* orientations of H-2/C4 and C1/H-3; therefore, two large $^3J_{C,H}$ values for H-2/C4 and C1/H-3 are enough to specify this rotamer. Alternatively, for example, because the *anti* arrangement of H-2/C4 in the A-2 rotamer is equivalent to the two *gauche* arrangements of H-2/H-3 and H-2/Y, the small ^3J(H-2,H-3) and large ^2J(H-2,C3) can compensate the lack of ^3J(H-2/C4). In this method, finding two antiperiplanar relations is the key to identifying a specific rotamer. With this notion in mind, it is possible to assign the relative configuration of a rotamer by the step-by-step interpretation of coupling constants, as exemplified in Fig. 6.4, without referring to Fig. 6.3.

This analysis is applicable to methylene-spaced stereocenters such as 1,3-dimethine systems when prochiral methylene protons are stereospecifically labeled according to their chemical shifts. All the possible rotamers for a methine–methylene

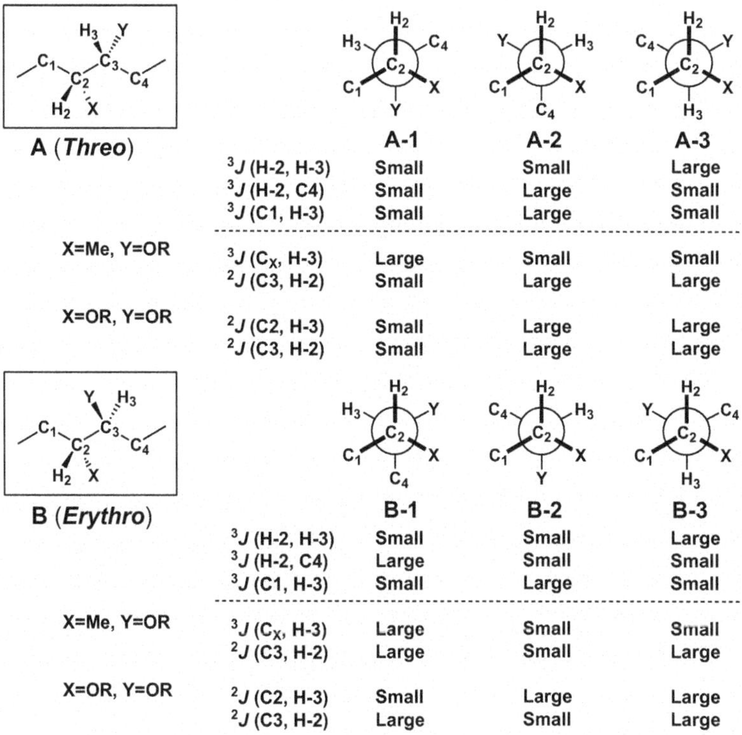

Fig. 6.3 JBCA for a 1,2-dimethine system. The expected magnitudes of coupling constants are listed for each rotamer. A combination of coupling constants can discriminate the rotamers except H/H-*anti* rotamers A-3 and B-3, which can be distinguished using NOEs. (Reproduced with permission from Ref. [2]. Copyright 1999 American Chemical Society)

bond can be unambiguously distinguished using $^{3}J_{H,H}$ and $^{2,3}J_{C,H}$. With one methine–methylene relationship in hand, the same examination for the other methine–methylene relationship leads to the diastereomeric determination of the 1,3-dimethine via the stereospecific assignment of methylene protons. An example is shown in Fig. 6.5. The methylene protons on the C3 position are labeled as $H3^h$ (high-field proton) and $H3^l$ (low-field proton), respectively. For the C2–C3 bond, the observed coupling constants can specify the single rotamer (Fig. 6.5a). In this rotamer, the two antiperiplanar relations are $H2/H3^h$ and $OH/H3^l$. Similarly, the rotamer for the C3–C4 bond was uniquely determined from the *anti*-orientation for $H3^h/H4$ and two *gauche* relations with respect to $H3^l$ (Fig. 6.5b). Then, the rotamers shown in Figs. 6.5a and b are connected at the C3 position so that the $H3^h$ and $H3^l$ in both the rotamers are consistent, thus establishing the diastereomeric relation between C2 and C4 as depicted in Fig. 6.5c.

The above analysis assumes a single and highly populated (>85%) rotamer in a C–C bond. However, this is not always the case for acyclic carbon chains. When multiple rotamers exist, averaged coupling constants, represented as medium *J* values, are

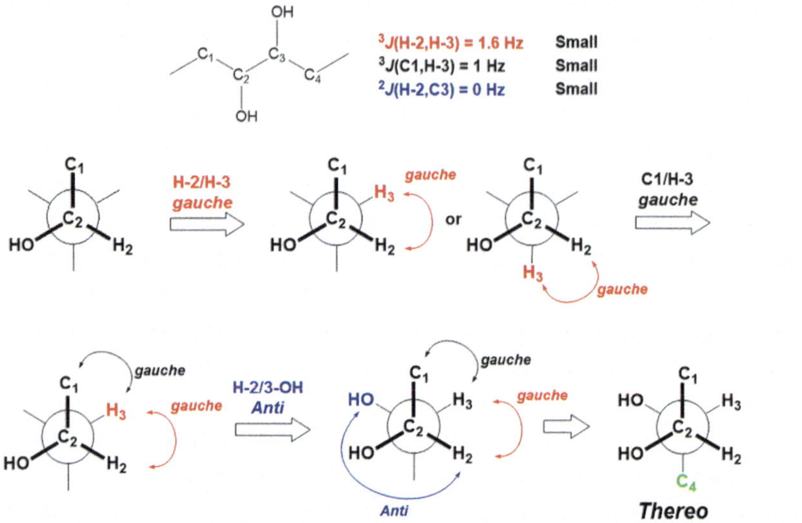

Fig. 6.4 An example of the configuration assignment of a 1,2-dimethine system by the step-by-step interpretation of observed coupling constants. The H-3 position is determined by the two *gauche* relations of H-2/H-3 and C1/H-3. The 3-OH position is determined by *anti*-H-2/3-OH relation derived from the small value of ^{2}J(H-2,C3). Accordingly, the remaining C4 should be uniquely placed in the *anti* position of C1, and the relative configuration is assigned as *threo*

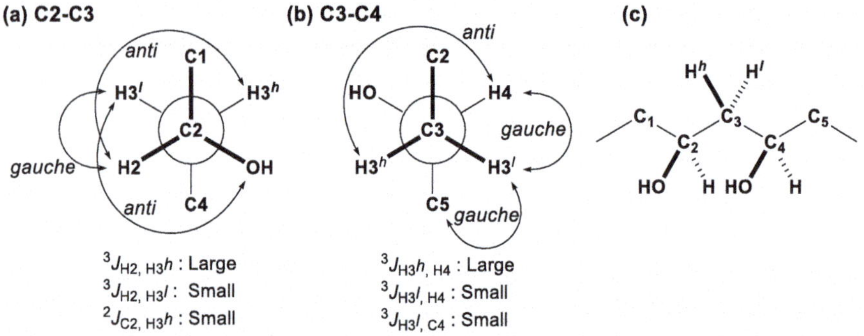

Fig. 6.5 An example of stereochemical assignment of 1,3-stereocenters. $H3^{h}$ and $H3^{l}$ represent the diastereotopic methylene protons on C3 with chemical shifts at higher and lower fields, respectively

observed owing to a conformational interconversion much faster than the NMR time scale. In most cases, the conformational equilibrium of a C–C bond is assumed to be the interconversion between two major staggered rotamers. Figure 6.6 shows an example of relative stereochemistry determination in exchanging rotamers. The medium magnitude of ^{3}J(H-2,H-3) indicates the *gauche*/*anti* equilibrium of H-2/H-3 dihedral angles. The medium magnitude of ^{3}J(H-2, Me) is also indicative of the equilibrium of *gauche*/*anti* between H2 and Me. The large value of ^{2}J(H-3,

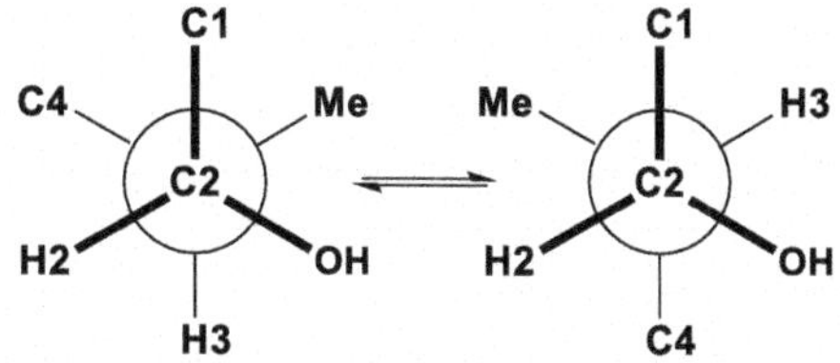

$^{3}J_{H2,H3}$: Medium	*gauche*	H2/H3	*anti*
$^{3}J_{H2,Me}$: Medium	*anti*	H2/Me	*gauche*
$^{2}J_{H3,C2}$: Large	*gauche*	H3/OH	*gauche*

Fig. 6.6 Stereochemical assignment of a C–C bond using the *J*-based method. A pair of equilibrating rotamers with *erythro* configuration can account for the observed coupling constants

C2) further indicates two concurrent *gauche* arrangements of H-3/OH. These data indicate the equilibrium of two exchanging rotamers shown in Fig. 6.6, thus leading to the configuration assignment of the C–C bond as *erythro*.

When three staggered rotamers have comparable populations, all the J values fall into an intermediate category, and the J-based method cannot be applied to the system. If the population of a minor rotamer is less than ca. 10%, its contribution would be negligible, and the relative stereochemistry can be determined as the above-discussed interconversion system between two major staggered rotamers.

Now that the introduction of $^{2,3}J_{C,H}$ values provides a method for the stereochemical assignment in acyclic natural products, measurements of $^{2,3}J_{C,H}$ are crucial for the success of this methodology. Since the 1990s, several practical methods have been proposed for the measurement of $^{2,3}J_{C,H}$ [15], thanks to the introduction of inverse detection and pulse-field-gradient (PFG) technique. In the original paper reporting JBCA [2], two methods were used: HETLOC (hetero half-filtered total correlation spectroscopy) [11, 12] and PS-HMBC (phase-sensitive heteronuclear multiple bond correlation) [16]. The sensitivities of those methods are almost comparable to or slightly lower than conventional HMBC. The HETLOC experiment provides a ^{1}H–^{1}H 2D TOCSY-like spectrum, where each cross-peak is split by $^{1}J_{C,H}$ along the F_1 direction and dislocated by $^{2,3}J_{C,H}$ along the F_2 direction. This cross-peak feature of HETLOC allows efficient measurements of small $^{2,3}J_{C,H}$ values. However, because HETLOC uses the TOCSY spin transfer, it cannot measure the coupling constants between protons and quaternary carbons or between protons and carbons belonging to different spin systems. To solve this problem, PS-HMBC experiments [16] were also introduced. In the PS-HMBC experiments, the relative cross-peak intensities arising from the same proton depend on the magnitude of $^{2,3}J_{C,H}$ [16]. Therefore, the combined use of HETLOC and PS-HMBC can increase the number of observed $^{2,3}J_{C,H}$ values. Several pulse sequences including J-HMBC [17], useful for estimating coupling constants between protons and quaternary carbons, were also reported for the estimation of $^{2,3}J_{C,H}$ values. The J-HMBC signal amplitude of spectra shows the characteristic $\sin(\pi\, ^{2,3}J_{C,H}\tau)$-dependence with increasing evolution time τ, and thus accurate C–H coupling constants can be obtained by fitting a sine curve to the experimental data or by performing a 3D-version of J-HMBC.

The J-based method was first applied to maitotoxin [18] (Fig. 6.1), the largest secondary metabolite known to date, followed by dysiherbaine [19] and amphidinol 3 [20]. Since then, this method has been used for the structure determination of diverse natural products [3]. During the course of its application, this method has been modified according to each compound, thus becoming a robust method. For example, the application of the J-based method has been extended from stereocenters bearing methyl and hydroxy groups to those bearing amine and chloride groups [3].

6.1.2 UDB Method

Around the same time when the JBCA method was reported, Kishi et al. reported the UDB method, providing another powerful tool for determining relative stereochemistry [4, 5]. In the 1980s and 1990s, Kishi's group achieved the stereochemical assignments of multifunctional and complex acyclic natural products such as palytoxin [21], AAL toxins/fumonisins [22], and maitotoxin [23] through stereocontrolled synthesis. These studies demonstrated the following facts [4]: The NMR profiles of these natural products show the specific stereochemistry of the small substituents on the carbon backbone, independent of the other moieties in the molecules, and the steric and stereoelectronic interactions between the structural clusters separated by a two/more methylene bridge are negligible. Based on these experimental discoveries, they developed the principle of the UDB method for assigning the stereochemistry of (acyclic) compounds. A highly functionalized molecule can be considered as the sum of independent clusters of stereocenters (stereocluster), and the stereochemistry of each independent stereocluster can be determined from the NMR data (^{13}C chemical shifts were mainly used in the initial publications of UDB) comparison of the libraries of model compounds with stereochemically unknown natural products. One of the original demonstrations of this technique was carried out on the oasomycins [5]. To determine the stereochemistry of the stereocluster in the rectangle shown in Fig. 6.7, they built a library of model compounds with eight possible diastereoisomers (database 1) and compared their ^{13}C chemical shifts with those of oasomycin B. According to Kishi's convention, the deviations ($\Delta\delta$) are presented in a bar chart form (Fig. 6.7). Apparently, the deviations in ^{13}C chemical shifts are the smallest in **1e**, indicating that the relative stereochemistry at the C6–C9 of oasomycin B is the same as that of **1e**.

Kishi's group created the libraries of stereoclusters (Databases 2–5) and determined the stereochemistry of oasomycin (Fig. 6.8) [24]. Although the creation of a database requires the laborious organic synthesis of all the possible stereoisomers, once the database is available, it is universally applicable to the stereochemical determination of various natural compounds containing the sterocluster.

Later, Kishi's group created databases for contiguous polyols such as 1,2,3-triols, 1,2,3,4-tetraols, and 1,2,3,4,5-pentaols and extended the UDB method using profiles consisting of two or three contiguous $^3J_{H,H}$ constants (Fig. 6.9) [25]. As exemplified for contiguous triols (Fig. 6.9), the internal and terminal $^3J_{H,H}$ pairs are distributed in

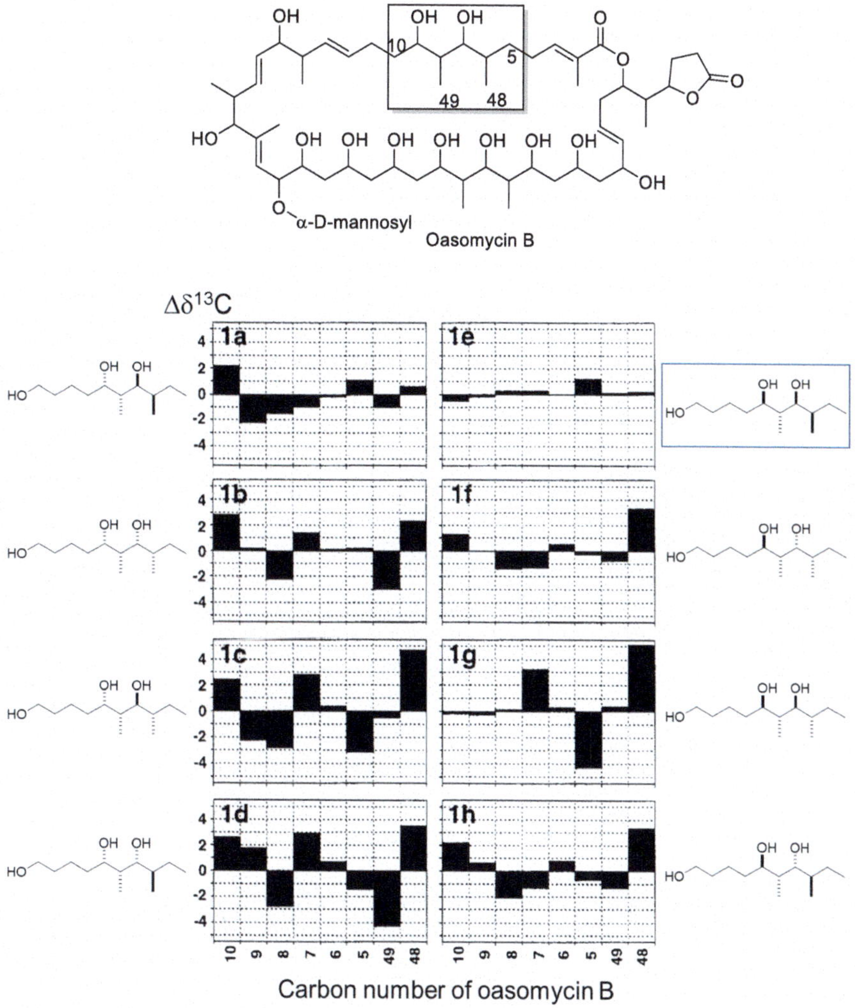

Fig. 6.7 Application of the UDB method to the stereochemical assignment of oasomycins. Difference in ^{13}C chemical shifts between oasomycin B and synthetic compounds **1a–h**. (Reproduced with permission from Ref. [5]. Copyright 1999 American Chemical Society)

four areas depending on their relative stereochemistry, cf, *syn/anti*, *syn/syn*, *anti/anti*, and *anti/syn*. This indicates that a $^3J_{H,H}$ profile can be used as a primary screening for predicting the stereochemistry of unknown polyols, whereas ^{13}C and ^{1}H chemical shift profiles are used to confirm the predicted stereochemistry.

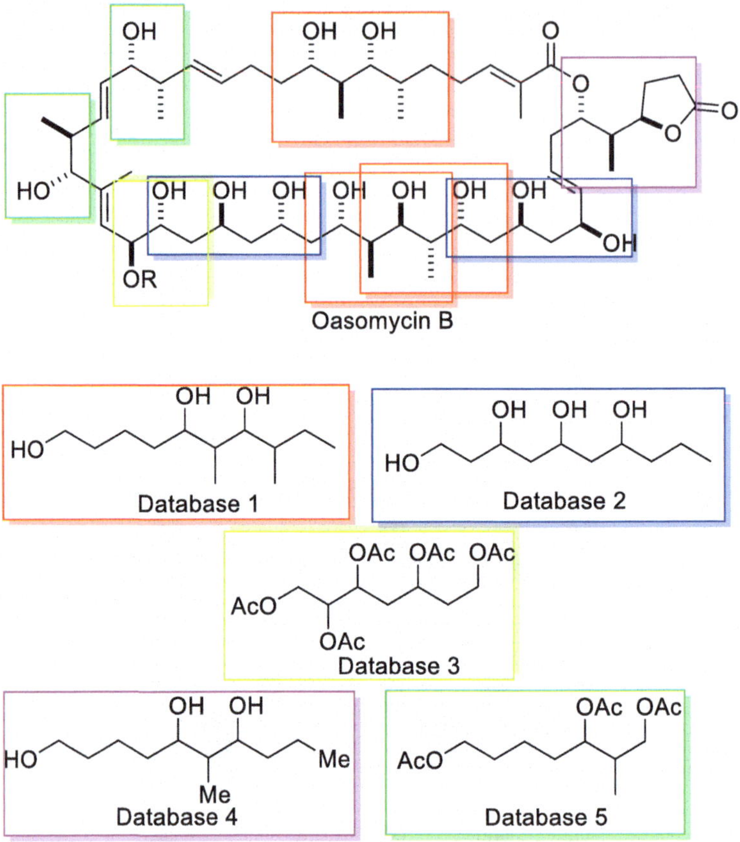

Fig. 6.8 Databases of stereoclusters to predict the stereochemistry of oasomycins

6.1.3 Calculations

One of the drawbacks of the JBCA method is that, unlike ${}^{3}J_{H,H}$ values that have well-documented empirical and theoretical correlation with the dihedral angles, the use of ${}^{2,3}J_{C,H}$ values for the analysis of relative configuration is accompanied by ambiguity in classifying a given value into large, medium, or small, because of the scarce accumulation of literature data, particularly for the stereocenters carrying less common substituents such as halogens. The NMR database method also faces the same difficult situation when NMR databases covering the stereocluster of interest are not available.

On the other hand, thanks to the developments in computational power and advances in quantum mechanical (QM) methods in the last two decades, experimental NMR parameters such as chemical shifts and coupling constants can be reproduced by calculations with high accuracy. Hence, it is no surprise that theoretical calculations are used to compensate not only the ambiguity in correlation

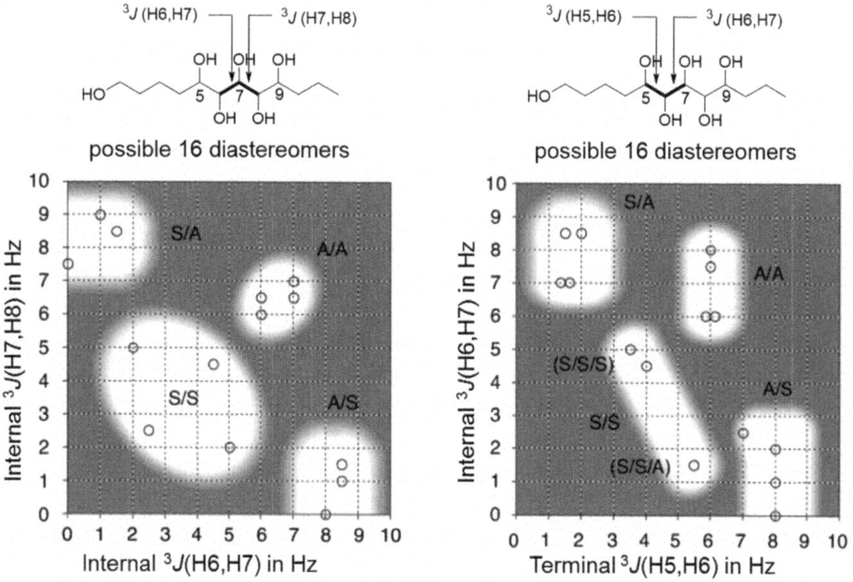

Fig. 6.9 Two-dimensional representation of the internal (left) and terminal (right) $^{3}J_{H,H}$ values observed for triols. The x and y axes represent the two contiguous $^{3}J_{H,H}$ values indicated by the arrows. S = *syn* and A = *anti*. (Reproduced with the permission from Ref. [25]. Copyright 2003 American Chemical Society)

between the experimental $^{2,3}J_{C,H}$ data and dihedral angles in the JBCA method but also the lack of experimental NMR databases for stereoclusters in the UDB method [3, 6].

The QM approach assumes that the possible stereoisomers show different NMR features (e.g., chemical shifts and *J* coupling constants). Specifically, the general workflow for determining the configuration using the QM method relies on the following two main phases: [27, 28]

(1) Generation of the ensembles of conformers, and subsequent calculation of chemical shifts and/or *J* values. This step starts with extensive conformational searches for all possible stereoisomers by molecular mechanics methods; subsequently, the sets of conformers are geometry-optimized by QM methods, and the gauge independent atomic orbital (GIAO) [29] was applied to calculate the chemical shifts and/or *J* couplings of the conformers. The Boltzmann-averaged values were computed based on the relative energy.
(2) Comparison of the Boltzmann-weighted final set of data with experimental values. They are quantitatively compared using statistical methods, such as by the mean absolute error (MAE), the root-mean-square deviation (RMSD), and the coefficient of determination R^2 and, as reported in recent studies, using emerging statistical tactics, such as the DP4 and the optimized DP4 + [30].

An example of QM calculation that compensates the traditional JBCA method is shown in Fig. 6.10 [31]. Since the absolute stereochemistry of the C4 and C10 of trichophycin C was determined as *S* and *R*, respectively, using the modified Mosher's method [9], the stereochemical assignment of C5 and C7 was carried out using the QM calculation at the DFT level. Hence, four different stereoisomers were considered, including C5*R*C7*R*, C5*S*C7*R*, C5*R*C7*S*, and C5SC7*S*, with C4 and C10 fixed to *S* and *R*, respectively. Briefly, 200 conformers generated through conformational sampling were submitted to the geometry optimization at the B3LYP/6-31g(d,p) level of theory. After redundant structures were discarded, the remaining conformers were ranked on the basis of the sum of electronic and thermal free energies. For those conformers whose energies were within the 2.2 kcal/mol window of the lowest energy conformer, *J* couplings were calculated using GIAO at the B3LYP/6–311 + g(d,p) level. Finally, the *J* values were averaged on the basis of their Boltzmann populations calculated from the energy gap from the lowest energy conformer, and the averaged *J* couplings are listed in Table 6.1 along with experimentally measured values. As is clearly shown in Fig. 6.10 and Table 6.1, the R^2 and RMSD between theoretically calculated and experimentally measured *J* couplings are the highest and lowest, respectively, for the stereoisomer C4*S*C5*R*C7*R*C10*R*.

This example adopted RMSD and R^2 as the statistical methods for comparing experimental and computed NMR parameters, but among the statistical tactics, the DP4 + method [32], which is based on a *Bayesian* probability, is the most popular and widely used today due to its overall satisfactory and reliable performance. The method has been applied to stereochemical assignments of a number of complicated natural products including karlotoxin [33] and gibbosols [34]. For more details on the DP4 + method, please refer to the papers reported by Sarotti et al. [32, 35].

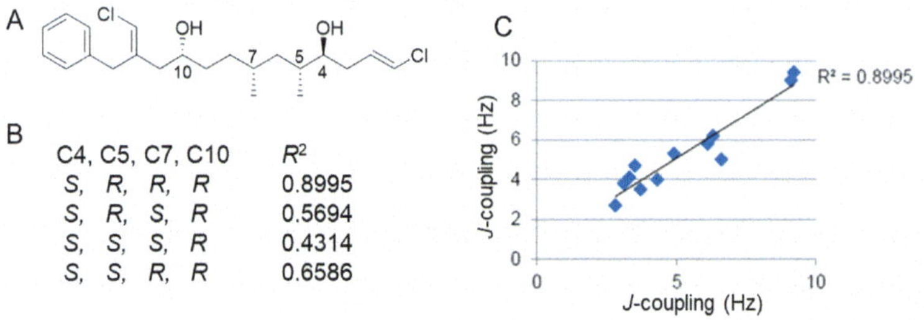

Fig. 6.10 **a** Four chiral centers of trichophycin C. The absolute stereochemistry at C4 and C10 was determined at *S* and *R*, respectively, using the modified Mosher's method. **b** R^2 values for correlations between computed and experimental *J*-couplings for stereochemistry at C4, C5, C7, and C10. **c** Plots of computed (horizontal) vs. experimental (vertical) *J*-coupling values for the isomer C4*S*C5*R*C7*R*C10*R*. (Reproduced with the permission from Ref. [31]. Copyright 2018 American Chemical Society)

Table 6.1 Calculated and observed coupling constants for trichophycin C

	Calcd.				Experimental
Coupling	C5*S*C7*S*	C5*R*C7*S*	C5*S*C7*R*	C5*R*C7*R*	
H6a–C8	2.4	2.4	7.1	6.1	5.8
H6b–C21	3.1	3	3.8	4.3	4.0
H6b–C22	2.7	6.1	2.6	6.3	6.2
H6a–H5	8.4	9.1	11.2	9.2	9.4
H6b–C8	5.1	5.1	2.2	3.1	3.8
H6a–H7	7.5	7.6	5.4	6.6	5.5
H6b–H7	8.0	8.4	10.9	9.1	9.0
H5–H4	4.5	3.6	3.0	4.9	5.3
H5–C3	2	3.6	1.5	3.3	4.1
H4–C6	2.1	2.9	2.1	2.8	2.7
H4–C22	4.6	4.3	5.3	3.7	3.5
H5–C4	1.4	2.8	0.2	3.5	4.7
RMSD (Hz)	2.1	1.6	2.2	0.7	

6.1.4 RDCs

As described above, the stereochemical assignments of organic molecules depend on chemical shifts, *J*-couplings, and NOEs, even though these NMR parameters do not provide long-range information. Therefore, it has been impossible to correlate distant stereocenters or stereocenters separated by a flexible linker, hindering the stereochemical determination of organic molecules with such stereocenters. This situation can be changed by introducing RDCs with angular information with respect to the direction of the external magnetic field, thus allowing to correlate the distant moieties of a molecule. The RDC techniques were mostly developed in the biomolecular NMR field, where RDCs are frequently used to elucidate and refine the 3D structures of peptides, proteins, and nucleic acids. Lately, this approach has been used to determine the relative configurations of organic molecules. The basic principle and applications of RDCs to organic molecules are briefly described below; readers should refer to review papers for more details [7, 8].

Dipolar couplings are directly proportional to $\langle \frac{3cos^2\theta - 1}{2} \rangle \gamma_I \gamma_S / r^3$, where θ is the angle between the I-S interspin vector and magnetic field direction; bracket < > indicates a time average of all the orientations sampled by the interspin vector; γ_I and γ_S are the magnetogyric ratios of the two spins I and S; *r* is the interspin distance. Thus, dipolar couplings provide structural information about the orientation and distance; those of C–H are typically in the order of several kHz in solid-state samples. In liquid samples, the dipolar couplings are generally averaged to zero, and thus the orientation and distance information is lost, because the isotropic fast tumbling motion of the molecules nulls the time average term denoted by the brackets. To access the dipolar couplings in liquid NMR spectra without a significant loss in signal resolution, the solute molecules should be weakly aligned, which can be attained using alignment media such as liquid crystals or stretched polymer gels. In such alignment media, the solute molecules are weakly oriented, and the dipolar couplings are not averaged to zero but scaled down from kHz to the order of several Hz. Such reduced dipolar couplings are known as RDCs.

For the application of RDCs to organic molecules, stretched polymer gels are most frequently used as the alignment media: polyacrylamide and polyacrylonitrile for water and polar organic solvents, and polystyrene for apolar organic solvents [8]. A preparation procedure involves the addition of a solvent to a dry polymer stick in an NMR tube. The swelling gel stretches along the NMR tube axis, and the polymer network also becomes anisotropic, preventing the solute samples from isotropic tumbling and consequently making the solute molecules weakly aligned (Fig. 6.11). To obtain RDCs, two measurements should be carried out: One in isotropic solution to obtain the scalar couplings J, and the other in anisotropic alignment media to obtain $|J + 2D|$. ${}^1J_{\mathrm{C,H}}/\,{}^1D_{\mathrm{C,H}}$ can be measured by conventional HMQC or HSQC experiments without carbon decoupling during the acquisition.

To understand the basic principle of RDC application to stereochemical determination, the distinction of axial versus equatorial protons on a six-membered ring is exemplified in Fig. 6.12. Because all the axial C–H bonds are parallel or antiparallel, they give the same C–H RDCs. In the α and β isomers of 4,6-*O*-ethylidene-D-glucopyranose [36], the RDC for the anomeric C–H bond in the β isomer should have the same size as those of the axial C–H bonds, whereas the α isomer produces a different size of RDC, thus making it possible to discriminate the stereoisomers based on the size of RDC. Although for practical purposes, such a simple problem can be solved by measuring NOEs and/or ${}^3J_{\mathrm{H,H}}$ values without referring to RDCs, this demonstrates how RDCs make it possible to correlate the stereochemistry of the anomeric position even with the most distant stereocenter in the acetal moiety.

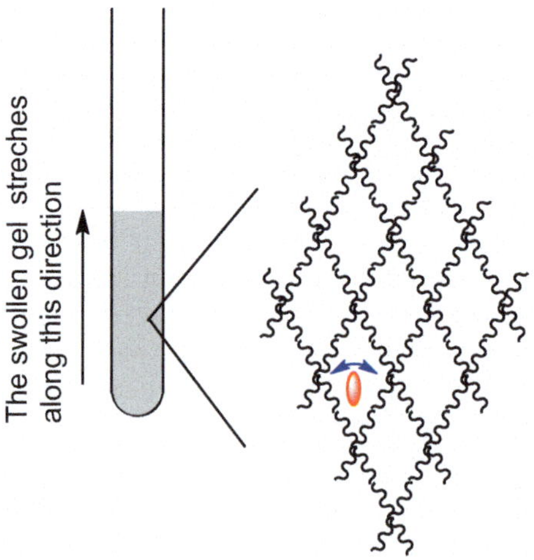

Fig. 6.11 Schematic drawing of an anisotropic gel network in an NMR tube and its prevention of the isotropic motion of solute molecules

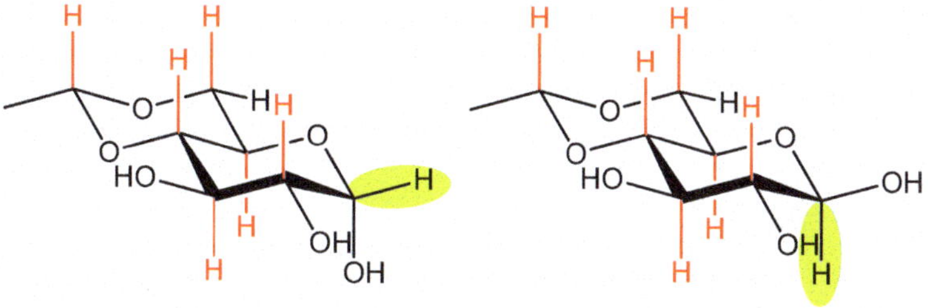

Fig. 6.12 α and β forms of 4,6-*O*-ethylidene-D-glucopyranose. The axial C–H bonds are highlighted in red, and the anomeric C–H bonds are shaded in yellow

To determine the relative stereochemistry of complex molecules using RDCs, the concept of alignment tensor is generally introduced. In the case of a rigid molecule, the alignment tensor represents the distribution of the external magnetic field vector in a reference molecular frame. When more than five RDCs are observed for a rigid molecule, its alignment tensor can be obtained by fitting a molecular structure to the experimental RDC data. From the obtained best-fitting alignment tensor and molecular structure, it is also possible to back-calculate the RDCs; this should be completely consistent with the experimental RDCs when the input molecular structure is correct. In contrast, a molecular structure with the wrong stereochemistry will cause a poor agreement between the back-calculated and experimental RDCs [37]. In this manner, RDCs can be used for the verification and falsification of relative configuration. Then, the application of RDCs to stereochemical analysis has extended from cyclic rigid compounds to flexible acyclic natural products as demonstrated by the report from Griesinger's group on the stereochemistry of sagittamide A (Fig. 6.13) [38]; the combined use of coupling constants, NOEs, and RDCs allowed the successful determination of all the stereochemistry. Recently, residual chemical shift anisotropies (RCSAs) of not only ^{13}C but also ^{1}H have also been applied to the stereochemical determination of organic molecules [39, 40]. In addition, more sophisticated methodologies combined with molecular dynamics simulations have been proposed to reduce the risk of incorrect structural assignments [41].

Although the application of RDCs and RCSAs to the stereochemical assignment of natural products has been gradually increasing, they have not been generalized yet, in particular for structurally unknown compounds. This is probably because the reliability of both the experimental results and the data analysis is still a question. In addition, the complexity of data analysis as well as the hesitation of mixing precious natural products with alignment gel media may limit the general application.

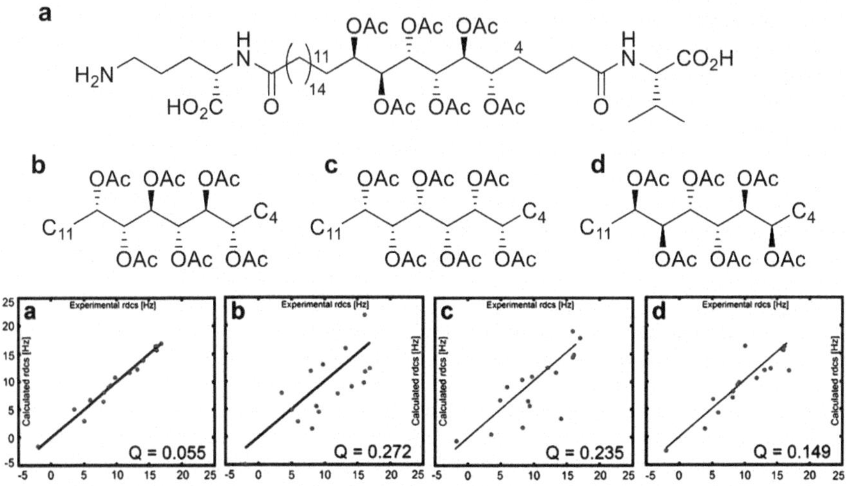

Fig. 6.13 Sagittamide A with possible stereoisomers (**b**-**d**) according to the *J*-coupling analysis (top) and experimental and back-calculated RDCs for Sagittamide A (bottom). From this data, stereochemistry **a** is the most probable. (Reproduced with the permission from Ref. [38]. Copyright 2007 American Chemical Society)

6.2 NMR Methods for Examining the Conformation and Intermolecular Interactions of Natural Products in Membranes

To gain a deeper insight into the biological activities of natural products, it is crucial to obtain more precise information regarding their interactions with membrane lipids, but it has been hardly obtained even with the current technologies. The fluid mosaic model was proposed by Singer and Nicolson in 1972 [42]; a plasma membrane was thought to be a two-dimensional homogeneous fluid, mainly consisting of rapidly diffusing lipids and proteins. This vision has been revised by the concept of "lipid rafts", where the lipid bilayers are heterogeneous in biological systems [43]. Some recent research results have demonstrated that certain physicochemical features of lipid rafts can be partly reproduced using artificial membranes in the absence of proteins [44, 45], significantly accelerating biophysical and chemical studies on the atomic-level mechanism of lipid–lipid interactions in membranes [46–48]. The methodologies developed in these studies for lipids also help examine the conformation of natural products associated with membranes. We highlight recent studies on natural products that interact with biological membranes in exerting their activities, some of which can modify the domain formation of biological membranes.

During our research on bioactive compounds, we learned that small molecules, particularly natural products, often show very high affinities for phospholipids and sterols in bilayer membranes when they exert membrane-permeabilizing activities [49]. To investigate large lipid-bilayer systems corresponding to cell membranes, therefore, solid-state NMR provides an important approach for the structural studies of complexes of natural products with membrane lipids. Solution NMR techniques can, however, be applied to characterize the structures of membrane-bound small molecules. Standard 2D ^{1}H–^{1}H experiments are often recorded with small lipid assemblies such as micelle suspensions and small bicelles [50, 51]. In this section, we describe four examples of our and others' NMR studies on small molecule–lipid complexes using solid-state and solution NMR techniques.

6.2.1 Saponins

6.2.1.1 Structural Features of Saponins and Their Interactions with Biological Model Membranes

As a first example, we chose saponin, a well-known membrane-active and hemolytic natural product. Since saponins are one of the natural products that could be potential drug candidates, biophysical studies have been conducted on the interaction of saponins with membrane lipids and membrane sterols. Although there is a wide diversity of saponin structures, the combination of key structures, such as the structure of sapogenin (aglycon of saponin), the position of the sugar chain substitution on the sapogenin, as well as the number and type of sugar chains, is responsible for a particular activity. Sapogenin structures can be grossly classified into two types, steroids and triterpenoids. There are extensive reviews on the cytotoxicity of these saponins, and much evidence has been gathered that monodesmosidic saponins are generally more cytotoxic than bisdesmosidic saponins [52]. Moreover, Li et al. reported that steroidal saponins exert hemolytic and cytotoxic activities by different mechanisms [53].

The biological effects of saponins as a surface-active molecules are attributed to their interaction with cell membranes. Saponins increase the permeability of biological membranes, which is the well-known cause of their hemolytic activity. The hemolytic potential of saponins often requires the presence of cholesterol (Chol) in the target membrane; Chol and saponin are believed to form insoluble complexes that subsequently accumulate into large micellar aggregates that disrupt the bilayer membrane and increase membrane permeability [54, 55]. According to the proposed mechanism, saponins first bind to the outer leaflet of the bilayer with hydrophobic sapogenins inserted along the acyl chains of phospholipids. Within the membrane, saponin binds to Chol to form a complex, which accumulates and develops into micelle clusters or plaques. Although this saponin-Chol interaction is widely accepted, some research results have revealed the membrane-disrupting activity of saponin on Chol-free membranes [56, 57]; in the absence of Chol in the bilayer, oleanolate saponin glycosylated at C3 inserts into the bilayer and thus increases the permeability of phosphatidylcholine (PC) bilayers.

Glycans of saponin are thought to be important in the membrane-disrupting activity of saponins. The length of the chain, the type of sugar, and the glycosidic linkage clearly affect the membrane activity of saponins and their affinity for Chol. However, a precisemode of structure-dependent interactions with lipids and Chol has not been elucidated. Sequential removal of the sugar units of digitonin [58], α-tomatin [59], and avenasin A1 [60], gradually weakened the membrane-disrupting ability of saponins, indicating that the number of sugar units is important for saponins to exhibit membrane-disrupting activity.

The diosgenyl saponins discussed in this subsection are the products of the *Dioscorea* species. They belong to the steroidal saponins and have a spirostanol sapogenin structure with oligosaccharide chains at the C-3 hydroxy group. Commonly isolated are dioscin (DSN), deltonin, gracilin, and trilin (TRL) (Fig. 6.14). Numerous studies have been conducted on the pharmacological activity of each saponin, suggesting their potential as an anticancer agent. Hydrolysis of saponin sugars yields sapogenin, diosgenin. This natural product is an industrially important precursor for the manufacture of steroid drugs.

There are not many examples of solid-state NMR studies of natural products that bind to and alter lipid bilayers except peptides. Relatively well-known are amphotericin B [49, 61], ionophore [62], and macrolide antibiotic [63], but examples of NMR studies on saponins in membranes are rather limited [64]. These NMR techniques are used to study the fluctuations of membrane-active molecules and their molecular orientation relative to the membrane using several NMR nuclei, ^{2}H, ^{13}C, ^{15}N, ^{31}P, etc. The main metrics are quadrupolar interactions, chemical shift anisotropy, magnetic dipole interactions [65], and relaxation times. Exactly the same techniques can be applied to membrane lipid molecules, allowing the effects of membrane-active molecules on lipid membranes to be investigated under the same conditions. These methods can also provide insight into the permeation of drugs through membranes and their sequestration in lipid pools, which are areas of great

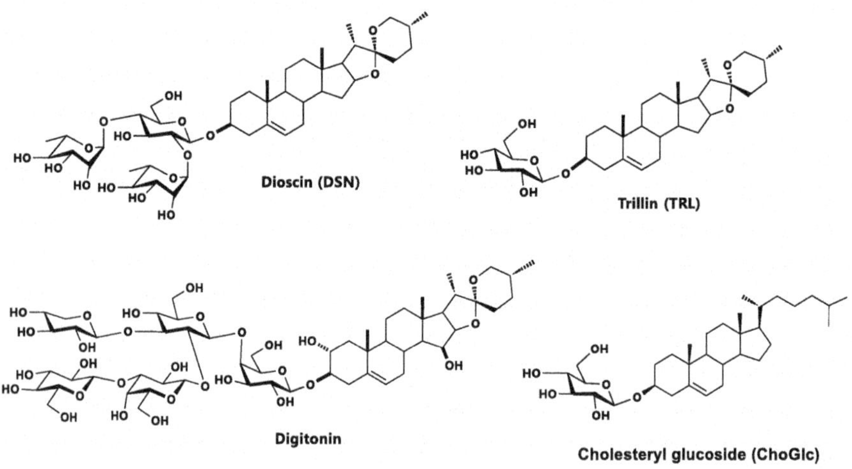

Fig. 6.14 Structures of the diosgenyl saponins (trillin and dioscin), digitonin, and ChoGlc discussed in this subsection

future importance. In this subsection, we will focus on diosgenyl saponins and discuss their influence on lipid-bilayer properties from the aspect of saponin structure, such as the orientation of sapogenin and the number and arrangement of sugar units. To examine the wobbling and orientation of the saponins in lipid bilayers and their interaction with sterols, five deuterated probes were used, 3*d*-Chol and 3*d*-diosgenin as sterol probes, 1'*d*-TRL and 3*d*-TRL as saponin probes, and 10′,10′-d_2-POPC as a phospholipid probe. Here, the orientation of saponin, sterol, and phospholipid molecules means the angle between the molecular axis of these compounds and the membrane normal, and wobbling means the range of motion of the C-2H bond from its mean orientation angle. Deuterium probes reproduce the orientation and wobbling of the original compounds without the effects of isotope labeling and their orientation and wobbling can be evaluated by quadrupole splitting width, $\Delta\nu$.

6.2.1.2 *Deuterium NMR Studies of Diosgenin as Diosgenyl Sapogenin*

Since diosgenin is a sapogenin common to diosgenyl saponins, it is interesting to examine the affinity between Chol and diosgenin as a model for the interaction between Chol and saponin in biological membranes. To this end, we measured the ^{2}H NMR of the ^{2}H-labeled sterols, deuterated 3*d*-Chol and 3*d*- diosgenin (Fig. 6.15) [66]. The quadrupole splitting values $\Delta\nu$ of these labeled sterols (10 mol %) in POPC bilayers were comparable (Table 6.2). Chol has a C8 alkyl tail and probably inserts between the acyl chains of phospholipids to enhance lipid ordering. Diosgenin bears a spirobicyclic tail and a very rigid structure throughout the molecule. The slight difference in $\Delta\nu$ may be due to the orientation difference between the tetracyclic cores of Chol and diosgenin. The same experiments with a one-to-one mixture of Chol and diosgenin gave interesting results. Panels C and D of Fig. 6.15 show the ^{2}H NMR spectra of POPC bilayers containing 9 mol% of each deuterated sterol (18 mol% of total sterols), and the $\Delta\nu$ values are strikingly similar (Table 6.2); $\Delta\nu$ of 3*d*-Chol significantly decreased with the addition of diosgenin (A $\rightarrow$ C), while the $\Delta\nu$ values of 3*d*-diosgenin increased (B $\rightarrow$ D). This suggests that the orientation of 3*d*-Chol and 3*d*-diosgenin changed due to the interaction of diosgenin and Chol, respectively, implying that heterophilicity is significantly stronger than homophilicity for these sterols, and that the Chol-dependence of membrane-activity of diosgenyl saponin is mediated by contact of the sapogenin with Chol.

The effect of diosgenin on membrane integrity was investigated by static ^{2}H and ^{31}P NMR spectra of POPC. In these spectra, the ^{2}H quadruple splitting width ($\Delta\nu$) and ^{31}P chemical shift anisotropy (CSA) provide physicochemical information about the ordering and orientation of hydrophobic acyl chains and a polar headgroup, respectively. In the experiment in Table 6.2, 10′,10′-d_2-POPC bearing a CD_2 group at the 10 position of a palmitoyl moiety and non-labeled POPC obtained from commercial sources were used for ^{2}H and ^{31}P spectral measurements. ^{2}H NMR showed comparable increases of $\Delta\nu$ by addition of Chol or diosgenin, indicating that both sterols have similar ordering effects on the acyl chains of POPC. As shown by ^{31}P chemical shift anisotropy (Table 6.2), an addition of diosgenin to POPC bilayers or

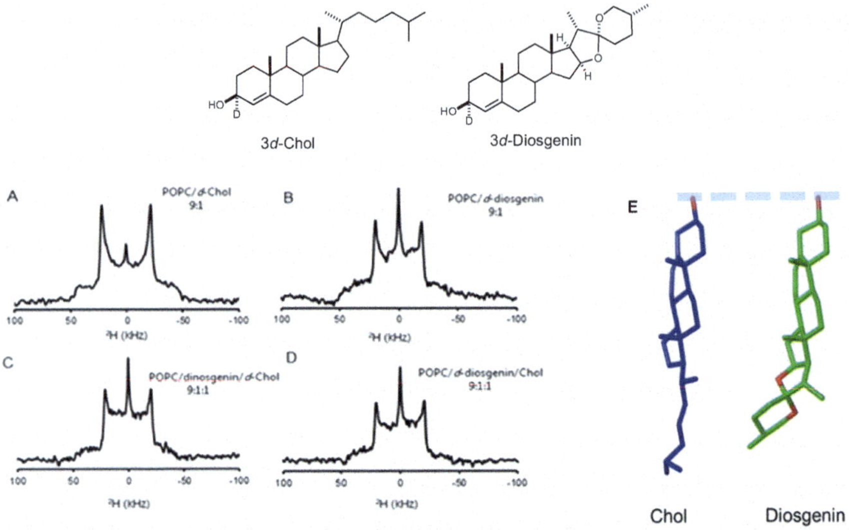

Fig. 6.15 ^{2}H NMR spectra of POPC-based MLVs containing different amounts of 3*d*-Chol **a** and 3*d*-diosgenin **b** at 30 °C. Those of POPC/Chol/diosgenin 9:1:1 containing 3*d*-Chol **c** and 3*d*-diosgenin **d**. The central peaks in the spectra are due to deuterated water and/or sterol in micelles formed during MLV preparation. E: Orientation of Chol and diosgenin deduced from ^{2}H NMR results. A broken line denotes the interfacial plane corresponding to the depth of the *sn-2* ester group of POPC. (Reproduced with the permission from Ref. [66]. Copyright 2021 Elsevier)

Table 6.2 Summary of NMR data for Chol, and diosgenin in POPC bilayers [67]

	Sterol in POPC (10 mol%)[a]			
^{2}H-observed sterol or lipid (kHz)	Non	Chol	Diosgenin	Chol + diosgenin
^{2}H Δv of 3*d*-Chol[a]		43.8[b]		42.0[b]
^{2}H Δv of 3*d*-diosgenin[a]			41.1[b]	42.1[b]
^{2}H $\Delta\nu$ of 10′,10′-d_2-POPC	19	25	23	–
^{31}P CSA ($\delta_{//}$ -$\delta_{\perp}$) of POPC (ppm)	45.9	45.4	45.1	44.5

[a] Data obtained from pure MLVs of POPC or Chol/diosgenin-containing MLVs at 30 °C
[b] Data obtained from the spectra in Fig. 6.15 in POPC MLVs

POPC-Chol bilayers did not significantly affect the wobbling or orientation of the POPC head group. This result means that the sugar chain is important for exerting the membrane activity of diosgenyl saponin.

6.2.1.3 *Deuterium NMR Studies on the Behavior of Deuterated Saponins in Lipid Bilayers*

To examine the effect of sugar substitution on the orientation of saponins in membranes, TRL was ^{2}H-labeled at the C3 position of the sapogenin (3*d*-TRL) and at the C1' position of the sugar moiety (1'*d*-TRL) (Fig. 6.16). These two site-specific probes were incorporated in POPC membranes in the presence or absence of Chol [67]. The ^{2}H NMR spectra of POPC bilayers with 10 mol% saponin probes showed typical Pake doublets, suggesting that the parallel orientation of saponin to the membrane is nearly perpendicular to the membrane plane. In the presence of Chol, the $\Delta\nu$ values increased for both 3*d*-TRL and 1'*d*-TRL probes (Fig. 6.17). The result suggests that the orientation of saponin in the membrane was somewhat affected by the contact with Chol, or by Chol-induced ordering of the acyl chains of POPC, resulting in the increase of $\Delta\nu$ [67].

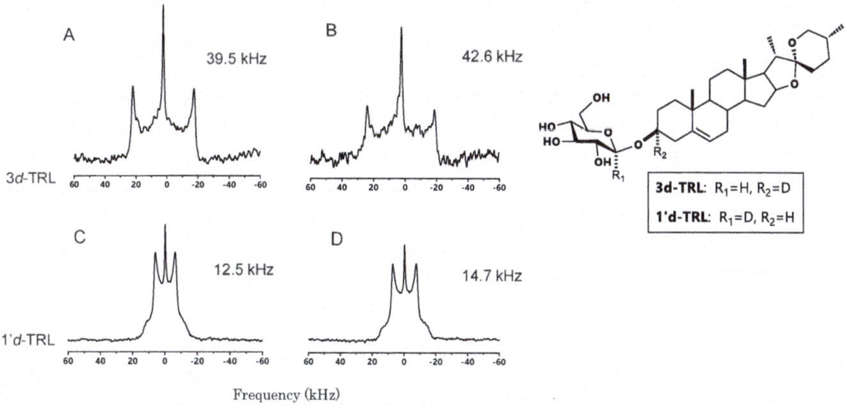

Fig. 6.16 Solid-state ^{2}H NMR spectra of 10 mol% deuterated TRLs in POPC and POPC–Chol (9:1) systems. **a** and **b**: 3*d*-TRL in the POPC and in POPC–Chol (9:1) bilayers, respectively. **c** and **d**: 1'*d*-TRL in unitary POPC and POPC–Chol 9:1 bilayers, respectively. The Δv values are shown in each spectrum. The central peak in each spectrum may be due to residual deuterated water

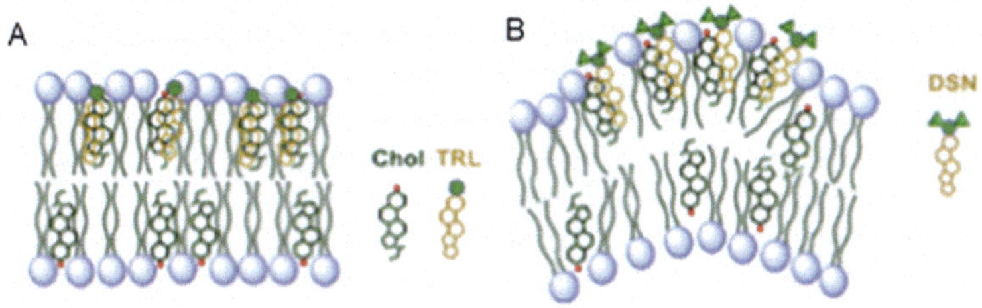

Fig. 6.17 Schematic illustrations of TRL and DSN interacting with Chol-containing bilayers, respectively

Table 6.3 Effect of saponin on membrane lipid and sterols obtained from NMR results in POPC bilayers

Sterol or lipid portion to be observed for its orientation/wobbling[a]	$^2H/^{31}P$-sterol or lipid for NMR observation	Sterol (10 mol%) in bilayers[b]	Ratio of the values in the presence and absence of 10 mol% saponin[c]		
			TRL	DSN	Digitonin
Chol tetracyclic core	^2H Δν of 3*d*-Chol	Chol	0.94	0.88	–
Diosgenin whole molecule	^2HΔν of 3*d*-diosgenin	Diosgenin	1.00	0.95	–
Acyl chain ordering of POPC	^2H *Δν* of 10′,10′d_2-POPC	Non	1.21	1.11	1.0
		Chol	1.20	0.95	0.80
Head group of POPC	^{31}P CSA ($\delta_{//} - \delta_{\perp}$) of POPC	Non	0.99	1.03	0.98
		Chol	1.00	0.88	0.97[d]
		Diosgenin	1.01	0.99	0.98

[a] Orientation of the labeled compounds denotes the angle between the molecular axis of the compound and the membrane normal, and wobbling means the range of motion of C-^2H bond (or CSA in an axially symmetric system) from its mean orientation angle. [b] Sterol was mixed with lipid at the beginning. [c] Relative ratio of NMR parameters in the presence of saponin to that in the absence of saponin, the latter of which is shown in Table 6.2. In these experiments, saponin was added to the vesicle suspension afterward. [d] Digitonin tends to not only raise bilayer curvature but also form rigid domains in the presence of Chol. Digitonin also causes the formation of small mixed micelles whose ^{31}P signal is detected at the isotropic chemical shift [69]

We recently examined the behavior of cholesteryl glucoside (Fig. 6.14) in bilayers using solid-state NMR [68]. The effect of the sugar unit on the ordering of the acyl chain was evaluated by using 10′,10′-d_2-POPC. The ordering effect of cholesteryl glucoside on the POPC acyl chain was 91% of Chol's effect, suggesting that sugar substitution on aglycon hardly affects bilayer order. In addition, *Δν* of Chol at the 3 position was decreased in the presence of TRL while that of diosgenin was unaffected by TRL (Table 6.3). The result further indicates that heterophilic interaction between Chol and TRL affect the orientation of both Chol and TRL as seen for the Chol and diosgenin case.

6.2.1.4 *Deuterium and Phosphorus Solid-State NMR Spectra of Saponin-Containing POPC Membranes*

The effects of TRL and DSN on membrane lipids and sterol were evaluated using three ^2H-labeled probes and ^{31}P of POPC [67] as shown in Table 6.3. The changes in Δν of 3*d*-Chol in the presence of TRL or DSN are slightly larger than those of 3*d*-diosgenin, also suggesting stronger interactions of saponin with Chol [69] as seen for 3*d*-TRL. The slightly smaller Δν values for 3*d*-diosgenin in the presence of 10 mol% DSN indicate that the ordering and orientation of diosgenin are somehow affected by the membrane-disturbing effects of DSN. Deuterated PC, 10',10'-d_2-POPC, was adopted to evaluate the membrane-disrupting activity of diosgenyl saponins since the 10' position of PC was known to be sensitive to Chol-induced ordering effects on acyl chains [70, 71] and the saponin's effect on POPC chain packing. The bilayers containing 10 mol% TRL increased the $\Delta\nu$ value of POPC by about 20% (Table 6.3) either in the presence or absence of Chol, indicating that TRL inserts deeper into POPC bilayers and enhances the packing of POPC acyl chains independent of Chol. A significant decrease in the $\Delta\nu$ value of POPC was observed when digitonin was added to POPC-Chol membranes, which may be a consequence of the disruption of membrane integrity by digitonin. On the contrary, the unchanged $\Delta\nu$ value for digitonin in the absence of Chol reveals that the saponin cannot bind efficiently to Chol-free POPC bilayers, probably due to its large sugar moiety (Table 6.3).

The NMR results further reveal that the depth position of saponins along the bilayer normally depends on the number of their sugar units [67]. The constant ^{31}P CSA values of POPC in the presence of TRL under the three conditions of Table 6.3 implies that TRL maintains the integrity of the POPC bilayer structure both in the presence and absence of Chol. The results further suggest that TRL sapogenin (diosgenin) is present at about the same depth as Chol, and its glucose portion seems to approach the phosphodiester of POPC. By contrast, the significantly reduced ^{31}P CSA values of DSN in the Chol-containing bilayers indicate that the sapogenin of DSN is located high along the bilayer normal and seems to lift the membrane lipids, causing higher membrane curvature (Fig. 6.17b). These results again suggest that the sapogenin portion (diosgenin) of DSN interacts with Chol with a high affinity inside the bilayer. This is because the oxygenated spiro-bicycles of sapogenin may weaken the homophilic interaction inside the hydrophobic bilayer while the *iso*-octyl side chain of Chol that has a strong ordering effect on lipid acyl chains [72] further promotes van der Waals (vdW) contacts between Chol and saponins in the bilayer environment. As a result, DSN alters the curvature of Chol-containing POPC bilayers and destabilizes the chain packing to some extent [67]. The potent membrane activity of DSN may be due to the DSN-induced formation of Chol-lipid mixed domains with high curvature, which weakens the barrier function of membrane lipids particularly around the domain boundaries.

Computational studies also provided a molecular view of saponin-Chol interactions. Lin and Wang reported that molecular dynamics (MD) simulation predicted possible binding modes for DSN with Chol; DSN has a stable orientation parallel to the bilayer normal, while Chol is quite oblique. Both sterol cores are presumed to contact each other near the vdW distance between their α faces, which is the side without angular methyl groups. The MD studies also estimated the energetics of complex formation and showed that the DSN-Chol complex is more stable than the Chol-Chol complex. This result is in line with the NMR experimental results mentioned above [73].

6.2.2 Erythromycin A

Erythromycin A (EA, Fig. 6.18) is an old antibiotic, but still widely prescribed. As a side effect known as phospholipidosis, the drug accumulates in lysosomal membranes and induces the excess accumulation of phospholipids [74, 75]. This is caused by the inhibition of lysosomal phospholipase activity by EA [74]. Isotropic bicelles have been applied to membrane-associated peptides and proteins [76, 77] and were used for the molecular characterization of the membrane-bound structure of EA [78]. First, ^{31}P NMR spectra of a large oriented bicelle containing dimyristoylphosphatidylcholine (DMPC), dihexanoylphosphatidylcholine (DHPC), dimyristoylphosphatidylglycerol (DMPG) and EA (the q value, [DMPC + DMPG]/[DMPC] = 3.3), were measured to confirm that the bicelle structure is maintained even at high contents of EA. The ^{1}H NMR spectra of EA in isotropic bicelles comprising DMPC-d_{54}, DHPC-d_{22}, and DMPG-d_{54} (q = 0.5) showed the conformation of bicelle-associated EA determined using NOEs and $^{3}J_{HH}$ data (Fig. 6.18b), consistent with the crystal structure of EA [79].

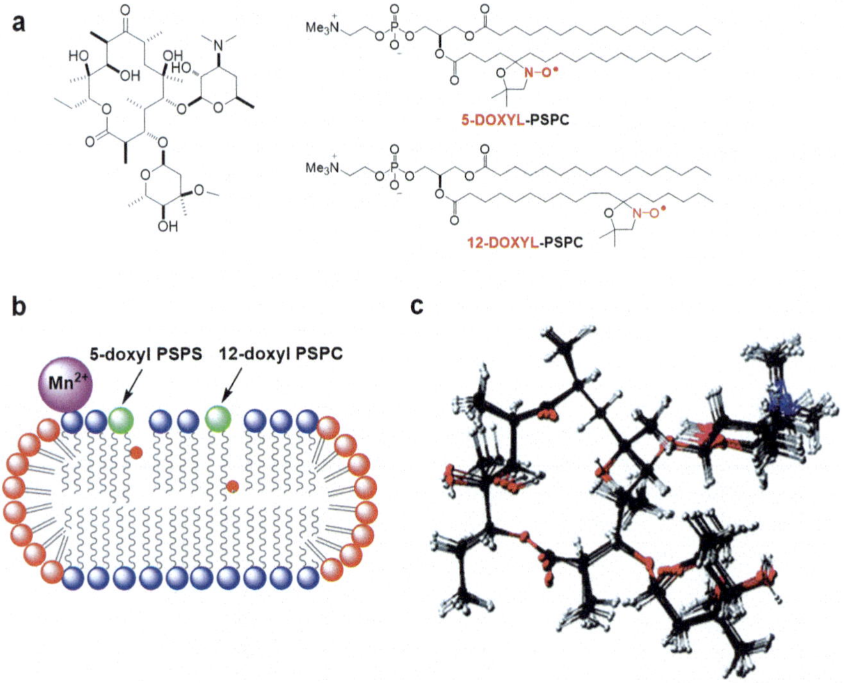

Fig. 6.18 Interactions of paramagnetic Doxyl-PC and Mn^{2+} with an isotropic bicelle. **a** Structures of EA and doxyl-PCs. **b** Isotropic bicelle containing doxyl-PCs in the presence of Mn^{2+}. **c** conformation of EA in bicelles [78]. (Reproduced from Ref. by permission of The Royal Society of Chemistry)

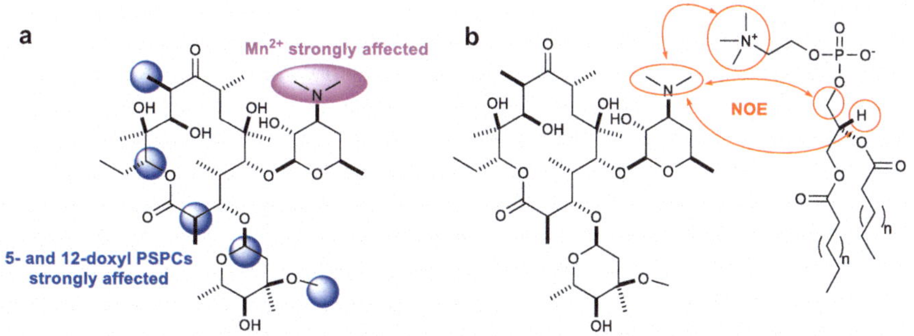

Fig. 6.19 **a** Results of paramagnetic relaxation experiments for EA and **b** results of NOE experiments on EA-containing phospholipids. [78] (Reproduced from Ref. by permission of The Royal Society of Chemistry)

The relaxation parameter T_{1M} of EA protons in bicelles containing paramagnetic Mn^{2+} ions (Fig. 6.19) was strongly shortened for the signals of the 7′,8′-dimethyl group, indicating that the ammonium group in aqueous media has ionic interactions with the phosphate groups of the bicelle and come close to the water–lipid interface. On the other hand, the protons affected by 12-doxyl-PSPC (Fig. 6.18) were positioned at 2, 13, 20, 1″, and 8″ (blue balls in Fig. 6.19), indicating that these protons reside in the relatively deeper portion of the bicelle interior. Taken together, EA resides in the shallow and middle parts of the membrane interior by placing the dimethylamine group close to the phosphate group and slightly immersing the macrolide portion in the lipid bilayer [78].

EA was shown to dominantly partition to the flat bilayer area of the bicelle to interact more preferentially with acidic DMPG than with neutral DMPC and DHPC, and DMPG preferentially resides in the DMPC-rich bilayer of the bicelles owing to the better matching of acyl chain length to that of DMPC. The position and conformation of EA in bicelles shown in Fig. 6.20 suggest that the drug is a potential inducer of phospholipidosis; the possible mechanism for the inhibition of phospholipase activity is the binding of EA to negatively charged lysosomal membranes [74, 75]. This NMR-based study showed that the large macrolide moiety of EA comes close to the ester groups of phospholipids (Fig. 6.20), thus probably hindering the approach of lysosomal phospholipase to the C1′ position of a lipid acyl chain. This led to the accumulation of phospholipids in the cell.

6.2.3 *Theonellamide A*

The next example is a cyclic peptide, which does not belong to typical membrane-active amphipathic peptides. In this section, we would like to describe another application of isotropic bicelles to membrane-bound molecules. Theonellamides (TNMs, Fig. 6.21) were isolated from the marine sponge *Theonella sp.* [80]. TNM

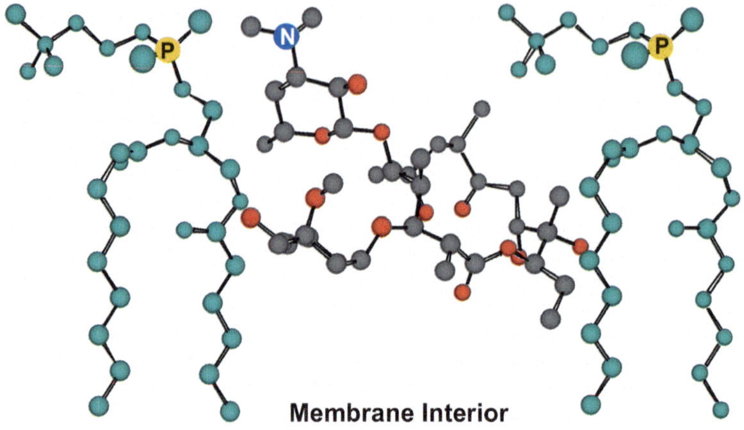

Fig. 6.20 Hypothetical interaction of erythromycin A (EA) and interactions with membrane phospholipids (light blue). It is assumed that electrostatic interactions exist between the phosphate group of lipids and the ammonium group of EA (see text). [78] (Reproduced from Ref. by permission of The Royal Society of Chemistry)

homologs possess a bismacrocyclic skeleton that encompasses nonribosomal amino acid residues and a histidinoalanine tether. In vitro binding assays using a fluorescent derivative showed that the peptide was specifically bound to 3β-hydroxysterols such as cholesterol (Chol) and ergosterol (Erg) [81]. The phenotypic changes in yeasts indicate that the membrane-related activity of TNM is distinct from that of polyene antifungals such as amphotericin B. The ^{2}H NMR spectra of deuterated-Chol and 3-*epi*-Chol showed that TNM-A significantly slows down the fast rotational motion of Chol and Erg, but not *epi*-cholesterol [82], indicating that 3β-hydroxysterol directly interacts with TNM-A in lipid bilayers.

To better understand the membrane-disrupting activity of TNM-A, solution NMR techniques were used for isotropic bicelles containing the peptide. In particular, several ^{1}H NMR experiments were conducted to examine the interaction mode of TNM-A in model membranes. First, the aggregate formation of TNM-A was evaluated by diffusion-ordered spectroscopy (DOSY) [83]. Using the accurate viscosity values of deuterated solvent systems, the results indicate that TNM-A has the oligomeric states of 2–9 molecules (depending on peptide concentration) in an aqueous environment, and this aggregation potentially plays a role in the membrane-disrupting activity of the peptide [84]. Further, TNM-A was mixed with small DMPC-d_{54}/ DHPC-d_{22} bicelles and subjected to ^{1}H NMR measurements in the presence of a paramagnetic quencher Mn^{2+} similar to erythromycin in the previous section (Fig. 6.21). In Chol-containing (10 mol%) bicelles, the ^{1}H signals of TNM-A were significantly broadened, indicating that the peptide efficiently associates with the Chol-containing bicelles. The addition of Mn^{2+} to these bicelles also elicited a concentration-dependent decrease in the relative intensity of TNM-A signals, further indicating that the peptide stays near the surface of the bilayers where the Mn^{2+}

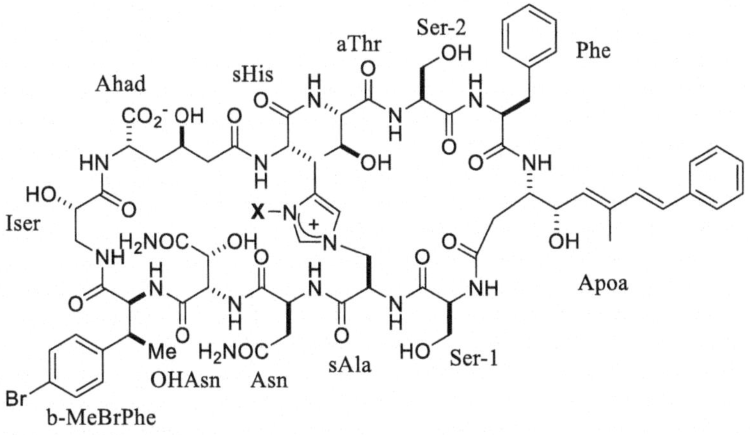

Fig. 6.21 Structure of Theonellamide-A (TNM-A)

residing near the phosphate group effectively affects the relaxation process of ^{1}H signals (Fig. 6.22). A comparison of the relative signal intensities with those of DMPC showed that the hydrophobic chains such as the Apoa and Phe groups of TNM-A bind at the lipid–water interface of the bilayer area of bicelles, close to the C2′ portion of the phospholipid acyl chain (Fig. 6.23). This shallow penetration of TNM-A to lipid bilayers can induce uneven membrane curvature and eventually disrupt the bilayer structure [84].

6.2.4 *Amphidinol 3*

The final example is amphidinol 3 (AM3). Dinoflagellates classified in the genus *Amphidinium* produce a unique class of polyhydroxy-polyene metabolites, which are collectively termed "amphidinols". More than 20 amphidinols have been reported since amphidinol 1 was discovered from *Amphidinium klebsii* in 1991 [85]. These natural products possess a conserved region sandwiched between two tetrahydropyran rings, and their structural variation of occurs predominantly in the polyhydroxy and polyene chains. Among amphidinols, AM3 (Fig. 6.24), which was also isolated from *A. klebsii* in 1996, has the most potent antifungal and hemolytic actions [86, 87].

The stereochemistry of AM3 was proposed in 1999 by combined use of JBCA, modified Mosher's method, and chemical degradation [88], but was revised several times by Oishi's synthetic studies of its partial structures [20, 89, 90] (Fig. 6.24). With regard to the elucidation of relative stereochemistry using the JBCA method, the assignments for the C38-C39 and C50-C51 bonds of AM3 failed. This can be

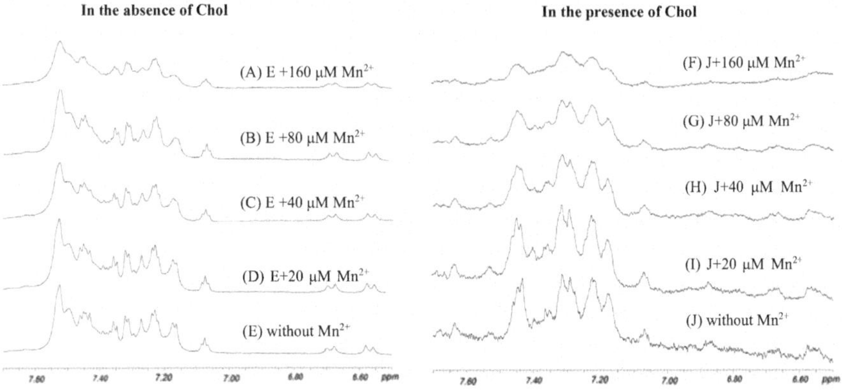

Fig. 6.22 ^{1}H NMR spectra of TNM-A incorporated in Chol-free **a**–**b** and Chol-containing (F–J) DMPC-d_{54}/DHPC-d_{22} bicelles (q = 0.5) in D_2O (100 mM NaCl) without Mn^{2+} **e** and **j** and with increasing concentrations of Mn^{2+} **a**–**d** and **f**–**i**. The amount of Chol added to bicelles was 10 mol% of DMPC-d_{54}. Mn^{2+}/TNM-A mol ratios: from the bottom to the top, 0:10, 1:10, 2:10, 4:10, and 8:10. The final concentrations of TNM-A and phospholipids (DMPC + DHPC) were 200 μM and 160 mM, respectively. All the spectra were recorded with 512 scans at 310 K. (Reproduced from Ref. [84], Copyright 2016, with permission from Elsevier)

attributable to the difficulty in determining the stereochemistry of the rotationally flexible and/or distorted bonds. In 2020, Oishi achieved the total synthesis of AM3 [91, 92], which unambiguously confirmed that the stereochemistry of AM3 revised by the partial synthesis is completely correct.

AM3 exhibits potent pore-forming activity on membranes, which depends on the presence of sterols such as Chol and ergosterol (Erg) [87, 88]. The direct interaction of AM3 with sterol in the membrane was verified using solid-state ^{2}H NMR experiments similar to the above TNM-A [93]. The quadrupole splitting patterns showed that 3*d*-Chol in lipid bilayers undergoes fast rotational motion in the absence of AM3, while the rotational motion became sluggish in its presence (Fig. 6.25a, b). A similar change was observed for 3*d*-Erg (Fig. 6.25c, d). In contrast, 3*d*-*epi*-Chol gave the Pake doublet signals with comparable intensity both in the absence and presence of AM3 (Fig. 6.25e, f). These spectral changes clearly show that the fast rotation of Chol and Erg was considerably attenuated because of their direct intermolecular interaction with AM3, while *epi*-Chol has a weaker interaction with AM3. This is consistent with its lack of pore formation of AM3 in the membrane containing *epi*-Chol. These ^{2}H NMR results explicitly demonstrate the direct intermolecular interaction between AM3 and 3β-hydroxysterols in membranes (Fig. 6.25g).

To elucidate the 3D structure of AM3 in membrane environments, the above-mentioned isotropic bicelles were employed for the conformation analysis using solution NMR [94]. The results show that the C20-C54 portion of AM3 takes a turn structure at the two tetrahydropyran rings. Note that this conformational study of AM3 was based on the originally proposed incorrect stereochemistry, because it was done before the revision of its configuration. Nevertheless, the hairpin-like

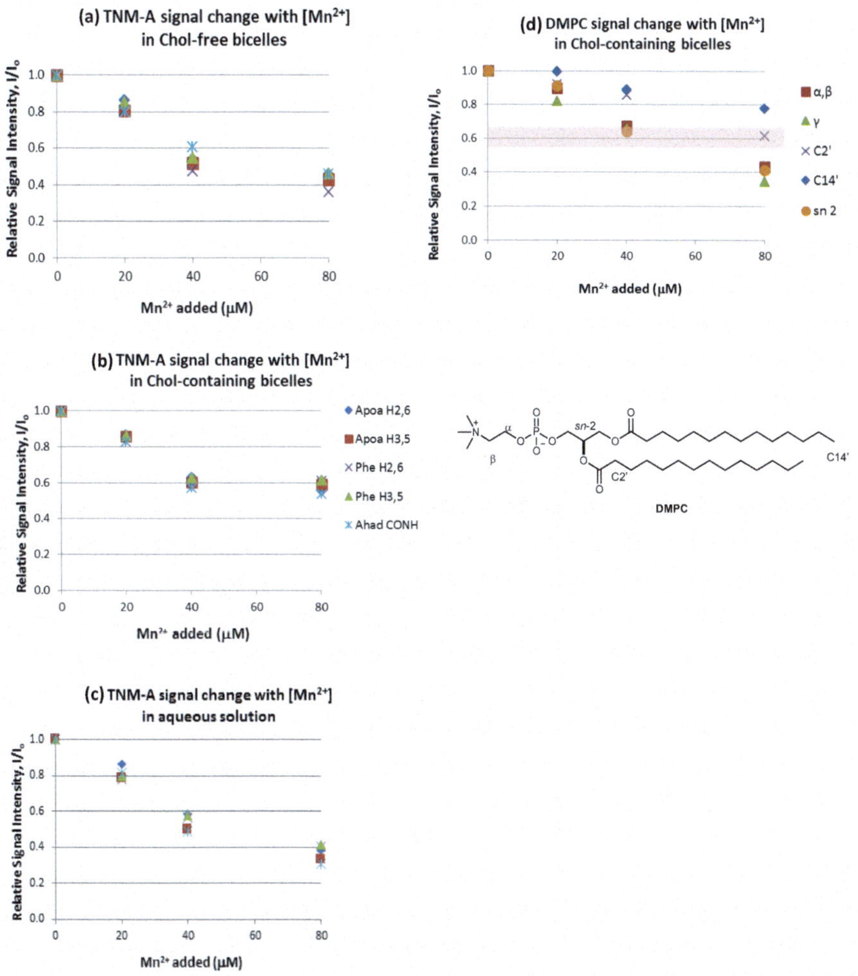

Fig. 6.23 **a** Relative intensities (I/I_0) of the ^1H signals of TNM-A incorporated in Chol-free bicelles with increasing concentrations of Mn^{2+} [84]. **b** Mn^{2+}-dependent I/I_0 values of the same signals in Chol-containing bicelles. **c** Mn^{2+}-dependent I/I_0 values of the same signals without bicelles. **d** Mn^{2+}-dependent I/I_0 values of DMPC signals in Chol-containing bicelles. The I/I_0 values were obtained from the ratio of the ^1H peak intensities in the presence and absence of Mn^{2+} at the indicated amino acid residue of TNM-A and depth-dependent fragments of DMPC as shown in the figure. The concentrations of TNM-A and phospholipids (DMPC + DHPC) in bicelles were 200 μM and 160 mM, respectively. A pink belt at $I/Io = 0.6$ in panel **d** depicts the depth of hydrophobic residue of TNM-A shown in panel B penetrating into the C2' position of POPC molecules of Chol-containing bicelles

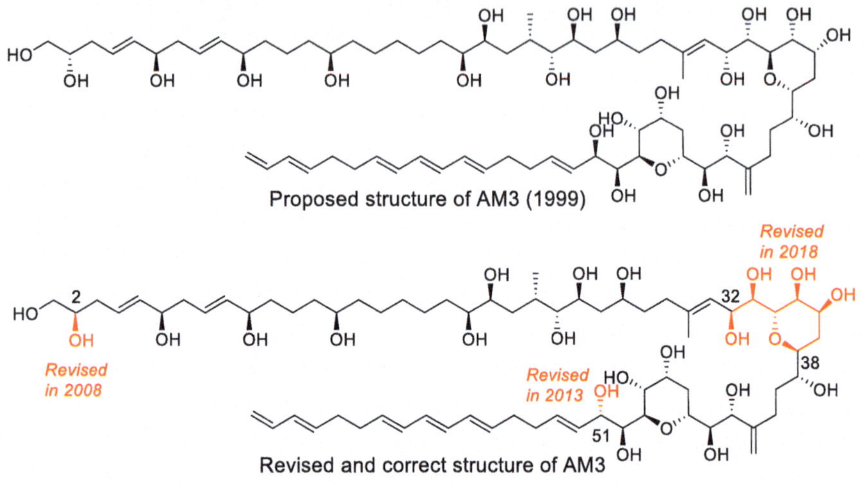

Fig. 6.24 Proposed and revised structures of AM3

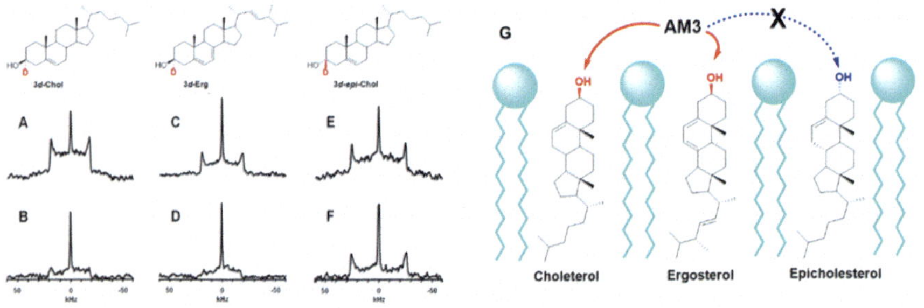

Fig. 6.25 ^{2}H NMR spectra of 3-*d*-sterols incorporated into lipid bilayers in the absence **a**, **c**, and **e** and presence **b**, **d**, and **f** of AM3. The membranes contained cholesterol-d_1 **a** and **b**, ergosterol-d_1 **c** and **d**, and epicholesterol-d_1 **e** and **f**. Isotropic signals at 0 ppm are mostly due to residual deuterium water. Schematic representation of the interaction between AM3 and sterols in bilayers **g**. (Reproduced with the permission from Ref. [93]. Copyright 2014 American Chemical Society)

turn structure at the two tetrahydropyran rings of AM3 is still valid even when the revised stereochemistry is used for the analysis. Unfortunately, the interaction of AM3 with sterol molecules was not elucidated from this experiment because the incorporation of Chol into isotropic bicelles resulted in a marked broadening of the signals, probably due to morphological changes in the isotropic bicelles.

More recently, we found that AM3 and cholestatrienol, a fluorescent sterol, undergo a short-range energy transfer within 1 nm, directly demonstrating a tight binding between AM3 polyene and the sterol alicyclic regions in lipid bilayers [52]. Based on this distance information, along with the conformation of AM3 determined using membrane-mimetic bicelles, we proposed a tight binding model between AM3 and the sterol in lipid bilayers (Fig. 6.26).

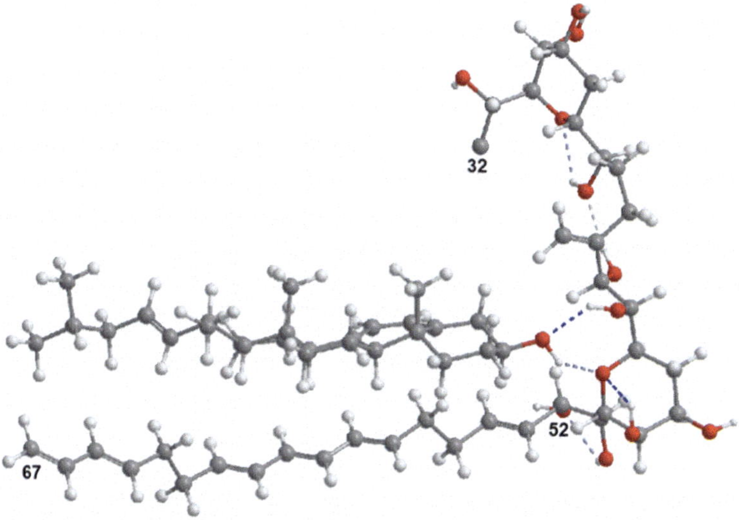

Fig. 6.26 A biding model of sterol (Erg) and AM3 (C32-C67). AM3 polyene and the sterol alicyclic region are in close contact within 1 nm

6.3 Summary and Outlook

In Sect. 6.1, we overviewed four important methods used for the stereochemical determination of natural products on the basis of solution NMR techniques. The basic concepts of the JBCA and UDB methods were reported in the 1990s, and since then both methods have been widely and frequently used in the field of natural products until today. The recent developments in computational methodology in predicting NMR data further complement the ambiguity of interpreting coupling constants in the original JBCA method and the lack of literature database in the UDB method, thus further increasing the reliability of these methods. The newly developed RDC method, still less generalized in the field of natural products, will be an important tool, particularly for the stereochemical correlation between distant stereogenic centers, even though only RDC analysis is not sufficient to determine all the stereochemistry of natural products. In this context, it is important to understand the advantages and disadvantages of each methodology and select the optimal method or combine multiple methods depending on the nature of the target organic molecules such as the molecular size, complexity, type and distribution of functional groups, and sample quantity.

In Sect. 6.2, we discussed the application of solid-state NMR techniques [49, 82] and solution NMR methods for bicelles [50, 78] to examine the membrane interactions of natural products and antibiotics; for reviews, see [48, 64]. However, membrane lipids are highly mobile, and thus the conformation of small molecules bound to membranes also rapidly change in contact with lipids. The time constants of these interactions are usually much faster than the timescale of the solid-state NMR experiments designed for weak interactions, such as dipole–dipole interaction, ^{2}H-qudrupole coupling, and ^{13}C/^{31}P chemical shift anisotropy; one exception is the Zeeman interaction, in which some of the nuclei have frequency similar to that of molecular motions. It is common to observe relaxation processes for this purpose, whereas it is difficult to capture the motion of lipid and drug molecules in membranes where multiple mobility modes coexist. Thus, site-specific isotope-labeling of the membrane-bound natural products, which allows us to select a specific mobility mode in a membrane-bound molecule, is adopted to detect these weak interactions. Therefore, lipid behavior should be deduced from these NMR data in combination with MD simulations as well as other modalities such as fluorescence microscopic techniques, X-ray/neutron scattering, and IR spectroscopy. Closer collaborations between biophysicists and synthetic chemists will surely accelerate fundamental research on the structural biology of membrane lipids and mechanism-of-action studies of natural products including drugs and antibiotics.

Acknowledgements We would like to thank Prof. Shigeru Matsuoka and Dr. Hiroshi Tsuchikawa, Oita University, Prof. Shinya Hanashima, Tottori University, and Dr. Yuichi Umegawa, Osaka University for their valuable comments. We are grateful to Drs. Raymond Malabed and Joan C. Ondevilla, De La Salle University for their efforts in the saponin project. Some of the research projects of the authors described in this review have been supported by JST ERATO lipid active structure projects (Grant Number JPMJER1005) and JSPS KAKENHI (Grant Numbers: 21H04707, 23K17373, 20H00405 and 22K19115).

References

1. Kwan, E.E.: Huang, S.G.: Structural Elucidation with NMR Spectroscopy: Practical Strategies for Organic Chemists. Eur J. Org. Chem. **2008,** 2671–2688 (2008)
2. Matsumori, N., Kaneno, D., Murata, M., Nakamura, H., Tachibana, K.: Stereochemical determination of acyclic structures based on carbon-proton spin-coupling constants. A method of configuration analysis for natural products. J. Org. Chem. **64**, 866–876 (1999)
3. Bifulco, G., Dambruoso, P., Lomez-Paloma, L., Riccio, R.: Determination of relative configuration in organic compounds by NMR spectroscopy and computational methods. Chem. Rev. **107**, 3744–3779 (2007)
4. Kobayashi, Y., Lee, J., Tezuka, K., Kishi, Y.: Toward creation of a universal NMR database for the stereochemical assignment of acyclic compounds: the case of two contiguous propionate units. Org. Lett. **1**, 2177–2180 (1999)
5. Lee, J., Kobayashi, Y., Tezuka, K., Kishi, Y.: Toward creation of a universal NMR database for the stereochemical assignment of acyclic compounds: proof of concept. Org. Lett. **1**, 2181–2184 (1999)
6. Micco, S.D., Chini, M.G., Riccio, R., Bifulco, G.: Quantum mechanical calculation of NMR parameters in the stereostructural determination of natural products. Eur. J. Org. Chem. **2010,** 1411–1434 (2010)
7. Thiele, C.M.: Use of RDCs in rigid organic compounds and some practical considerations concerning alignment media. Concept Magn. Reson. A **30A**, 65–80 (2007)
8. Kummerlöwe, G., Luy, B.: Residual dipolar couplings as a tool in determining the structure of organic molecules. Trends Analyt. Chem. **28**, 483–493 (2009)

9. Ohtani, I., Kusumi, T., Kashman, Y., Kakisawa, H.: High-field FT NMR application of Mosher's method. The absolute configurations of marine terpenoids. J. Am. Chem. Soc. **113**, 4092–4096 (1991)
10. (a) Seco, J.M., Quiñoá, E., Riguera, R.: The assignment of absolute configuration by NMR. Chem. Rev. **104**, 17–118 (2004); (b) Kong, L.Y., Wang, P.: Determination of the absolute configuration of natural products. Chin. J. Nat. Med. **11**, 193–198 (2013)
11. Kurz, M., Schmieder, P., Kessler, H.: HETLOC, an efficient method for determining heteronuclear long-range couplings with heteronuclei in natural abundance. Angew. Chem. Int. Ed. **30**, 1329–1331 (1991)
12. Wollborn, U., Leibfritz, D.: Measurement of heteronuclear long-range coupling constants from inverse homonuclear 2D NMR spectra. J. Magn. Reson. **98**, 142–146 (1992)
13. For a review, see: Hansen, P.E.: Carbon-hydrogen spin-spin coupling constants. Prog. NMR Spectrosc. **14**, 175295 (1981)
14. (a) Schwarcz, J.A., Cyr, N., Perlin, A.S.: Orientation effects and the sign of two-bond ^{13}C–^{1}H coupling. Can J. Chem. **53**, 1872–1875 (1994); (b) Parella, T., Sanchez-Ferrando, F., Virgili, A.: Effect of hydroxyl and carbonyl groups on long-range proton-carbon coupling constants. Magn. Reson. Chem. **32**, 657–664 (1994); (c) Podlasek, C.A., Wu, J., Stripe, W.A., Bondo, P.B., Serianni, A.S.: [^{13}C]Enriched methyl aldopyranosides: structural interpretations of ^{13}C-^{1}H spin-coupling constants and ^{1}H chemical shifts. J. Am. Chem. Soc. **117**, 8635–8644 (1995)
15. Marquez, B.L., Gerwick, W.H., Williamson, R.T.: Survey of NMR experiments for the determination of nJ(C, H) heteronuclear coupling constants in small molecules. Magn. Reson. Chem. **39**, 499–530 (2001)
16. (a) Zhu, G., Bax, A.: Measurement of long-range ^{1}H-^{13}C coupling constants from quantitative 2D heteronuclear multiple-quantum correlation spectra. J. Magn. Reson. **104A**, 353–357 (1993); (b) Zhu, G., Live, D., Bax, A.: Analysis of sugar puckers and glycosidic torsion angles in a DNA G-tetrad structure by heteronuclear three-bond J coupling. J. Am. Chem. Soc. **116**, 8370–8371 (1994)
17. Willker, W., Leibfritz, D.: Determination of heteronuclear long-range H, X coupling constants from gradient-selected HMBC spectra. Magn. Reson. Chem. **33**, 632–638 (1995)
18. (a) Sasaki, M., Matsumori, N., Maruyama, T., Nonomura, T., Murata, M., Tachibana, K., Yasumoto, T.: The complete structure of maitotoxin .1. Configuration of the C1-C14 side chain. Angew. Chem. Int. Ed. **35**, 1672–1675 (1996); (b) Nonomura, T., Sasaki, M., Matsumori, N., Murata, M., Tachibana, K., Yasumoto, T.: The complete structure of maitotoxin 2. Configuration of the C135–C142 side chain and absolute configuration of the entire molecule. Angew. Chem. Int. Ed. **35**, 1675–1678 (1996); (c) Matsumori, N., Nonomura, T., Sasaki, M., Murata, M., Tachibana, K., Satake, M., Yasumoto, T.: Long-range carbon-proton coupling constants for stereochemical assignment of acyclic structures in natural products: Configuration of the C5-C9 portion of maitotoxin. Tetrahedron Lett. **37**, 1269–1272 (1996)
19. Sakai, R., Kamiya, H., Murata, M., Shimamoto, K.: Dysiherbaine: A new neurotoxic amino acid from the Micronesian marine sponge Dysidea herbacea. J. Am. Chem. Soc. **119**, 4112–4116 (1997)
20. Murata, M., Matsuoka, S., Matsumori, N., Paul, G.K., Tachibana, K.: Absolute configuration of amphidinol 3, the first complete structure determination from amphidinol homologues: Application of a new configuration analysis based on carbon-hydrogen spin-coupling constants. J. Am. Chem. Soc. **121**, 870–871 (1999)
21. Cha, J.K., Christ, W.J., Finan, J.M., Fujioka, H., Kishi, Y., Klein, L.L., Ko, S.S., Leder, J., McWhorter, W.W. Jr., Pfaff, K.-P., Yonaga, M., Uemura, D., Hirata, Y.: Stereochemistry of palytoxin .4. Complete structure. J. Am. Chem. Soc. **104**, 7369–7371 (1982)
22. (a) Boyle, C.D., Harmange, J.-C., Kishi, Y.: Novel structure elucidation of AAL toxin t-a backbone. J. Am. Chem. Soc. **116**, 4995–4996 (1994) (b) Boyle, C.D., Kishi, Y.: Absolute-configuration at the tricarballylic acid moieties of fumonisin b-1 and AAL toxin t-a. Tetrahedron Lett. **36**, 5695–5698 (1995)
23. (a) Zheng, W., DeMattei, J.A., Wu, J.-P., Duan, J. J.-W., Cook, L.R., Oinuma, H., Kishi, Y.: Complete relative stereochemistry of maitotoxin. J. Am. Chem. Soc. **118**, 7946–7968 (1996);

(b) Cook, L.R., Oinuma, H., Semones, M.A., Kishi, Y.: The stereochemical assignment and conformational analysis of the V/W-ring juncture of maitotoxin. J. Am. Chem. Soc. **119**, 7928–7937 (1997)
24. Kobayashi, Y., Tan, C.-H., Kishi, Y.: Toward creation of a universal NMR database for stereochemical assignment: Complete structure of the desertomycin/oasomycin class of natural products. J. Am. Chem. Soc. **123**, 2076–2078 (2001)
25. Higashibayashi, S., Czechtizky, W., Kobayashi, Y., Kishi, Y.: Universal NMR databases for contiguous polyols. J. Am. Chem. Soc. **125**, 14379–14393 (2003)
26. Plaza, A., Bifulco, G., Keffer, J.L., Lloyd, J.R., Baker, H.L., Bewley, C.A.: Celebesides A-C and theopapuamides B-D, depsipeptides from an Indonesian sponge that inhibit HIV-1 entry. J. Org. Chem. **74**, 504–512 (2009)
27. Bifulco, G., Bassarello, C., Riccio, R., Gomez-Paloma, L.: Quantum mechanical calculations of NMR J coupling values in the determination of relative configuration in organic compounds. Org. Lett. **6**, 1025–1028 (2004)
28. (a) Barone, G., Gomez- Paloma, L., Duca, D., Silvestri, A., Riccio, R., Bifulco, G.: Determination of the relative stereochemistry of flexible organic compounds by ab initio methods: conformational analysis and Boltzmann-averaged GIAO ^{13}C NMR chemical shifts. Chem. Eur. J. **8**, 3233–3239 (2002); (b) Barone, G., Duca, D., Silvestri, A., Gomez-Paloma, L., Riccio, R., Bifulco, G.: Determination of the relative stereochemistry of flexible organic compounds by ab initio methods: Conformational analysis and Boltzmann-averaged GIAO ^{13}C NMR chemical shifts. Chem. Eur. J. **8**, 3240–3245 (2002)
29. (a) Ditchfield, R.: Molecular-orbital theory of magnetic shielding and magnetic susceptibility. J. Chem. Phys. **56**, 5688–5691 (1972); (b) Wolinski, K., Hinton, J.F., Pulay, P.: Efficient implementation of the gauge-independent atomic orbital method for NMR chemical-shift calculations. J. Am. Chem. Soc. **112**, 8251–8260 (1990)
30. Lauro, G., Das, P., Riccio, R., Reddy, D.S., Bifulco, G.: DFT/NMR approach for the configuration assignment of groups of stereoisomers by the combination and comparison of experimental and predicted sets of data. J. Org. Chem. **85**, 3297–3306 (2020)
31. Bertin, M.J., Saurí, J., Liu, Y., Via, C.W., Roduit, A.F., Williamson, R.T.: Trichophycins B-F, chlorovinylidene-containing polyketides isolated from a cyanobacterial bloom. J. Org. Chem. **83**, 13256–13266 (2018)
32. Grimblat, N., Zanardi, M.M., Sarotti, A.M.: Beyond DP4: An improved probability for the stereochemical assignment of isomeric compounds using quantum chemical calculations of NMR shifts. J. Org. Chem. **80**, 12526–12534 (2015)
33. Waters, A.L., Oh, J., Place, A.R., Hamann, M.T.: Stereochemical studies of the karlotoxin class using NMR spectroscopy and DP4 chemical-shift analysis: Insights into their mechanism of action. Angew. Chem. Int. Ed. **54**, 15705–15710 (2015)
34. Li, W.S., Yan, R.J., Yu, Y., Shi, Z., Mándi, A., Shen, L., Kurtán, T., Wu, J.: Determination of the absolute configuration of super-carbon-chain compounds by a combined chemical, spectroscopic, and computational approach: Gibbosols A and B. Angew. Chem. Int. Ed. **59**, 13028–13036 (2020)
35. Marcarino, M.O., Cicetti, S., Zanardi, M.M., Sarotti, A.M.: A critical review on the use of DP4+ in the structural elucidation of natural products: the good, the bad and the ugly. A practical guide. Nat. Prod. Rep. **39**, 58–76 (2022)
36. Yan, J., Kline, A.D., Mo, H., Shapiro, M.J., Zartler, E.R.: A novel method for the determination of stereochemistry in six-membered chairlike rings using residual dipolar couplings. J. Org. Chem. **68**, 1786–1795 (2003)
37. Freudenberger, J.C., Spiteller, P., Bauer, R., Kessler, H., Luy, B.: Stretched poly(dimethylsiloxane) gels as NMR alignment media for apolar and weakly polar organic solvents: an ideal tool for measuring RDCs at low molecular concentrations. J. Am. Chem. Soc. **126**, 14690–14691 (2004)
38. Schuetz, A., Junker, J., Leonov, A., Lange, O.F., Molinski, T.F., Griesinger, C.: Stereochemistry of sagittamide A from residual dipolar coupling enhanced NMR. J. Am. Chem. Soc. **129**, 15114–15115 (2007)

39. Nath, N., Schmidt, M., Gil, R.R., Williamson, R.T., Martin, G.E., Navarro-Vázquez, A., Griesinger, C., Liu, Y.: Determination of relative configuration from residual chemical shift anisotropy. J. Am. Chem. Soc. **138**, 9548–9556 (2016)
40. Nath, N., Fuentes-Monteverde, J.C., Pech-Puch, D., Rodríguez, J., Jiménez, C., Noll, M., Kreiter, A., Reggelin, M., Navarro-Vázquez, A., Griesinger, C.: Relative configuration of micrograms of natural compounds using proton residual chemical shift anisotropy. Nat. Commun. **11**, 4372 (2020). https://doi.org/10.1038/s41467-020-18093-5
41. Tzvetkova, P., Sternberg, U., Gloge, T., Navarro-Vázquez, A., Luy, B.: Configuration determination by residual dipolar couplings: accessing the full conformational space by molecular dynamics with tensorial constraints. Chem. Sci. **10**, 8774–8791 (2019)
42. Singer, S.J., Nicolson, G.L.: The fluid mosaic model of the structure of cell membranes. Science **175**, 720–731 (1971)
43. Simons, K., Ikonen, E.: Functional rafts in cell membranes. Nature **387**, 569–572 (1997)
44. Maulik, P.R., Shipley, G.G.: *N*-palmitoyl sphingomyelin bilayers: structure and interactions with cholesterol and dipalmitoylphosphatidylcholine. Biochemistry **35**, 8025–8034 (1996)
45. Veach, S., Keller, S.: Separation of liquid phases in giant vesicles of ternary mixtures of phospholipids and cholesterol. Biophys. J. **85**, 3074–3083 (2003)
46. Quinn, P.J.: Lipid-lipid interactions in bilayer membranes: married couples and casual liaisons. Prog. Lipid Res. **51**, 179–198 (2012)
47. Simons, K., Sampaio, J.L.: Membrane organization and lipid rafts. Cold Spring Harb. Perspect. Biol. **3**, a004697 (2011)
48. Murata, M., Sugiyama, S., Matsuoka, S., Matsumori, N.: Bioactive structure of membrane lipids and natural products elucidated by a chemistry-based approach. Chem. Rec. **15**, 675–690 (2015)
49. Umegawa, Y., Yamamoto, T., Dixit, M., Funahashi, K., Seo, S., Nakagawa, Y., Suzuki, T., Matsuoka, S., Tsuchikawa, H., Hanashima, S., Oishi, T., Matsumori, M., Shinoda, W., Murata, M.: Amphotericin B assembles into seven-molecule ion channels: An NMR and molecular dynamics study. Sci. Adv. **8** eabo2658 (2022).
50. Matsumori, N., Morooka, A., Murata, M.: Conformation and location of membrane-bound salinomycin-sodium complex deduced from NMR in isotropic bicelles. J. Am. Chem. Soc. **129**, 14989–14995 (2007)
51. Houdai, T., Matsumori, N., Murata, M.: Structure of membrane-bound amphidinol 3 in isotropic small bicelles. Org. Lett. **10**, 4191–4194 (2008)
52. Malabed, R., Hanashima, S., Murata, M., Sakurai, K.: Interactions of OSW-1 with lipid bilayers in comparison with digitonin and soyasaponin. Langmuir **36**, 3600–3610 (2020)
53. Li, W., Qiu, Z., Wang, Y., Zhang, Y., Li, M., Yu, J., Zhang, L., Zhu, Z., Yu, B.: Synthesis, cytotoxicity, and hemolytic activity of 6'-O-substituted dioscin derivatives. Carbohydr. Res. **343**, 2705–2715 (2007)
54. Frenkel, N., Makky, A., Sudji, I.R., Wink, M., Tanaka, M.: Mechanistic investigation of interactions between steroidal saponin digitonin and cell membrane models. J. Phys. Chem. B **118**, 14632–14639 (2014)
55. Glauert, A.M., Dingle, J.T., Lucy, J.A.: Action of saponin on biological cell membranes. Nature **196**(4858), 953–955 (1962)
56. Keukens, E.A.J., de Vrije, T., van den Boom, C., de Waard, P., Plasman, H.H., Thiel, F., Chupin, V., Jongen, W.M.F., de Kruijff, B.: Molecular basis of glycoalkaloid induced membrane disruption. Biochim. Biophys. Acta **1240**, 216–228 (1995)
57. Hu, M., Konoki, K., Tachibana, K.: Cholesterol-independent membrane disruption caused by triterpenoid saponins. Biochim. Biophys. Acta **1299**, 252–258 (1996)
58. Sudji, I.R., Subburaj, Y., Frenkel, N., García-Sáez, A.J., Wink, M.: Membrane disintegration caused by the steroid saponin digitonin is related to the presence of cholesterol. Molecules **20**, 20146–20160 (2015)
59. Stine, K.J., Hercules, R.K., Duff, J.D., Walker, B.W.: Interaction of the glycoalkaloid tomatine with DMPC and sterol monolayers studied by surface pressure measurements and brewster angle microscopy. J. Phys. Chem. B **110**, 22220–22229 (2006)

60. Armah, C.N., Mackie, A.R., Roy, C., Price, K., Osbourn, A.E., Bowyer, P., Ladha, S.: The membrane-permeabilizing effect of avenacin A-1 involves the reorganization of bilayer cholesterol. Biophys. J. **76**, 281–290 (1999)
61. Lewandowska, A., Soutar, C.P., Greenwood, A.I., Nimerovsky, E., De Lio, A.M., Jordan, H.T., Hisao, G.S., Khandelwal, A., Zhang, J., SantaMaria, A.M., Schwieters, C.D., Pogorelov, T.V., Burke, M.D., Rienstra, C.M.: Fungicidal amphotericin B sponges are assemblies of staggered asymmetric homodimers encasing large void volumes. Nat. Struc. Mol. Biol. **28**, 972–981 (2021)
62. Saitô, H., Tabeta, R., Motoko, Y.: High-resolution ^{13}C NMR study of free and metal-complexed ionophores in the solid state: Conformation and dynamics of the macrocyclic ring and the effect of intermolecular short contact of methyl groups on spin–lattice relaxation times and displacements of chemical shifts. Magn. Reson. Chem. **26**, 775–786 (1988)
63. Hayashi, T., Tsuchikawa, T., Umegawa, Y., Murata, M. Small structural alterations greatly influence the membrane affinity of lipophilic ligands: Membrane interactions of bafilomycin A1 and its desmethyl derivative bearing ^{19}F-labeling., Bioog. Med. Chem. **27,** 1677–1682 (2019)
64. Murata, M., Umegawa, Y., Hanashima, S. NMR studies on the structure and functions of small biomolecules with highly flexible conformation. Bull. Chem. Soc. Jpn. (2023) https://doi.org/10.1246/bcsj.20230166
65. Matsuoka, S., Inoue, M.: Application of REDOR NMR in natural product chemistry. Chem. Commun. 5664–5675 (2009)
66. Ondevilla, J.C., Hanashima, S., Mukogawa, A., Umegawa, Y., Murata, M.: Diosgenin-induced physicochemical effects on phospholipid bilayers in comparison with cholesterol. Bioorg. Med. Chem. Lett. **36**, 127816 (2021)
67. Ondevilla, J.C., Hanashima, A., Mukogawa, A., Garza-Miyazato, D., Umegawa, Y., Murata, M.: Effect of the number of sugar units on the interaction between diosgenyl saponin and membrane lipids. Biochim. Biophys. Acta **1865**, 184145 (2023)
68. Hanashima, S., Fukuda, N., Malabed, R., Murata, M., Kinoshita, M., Greimel, P., Hirabayashi, Y.: β-Glucosylation of cholesterol reduces sterol-sphingomyelin interactions. Biochim. Biophys. Acta **1863**, 183496 (2021)
69. Malabed, R., Hanashima, S., Murata, M., Sakurai, K.: Sterol-recognition ability and membrane-disrupting activity of ornithogalum saponin OSW-1 and usual 3-O-glycosyl saponins. Biochim. Biophys. Acta Biomembrane **1859**, 2516–2515 (2017)
70. Matsumori, M., Yasuda, T., Okazaki, H., Suzuki, T., Yamaguchi, T., Tsuchikawa, H., Doi, M., Oishi, T., Murata, M.: Comprehensive molecular motion capture for sphingomyelin by site-specific deuterium labeling. Biochemistry **51**, 8363–8370 (2012)
71. Yasuda, T., Tsuchikawa, H., Murata, M., Matsumori. N.: Deuterium NMR of raft model membranes reveals domain-specific order profiles and compositional distribution. Biophys. J. **108,** 2502–2506 (2015).
72. Scheidt, H.A., Meyer, T., Nikolaus, J., Baek, D.J., Haralampiev, I., Thomas, L., Bittman, R., Müller, P., Herrmann, A., Huster, D.: Angew. Chem. Int. Ed. **52**, 12848 (2013)
73. Lin, F., Wang, R.: Hemolytic mechanism of dioscin proposed by molecular dynamics simulations. J. Mol. Model. **16**, 107–118 (2010)
74. Montenez, J.P., Van Bambeke, F., Piret, J., Brasseur, R., Tulkens, P.M., Mingeot-Leclercq, M.P.: Interactions of macrolide antibiotics (Erythromycin A, roxithromycin, erythromycylamine [Dirithromycin], and azithromycin) with phospholipids: computer-aided conformational analysis and studies on acellular and cell culture models. Toxicol. Appl. Pharmacol. **156**, 129–140 (1999)
75. Reasor, M.J., Kacew, S.: Drug-induced phospholipidosis: are there functional consequences? Exp. Biol. Med. **226**, 825–830 (2001)
76. Andersson, A., Biverståhl, H., Nordin, J., Danielsson, J., Lindahl, E., Mäler, L.: The membrane-induced structure of melittin is correlated with the fluidity of the lipids. Biochim. Biophys. Acta **1768**, 115–121 (2007)

77. Biverståhl, H., Lind, L., Bodor, A., Mäler, L.: Biophysical studies of the membrane location of the voltage-gated sensors in the HsapBK and KvAP K(+) channels. Biochim. Biophys. Acta **1788**, 1976–1986 (2009)
78. (a) Matsumori, N., Morooka, A., Murata, M.: Detailed description on conformation and location of membrane-bound erythromycin A using small bicelle. *J Med. Chem* **49**, 3501–3508 (2006); (b) Matsumori, N., Murata, M.: 3D structures of membrane-associated small molecules as determined in isotropic bicelles. Nat Prod. Rep **27**, 1480–1492 (2010)
79. Harris, D.R., McGeachin, S.G., Mills, H.H.: The structure and stereochemistry of erythromycin A. Tetrahedron Lett. **6**, 679–685 (1965)
80. Matsunaga, S., Fusetani, N., Hashimoto, K., Wälchli, M.: Theonellamide F. A novel antifungal bicyclic peptide from a marine sponge *Theonella* sp. J. Am. Chem. Soc. **111**, 2582–2588 (1989)
81. Nishimura, S., Arita, Y., Honda, M., Iwamoto, K., Matsuyama, A., Shirai, A., Kawasaki, H., Kakeya, H., Kobayashi, T., Matsunaga, S., Yoshida, M.: Marine antifungal theonellamides target 3beta-hydroxysterol to activate Rho1 signaling. Nat. Chem. Biol. **6**, 519–526 (2010)
82. Espiritu, R., Matsumori, N., Murata, M., Nishimura, S., Kakeya, H., Matsunaga, S., Yoshida, M.: Interaction between the marine sponge cyclic peptide theonellamide A and sterols in lipid bilayers as viewed by surface plasmon resonance and solid-state ^{2}H nuclear magnetic resonance. Biochemistry **52**, 2410–2418 (2013)
83. Morris, K.F., Johnson, C.S., Jr.: Diffusion-ordered two-dimensional nuclear magnetic resonance spectroscopy. J. Am. Chem. Soc. **114**, 3139–3141 (1992)
84. Cornelio, K., Espiritu, R.A., Todokoro, Y., Hanashima, S., Kinoshita, M., Matsumori, N., Murata, M., Nishimura, S., Kakeya, H., Yoshida, M., Matsunaga, S.: Sterol-dependent membrane association of the marine sponge-derived bicyclic peptide theonellamide A as examined by ^{1}H NMR. Bioorg. Med. Chem. **24**, 5235–5242 (2016)
85. Satake, M., Murata, M., Yasumoto, T., Fujita, T., Naoki, H.: Amphidinol, a polyhydroxy-polyene antifungal agent with an unprecedented structure, from a marine dinoflagellate. Amphidinium klebsii. J. Am. Chem. Soc. **113**, 9859–9861 (1991)
86. Paul, G.K., Matsumori, N., Konoki, K., Sasaki, M., Murata, M., Tachibana, K.: Structure of amphidinol 3 and its cholesterol-dependent membrane perturbation: a strong antifungal metabolite produced by dinoflagellate, *Amphidinium klebsii.* In Harmful and Toxic Algal Blooms; Yasumoto, T., Oshima, Y., Fukuyo, Y., Eds.; UNESCO: Paris, pp 503– 506 (1996)
87. Morsy, N., Houdai, T., Konoki, K., Matsumori, N., Oishi, T., Murata, M.: Effects of lipid constituents on membrane-permeabilizing activity of amphidinols. Bioorg. Med. Chem. **16**, 3084–3090 (2008)
88. Oishi, T., Kanemoto, M., Swasono, R., Matsumori, N., Murata, M.: Combinatorial synthesis of the 1,5-polyol system based on cross metathesis: Structure revision of amphidinol 3. Org. Lett. **10**, 5203–5206 (2008). https://doi.org/10.1021/ol802168r
89. Ebine, M., Kanemoto, M., Manabe, Y., Konno, Y., Sakai, K., Matsumori, N., Murata, M., Oishi, T.: Synthesis and structure revision of the C43–C67 part of amphidinol 3. Org. Lett. **15**, 2846–2849 (2013)
90. Wakamiya, Y., Ebine, M., Murayama, M., Omizu, H., Matsumori, N., Murata, M., Oishi, T.: Synthesis and stereochemical revision of the C31–C67 fragment of amphidinol 3. Angew. Chem. Int. Ed. **57**, 6060–6064 (2018)
91. Wakamiya, Y., Ebine, M., Matsumori, N., Oishi, T.: Total synthesis of amphidinol 3: A general strategy for synthesizing amphidinol analogues and structure–activity relationship study. J. Am. Chem. Soc. **142**, 3472–3478 (2020)
92. Oishi, T.: Structure Determination, Chemical synthesis, and evaluation of biological activity of super carbon chain natural products. Bull. Chem. Soc. Jpn **93**, 1350–1360 (2020)
93. Espiritu, R.A., Matsumori, N., Tsuda, M., Murata, M.: Direct and stereospecific interaction of amphidinol 3 with sterol in lipid bilayers. Biochemistry **53**, 3287–3293 (2014)
94. Hieda, M., Tsujimura, K., Kinoshita, M., Matsumori, N.: Formation of a tight complex between amphidinol 3 and sterols in lipid bilayers revealed by short-range energy transfer. Bull. Chem. Soc. Jpn **95**, 1753–1759 (2022)

Chapter 7
Technical Basis for Nuclear Magnetic Resonance Approach for Glycoproteins

Koichi Kato, Saeko Yanaka, and Hirokazu Yagi

Abstract Significant strides have been achieved in unraveling the 3D structures of proteins; however, the structural investigation of glycoproteins presents distinctive challenges due to the unpredictable and heterogeneous nature of protein glycosylation, coupled with the intricate and flexible structures of glycoprotein glycans. Conventional bacterial expression systems are inadequate for producing glycoproteins, posing a hurdle for structural biology endeavors. Confronting these obstacles, nuclear magnetic resonance (NMR) spectroscopy, in synergy with various analytical and preparative techniques, stands out as a potent resolution. Recombinant glycoproteins can now be expressed using stable isotope labeling through various eukaryotic production platforms. High-performance liquid chromatography mapping and mass

K. Kato (✉)
Graduate School of Pharmaceutical Sciences, Nagoya City University, 3-1 Tanabe-dori, Mizuho-ku 467-8603, Nagoya, Japan
e-mail: kkatonmr@ims.ac.jp

Exploratory Research Center on Life and Living Systems (ExCELLS) and Institute for Molecular Science, National Institutes of Natural Sciences, 5-1 Higashiyama Myodaiji, Okazaki 444-8787, Japan

S. Yanaka
Exploratory Research Center on Life and Living Systems (ExCELLS), National Institutes of Natural Sciences, 5-1 Higashiyama Myodaiji, Okazaki 444-8787, Japan

Graduate School of Pharmaceutical Sciences, Kyushu University, 3-1-1 Maidashi, Higashi 812-8582, Fukuoka, Japan

H. Yagi
Graduate School of Pharmaceutical Sciences, Nagoya City University, 3-1 Tanabe-dori, Mizuho-ku 467-8603, Nagoya, Japan
e-mail: hyagi@phar.nagoya-cu.ac.jp

Exploratory Research Center on Life and Living Systems (ExCELLS), National Institutes of Natural Sciences, 5-1 Higashiyama Myodaiji, Okazaki 444-8787, Japan

S. Yanaka
e-mail: saeko-yanaka@ims.ac.jp

The NMR Society of Japan (ed.), *Experimental Approaches of NMR Spectroscopy II*,
https://doi.org/10.1007/978-981-97-6830-1_7

spectrometry enable the analysis of glycoforms, while genetic engineering of production vehicles and in vitro enzymatic reactions facilitate glycoprotein glycan remodeling. Stable-isotope-assisted NMR techniques offer a robust approach, furnishing detailed insights into the conformational dynamics and interactions of carbohydrate chains in solution. This information provides a valuable understanding of the functional mechanisms of glycoprotein glycans. Moreover, the synergy of NMR spectroscopy and computational approaches empowers us to enhance the functionality of carbohydrate chains by designing and optimizing their conformational spaces. This strategic approach lays the structural foundation for crafting glycoproteins with tailored, and even artificial, glycoforms.

Keywords Glycoprotein · Oligosaccharide · Stable isotope labeling · Eukaryotic expression system · NMR spectroscopy

7.1 Introduction

Advancements in biomolecular NMR spectroscopy have been driven by the constant development of methods for the expression of isotope-labeled biomacromolecules and novel hardware, exemplified by higher-field-strength magnets along with sophisticated measurement techniques. This progress has enabled high-throughput three-dimensional (3D)-structure determination of proteins, quantitative characterization of conformational dynamics of biomolecules on a broad range of timescales [1–3], depiction of huge protein complexes beyond 1 MDa [4], and observation of protein behaviors in intracellular environments [5]. Advances in biomolecular NMR spectroscopy have also provided a technical basis for high-throughput analyses of protein–ligand interactions using chemical libraries [6, 7]. To date, proteins subjected to NMR-based structural analysis have been conventionally produced using bacterial protein expression systems or cell-free expression systems [8]. Although more than half of proteins in nature are estimated to be modified with carbohydrate moieties [9], these recombinant proteins are not glycosylated by the aforementioned expression systems.

Carbohydrate chains have a significant impact on the physicochemical and biofunctional properties of the carrier proteins [10]. Thermal stability and solubility of proteins are governed by the carbohydrate chains, which can also affect protein conformations, and thereby control their biological functions. Moreover, glycans directly involved in interactions with other proteins, which are collectively termed lectins, mediate a variety of biological processes involved in cell–cell communications, viral and bacterial infections, immune responses, and protein-fate determination. Therefore, without knowledge of the glycan structures, one can barely understand the functional mechanisms of glycoproteins in the biological context. However, despite their biological importance, structural biology studies of glycoproteins have been falling behind. Indeed, atomic coordinates of glycoproteins represent only 5%

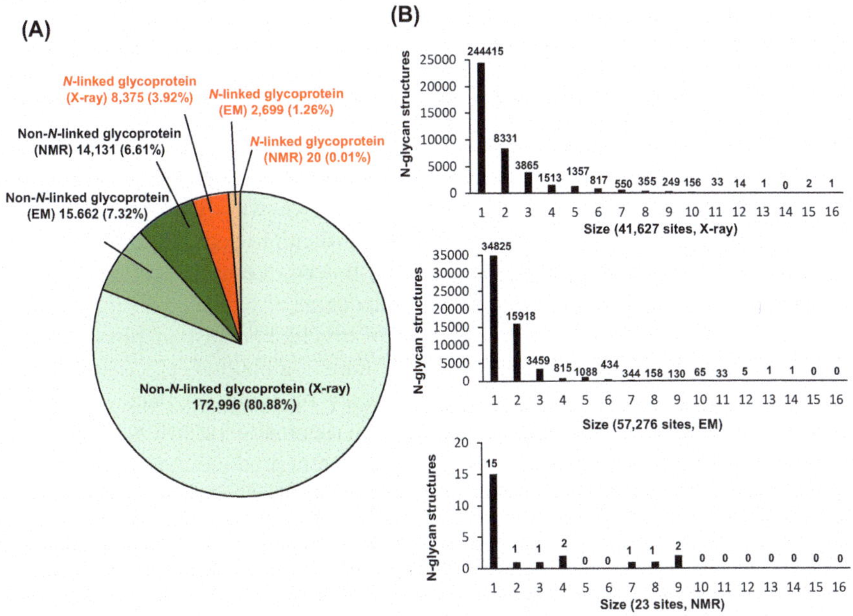

Fig. 7.1 Current statistics regarding Protein Data Bank entries on *N*-linked glycoprotein structures (data were extracted on Jan 4, 2024). **a** Number and percentage of X-ray crystallography, EM, and NMR structures of *N*-linked glycoproteins. **b** Size distribution of *N*-glycans among glycoprotein structures revealed by X-ray crystallography (top), EM (middle), and NMR spectroscopy (bottom)

of the total entries in the Protein Data Bank, and most of these provide structural information on only one or two glycan residues (Fig. 7.1).

Besides the traditional application for determination of the covalent structures of oligosaccharides [11–15], NMR spectroscopy can play versatile roles in characterizing dynamic conformations and interactions of oligosaccharides. In this chapter, we will outline the major obstacles in studying glycoproteins via NMR spectroscopy and the current state-of-the-art methodology for NMR-based structural glycobiology.

7.2 Enigmatic Aspects of Carbohydrate Structures

Needless to say, any structural biology studies presuppose the covalent structures of the target biomolecules, e.g., the amino acid sequences of proteins. An obvious example can be seen in structural genomics, in which accumulating genome sequence data has prompted comprehensive structural determination of the encoded proteins. Employing the extensive repository of accumulated 3D structural data, neural network-based machine learning methods have significantly improved the accuracy of predicting protein 3D structures based solely on amino acid sequences

[16, 17]. From this perspective, one of the biggest problems in the structural biology of glycoproteins is that the covalent structures of glycans are not directly encoded in the genome. Although the necessary sequence for eukaryotic *N*-glycosylation, i.e., Asn-X-Ser/Thr(/Cys) in which X is not proline, is widely known, this is insufficient [10]. Furthermore, no consensus motif has been established for other glycosylation sites, despite a certain tendency of amino acid sequences in *O*-glycosylation sites [10]. Hence, glycosylation sites of a protein with a known amino acid sequence remain unpredictable. Moreover, it is much more difficult to deduce the sequence of carbohydrate moiety at each glycosylation site. Indeed, the carbohydrate moieties displayed at one site of protein often have a high degree of sequence divergence and heterogeneity, depending not only on the protein structure but also on physiological and pathological conditions [10]. In the case of immunoglobulin G (IgG), which is a major serum glycoprotein, its glycosylation depends on aging, pregnancy, delivery, and onset of connective tissue disease typified by rheumatoid arthritis [18–25]. Of course, recombinant glycoproteins produced in different production vehicles exhibit different glycoforms and even cell culture conditions affect the glycosylation profiles [26–29].

The intrinsic complexity of oligosaccharide structures brings further difficulty in structural analysis of glycoproteins. The building blocks of sugar chains, i.e., monosaccharides, are structurally simple with little variation in their functional groups in comparison with amino acids [10]. However, unlike polypeptide chains and nucleic acids, sugar residues are connected through multiple types of glycosidic linkages, potentially giving rise to a wide variety of isomeric and branching structures. Indeed, the number of all possible isomers of a tetrasaccharide composed of four different monosaccharides was calculated to be 34,560, whereas only 24 different tetrapeptides are made from four different amino acid residues [30]. Glycosidic linkage isomerism is critical to glycoprotein functions, as best exemplified by infections with influenza viruses, which bind host cells by recognizing specific sialyl linkage isomers [31, 32]. Paradoxically, the structural simplicity (or similarity) of sugar residues precludes structural analyses: isomeric structures such as glucose, mannose, and galactose cannot be distinguished by conventional mass spectrometry (MS), whereas the little diversity of the functional groups causes a chemical shift degeneracy in NMR spectroscopy.

Another difficulty for structural analysis of glycoproteins is associated with the motional freedom of glycosidic linkages in the carbohydrate moieties, giving rise to conformational fluctuations. This flexible property of glycans, along with their structural heterogeneity, hinders glycoprotein crystallization, leading to ambiguities in interpreting electron densities even if successful [33], a concern also present in structural analyses by cryo-electron microscopy (cryo-EM) [34] (Fig. 7.1). Moreover, a tricky problem is that mutational deglycosylation can cause unexpected or even undesirable modifications of the proteins, such as sulfation and nicks [35, 36].

All these characteristics of the carbohydrate chains make structural analysis of glycoproteins (and even oligosaccharides) a challenging task. Namely, one who attempts to perform structural glycobiology must conduct a series of analyses, from the determination of carbohydrate sequences to characterization of the conformational dynamics of target glycoproteins including their glycans. NMR spectroscopy plays a central role in addressing these issues because it can, in principle, characterize dynamic structures of heterogenous biomacromolecules in solution. However, it should be noted that a variety of preparative and analytical techniques that complement the NMR approach are indispensable for conducting structural analysis of glycoproteins. In the following sections, we will describe how one can perform structural glycobiology by NMR spectroscopy aided by these complementing methods.

7.3 Expression of Isotope-Labeled Glycoproteins

As glycoproteins cannot be produced by conventional protein expression systems such as *Escherichia coli*, appropriate eukaryotic vehicles have to be chosen for the production of NMR quantities of glycoproteins with stable isotope labeling. Typical choices would be cell lines derived from mammals. Several groups developed metabolic isotope labeling of recombinant proteins using mammalian cells, including Chinese hamster ovary (CHO) cells, mouse hybridomas, and adenovirus vector-based mammalian cells [37–41]. In particular, uniform and amino acid-selective ^{13}C and/or ^{15}N labeling techniques were established using IgG as a model glycoprotein, yielding several NMR applications [39, 41]. The composition of the serum-free medium used for the production of isotopically labeled IgG is shown in Table 7.1. Amino acid-selective labeling is accomplished by replacing specific amino acid(s) with their isotope-labeled counterpart(s), although this often suffers from metabolic isotope scrambling and dilution. To suppress the undesirable metabolic processes, metabolic inhibitors such as β-chloro-L-alanine are often added to the medium [42].

For uniform ^{13}C/^{15}N labeling of a glycoprotein, it is necessary to replace all the metabolic precursors with the isotope-labeled ones, which are commercially very expensive. A more practical approach is to use an isotope-labeled amino acid mixture derived from algae with a modification of the medium composition (Table 7.1). Using this protocol, uniformly ^{13}C/^{15}N-labeled IgG glycoproteins were produced through mammalian cell culture. Figure 7.2A shows a ^{1}H-^{15}N HSQC spectrum of ^{13}C/^{15}N-labeled human IgG-Fc obtained from CHO cells [43]. A significant limitation in employing mammalian cells is the high production costs attributed to the substantial amount of labeled L-glutamine necessary for their cultivation. To tackle this challenge, a cost-effective method has been devised for uniform isotope labeling by cultivating mammalian cells under glutamine-free conditions, achieved through the co-expression of glutamine synthase [44].

Isotope labeling of the carbohydrate moieties can be simply achieved by using ^{13}C-labeled glucose as a metabolic precursor [45–47]. ^{13}C- and/or ^{15}N-labeled glucosamine (but not *N*-acetylglucosamine) can be used for the metabolic labeling of GlcNAc residues in the carbohydrate moieties of glycoproteins [45, 48].

Table 7.1 Composition of the serum-free medium used for metabolic labeling of IgG glycoproteins (mg/L)

		Algal mixture	–	+
NaCl	6208	L-Arginine · HCl	76.1	–
KCl	388	L-Arginine	97.0	72.0
$CaCl_2$ (unhyd)	97	L-Asparagine · H_2O	42.5	42.5
$Ca(NO_3)_2$ (unhyd)	33.7	L-Aspartic acid	9.7	–
$MgSO_4$ (unhyd)	69.0	L-Cystine · $2H_2O$	31.5	31.5
Na_2HPO_4 (unhyd)	388.5	L-Cysteine · HCl · H_2O	16.7	16.7
NaH_2PO_4 (unhyd)	55.8	L-Glutamic acid	9.7	–
D-Biotin	0.11	L-Glutamine	450	450
D-Ca Pantothenate	0.61	L-Histidine	27.7	15.5
Choline chloride	26.5	Glycine	9.9	–
Folic acid	0.97	L-Hydroxyproline	9.7	–
i-Inositole	18.0	L-Isoleucine	49.5	–
Nicotinamide	0.97	L-Leucine	49.5	13.1
p-aminobenzoic acid	0.485	L-Lysine · HCl	54.8	–
Pyridoxal · HCl	0.485	L-Methionine	14.6	–
Pyridoxine · HCl	0.485	L-Proline	14.7	–
Riboflavin	0.146	L-Serine	29.6	–
Thiamin · HCl	0.97	L-Threonine	48.0	–
Vitamine B12	0.0037	L-Valine	47.0	–
Glutathione (red)	0.485	L-Phenylalanine	22.8	–
Choline Bitartrate	0.873	L-Tryptophan	7.3	7.3
Putrescine · $2H_2O$	0.0125	L-Tyrosine	27.2	16.9
Sodium pyruvate	110.0	Algal mixture	–	867
Thymidine	0.0125			
Hypoxanthine	0.025			
Monoethanolamine	20.0			
Sodium Selenite	0.0017			
Kanamycin sulfate	60.0			
Phenol red	5.0			
Glucose	2000			
Sodium Succinate	50.0			
Succinic acid	37.5			
Bovine Transferrin (holo)	5.0			
Bovine Insulin	10.0			
HEPES	3570			
Sodium Bicarbonate	1400			

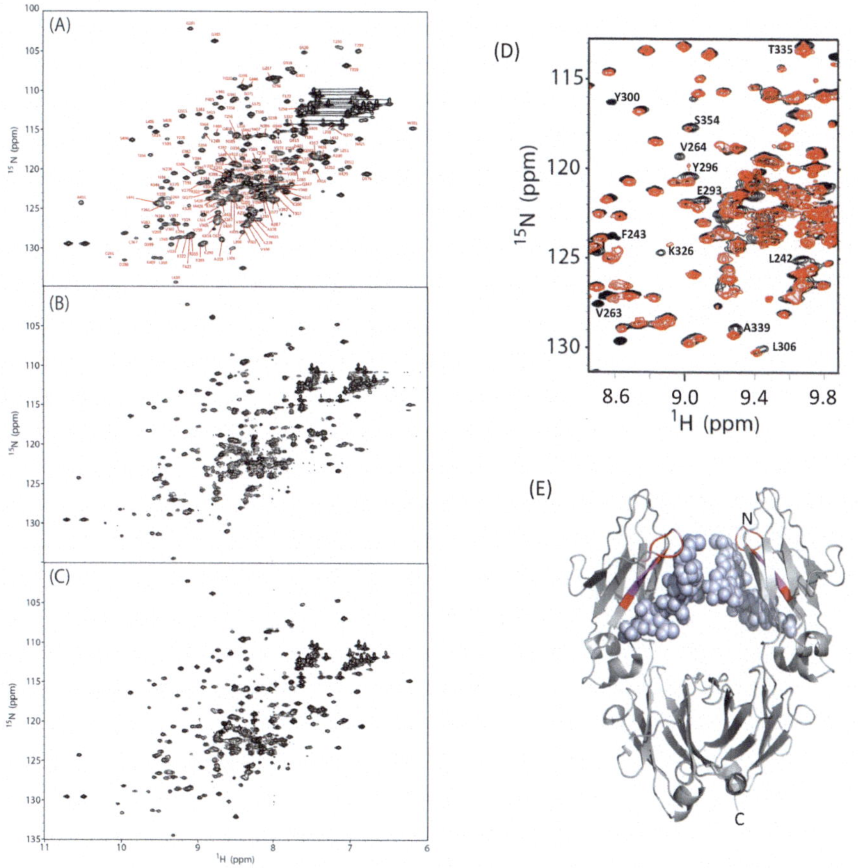

Fig. 7.2 ^{1}H-^{15}N HSQC spectra of isotope-labeled human IgG1-Fc are shown obtained from **a** CHO cells, **b** transgenic tobacco plants, and **c** silkworms. **d** Superposition of ^{1}H-^{15}N HSQC spectra of the uniformly ^{15}N/^{13}C-labeled IgG1-Fc expressed by transgenic silkworm (red) and CHO cells (black). **e** Mapping of the amino acid residues with observable chemical shift changes [$(0.2\delta_N{}^2 + \delta_H{}^2)^{1/2}$ > 0.1 ppm and > 0.2 ppm in pink and red, respectively] on the crystal structure [PDB code: 3AVE] [79]. The *N*-glycans are represented by gray sphere. The molecular graphics were generated with PyMOL [132]. In A, backbone assignments are annotated by the resonance peaks with one-letter amino acid codes and the sequence numbers. Side-chain resonances corresponding to NH_2 amides are connected by horizontal lines. Adapted from the references [43, 49, 50] with the modification

In addition to mammalian cells, metabolic isotope labeling techniques have been developed for NMR applications in various eukaryotic expression systems including yeast, *Dictyostelium discoideum*, insects, and plants [49–54]. As an insect cell-based production vehicle, Sf9 cells infected with recombinant baculoviruses are widely used for the expression of membrane glycoproteins for NMR as well as crystallographic analyses [51, 55, 56]. Silkworms have also been demonstrated as promising production vehicles for isotope-labeled recombinant glycoproteins [50]. In this approach, baculovirus-infected fifth instar larvae were reared with an artificial diet containing a protein mixture derived from *Candida utilis* grown in a culture medium containing ^{15}N-labeled ammonium sulfate as a nitrogen source. ^{15}N-enrichment level of recombinant human IgG achieved in this system was approximately 80%. By strategically timing the infection and harvesting from silkworm pupae, it is also possible to enhance the yield of isotope-labeled recombinant glycoproteins [57].

Plant expression systems are now attracting attention from pharmaceutical interests. Transgenic tobacco (*Nicotiana benthamiana*) was tested for expression of ^{15}N-labeled IgG [49]. Four-week-old seedlings of transgenic tobacco were placed in a drip hydroponic system and cultivated for 49 days with isotopically labeled Murashige and Skoog medium containing potassium nitrate ($K^{15}NO_3$) and ammonium nitrate ($^{15}NH_4{}^{15}NO_3$) as the major nitrogen sources. The degree of ^{15}N-enrichment of recombinant human IgG thus produced was in the approximately 50%–60% range.

Figure 7.2 compares the ^{1}H-^{15}N HSQC spectra of recombinant human IgG1-Fc expressed by transgenic tobacco and silkworm as well as CHO cells [43, 49, 50]. Although the spectra are similar, significant chemical shift differences were observed for the amino acid residues in close proximity to *N*-glycans. The plant-derived IgG1-Fc exhibits α1,3-linked fucose and β1,2-linked xylose residues in the common conserved $Man_3GlcNAc_2$ core, whereas the insect cell-derived glycoprotein expresses paucimannose-type glycans as the major components (Fig. 7.3). Because these glycoforms are not expressed in mammalian vehicles, the observed spectral differences could be attributed to microenvironmental alterations caused by differential glycosylation.

Yeast can also be employed for expression of isotope-labeled recombinant glycoproteins [52, 53, 58]. For example, *Pichia pastoris* strain GS115 was used for producing ^{15}N-labeled human chorionic gonadotropin and human follicle-stimulating hormone [52]. From cultures of *P. pastoris* using $^{15}NH_4Cl$/glucose-glycerol-methanol as nitrogen and carbon sources, human chorionic gonadotropin was obtained with a ^{15}N enrichment of 70%, with modification by the yeast-strain specific high-mannose type oligosaccharides ranging from $Man_8GlcNAc_2$ to $Man_{11}GlcNAc_2$. Kamiya et al. described a method for the overexpression of uniformly ^{13}C-labeled decasaccharides M8B and undecasaccharide M9 using genetically engineered *Saccharomyces cerevisiae* cells grown in a medium containing ^{13}C-labeled glucose [53, 58, 59].

Deuteration is effective to suppress dipolar relaxation for dealing with larger biomacromolecules [60, 61]. However, one major problem in eukaryotic expression systems is difficulty in metabolic deuteration using heavy water because the cells cannot survive in such a milieu. Successful protocols have been reported for the deuteration of recombinant proteins with high deuteration degrees and sufficient expression yields [56, 62, 63]. These protocols involve using insect cells cultured in a medium containing deuterated yeast extract or deuterated algal amino acid mixture. This line of study will open up new avenues for NMR analyses of larger glycoproteins expressed by eukaryotic expression systems.

Although *E. coli* lacks a protein glycosylation pathway, protein *N*-glycosylation has been found in a variety of bacteria including *Campylobacter jejuni*, which

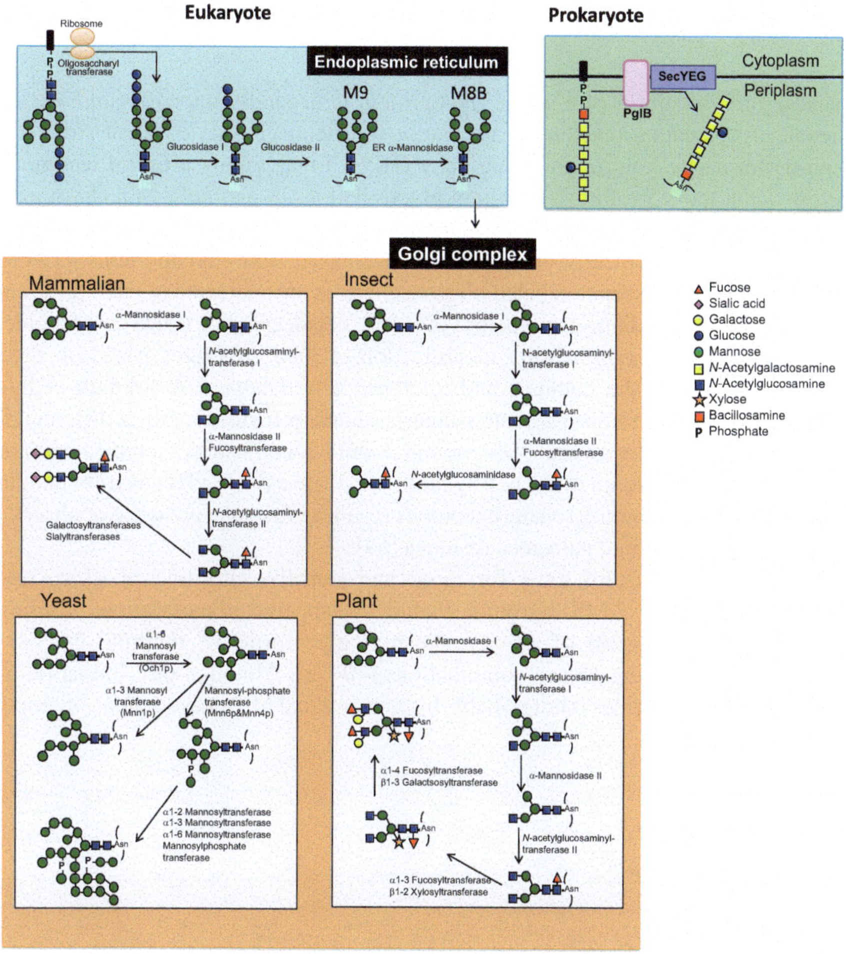

Fig. 7.3 *N*-glycan processing pathways as the basis of glycoform-engineering strategies. Schematic drawing of the *N*-glycan processing pathway in the ER common to mammalian cells, insects, yeast, and plants, and the distinct pathways in the Golgi complex along with a prokaryotic *N*-glycan pathway exemplified by that in *C. jejuni*. Adapted from the reference [127] with the modification

expresses proteins modified with the *N*-linked heptasaccharide, GalNAcα1-4GalNAcα1-4(Glcβ1-3)GalNAcα1-4GalNAcα1-4GalNAcα1-3Bac [64] (Fig. 7.3). A $^{13}C/^{15}N$-labeled recombinant protein produced in *E. coli* was successfully modified in vitro with this oligosaccharide by using enzymes from *C. jejuni* [65].

7.4 Glycosylation Profiling

Because the glycoform of a given glycoprotein is generally unpredictable, experimental glycosylation profiling is necessary for identifying its covalent structure. High-performance liquid chromatography (HPLC) mapping is a useful technique for the quantitative profiling of protein glycosylation. In particular, Takahashi and coworkers have developed 3D HPLC mapping using pyridyl-2-amination of *N*-linked oligosaccharides chemically or enzymatically cleaved from glycoproteins [66–68] (Fig. 7.4). The fluorescence-labeled oligosaccharides are subjected to three different HPLC columns, i.e., diethylaminoethyl (DEAE), octadecylsilica (ODS), and amide, in a sequential manner, enabling identification of their structures based on their elution positions on the columns and reference to a database. A strength of this method is the ability to discriminate isomeric oligosaccharides by their differential elution times, primarily on the ODS column. Figure 7.4a illustrates a typical elution profile on an ODS column of the pyridylamino derivatives of *N*-glycans from human IgG1-Fc, which exhibit microheterogeneities resulting from the presence or absence of non-reducing terminal galactose residues [69].

MS also offers useful tools for rapid and sensitive profiling of glycosylation of glycoproteins. MS analyses enable site-specific glycosylation profiling as well as determination of entire molecular glycoforms of multiple glycosylated proteins including glycan combinations [70–75]. Although discrimination of isomeric oligosaccharides is not possible by conventional MS measurement, attempts

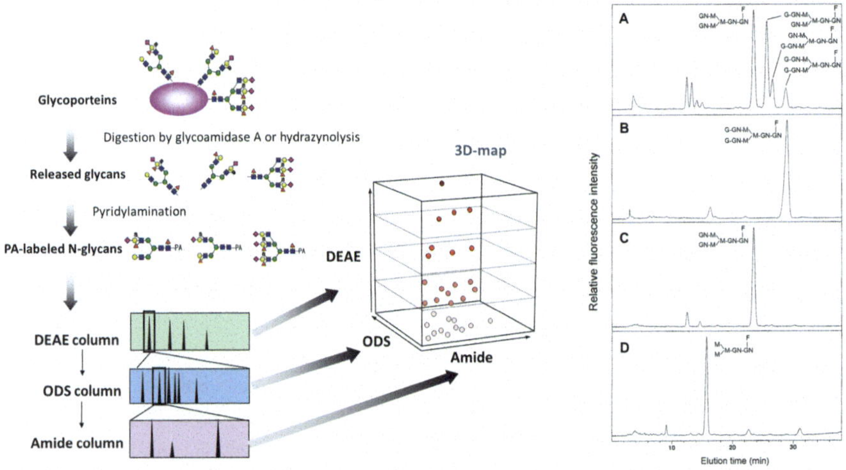

Fig. 7.4 Scheme of *N*-glycosylation profiling based on the 3D HPLC map (left) and elution profiles on an ODS column of pyridylamino derivatives of the *N*-linked oligosaccharides released from the human IgG1-Fc glycoforms (right). **a** nontreated Fc, **b** Fc(G2), **c** Fc(G0), and **d** Fc(M3). Key: F, fucose; G, galactose; GN, *N*-acetylglucosamine; M, mannose. Partially adapted from the reference [69] with the modification

have been made to achieve it by analyzing the fragmentation patterns observed in multistage tandem mass (MS^n) spectra [76–78].

7.5 Remodeling of Glycoprotein Glycoforms

Heterogeneous glycosylation hampers unambiguous structural studies of glycoproteins, making structural biologists *glycophobics*. To overcome this problem, various techniques have been applied for remodeling glycoprotein glycoforms to reduce glycan heterogeneity. Glycoform remodeling is also useful for systematic analyses of the structure–function relationships of glycoproteins, addressing the functional properties of individual glycoforms.

Glycosidase treatment has been traditionally used for trimming glycoprotein glycans. Glycoproteins with a uniform glycan structure can also be prepared by in vitro enzymatic attachment of non-reducing terminal sugar residue(s) [46, 69]. Figure 7.4b and c shows elution profiles of *N*-glycans derived from digalactosyl (G2) and agalactosyl (G0) glycoforms of human IgG1-Fc prepared by enzymatic galactosylation and degalactosylation, respectively. The IgG1-Fc glycans could be further trimmed into thc trimannosyl core $FucGlcNAc_2Man_3$ by *N*-acetylhexosaminidase treatment (Fig. 7.4d) and subsequently into the Fucα1 $\rightarrow$ 6GlcNAc disaccharide (FGN) with Endo D treatment. The glycan could be completely eliminated by PNGase F treatment. Figure 7.5 compares the ^{1}H-^{15}N HSQC spectra of a series of Fc glycoforms (G2, G0, M3, FGN, and deglycosylation), displaying progressive spectral changes of the polypeptide backbone of Fc upon stepwise trimming of the glycans [69]. The results indicated that removal of the cleavage at the GlcNAcβ1-4GlcNAc linkage of the Fc glycans conformationally affected areas proximal to the hinge region, including the Fc γ receptor-binding sites, thereby impairing receptor binding. Therefore, the carbohydrate moieties of Fc play crucial roles in maintaining the structural integrity of the functional sites of IgG.

Another approach for controlling glycosylation is the genetic engineering of glycoprotein-producing cells by knocking out genes responsible for a specific process(es) during glycan formation. For example, CHO cells lacking FUT8, an enzyme catalyzing the transfer of fucose from GDP-fucose to GlcNAc in an α-1,6 linkage, produce a completely non-fucosylated glycoform of IgG [44, 79], which has a higher affinity for Fcγ receptor III on natural killer cells and consequently exerts an increased magnitude of antibody-dependent cell-mediated cytotoxicity (ADCC) [80–82]. Figure 7.6 compares the ^{1}H-^{15}N HSQC spectra of the fucosylated and non-fucosylated forms of IgG-Fc labeled with ^{15}N at the amide groups of tyrosine residues, showing chemical shift perturbation and line-broadening of the peak originating from Tyr296, which is proximal to the fucose residue (vide infra) [79].

Genetic engineering of *S. cerevisiae* by deletion of genes involved in *N*-glycan processing has enabled overexpression of homogeneous high-mannose-type oligosaccharides M9 and M8B with ^{13}C labeling [53, 58] (Fig. 7.7). Several attempts have been made to produce recombinant proteins with human-like glycosylation in

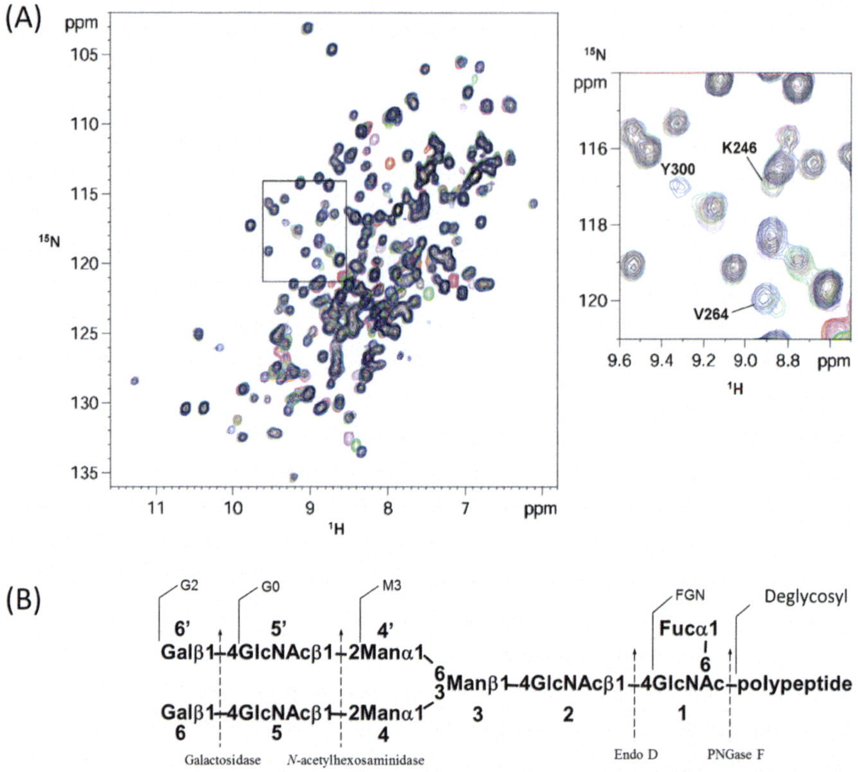

Fig. 7.5 **a** ^{1}H-^{15}N HSQC spectral changes upon trimming of the carbohydrate chains of uniformly ^{13}C/^{15}N-labeled human IgG1-Fc. The spectra of the different glycoforms of Fc with different colors were superimposed. Black, Fc(G2); blue, Fc(G0); green, Fc(M3); orange, Fc(FGN); and red, deglycosylated Fc. **b** The carbohydrate sequence attached to Asn297 of human IgG1-Fc. The arrows indicate the cleavage sites of the glycosidases used and the resultant glycoforms are also shown. Adapted from the reference [69] with permission

yeast [83–85]. Currently, sialylated biantennary *N*-glycans can be expressed by engineered *P. pastoris*, in which four genes responsible for yeast-specific glycosylation were knocked out and 14 heterologous genes involved in the biosynthesis of human-type glycosylation were introduced [84]. Furthermore, IgG modified with complex-type *N*-glycans could be expressed in silkworms by co-expression of human β-1,2-*N*-acetylglucosaminyltransferase II and β1,4-galactosyltransferase [86]. This genetically engineered production vehicle may be used for expression of isotope-labeled recombinant glycoproteins for NMR studies.

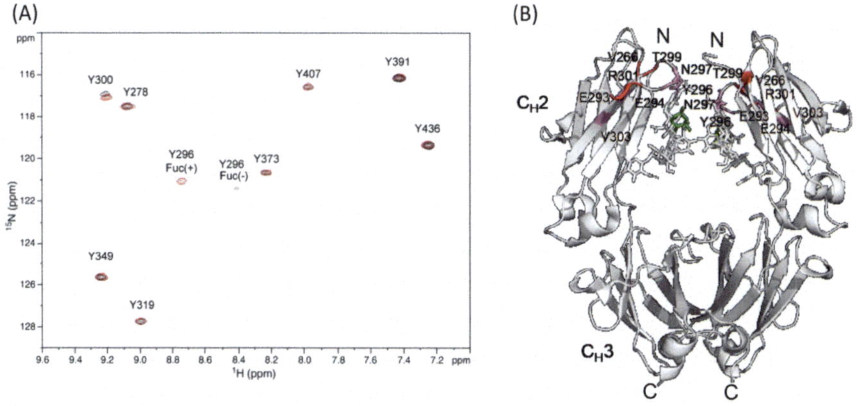

Fig. 7.6 **a** ^{1}H-^{15}N HSQC spectral comparison between the fucosylated (red) and Fuc nonfucosylated (black) glycoforms of ^{15}N-Tyr-labeled human IgG1-Fc. Tyr296 of non-fucosylated Fc exhibits a significant broadening of the HSQC peak. **b** Mapping on the crystal structure of fucosylated IgG-Fc of the amino acid residues showing the chemical shift difference between fucosylated and non-fucosylated IgG-Fc. The chemical shift differences are quantified for each residue according to the Eq. $(0.2\delta_N{}^2 + \delta_H{}^2)^{1/2}$, where δ_N and δ_H represent the differences in nitrogen and proton chemical shifts between fucosylated and non-fucosylated IgG-Fc. The amino acid residues showing chemical shift differences [$\Delta\delta$> 0.1 ppm] and [$\Delta\delta$> 0.05 ppm] are colored in red and pink, respectively, on the crystal structure of IgG1-Fc [PDB code: 3AVE] [79], respectively. Fucose residues are shown in green. The *N*-glycans are represented by stick models. The molecular graphics were generated with PyMOL [132]. Adapted from the references [44, 79] with permission

7.6 Spectral Observations and Assignments

Carbohydrate NMR spectral analyses are often hampered by peak overlapping, particularly when dealing with liberated oligosaccharides or solvent-exposed glycans. A straightforward approach for solving this problem is to employ a higher magnetic field so as to increase spectral resolution [15, 87, 88]. Line narrowing of oligosaccharide CH peaks is expected in a higher magnetic field, because in systems where dipole–dipole interaction is a dominant relaxation mechanism, T_2 longevity of the molecules with an appropriate motional correlation time increases in the higher field [89, 90]. Paramagnetic lanthanide probes can also be used for improving chemical shift dispersion. Figure 7.7c illustrates HSQC spectra of high-mannose-type undecasaccharide M9 with a lanthanide chelating tag covalently attached at its reducing terminus with La^{3+} and Tm^{3+} [91]. The overlapping peaks originating from the three non-reducing terminal mannose residues exhibit different pseudo-contact shifts (PCSs), dissolving the chemical shift degeneracy because their spatial arrangements are different with respect to the reducing terminus. PCS thus provides conformational information on the oligosaccharides (vide infra).

Although scalar connectivity-based techniques are applicable for spectral assignments of glycoprotein polypeptide chains, sequential assignments of the glycan remain partially dependent on nuclear Overhauser effect (NOE) connectivities.

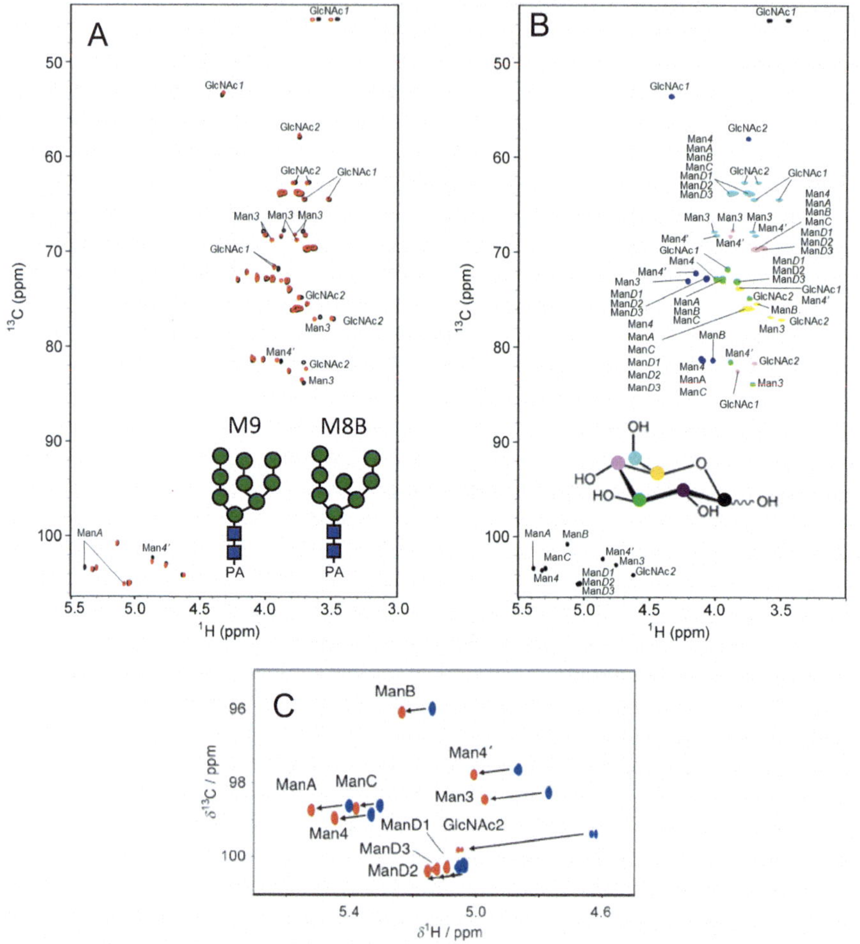

Fig. 7.7 NMR spectra of the high-mannose-type undecasaccharide M9. **a** ^{1}H-^{13}C HSQC spectra of M9 (black) uniformly labeled with ^{13}C compared with that of M8B (red). **b** ^{1}H-^{13}C HSQC spectra of M9, metabolically ^{13}C-labeled with D-[1-^{13}C]glucose (black), D-[2-^{13}C]glucose (blue), D-[3-^{13}C]glucose (green), D-[4-^{13}C]glucose (magenta), D-[5-^{13}C]glucose (orange) or D-[6-^{13}C]glucose (cyan). The six spectra were superposed and the ^{13}C-labeled positions in the glucose isotopomers used as metabolic precursors are shown with circles in the same colors as the corresponding spectra. **c** ^{1}H-^{13}C HSQC spectra of M9, modified with a lanthanide chelating tag with Tm^{3+} (red) and La^{3+} (blue). Chemical shift differences induced by PCS are indicated by arrows. Adapted from the references [53, 58, 91] with the permission

Therefore, selective isotope labeling of the carbohydrate moieties facilitates spectral assignments. Metabolic labeling using appropriate isotope-labeled metabolic precursors enables residue-specific and/or position-specific ^{13}C incorporation [40, 41]. Figure 7.7b shows the superposition of six ^{1}H-^{13}C HSQC spectra of M9 labeled with ^{13}C at different positions using six different glucose isotopomers as metabolic

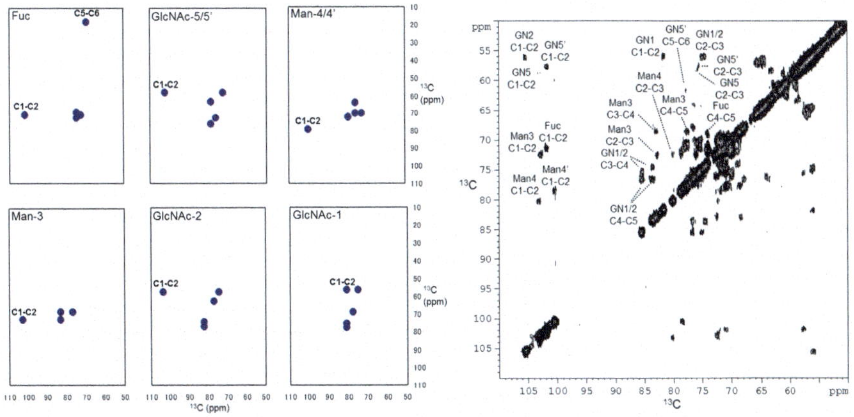

Fig. 7.8 Two-dimensional ^{13}C–^{13}C NOESY patterns expected for the sugar residues in a biantennary complex-type oligosaccharide (left) and the oligosaccharide region of the two-dimensional ^{13}C–^{13}C NOESY spectrum of ^{13}C-labeled mouse IgG2b–Fc (right). Adapted from the reference [96] with the modification

precursors [58]. Peak classification into the sugar residue types can be made based on differential protonation levels during metabolic labeling using perdeuterated ^{13}C-glucose as a metabolic precursor [53, 92]. In vitro enzymatic attachment of an isotope-labeled sugar onto the non-reducing end of the glycan is also useful for selective observation of NMR signals originating from the terminal sugar residues [45, 93–95]. Yamaguchi et al. proposed that ^{13}C-detected ^{13}C-^{13}C NOE spectroscopy (NOESY) experiments are useful for spectral assignments of large glycoproteins because the observed peaks can be easily classified into monosaccharide types with specific linkage configurations by inspecting the intra-residue ^{13}C-^{13}C connectivity patterns (Fig. 7.8) [96].

Although carbohydrate hydroxyl groups are functionally crucial, as exemplified by hydrogen bonding, direct observation of their ^{1}H NMR signals is extremely difficult in aqueous solution. An indirect approach utilizes the H/D isotope effect on the ^{13}C chemical shift at ^{13}C-OH groups to estimate the exchange rate for OH with the solvent in a 50/50 H_2O/D_2O solution at steady state [97, 98]. This technique is applicable for the estimation of the NH exchange rate of the acetamide group of *N*-acetylglucosamine residues of a glycoprotein with appropriate ^{13}C labeling (Fig. 7.9) [39]. Freedberg and coworkers used a ^{1}H-^{13}C HSQC – total correlated spectroscopy (TOCSY)-based technique to measure OH exchange rates in a 50/50 H_2O/D_2O solution, enabling quantitative characterization of transient hydrogen bonds involving the hydroxyl groups [99, 100]. Significant advancements have also been achieved in NMR studies focused on directly observing the hydroxyl groups of glycans, enabling the capture and examination of hydrogen bonds formed within glycans or between oligosaccharides and lectins [100, 101].

Upon secretion, glycoproteins frequently operate in the extracellular fluid. The application of stable isotope labeling enables the selective observation of NMR

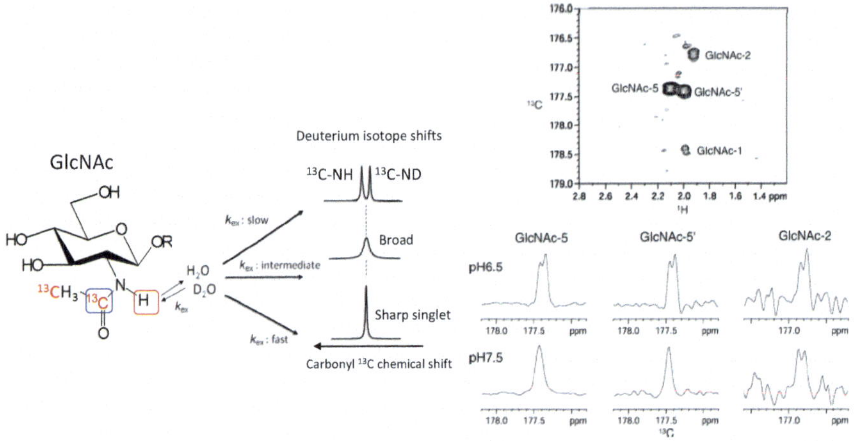

Fig. 7.9 Schematic view of deuterium—hydrogen exchange of the amide group of GlcNAc residue on the line shape of the carbonyl ^{13}C resonance deu to deuterium isotope shift and (left) and 400 MHz 2D H(C)CO spectra of IgG-Fc showing the peaks from the GlcNAc acetamide groups. The Fc sample was dissolved in a sodium phosphate buffer H_2O:D_2O = 1:1. The nomenclature of oligosaccharide residues is represented in Fig. 7.5b. Adapted from the reference [39] with permission

signals originating from the target glycoprotein within multimolecular crowding environments, such as serum. This capability has been employed to explore interactions between antibodies and serum components [102–104] On the flip side, glycoproteins, exemplified by antibodies, play a crucial role in the biopharmaceutical field. There is an expectation to utilize NMR for assessing higher-order structures of therapeutic antibodies. Achieving this objective requires conducting NMR measurements on non-labeled samples, and ongoing investigations are being carried out in this direction [105]

7.7 NMR Analyses of Dynamic Conformations and Interactions of Oligosaccharides

As in the case of other biomolecules, conformations of oligosaccharides are primarily characterized based on NOE data complemented with those of *J* coupling and residual dipolar coupling [106–108]. NOE data are also used for probing carbohydrate-protein interactions. Figure 7.10a shows intramolecular NOE connectivities observed between the core fucose of biantennary complex-type glycan and the aromatic ring of its spatially proximal tyrosine residue (Tyr296) in human IgG1-Fc. Core fucosylation precludes optimum binding of IgG1 to Fcγ receptor IIIa, which is mediated by the Tyr296 side chain, and thereby compromises ADCC activity [80, 109]. This is, at least partially, attributed to steric masking of the functionally important tyrosine by the fucose. In the absence of the core fucose, Tyr296 gains motional freedom as exemplified by the exchange line broadening of the ^{1}H-^{15}N HSQC peak originating from this tyrosine [79].

Glycoprotein recognition by lectins can be investigated by observing intermolecular NOEs using isotope-labeled oligosaccharides or glycopeptides [110–112]. Figure 7.10b shows a ^{15}N-edited NOESY spectrum indicating intermolecular NOE connectivities between the carbohydrate-binding domain of Fbs1, a cytosolic lectin that functions as the substrate recognition part of a multi-subunit ubiquitin ligase and its cognate glycopeptide cleaved from metabolically ^{13}C/^{15}N-labeled IgG [110]. In this specific case, the lectin contacts not only the innermost part of the glycan but also the side chain of the glycosylated asparagine. A glycan-polypeptide junction of glycoprotein is usually shielded from the solvent by spatially surrounding amino acid residues but can be exposed in its misfolded state and targeted for ubiquitin-mediated proteasomal degradation. Fbs1 captures the junction and protects it against

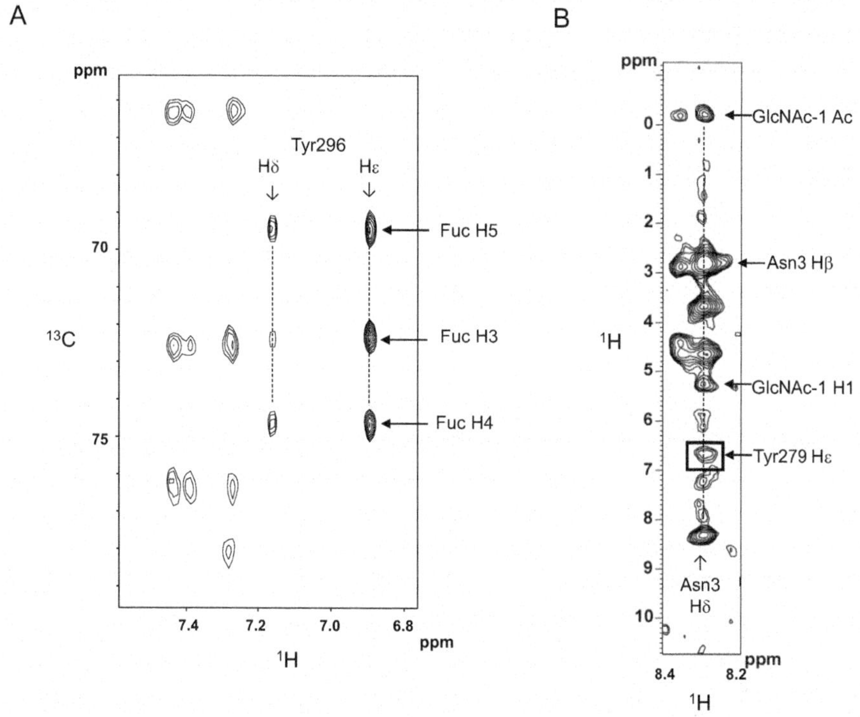

Fig. 7.10 Parts of **a** 2D HSQC-NOESY spectrum of human IgG1-Fc (metabolically labeled with [$^{13}C_6$]glucose) exhibiting intramolecular NOE connectivities between Tyr296 and the core fucose residue and **b** ^{15}N-edited NOESY spectrum of the ^{15}N-labeled Fc-derived heptapeptide carrying $Man_3GlcNAc_2$ bound to the carbohydrate-binding domain of Fbs1 exhibiting intermolecular NOE peak (boxed) between Tyr279 Hε (Fbs1) and Asn3 Hδ (glycopeptide). Adapted from the references [110] with permission

attack by the cytosolic deglycosylation enzyme during ubiquitination of glycoprotein substrates. Because carbohydrate-lectin interactions are typically weak, with dissociation constants in the submillimolar range, transferred NOE and saturation transfer difference experiments are effective to characterize such weak interactions [113–115].

NOE-based conformational analysis of carbohydrate chains is hampered by low proton density, in comparison with that of protein. Moreover, the conformational dynamics of carbohydrate chains hamper the quantitative interpretation of NOE data. In consideration of these issues, paramagnetism-assisted NMR approaches have been applied to the conformational analysis of oligosaccharides [116–118]. In particular, PCS data, obtained employing paramagnetic lanthanide probes, are used as a source of long distance (as far as 40 Å) information independent of the correlation time of the heterogeneous internal motion of oligosaccharides. Paramagnetic relaxation enhancements have also been used to characterize the conformation and metal binding of oligosaccharides and their clusters [119–121].

Because oligosaccharides are extremely flexible molecules, NMR provides information averaged over their dynamic conformational ensembles. To address this issue, NMR data are treated as population-weighted averages of the possible conformers and reproduced by a population-optimized combination of selected low-energy conformers. In another approach, NMR data are used for the validation of conformational ensembles derived from molecular dynamics (MD) simulation of flexible oligosaccharides that exhibit shallow and broad energy minima in their conformational space [91, 122–124]. In this approach, experimentally observed PCS data are compared with those back-calculated from the MD-derived conformational ensembles (Fig. 7.11). NMR validation showed that conformational spaces of the triantennary high-mannose-type oligosaccharides could be sufficiently explored by replica-exchange MD simulation with a total simulation time of several microseconds [91]. By employing non-linear multivariate analysis of NMR-validated conformational ensembles, oligosaccharide conformers can be classified in relation to their conformational free-energy landscape [125]. In conjunction with crystallographic data of lectin-bound oligosaccharides, carbohydrate recognition by lectins is characterized by conformational selection as well as induced-fit mechanisms [118, 124].

Oligosaccharides accommodated in a protein can be less mobile than liberated ones. As a typical example, the pair of biantennary oligosaccharides packed within the quaternary structure of IgG-Fc is visualized for the most part in the crystalline state [79, 80]. However, the terminal galactose residues exhibit different dynamic properties, which have been characterized on the basis of relaxation data [45, 93]. Therefore, the glycans attached to proteins can exhibit heterogeneous properties in terms of conformational dynamics.

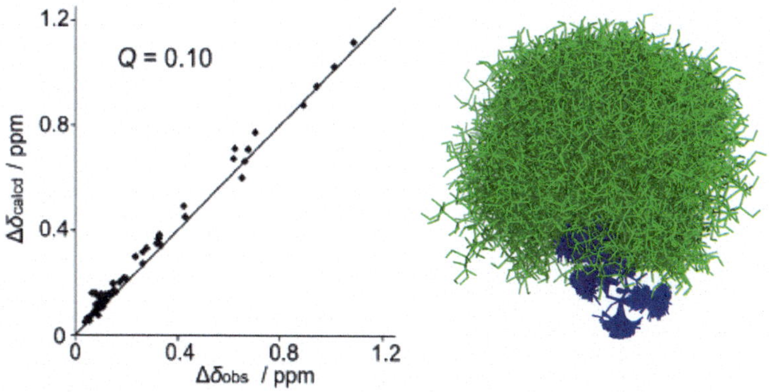

Fig. 7.11 Conformational ensemble of M9 obtained from replica-exchange MD calculation validated by the PCS data (left) and the correlation between experimentally observed PCS values with Tm^{3+} and theoretical PCS data back-calculated from the ensemble model (right). Adapted from the reference [91] with permission

7.8 Perspectives

NMR spectroscopy in conjunction with computational approaches has now successfully characterized dynamic conformational ensembles of oligosaccharides and glycoprotein glycans. This prompts us to improve the functionality of carbohydrate chains by designing and optimizing their conformational spaces [126]. To realize this glycoengineering, it is obviously necessary to create glycoproteins with tailored and even artificial glycoforms.

A hybrid approach combining cell engineering and chemoenzymatic techniques has the ability to deal with glycoform heterogeneity of glycoproteins in NMR structural glycobiology [127]. This line of technical development will realize isotope-labeled glycoproteins with tailored glycoforms. By installing glycosylation systems of *C. jejuni* and *S. cerevisiae* into *E. coli*, bacterial expression of recombinant glycoproteins having the core structure of eukaryotic glycans, i.e., $Man_3GlcNAc_2$, has been achieved [128]. This glycan can be substituted with synthetic donor oligosaccharides by in vitro enzymatic transglycosylation catalyzed by endo-β-*N*-acetylglucosaminidases, leaving the innermost GlcNAc residue as acceptor site [129].

Furthermore, synthetic approaches have been successfully developed to prepare glycoproteins involving chemical ligation of recombinant non-glycosylated peptides and chemically synthesized glycopeptides (or glycosyl asparagine thioacid) [130, 131]. These technical advancements will open up new possibilities to create glycoproteins with artificial glycoforms. NMR spectroscopy along with computational approaches will play critical roles in the detailed characterization of dynamic conformations and interactions of those *neo-glycoproteins* for a better understanding of the functional mechanisms of carbohydrate chains.

Acknowledgements We thank Dr. Issaku Yamada (The Noguchi Institute) for the statistical analysis of PDB data and Drs. Takumi Yamaguchi (School of Material Science, Japan Advanced Institute of Science and Technology) and Maho Yagi-Utsumi (Nagoya City University) for useful discussion. This work was partly supported by the Nanotechnology Platform Program (Molecule and Material Synthesis) of MEXT, MEXT/JSPS Grants in Aid for Scientific Research (JP25102008, JP15K07935, JP17H05893, JP19H01017, JP20K15981, JP21H02625, JP22H02755, JP23H03838, and JP24H00599), the Japan Agency for Medical Research and Development (JP21ae0121020h0001 and JP21ae0121013h0301), MEXT Promotion of Development of a Joint Usage/ Research System Project: Coalition of Universities for Research Excellence Program (CURE) Grant Number JPMXP1323015488, JST-CREST (Grant Number JP MJCR21E3 to K.K.), Human Glycome Atlas Project, and the Cooperative Research Program of Institute for Protein Research, Osaka University (NMRCR-16-05,-17-05,-18-05, -19-05, -20-05, and -21-05, and VFCR-22-05 and -23-05).

References

1. Boehr, D.D., Nussinov, R., Wright, P.E.: The role of dynamic conformational ensembles in biomolecular recognition. Nat. Chem. Biol. **5**(11), 789–796 (2009). https://doi.org/10.1038/nchembio.232
2. Kovermann, M., Rogne, P., Wolf-Watz, M.: Protein dynamics and function from solution state NMR spectroscopy. Q. Rev. Biophys. **49**, e6 (2016). https://doi.org/10.1017/S0033583516000019.7
3. Ikeya, T., Ban, D., Lee, D., Ito, Y., Kato, K., Griesinger, C.: Solution NMR views of dynamical ordering of biomacromolecules. Biochim. Biophys. Acta Gen. Subj. **1862**(2), 287–306 (2018). https://doi.org/10.1016/j.bbagen.2017.08.020
4. Rosenzweig, R., Kay, L.E.: Bringing Dynamic Molecular Machines into Focus by Methyl-TROSY NMR. Annu. Rev. Biochem. **83**, 291–315 (2014). https://doi.org/10.1146/annurev-biochem-060713-035829
5. Freedberg, D.I., Selenko, P.: Live cell NMR. Annu. Rev. Biophys. **43**, 171–192 (2014). https://doi.org/10.1146/annurev-biophys-051013-023136
6. Skinner, A.L., Laurence, J.S.: High-Field solution NMR spectroscopy as a tool for assessing protein interactions with small molecule ligands. J Pharm Sci-Us. **97**(11), 4670–4695 (2008). https://doi.org/10.1002/jps.21378
7. Dias, D.M., Ciulli, A.: NMR approaches in structure-based lead discovery: recent developments and new frontiers for targeting multi-protein complexes. Prog. Biophys. Mol. Biol. **116**(2–3), 101–112 (2014). https://doi.org/10.1016/j.pbiomolbio.2014.08.012
8. Carlson, E.D., Gan, R., Hodgman, C.E., Jewett, M.C.: Cell-free protein synthesis: applications come of age. Biotechnol. Adv. **30**(5), 1185–1194 (2012). https://doi.org/10.1016/j.biotechadv.2011.09.016
9. Apweiler, R., Hermjakob, H., Sharon, N.: On the frequency of protein glycosylation, as deduced from analysis of the SWISS-PROT database. Biochem. Biophys. Acta. **1473**(1), 4–8 (1999)
10. Essentials of Glycobiology. 4th edition ed. Cold Spring Harbor Laboratory Pres. (2009)
11. Bubb, W.A.: NMR spectroscopy in the study of carbohydrates: Characterizing the structural complexity. Concepts Magn. Reson. Part A.19A(1):1–7. (2003) https://doi.org/10.1002/cmr.a.10080.
12. Prestegard, J.H., Koerner, T.A.W., Demou, P.C., Yu, R.K.: Complete analysis of oligosaccharide primary structure using two-dimensional high-field proton NMR. J. Am. Chem. Soc. **104**(18), 4993–4995 (1982). https://doi.org/10.1021/ja00382a063

13. Vliegenthart, J.F.: High resolution ^{1}H-NMR spectroscopy of carbohydrate structures. Adv. Exp. Med. Biol. **125**, 77–91 (1980)
14. Gray, C.J., Migas, L.G., Barran, P.E., Pagel, K., Seeberger, P.H., Eyers, C.E., et al.: Advancing solutions to the carbohydrate sequencing challenge. J. Am. Chem. Soc. **141**(37), 14463–14479 (2019). https://doi.org/10.1021/jacs.9b06406
15. Fontana, C., Widmalm, G.: Primary structure of Glycans by NMR spectroscopy. Chem. Rev. **123**(3), 1040–1102 (2023). https://doi.org/10.1021/acs.chemrev.2c00580
16. Jumper, J., Evans, R., Pritzel, A., Green, T., Figurnov, M., Ronneberger, O., et al.: Highly accurate protein structure prediction with AlphaFold. Nature **596**(7873), 583–589 (2021). https://doi.org/10.1038/s41586-021-03819-2
17. Baek, M., DiMaio, F., Anishchenko, I., Dauparas, J., Ovchinnikov, S., Lee, G.R., et al.: Accurate prediction of protein structures and interactions using a three-track neural network. Science **373**(6557), 871–876 (2021). https://doi.org/10.1126/science.abj8754
18. Albrecht, S., Unwin, L., Muniyappa, M., Rudd, P.M.: Glycosylation as a marker for inflammatory arthritis. Cancer Biomark. **14**(1), 17–28 (2014). https://doi.org/10.3233/CBM-130373
19. Adamczyk, B., Albrecht, S., Stockmann, H., Ghoneim, I.M., Al-Eknah, M., Al-Busadah, K.A., et al.: Pregnancy-Associated Changes of IgG and Serum N-Glycosylation in Camel (Camelus dromedarius). J. Proteome Res. 15(9). pp. 3255–65. (2016). https://doi.org/10.1021/acs.jproteome.6b00439.
20. Parekh, R.B., Dwek, R.A., Sutton, B.J., Fernandes, D.L., Leung, A., Stanworth, D., et al.: Association of rheumatoid arthritis and primary osteoarthritis with changes in the glycosylation pattern of total serum IgG. Nature **316**(6027), 452–457 (1985)
21. Yamada, E., Tsukamoto, Y., Sasaki, R., Yagyu, K., Takahashi, N.: Structural changes of immunoglobulin G oligosaccharides with age in healthy human serum. Glycoconjugate J. **14**(3), 401–405 (1997)
22. Kibe, T., Fujimoto, S., Ishida, C., Togari, H., Wada, Y., Okada, S., et al.: Glycosylation and placental transport of immunoglobulin G. J. Clin. Biochem Nutr. **21**(1), 57–63 (1996)
23. Cobb, B.A.: The history of IgG glycosylation and where we are now. Glycobiology **30**(4), 202–213 (2020). https://doi.org/10.1093/glycob/cwz065
24. Flevaris, K., Kontoravdi, C.: Immunoglobulin G N-glycan Biomarkers for Autoimmune Diseases: Current State and a Glycoinformatics Perspective. Int. J. Mol. Sci. 23(9). (2022). https://doi.org/10.3390/ijms23095180.
25. Shkunnikova, S., Mijakovac, A., Sironic, L., Hanic, M., Lauc, G., Kavur, M.M.: IgG glycans in health and disease: Prediction, intervention, prognosis, and therapy. Biotechnol. Adv. **67**, 108169 (2023). https://doi.org/10.1016/j.biotechadv.2023.108169
26. Park, S.S., Park, J., Ko, J., Chen, L., Meriage, D., Crouse-Zeineddini, J., et al.: Biochemical assessment of erythropoietin products from Asia versus US Epoetin alfa manufactured by Amgen. J. Pharm. Sci. **98**(5), 1688–1699 (2009). https://doi.org/10.1002/jps.21546
27. Zeck, A., Pohlentz, G., Schlothauer, T., Peter-Katalinic, J., Regula, J.T.: Cell type-specific and site directed *N*-glycosylation pattern of FcγRIIIa. J. Proteome Res. **10**(7), 3031–3039 (2011). https://doi.org/10.1021/pr1012653
28. Goh, J.B., Ng, S.K.: Impact of host cell line choice on glycan profile. Crit. Rev. Biotechnol. **38**(6), 851–867 (2018). https://doi.org/10.1080/07388551.2017.1416577
29. Dumont, J., Euwart, D., Mei, B., Estes, S., Kshirsagar, R.: Human cell lines for biopharmaceutical manufacturing: history, status, and future perspectives. Crit. Rev. Biotechnol. **36**(6), 1110–1122 (2016). https://doi.org/10.3109/07388551.2015.1084266
30. Lindhorst, T,K.: Essentials of carbohydrate chemistry and biochemistry. 3rd Completely Revised and Enlarged Edition ed. Wily. p. 1–332 (2007)
31. de Graaf, M., Fouchier, R.A.: Role of receptor binding specificity in influenza A virus transmission and pathogenesis. EMBO J. **33**(8), 823–841 (2014). https://doi.org/10.1002/embj.201387442
32. Zhao, C., Pu, J.: Influence of host sialic acid receptors structure on the host specificity of influenza viruses. Viruses. 14(10). (2022) https://doi.org/10.3390/v14102141

33. Wormald, M.R., Petrescu, A.J., Pao, Y.L., Glithero, A., Elliott, T., Dwek, R.A.: Conformational studies of oligosaccharides and glycopeptides: complementarity of NMR, X-ray crystallography, and molecular modelling. Chem. Rev. **102**(2), 371–386 (2002)
34. Atanasova, M., Bagdonas, H., Agirre, J.: Structural glycobiology in the age of electron cryomicroscopy. Curr. Opin. Struct. Biol. **62**, 70–78 (2020). https://doi.org/10.1016/j.sbi.2019.7.12.003
35. Masuda, K., Yamaguchi, Y., Takahashi, N., Jefferis, R., Kato, K.: Mutational deglycosylation of the Fc portion of immunoglobulin G causes O-sulfation of tyrosine adjacently preceding the originally glycosylated site. FEBS Lett. **584**(15), 3474–3479 (2010)
36. Shibata-Koyama, M., Iida, S., Okazaki, A., Mori, K., Kitajima-Miyama, K., Saitou, S., et al.: The N-linked oligosaccharide at FcγRIIIa Asn-45: an inhibitory element for high FcγRIIIa binding affinity to IgG glycoforms lacking core fucosylation. Glycobiology **19**(2), 126–134 (2009)
37. Yamaguchi, Y., Kato, K.: Dynamics and interactions of glycoconjugates probed by stable-isotope-assisted NMR spectroscopy. Methods Enzymol. **478**, 305–322 (2010)
38. Sastry, M., Xu, L., Georgiev, I.S., Bewley, C.A., Nabel, G.J., Kwong, P.D.: Mammalian production of an isotopically enriched outer domain of the HIV-1 gp120 glycoprotein for NMR spectroscopy. J. Biomol. NMR **50**(3), 197–207 (2011). https://doi.org/10.1007/s10858-011-9506-4
39. Kato, K., Yamaguchi, Y., Arata, Y.: Stable-isotope-assisted NMR approaches to glycoproteins using immunoglobulin G as a model system. Prog. Nucl. Magn. Reson. Spectrosc. **56**(4), 346–359 (2010). https://doi.org/10.1016/j.pnmrs.2010.03.001
40. Yamaguchi, Y., Yagi, H., Kato, K.: Stable Isotope Labeling of Glycoproteins for NMR Study. In: Kato, K., Peters, T. (eds.) NMR in Glycoscience and Glycotechnology, pp. 194–207. RSC Publishing, Cambridge (2017)
41. Kato, K., Yamaguchi, Y.: Glycoproteins and Antibodies: Solution NMR studies. Encyclopedia of Magnetic Resonance. Chichester: John Wiley & Sons. pp. 1779–90 (2012)
42. Kato, K., Matsunaga, C., Igarashi, T., Kim, H., Odaka, A., Shimada, I., et al.: Complete assignment of the methionyl carbonyl carbon resonances in switch variant anti-dansyl antibodies labeled with [1-^{13}C]methionine. Biochemistry **30**(1), 270–278 (1991). https://doi.org/10.1021/bi00215a037
43. Yagi, H., Zhang, Y., Yagi-Utsumi, M., Yamaguchi, T., Iida, S., Yamaguchi, Y., et al.: Backbone ^{1}H, ^{13}C, and ^{15}N resonance assignments of the Fc fragment of human immunoglobulin G glycoprotein. Biomol. NMR Assign. **9**(2), 257–260 (2015). https://doi.org/10.1007/s12104-014-9586-7
44. Yanaka, S., Yagi, H., Yogo, R., Onitsuka, M., Kato, K.: Glutamine-free mammalian expression of recombinant glycoproteins with uniform isotope labeling: an application for NMR analysis of pharmaceutically relevant Fc glycoforms of human immunoglobulin G1. J. Biomol. NMR **76**(1–2), 17–22 (2022). https://doi.org/10.1007/s10858-021-00387-5
45. Yamaguchi, Y., Kato, K., Shindo, M., Aoki, S., Furusho, K., Koga, K., et al.: Dynamics of the carbohydrate chains attached to the Fc portion of immunoglobulin G as studied by NMR spectroscopy assisted by selective ^{13}C labeling of the glycans. J. Biomol. NMR **12**(3), 385–394 (1998)
46. Yanaka, S., Yamaguchi, Y., Takizawa, T., Miyanoiri, Y., Yogo, R., Shimada, I., et al.: NMR assignments of the N-glycans of the Fc fragment of mouse immunoglobulin G2b glycoprotein. Biomol. NMR Assign. **15**(1), 187–192 (2021). https://doi.org/10.1007/s12104-020-10004-5
47. Rogals, M.J., Yang, J.Y., Williams, R.V., Moremen, K.W., Amster, I.J., Prestegard, J.H.: Sparse isotope labeling for nuclear magnetic resonance (NMR) of glycoproteins using ^{13}C-glucose. Glycobiology **31**(4), 425–435 (2021). https://doi.org/10.1093/glycob/cwaa071
48. Kato, K., Yamaguchi, Y.: Structural Analyses of Glycoconjugates by NMR. In: Taniguchi N, Suzuki A, Ito Y, Narimatsu H, Kawasaki T, Hase S, editors. Experimental Glycoscience. Springer Japan. pp. 45–50 (2008)
49. Yagi, H., Fukuzawa, N., Tasaka, Y., Matsuo, K., Zhang, Y., Yamaguchi, T., et al.: NMR-based structural validation of therapeutic antibody produced in Nicotiana benthamiana. Plant Cell Rep. **34**(6), 959–968 (2015). https://doi.org/10.1007/s00299-015-1757-1

50. Yagi, H., Nakamura, M., Yokoyama, J., Zhang, Y., Yamaguchi, T., Kondo, S., et al.: Stable isotope labeling of glycoprotein expressed in silkworms using immunoglobulin G as a test molecule. J. Biomol. NMR **62**(2), 157–167 (2015). https://doi.org/10.1007/s10858-015-9930-y
51. Walton, W.J., Kasprzak, A.J., Hare, J.T., Logan, T.M.: An economic approach to isotopic enrichment of glycoproteins expressed from Sf9 insect cells. J. Biomol. NMR **36**(4), 225–233 (2006). https://doi.org/10.1007/s10858-006-9086-x
52. Blanchard, V., Gadkari, R.A., George, A.V., Roy, S., Gerwig, G.J., Leeflang, B.R., et al.: High-level expression of biologically active glycoprotein hormones in Pichia pastoris strains–selection of strain GS115, and not X-33, for the production of biologically active N-glycosylated 15N-labeled phCG. Glycoconj. J. **25**(3), 245–257 (2008). https://doi.org/10.1007/s10719-007-9082-8
53. Kamiya, Y., Yamamoto, S., Chiba, Y., Jigami, Y., Kato, K.: Overexpression of a homogeneous oligosaccharide with ^{13}C labeling by genetically engineered yeast strain. J. Biomol. NMR **50**(4), 397–401 (2011). https://doi.org/10.1007/s10858-011-9525-1
54. Cubeddu, L., Moss, C.X., Swarbrick, J.D., Gooley, A.A., Williams, K.L., Curmi, P.M., et al.: Dictyostelium discoideum as expression host: isotopic labeling of a recombinant glycoprotein for NMR studies. Protein Expr. Purif. **19**(3), 335–342 (2000). https://doi.org/10.1006/prep.2000.1269
55. Nettleship, J.E., Assenberg, R., Diprose, J.M., Rahman-Huq, N., Owens, R.J.: Recent advances in the production of proteins in insect and mammalian cells for structural biology. J. Struct. Biol. **172**(1), 55–65 (2010). https://doi.org/10.1016/j.jsb.2010.02.006
56. Franke, B., Opitz, C., Isogai, S., Grahl, A., Delgado, L., Gossert, A.D., et al.: Production of isotope-labeled proteins in insect cells for NMR. J. Biomol. NMR **71**(3), 173–184 (2018). https://doi.org/10.1007/s10858-018-0172-7
57. Yagi, H., Yanaka, S., Yogo, R., Ikeda, A., Onitsuka, M., Yamazaki, T., et al.: Silkworm pupae function as efficient producers of recombinant glycoproteins with Stable-Isotope Labeling. Biomolecules. 10(11). (2020) https://doi.org/10.3390/biom10111482
58. Kamiya, Y., Yanagi, K., Kitajima, T., Yamaguchi, T., Chiba, Y., Kato, K.: Application of metabolic ^{13}C labeling in conjunction with high-field nuclear magnetic resonance spectroscopy for comparative conformational analysis of high mannose-type oligosaccharides. Biomolecules **3**(1), 108–123 (2013). https://doi.org/10.3390/biom3010108
59. Shenoy, A., Davis, A.R., Roberts, E.T., Amster, I.J., Barb, A.W.: Metabolic ^{15}N labeling of the N-glycosylated immunoglobulin G1 Fc with an engineered Saccharomyces cerevisiae strain. J. Biomol. NMR **76**(4), 95–105 (2022). https://doi.org/10.1007/s10858-022-00397-x
60. Sattler, M., Fesik, S.W.: Use of deuterium labeling in NMR: Overcoming a sizeable problem. Structure. **4**(11), 1245–1249 (1996). https://doi.org/10.1016/S0969-2126(96)00133-5
61. Kainosho, M., Guntert, P.: SAIL–stereo-array isotope labeling. Q. Rev. Biophys. **42**(4), 247–300 (2009). https://doi.org/10.1017/S0033583510000016
62. Opitz, C., Isogai, S., Grzesiek, S.: An economic approach to efficient isotope labeling in insect cells using homemade ^{15}N-, ^{13}C- and ^{2}H-labeled yeast extracts. J. Biomol. NMR **62**(3), 373–385 (2015). https://doi.org/10.1007/s10858-015-9954-3
63. Kofuku, Y., Ueda, T., Okude, J., Shiraishi, Y., Kondo, K., Mizumura, T., et al.: Functional dynamics of deuterated beta2 -adrenergic receptor in lipid bilayers revealed by NMR spectroscopy. Angew. Chem. Int. Ed. Engl. **53**(49), 13376–13379 (2014). https://doi.org/10.1002/anie.201406603
64. Guarino, C., DeLisa, M.P.: A prokaryote-based cell-free translation system that efficiently synthesizes glycoproteins. Glycobiology **22**(5), 596–601 (2012). https://doi.org/10.1093/glycob/cwr151
65. Slynko, V., Schubert, M., Numao, S., Kowarik, M., Aebi, M., Allain, F.H.: NMR structure determination of a segmentally labeled glycoprotein using in vitro glycosylation. J. Am. Chem. Soc. **131**(3), 1274–1281 (2009). https://doi.org/10.1021/ja808682v
66. Takahashi, N., Nakagawa, H., Fujikawa, K., Kawamura, Y., Tomiya, N.: Three-dimensional elution mapping of pyridylaminated N-linked neutral and sialyl oligosaccharides. Anal. Biochem. **226**(1), 139–146 (1995)

67. Takahashi, N., Kato, K.: GALAXY(Glycoanalysis by the Three Axes of MS and Chromatography):a web application that assists structural analyses of *N*-Glycans. Trends Glycosci Glycotech. **15**, 235–251 (2003)
68. Yagi, H., Amagasa, E., Shiota, M., Yamada, I., Aoki-Kinoshita, K.F., Kato, K.: GALAXY ver3: updated web application for glycosylation profiling based on 3D HPLC map. Glycobiology **32**(8), 646–650 (2022). https://doi.org/10.1093/glycob/cwac025
69. Yamaguchi, Y., Nishimura, M., Nagano, M., Yagi, H., Sasakawa, H., Uchida, K., et al.: Glycoform-dependent conformational alteration of the Fc region of human immunoglobulin G1 as revealed by NMR spectroscopy. Biochim. Biophys. Acta **1760**(4), 693–700 (2006)
70. Yagi, H., Kuo, C.W., Obayashi, T., Ninagawa, S., Khoo, K.H., Kato, K.: Direct mapping of additional modifications on phosphorylated O-glycans of alpha-Dystroglycan by mass spectrometry analysis in conjunction with knocking out of causative genes for dystroglycanopathy. Mol. Cell. Proteomics **15**(11), 3424–3434 (2016). https://doi.org/10.1074/mcp.M116.062729
71. Masuda, K., Yamaguchi, Y., Kato, K., Takahashi, N., Shimada, I., Arata, Y.: Pairing of oligosaccharides in the Fc region of immunoglobulin G. FEBS Lett. **473**(3), 349–357 (2000)
72. Thaysen-Andersen, M., Packer, N.H.: Advances in LC-MS/MS-based glycoproteomics: getting closer to system-wide site-specific mapping of the N- and O-glycoproteome. Biochem. Biophys. Acta. **1844**(9), 1437–1452 (2014). https://doi.org/10.1016/j.bbapap.2014.05.002
73. Terral, G., Beck, A., Cianferani, S.: Insights from native mass spectrometry and ion mobility-mass spectrometry for antibody and antibody-based product characterization. J. Chromatogr. B Analyt. Technol. Biomed. Life Sci. **1032**, 79–90 (2016). https://doi.org/10.1016/j.jchromb.2016.03.044
74. Yang, Y., Franc, V., Heck, A.J.R.: Glycoproteomics: a balance between high-throughput and in-depth analysis. Trends Biotechnol. **35**(7), 598–609 (2017). https://doi.org/10.1016/j.tibtech.2017.04.010
75. Reiding, K.R., Bondt, A., Franc, V., Heck, A.J.R.: The benefits of hybrid fragmentation methods for glycoproteomics. Trac-Trends in Analytical Chemistry. **108**, 260–268 (2018). https://doi.org/10.1016/j.trac.2018.09.007
76. Ito, H., Kameyama, A., Sato, T., Narimatsu, H.: Preparation of a glycan library using a variety of glycosyltrasferases. Methods Mol. Biol. **534**, 283–291 (2009). https://doi.org/10.1007/978-1-59745-022-5_20
77. Kato, K., Yamaguchi, Y., Takahashi, N., Nishimura, M., Iwamoto, S., Sekiya, S., et al.: Discrimination of isomeric fragment ions observed in tandem mass spectra of biantennary oligosaccharides by use of selective isotope labeling. J Mass Spectrom Soc Japan. **52**, 284–288 (2004)
78. Hsiao, C.T., Wang, P.W., Chang, H.C., Chen, Y.Y., Wang, S.H., Chern, Y., et al.: Advancing a High Throughput Glycotope-centric Glycomics Workflow Based on nanoLC-MS(2)-product Dependent-MS(3) Analysis of Permethylated Glycans. Mol. Cell. Proteomics **16**(12), 2268–2280 (2017). https://doi.org/10.1074/mcp.TIR117.000156
79. Matsumiya, S., Yamaguchi, Y., Saito, J., Nagano, M., Sasakawa, H., Otaki, S., et al.: Structural comparison of fucosylated and nonfucosylated Fc fragments of human immunoglobulin G1. J. Mol. Biol. **368**(3), 767–779 (2007)
80. Mizushima, T., Yagi, H., Takemoto, E., Shibata-Koyama, M., Isoda, Y., Iida, S., et al.: Structural basis for improved efficacy of therapeutic antibodies upon defucosylation of their Fc glycans. Genes Cells **16**(11), 1071–1080 (2011)
81. Shields, R.L., Lai, J., Keck, R., O'Connell, L.Y., Hong, K., Meng, Y.G., et al.: Lack of fucose on human IgG1 N-linked oligosaccharide improves binding to human FcγRIII and antibody-dependent cellular toxicity. J. Biol. Chem. **277**(30), 26733–26740 (2002)
82. Shinkawa, T., Nakamura, K., Yamane, N., Shoji-Hosaka, E., Kanda, Y., Sakurada, M., et al.: The absence of fucose but not the presence of galactose or bisecting N-acetylglucosamine of human IgG1 complex-type oligosaccharides shows the critical role of enhancing antibody-dependent cellular cytotoxicity. J. Biol. Chem. **278**(5), 3466–3473 (2003)
83. Chiba, Y., Akeboshi, H.: Glycan engineering and production of 'humanized' glycoprotein in yeast cells. Biol. Pharm. Bull. **32**(5), 786–795 (2009)

84. Hamilton, S.R., Davidson, R.C., Sethuraman, N., Nett, J.H., Jiang, Y., Rios, S., et al.: Humanization of yeast to produce complex terminally sialylated glycoproteins. Science **313**(5792), 1441–1443 (2006). https://doi.org/10.1126/science.1130256
85. Irani, Z.A., Kerkhoven, E.J., Shojaosadati, S.A., Nielsen, J.: Genome-scale metabolic model of Pichia pastoris with native and humanized glycosylation of recombinant proteins. Biotechnol. Bioeng. **113**(5), 961–969 (2016). https://doi.org/10.1002/bit.25863
86. Kato, T., Kako, N., Kikuta, K., Miyazaki, T., Kondo, S., Yagi, H., et al.: N-Glycan modification of a recombinant protein via coexpression of human glycosyltransferases in silkworm pupae. Sci. Rep. **7**(1), 1409 (2017). https://doi.org/10.1038/s41598-017-01630-6
87. Kato, K., Sasakawa, H., Kamiya, Y., Utsumi, M., Nakano, M., Takahashi, N., et al.: 920 MHz ultra-high field NMR approaches to structural glycobiology. Biochem. Biophys. Acta. **1780**(3), 619–625 (2008). https://doi.org/10.1016/j.bbagen.2007.11.014
88. Blundell, C.D., Reed, M.A., Overduin, M., Almond, A.: NMR spectra of oligosaccharides at ultra-high field (900 MHz) have better resolution than expected due to favourable molecular tumbling. Carbohydr. Res. **341**(12), 1985–1991 (2006). https://doi.org/10.1016/j.carres.2006.05.017
89. Canet, D., Robert, J.B.: Behaviour of the NMR Relaxation Parameters at High Fields. In: Robert, J.B. (ed.) NMR at Very High Field, pp. 45–89. Springer-Verlag (1991)
90. Meersmann, T., Schwager, M., Varma, V., Bodenhausen, G.: Little-known advantages of very high fields in NMR. J Magn Reson Ser A **119**, 275–279 (1996)
91. Yamaguchi, T., Sakae, Y., Zhang, Y., Yamamoto, S., Okamoto, Y., Kato, K.: Exploration of conformational spaces of high-mannose-type oligosaccharides by an NMR-validated simulation. Angew. Chem. Int. Ed. Engl. **53**(41), 10941–10944 (2014). https://doi.org/10.1002/anie.201406145
92. Yamaguchi, Y., Takizawa, T., Kato, K., Arata, Y., Shimada, I.: ^{1}H and ^{13}C NMR assignments for the glycans in glycoproteins by using $^{2}H/^{13}C$-labeled glucose as a metabolic precursor. J. Biomol. NMR **18**(4), 357–360 (2000)
93. Barb, A.W., Prestegard, J.H.: NMR analysis demonstrates immunoglobulin G N-glycans are accessible and dynamic. Nat. Chem. Biol. **7**(3), 147–153 (2011). https://doi.org/10.1038/nchembio.511
94. Gilhespy-Muskett, A.M., Partridge, J., Jefferis, R., Homans, S.W.: A novel ^{13}C isotopic labelling strategy for probing the structure and dynamics of glycan chains in situ on glycoproteins. Glycobiology **4**(4), 485–489 (1994)
95. Zhu, T., Yamaguchi, T., Satoh, T., Kato, K.: A Hybrid Strategy for the Preparation of ^{13}C-Labeled High-mannose-type Oligosaccharides with Terminal Glucosylation for NMR Study. Chem. Lett. (2015). https://doi.org/10.1246/cl.150898
96. Yamaguchi, Y., Walchli, M., Nagano, M., Kato, K.: A C-13-detection NMR approach for large glycoproteins. Carbohyd. Res. **344**(4), 535–538 (2009). https://doi.org/10.1016/j.carres.2008.12.003
97. Hanashima, S., Kato, K., Yamaguchi, Y.: ^{13}C-NMR quantification of proton exchange at LewisX hydroxyl groups in water. Chem. Commun. (Camb.) **47**(38), 10800–10802 (2011). https://doi.org/10.1039/c1cc13310a
98. Pfeffer, P.E., Valentine, K.M., Parrish, F.W.: Deuterium-induced differential isotope shift C-13 Nmr .1. Resonance Re-Assignments of Monosaccharides and Disaccharides. J. Am. Chem. Soc. **101**(5):1265–74. (1979) https://doi.org/10.1021/Ja00499a036
99. Battistel, M.D., Azurmendi, H.F., Freedberg, D.I.: Glycan OH exchange rate determination in aqueous solution: seeking evidence for transient hydrogen bonds. (2017). https://doi.org/10.1021/acs.jpcb.6b10594
100. Battistel, M.D., Azurmendi, H.F., Freedberg, D.I.: Intramolecular hydrogen bonding in glycans in aqueous solution. NMR in Glycoscience and Glycotechnology. pp. 1–7 (2017)
101. Nestor, G., Anderson, T., Oscarson, S., Gronenborn, A.M.: Direct observation of carbohydrate hydroxyl protons in hydrogen bonds with a protein. J. Am. Chem. Soc. **140**(1), 339–345 (2018). https://doi.org/10.1021/jacs.7b10595

102. Yanaka, S., Yogo, R., Yagi, H., Onitsuka, M., Wakaizumi, N., Yamaguchi, Y., et al.: Negative interference with antibody-dependent cellular cytotoxicity mediated by rituximab from its interactions with human serum proteins. Front. Immunol. 14. (2023). https://doi.org/10.3389/fimmu.2023.1090898.
103. Yanaka, S., Yamazaki, T., Yogo, R., Noda, M., Uchiyama, S., Yagi, H., et al.: NMR detection of semi-specific antibody interactions in serum environments. Mol. **22**(10). (2017). https://doi.org/10.3390/molecules22101167.
104. Yanaka, S., Yagi, H., Yogo, R., Yagi-Utsumi, M., Kato, K.: Stable isotope labeling approaches for NMR characterization of glycoproteins using eukaryotic expression systems. J. Biomol. NMR **71**(3), 193–202 (2018). https://doi.org/10.1007/s10858-018-0169-2
105. Brinson, R.G., Marino, J.P., Delaglio, F., Arbogast, L.W., Evans, R.M., Kearsley, A., et al.: Enabling adoption of 2D-NMR for the higher order structure assessment of monoclonal antibody therapeutics. MAbs **11**(1), 94–105 (2019). https://doi.org/10.1080/19420862.2018.1544454
106. Hadad, M.J., Zhang, W., Turney, T., Sernau, L., Wang, X., Woods, R.J., et al.: NMR Spin-Couplings in Saccharides: Relationships Between Structure, Conformation and the Magnitudes of JHH, JCH and JCC Values. In: Kato K, Peters T, editors. NMR in Glycoscience and Glycotechnology. Cambridge: RSC Publishing; 2017. pp. 20–100 (2017)
107. Prestegard, J.H., Eletsky, A.: Paramagnetic, RDC and RCSA Constraints in the Structural Analysis of Glycans. In: Kato, K., Peters, T. (eds.) NMR in Glycoscience and Glycotechnology, pp. 123–149. RSC Publishing, Cambridge (2017)
108. Schubert, M.: Insights into Carbohydrate Recognition by 3D Structure Determination of Protein-Carbohydrate Complexes Using NMR. In: Kato, K., Peters, T. (eds.) NMR in Glycoscience and Glycotechnology, pp. 101–122. RSC Publishing, Cambridge (2017)
109. Yanaka, S., Yogo, R., Inoue, R., Sugiyama, M., Itoh, S,G., Okumura, H., et al.: Dynamic views of the Fc region of immunoglobulin G provided by experimental and computational observations. Antibodies (Basel). **8**(3). (2019). https://doi.org/10.3390/antib8030039
110. Yamaguchi, Y., Hirao, T., Sakata, E., Kamiya, Y., Kurimoto, E., Yoshida, Y., et al.: Fbs1 protects the malfolded glycoproteins from the attack of peptide:N-glycanase. Biochem. Biophys. Res. Commun. **362**(3), 712–716 (2007). https://doi.org/10.1016/j.bbrc.2007.08.056
111. Nestor, G., Anderson, T., Oscarson, S., Gronenborn, A.M.: Exploiting Uniformly ^{13}C-Labeled Carbohydrates for Probing Carbohydrate-Protein Interactions by NMR Spectroscopy. J. Am. Chem. Soc. **139**(17), 6210–6216 (2017). https://doi.org/10.1021/jacs.7b01929
112. Nestor, G., Ruda, A., Anderson, T., Oscarson, S., Widmalm, G., Gronenborn, A.M.: A detailed picture of a protein-carbohydrate hydrogen-bonding network revealed by NMR and MD simulations. Glycobiology **31**(4), 508–518 (2021). https://doi.org/10.1093/glycob/cwaa081
113. Yamaguchi, Y., Kato, K.: Analyses of Sugar-Protein Interactions by NMR. In: Taniguchi, N., Suzuki, A., Ito, Y., Narimatsu, H., Kawasaki, T., Hase, S. (eds.) Experimental Glycoscience, pp. 121–123. Springer Japan, Tokyo (2008)
114. Mayer, M., Meyer, B.: Group epitope mapping by saturation transfer difference NMR to identify segments of a ligand in direct contact with a protein receptor. J. Am. Chem. Soc. **123**(25), 6108–6117 (2001)
115. Jimenes-Barbero, J., Peters, T.: TR-NOE experiments to study carbohydrate-protein interactions. In: Jimenes-Barbero, J., Peters, T. (eds.) NMR spectroscopy of glycoconjugates, pp. 289–309. Wiley-VCH, Weinheim (2006)
116. Canales, A., Mallagaray, A., Pérez-Castells, J., Boos, I., Unverzagt, C., André, S., et al.: Breaking Pseudo-Symmetry in Multiantennary Complex N-Glycans Using Lanthanide-Binding Tags and NMR Pseudo-Contact Shifts. Angew. Chem. Int. Ed. **52**(51), 13789–13793 (2013). https://doi.org/10.1002/anie.201307845
117. Kato, K., Yamaguchi, T.: Paramagnetic NMR probes for characterization of the dynamic conformations and interactions of oligosaccharides. Glycoconjugate J. **32**(7), 505–513 (2015). https://doi.org/10.1007/s10719-015-9599-1
118. Kato, K., Yagi, H., Yamaguchi, T.: NMR characterization of dynamic conformations and interactions of oligosaccharides. Mod. Magn. Reson.. Springer International Publishing. pp. 737–754. (2018). https://doi.org/10.1007/978-3-319-28388-3_35

119. Yamaguchi, T, Kamiya, Y., Y.M., C., Yamamoto, S., Kato, K.: Terminal Spin Labeling of a High-mannose-type Oligosaccharide for Quantitative NMR Analysis of Its Dynamic Conformation. Chem. Lett. **42**(5):544–6. (2013) https://doi.org/10.1246/cl.130040
120. Yan, G., Yamaguchi, T., Suzuki, T., Yanaka, S., Sato, S., Fujita, M., et al.: Hyper-Assembly of Self-Assembled Glycoclusters Mediated by Specific Carbohydrate-Carbohydrate Interactions. Chem. Asian J. **12**(9), 968–972 (2017). https://doi.org/10.1002/asia.201700202
121. Demarco, M.L., Woods, R.J., Prestegard, J.H., Tian, F.: Presentation of membrane-anchored glycosphingolipids determined from molecular dynamics simulations and NMR paramagnetic relaxation rate enhancement. J. Am. Chem. Soc. **132**(4), 1334–1338 (2010). https://doi.org/10.1021/ja907518x
122. Zhang, Y., Yamamoto, S., Yamaguchi, T., Kato, K.: Application of paramagnetic NMR-Validated molecular dynamics simulation to the analysis of a conformational ensemble of a branched oligosaccharide. Molecules **17**(6), 6658 (2012)
123. Yamamoto, S., Zhang, Y., Yamaguchi, T., Kameda, T., Kato, K.: Lanthanide-assisted NMR evaluation of a dynamic ensemble of oligosaccharide conformations. Chem. Commun. **48**(39), 4752–4754 (2012). https://doi.org/10.1039/C2CC30353A
124. Suzuki, T., Kajino, M., Yanaka, S., Zhu, T., Yagi, H., Satoh, T., et al.: Conformational analysis of a High-Mannose-Type oligosaccharide displaying Glucosyl determinant recognised by molecular chaperones using NMR-Validated molecular dynamics simulation. Chem. Bio. Chem. (2017). https://doi.org/10.1002/cbic.201600595
125. Watanabe, T., Yagi, H., Yanaka, S., Yamaguchi, T., Kato, K.: Comprehensive characterization of oligosaccharide conformational ensembles with conformer classification by free-energy landscape via reproductive kernel Hilbert space. Phys. Chem. Chem. Phys. **23**(16), 9753–9760 (2021). https://doi.org/10.1039/d0cp06448c
126. Suzuki, T., Yanaka, S., Watanabe, T., Yan, G., Satoh, T., Yagi, H., et al.: Remodeling of the oligosaccharide conformational space in the prebound state to improve lectin-binding affinity. Biochem. **207**. https://doi.org/10.1021/acs.biochem.9b00594
127. Kamiya, Y., Satoh, T., Kato, K.: Recent advances in glycoprotein production for structural biology: toward tailored design of glycoforms. Curr. Opin. Struct. Biol. **26**, 44–53 (2014). https://doi.org/10.1016/j.sbi.2014.03.008
128. Valderrama-Rincon, J.D., Fisher, A.C., Merritt, J.H., Fan, Y.Y., Reading, C.A., Chhiba, K., et al.: An engineered eukaryotic protein glycosylation pathway in Escherichia coli. Nat. Chem. Biol. **8**(5), 434–436 (2012). https://doi.org/10.1038/nchembio.921
129. Schwarz, F., Huang, W., Li, C., Schulz, B.L., Lizak, C., Palumbo, A., et al.: A combined method for producing homogeneous glycoproteins with eukaryotic N-glycosylation. Nat. Chem. Biol. **6**(4), 264–266 (2010). https://doi.org/10.1038/nchembio.314
130. Kajihara, Y., Tanabe, Y., Sasaoka, S., Okamoto, R.: Homogeneous human complex-type oligosaccharides in correctly folded intact glycoproteins: evaluation of oligosaccharide influence on protein folding, stability, and conformational properties. Chemistry **18**(19), 5944–5953 (2012). https://doi.org/10.1002/chem.201103428
131. Nomura, K., Maki, Y., Okamoto, R., Satoh, A., Kajihara, Y.: Glycoprotein semisynthesis by chemical insertion of glycosyl asparagine using a bifunctional thioacid-mediated strategy. J. Am. Chem. Soc. **143**(27), 10157–10167 (2021). https://doi.org/10.1021/jacs.1c02601
132. DeLano, W.L.: The PyMOL molecular graphics system DeLano scientific. San Carlos, CA, USA. (2002)

Chapter 8
NMR Studies on RNA

Taiichi Sakamoto, Maina Otsu, Rika Ichijo, and Gota Kawai

Abstract Basic methods for NMR analysis of RNA are presented for researchers who recently joined or want to join this field. This chapter includes the design of RNA sequences, sample preparation, NMR measurements, and analysis of exchangeable and non-exchangeable proton signals as well as structure calculation. It is usually required to design shorter RNA fragments to be analyzed by NMR. RNA samples are prepared by in vitro transcription or chemical synthesis. In addition, the enzymatic ligation method and artificial base pair systems can be used. Because proton signals of RNA are usually overlapped, stable-isotopic labeling is required to separate the overlapped signals. Site-specific 10% $^{13}C/^{15}N$-labeled RNAs are useful to identify signals for the labeled residue. Because RNAs are usually elongated molecules, long-range constraints such as residual dipolar couplings are useful for obtaining information on the global conformation of RNAs. Pf1 filamentous phage has been generally used for the alignment of RNAs. Some examples of Interaction analysis and structure screening are also shown. Furthermore, analysis for longer RNAs and in-cell NMR are also introduced.

Keywords RNA · *In vitro* transcription · Imino proton · Residual dipolar coupling · Stable isotopic labeling

In this chapter, general methods for RNA structure determination as well as some examples are shown. In the case of protein, the method for sequential assignment by using $^{13}C/^{15}N$-labeled protein samples and triple-resonance experiments is established and, furthermore, most of the processes are automated by the State-of-Art programs. In contrast, the assignment method for RNA is still under development. Here, we focused on rather a basic method for NMR analysis of RNA for researchers who recently joined or want to join this field. Please refer to the previous review articles for NMR analysis of RNA [1, 2]. For recent progress, please refer to Ref. [3].

T. Sakamoto · M. Otsu · R. Ichijo · G. Kawai (✉)
Department of Life Science, Chiba Institute of Technology, Narashino 275-0016, Chiba, Japan
e-mail: gkawai@sea.it-chiba.ac.jp

The NMR Society of Japan (ed.), *Experimental Approaches of NMR Spectroscopy II*,
https://doi.org/10.1007/978-981-97-6830-1_8

8.1 Design of RNA Sequences

Because RNA consists of only four kinds of nucleotides, signal overlaps are always to be overcome. Thus, it is usually required to design shorter RNA fragments to be analyzed by NMR. In many cases for stable isotopic labeling, the in vitro transcription using the T7 RNA polymerase is used. To increase the transcription efficiency, the 5' terminal must be a G residue, hopefully GGG. This requirement limits the flexibility of the fragment design.

To design shorter fragments, it is recommended to check the possible secondary structure of the target RNA by the secondary structure prediction programs such as CentroidFold [4] or vsfold5 [5]. Figure 8.1 shows an example of the secondary structure predicted by vsfold5. For this RNA, two GC base pairs were added to its terminal to stabilize the structure as well as make the 5' terminal start with GG. Sometimes, an A residue is added to the 3' terminal to stabilize the stem structure.

For larger RNA, it may be required to divide into several pieces to assign NMR signals and obtain structural restraints. Figure 8.2 shows an example of such division; to determine the structures of two types of dimers of 39nt RNA, two fragments are prepared [6]. By using the structures of each part, the overall structures were constructed. The constructed overall structure can be refined by RDC or other long-range information.

A

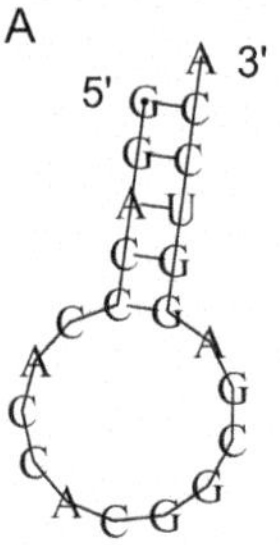

B

5'-CTAATACGACTCACTATA-3'
3'-GATTATGCTGAGTGATATCCTGGGTGGTGCCGCTCCAGGT-5'

Fig. 8.1 Design of RNA and DNA template. **a** The predicted secondary structure of a designed RNA. **b** An example of the template DNA for the in vitro transcription by the T7 RNA polymerase. Underlines indicate the T7 promoter

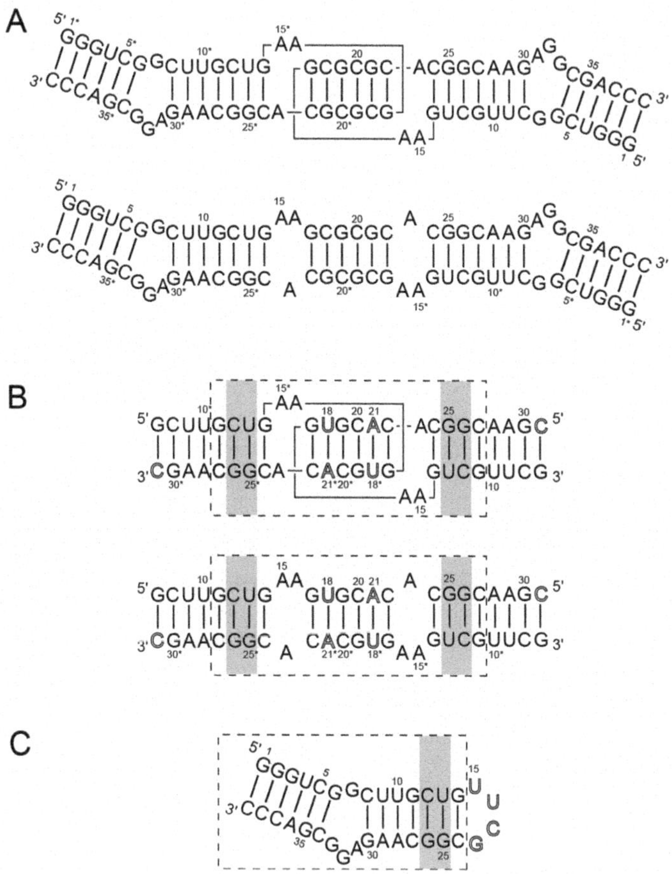

Fig. 8.2 Design of RNA fragments for the NMR analysis of larger RNA. **a** Secondary structure of the target RNA, HIV-1 DIS, in the kissing dimer (upper) and extended-duplex dimer (lower) conformation. **b** A fragment containing the self-complementary loop. **c** A fragment containing the internal loop. Residues in the grey box will be superimposed for structure determination of the whole RNA. [5] Reproduced from ref. by permission of Oxford University Press

8.2 Sample Preparation

To prepare a large amount of RNA samples for NMR (100 nmol, for example), two methods, chemical synthesis, and enzymatic synthesis (in vitro transcription) are available. Chemical synthesis is useful for the preparation of short RNAs (approximately less than 30 nucleotides), while in vitro transcription is suitable for long RNAs. Furthermore, chemical synthesis is useful for site-specific labeling with stable isotopes, and in vitro transcription is available for uniform labeling or nucleotide-specific labeling.

8.2.1 In Vitro Transcription

Although T3 RNA polymerase, SP6 RNA polymerase, and T7 RNA polymerase are commercially available for in vitro transcription, the T7 RNA polymerase-based in vitro transcription system is predominant for the preparation of NMR samples. The experimental procedure of in vitro transcription using T7 RNA polymerase was described previously [7], and some commercial kits, in which reaction buffers are optimized, are available (Epicentre Technologies Corp.). The promoter sequence of the template DNA is critical for recognition by the polymerase. Figure 8.1b shows an example of the DNA template; antisense strand (lower) and sense strand (upper).

Although the transcription efficiency depends on RNA sequence, the T7 RNA polymerase-based in vitro transcription system is efficient, simple, and cost-effective for the preparation of RNA samples. Usually, 1–10 ml of reaction is needed to prepare 100 nmol of RNA sample which is the standard quantity of RNA for structure determination using NMR. On the other hand, it has some drawbacks as follows. Short RNAs with less than 10 nucleotides are produced with only very low efficiency. The polymerase requires a guanosine at the 5' terminal of the transcript, and other nucleotides A, C, and U cannot be incorporated. It prefers purine-rich sequences near the 5' end of RNA, such as GGG and GGA. Furthermore, 1 or 2 nucleotides of random sequences are added beyond the desired sequence, and it is difficult to purify the desired product away from these by-products. Thus, heterogeneity of the sample often leads to multiple sets of NMR signals and makes analysis of the spectra complicated. To avoid heterogeneity, ribozymes can be applied for the removal of the additional nucleotides [8].

8.2.2 Chemical Synthesis

Although the efficiency of chemical synthesis of RNA is not as high as that of DNA, solid-phase synthesis by the phosphoramidite method has been established to produce large amounts of RNA. To synthesize the desired RNA, the phosphoramidite building blocks with 2'-O-butyldimethylsilyl as a 2'-hydroxyl protecting group are sequentially coupled to the growing oligonucleotide chain in the order. The process is fully automated using a DNA/RNA synthesizer and purified RNAs that have desired sequences can be purchased from commercial companies. By using the $^{13}C/^{15}N$-labeled phosphoramidite building blocks, labeled residues can be incorporated into specific positions of the RNA sequence.

Fig. 8.3 Scheme of the enzymatic ligation of RNA by T4 RNA ligase. Y: pyrimidine, R: purine, N: any of A, C, G, U

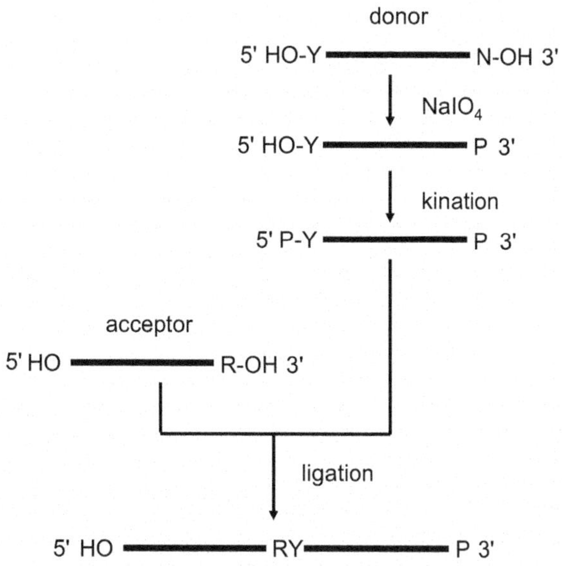

8.2.3 Enzymatic Ligation

To incorporate labeled or modified nucleotides into RNA fragments at the specific positions in the sequence, enzymatic ligation using T4 RNA ligase can be used. For shorter RNA, typically shorter than 50nt, residue-specific labeling can be introduced by the chemical synthesis. Then, longer RNA with labeled or modified nucleotides can be prepared by ligating fragments.

To design the fragments to be used for ligation with T4 RNA ligase, it is required that the 5' fragment (acceptor) have a purine residue at the 3' terminal with 3'-OH and the 3' fragment (donor) have a pyrimidine residue at the 5' terminal with 5'-phosphate. To avoid the self-ligation of each fragment, the acceptor and donor should have 5'-OH and 3'-phosphate, respectively, as shown in Fig. 8.3. The 5'-phosphate can be added by kinase, and 3'-phosphate can be generated by the $NaIO_4$ treatment [9]. This method was used for specific $^{13}C/^{15}N$-labeling of tRNA [10].

8.2.4 Artificial Base Pair System

The artificial base pair system is an alternative way to incorporate modified nucleotides into RNA fragments at the specific positions in the sequence [11]. The artificial deoxynucleotide triphosphates, dDsTP and dPxTP, can be used for conventional PCR reactions in which the Ds-Px base pair is formed and starting from a small amount of the chemically synthesized DNA fragment with Ds in the specific position, fragments with Ds or Px were amplified. The amplified double-stranded fragment is used for in vitro transcription by T7 RNA polymerase, for example, and the artificial ribonucleotide triphosphate, N_3-PaTP, is incorporated into the transcript. The N_3-Pa base can react with a variety of compounds by the click reaction. For example, spin-label such as (2,2,6,6-Tetramethylpiperidin-1-yl)oxyl or TEMPO can be introduced into the specific position for the paramagnetic relaxation enhancement (PRE) measurement.

8.2.5 *Stable Isotopic Labeling*

Because proton signals of RNA are usually overlapped, stable isotopic labeling is required to separate the overlapped signals. Using commercially available $^{13}C/^{15}N$-labeled NTPs (Taiyo Nippon Sanso Co.), stable isotopic labeled RNAs can be synthesized by using an in vitro transcription system. The in vitro transcription system is useful for uniform labeling and nucleotide-specific labeling. The nucleotide-specific labeling is required for long RNAs to separate the overlapped signals. On the other hand, chemical synthesis with $^{13}C/^{15}N$-labeled phosphoramidite building blocks has been carried out for site-specific labeling although chemical synthesis is available for short RNAs. To avoid the overlap of signals, site-specific labeling or segmental labeling is required for long RNAs. Thus, for the site-specific labeling and segmental labeling, preparative ligation methods of a short RNA fragment containing isotope-labeled nucleotides with the other RNA fragments have been developed [12–15]. Recently, a solid–liquid phase transcription method using an automated robotic platform was developed for the synthesis of RNAs with site-selective labeling [16]. Furthermore, methods to incorporate labels at specific atomic positions within each nucleotide, such as 1',5',8-^{13}C3 guanosine residue, have been developed using recombinantly expressed enzymes and chemically synthesized ribose and nucleobase [17]. As described below, site-specific 10% $^{13}C/^{15}N$-labeled RNAs can be used to identify signals for the labeled residues and, at the same time, obtain structural information for whole RNA by using conventional experiments such as ^{1}H-^{1}H NOESY with a single RNA sample. Because the 10% $^{13}C/^{15}N$-labeled nucleotide unit is a mixture of the uniformly $^{13}C/^{15}N$-labeled unit and the unlabeled unit with the molar ratio of 1:9, the cost for the labeled residues can be saved by using 10% $^{13}C/^{15}N$-labeled units. It is not recommended to introduce 10% $^{13}C/^{15}N$-labeling to the residues with broad signals because a broad 10% signal can be weaker than overlapped natural abundance signals.

8.2.6 *Purification*

In the case of in vitro transcription, long by-products such as n + 1 and n + 2 products are synthesized. On the other hand, short by-products such as n-1 and n-2 products are produced in the case of chemical synthesis. For the purification of the desired RNA away from these by-products, preparative polyacrylamide gel electrophoresis (PAGE) under denaturing conditions with 7 M urea is widely used. This method has the advantage that single nucleotide resolution can be achieved for approximately less than 30 mer RNAs. After electrophoresis, RNA bands can be visualized by UV shadowing and cut out from the gel. Following the elution of the desired RNAs from the crushed gel (crush and soak procedure), the RNAs are purified using ethanol precipitation. For the folding of RNA conformation, the RNAs are annealed as follows. To obtain single-stranded RNAs, RNA samples are incubated at around 95 °C and then snap-cooled on ice under low salt concentration. On the other hand, to obtain double-stranded conformation, the samples are cooled gradually after incubation at around 95 °C under high salt concentration. The confirmation can be confirmed by native PAGE. To exchange buffer solution for NMR measurement, an ultrafiltration membrane is useful, and the concentration of RNAs is estimated by UV absorption.

An alternative approach to purification is gel filtration chromatography. Although gel filtration chromatography does not have single nucleotide resolution, it is useful to separate the desired long RNAs from the unincorporated nucleotides and short abortive by-products without denaturing RNA conformation [18, 19]. It is noted that it may be difficult to refold the conformation of long RNAs.

8.3 Measurements

Exchangeable imino protons and amino protons can be observed only in H_2O containing 5% D_2O, while non-exchangeable protons can be observed even in 100% D_2O.

To obtain NMR restraints precisely and as much as possible, it is recommended to measure NMR spectra in 100% D_2O because many ribose protons overlap with the water signal. However, it is also possible to analyze NMR signals of RNA only with spectra measured in H_2O containing 5% D_2O, especially for RNA with stable isotopic labeling.

8.3.1 Exchangeable Protons

Imino proton signals of guanosine and uridine residues observed between 10–15 ppm contain valuable information about base pairing in the RNA molecule, because these signals are observable when the imino protons are involved in hydrogen bonding or protected from exchange with the bulk solvent water. Because of the fast exchange of imino and amino protons with water protons, presaturation of the H_2O signal cannot be used for solvent suppression to observe exchangeable base proton signals in RNA in contrast to NMR of proteins. Despite the undesirable baseline distortions, the jump-and-return pulse sequence is useful for solvent suppression of RNA samples

[20]. Furthermore, WATERGATE or similar solvent suppression schemes can be used to overcome the drawbacks of the jump-and-return pulse sequence [21].

The Imino proton signals of A:U and G:C base pairs tend to be observed at 13–15 ppm and 12–14 ppm, respectively, whereas imino protons of noncanonical base pairs (ex. G:U base pair) often show upfield chemical shifts (10–12 ppm) (Fig. 8.4). The chemical shifts of the imino proton signals depend strongly on the sequence of RNA. The analysis of imino proton signals is useful to elucidate the secondary structure of RNA. In addition, the conformational stability of RNA molecules can be investigated by following temperature-induced changes in the imino proton spectra. Amino proton signals of cytidine are observable when they are involved in hydrogen bonding, whereas those of guanosine and adenosine are hardly observed. By HNN-COSY experiment of ^{15}N-labeled RNA, scalar couplings across an N–H..N-type hydrogen bond can be observed for the nitrogen and proton of the hydrogen-bond donor and the nitrogen of the hydrogen-bond acceptor [22].

8.3.2 Non-Exchangeable Protons

For the measurement of NMR for non-exchangeable protons, D_2O is a useful solvent because signals around water can be easily observed without a water suppression pulse scheme. Solvent water H_2O can be replaced with D_2O by lyophilization or evaporation, because shorter RNA does not aggregate during such treatment in general. As shown in the NOESY spectrum (Fig. 8.5), signals of H2', H3', H4', H5', and H5" are overlapped at 3.5–5 ppm, while H1' and base protons can be observed separately.

In the case of ^{13}C/^{15}N-labeled RNA, the water signal can be suppressed effectively by the ^{13}C-edited experiments and the NMR analysis can be done with spectra measured in H_2O.

8.3.3 Residual Dipolar Coupling

Residual dipolar couplings (RDCs) between spins i and j provide long-range constraints on the average orientation (θ) on the inter-nuclear bond vector relative to the applied magnetic field (B_0) [23]. Because RNAs are usually elongated molecules, the long-range constraints are extremely useful for obtaining information on the global conformation of RNAs. Although several media can be applied, Pf1 filamentous phage has been generally used for the alignment of RNAs (Fig. 8.6) [24, 25]. The negatively charged Pf1 phage is aligned in the magnetic field, which induces a slight alignment of RNA molecules. For the measurement of RDCs, 5–15 mg/ml of Pf1 phage is added to the stable isotope-labeled RNAs. RDC can be measured as the difference of coupling width (Hz) between with and without phage by undecoupled ^{15}N-^{1}H SQC or ^{13}C-^{1}H SQC spectrum. To avoid the overlap of signals, in-phase and

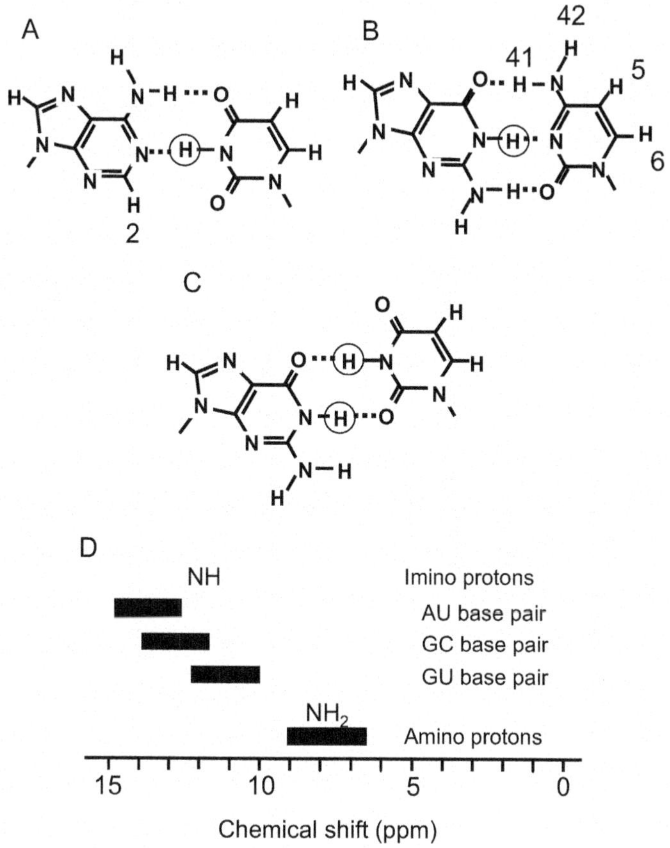

Fig. 8.4 Imino protons in the typical base pairs of RNA. **a** A:U base pair. **b** G:C base pair. **c** G:U base pair. Imino protons are indicated by circles. **d** Distributions of proton chemical shifts for the exchangeable protons in RNA

anti-phase heteronuclear single quantum coherence (IPAP-HSQC) [26] or transverse relaxation optimized spectroscopy (TROSY) [27] can be used.

Furthermore, Al-Hashimi et al. have applied RDC for studying RNA structural dynamics [28, 29]. Inherent flexibility involving local motions of nucleotide residues and collective helix motions can be characterized over a wide range of timescales (< milliseconds).

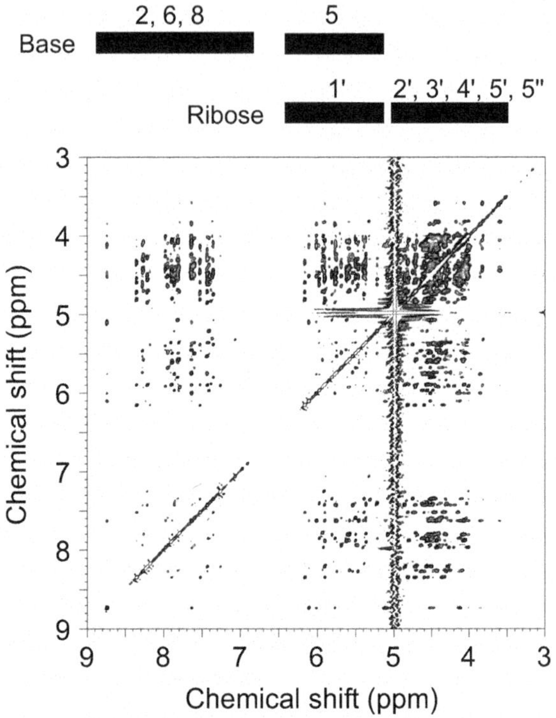

Fig. 8.5 A 2D NOESY spectrum of a 22-mer RNA in D_2O. Distributions of proton chemical shifts for the non-exchangeable protons in RNA were indicated at the top

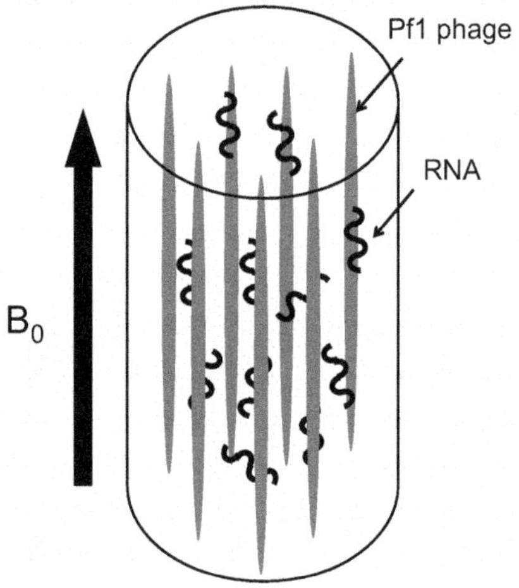

Fig. 8.6 Schematic representation of the alignment of RNAs using pf1 phage for RDC measurement. Pf1 phage is aligned in the magnetic field, which induces a slight alignment of RNA molecules

8.4 Signal Assignments

8.4.1 Exchangeable Protons

Based on the predicted secondary structure of RNA, the imino proton signals can be sequentially assigned along the stem sequence using the 2D NOESY spectrum (Fig. 8.7). A strong NOE cross peak between the uridine H3 and the adenosine H2 can be observed in A:U base pairs. In G:C base pairs, NOE cross peaks can be observed between the guanosine H1 and the amino protons of cytidine. Because one of the amino protons of cytidine in G:C base pair is close to H5 of the cytidine, H5/H6 signals can be assigned by following the NOE connectivity as H1(G)-H41/42(C)-H6(C). G:U base pairs, whose imino proton signals resonate higher field around 12–10 ppm, can be easily identified by a strong NOE cross-peak between the guanosine H1 and the uridine H3.

Heteronuclear multi-quantum coherence (HMQC) spectrum of ^{15}N-labeled RNA is useful to assign the imino proton signals. Because the ^{15}N chemical shift of guanosine N1 is around 150 ppm and that of uridine N3 is around 160 ppm, imino proton signals of guanosine and uridine can be clearly discriminated.

8.4.2 Non-Exchangeable Protons

Although signals of ribose protons are highly overlapped, H1' signals can be observed separately from other ribose signals. Thus, sequential assignment of base and ribose H1' signals can be accomplished by the analysis of NOE connectivities in helical A-form RNA (Fig. 8.8). The H1' of residue i shows NOE cross peaks to the H8/H6 of residue i (intra-residue) and residue i + 1 (inter-residue), which enables sequential assignment of these signals along the RNA sequence. Figure 8.9 shows the region where NOEs used for the sequential assignment, NOEs between H2/H8/H6 and H5/H1', are observed. Because the inter-residue H1'- H8/H6 distance is larger than 4 Å in helical A-form RNA, spin-diffusion-mediated NOEs, through strong NOE between H6/H8 and H2'/H3' protons, caused by long mixing time (> 400 ms) of NOESY experiments may be useful for the sequential assignment. Furthermore, the inter-residue NOE cross-peaks between base protons are helpful for sequential assignment. Although NOEs between base protons and the sugar protons H4', H5', and H5'' can be observed, their assignment is difficult because of the low dispersion in most RNA sequences.

To perform the NOE-based assignment procedure, connectivities between non-exchangeable base protons and imino protons are important. The H5 proton of cytidine can be identified by NOE connectivities to the amino protons of cytidine in H_2O as described above. The pyrimidine H5 and H6 signals are easy to identify because of the strong NOE cross-peaks observed in the NOESY experiment, which can be confirmed by the TOCSY experiment. Signals of cytidine and uridine can be

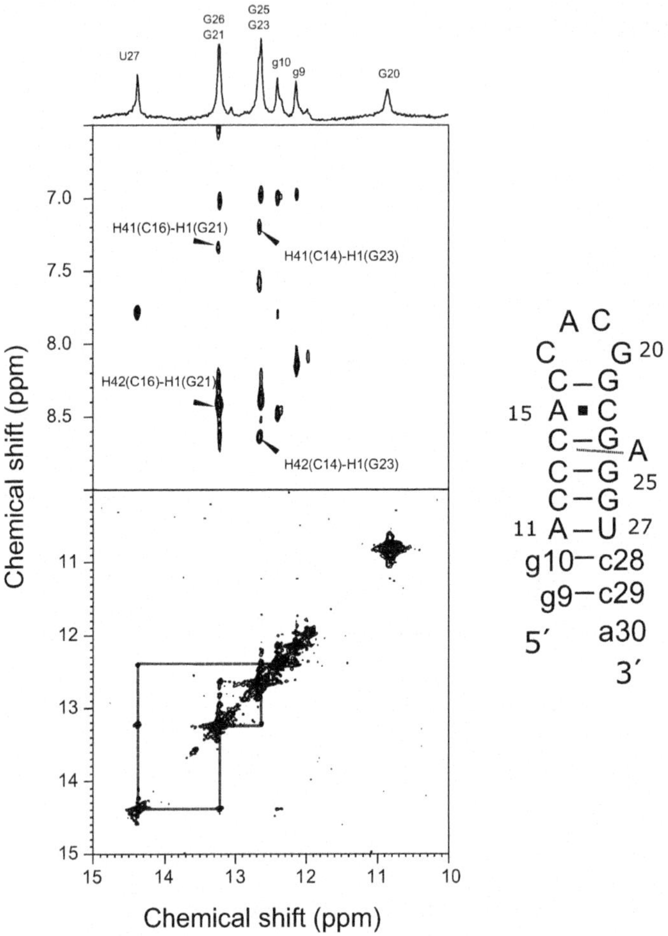

Fig. 8.7 Assignment of imino proton signals of a 22-mer RNA. Sequential connectivities are indicated by lines in the 2D NOESY spectrum and assignments are shown on the 1D spectrum. Secondary structure of the 22-mer RNA is shown. [29] Reproduced from ref. by permission of Oxford University Press

distinguished by the difference between the chemical shifts of the C5 resonances in a ^{13}C-^{1}H SQC (Fig. 8.10). Furthermore, H2 signals of adenosine can be identified by NOE contacts to the uridine imino protons in H_2O. The C2H2 region (150 ppm in ^{13}C) is well resolved in a ^{13}C-^{1}H SQC. In helical A-form RNA, the distance between H2 (i) and the H1' (i + 1) is shorter than 4 Å, and, thus, strong NOEs can be observed between them, which enables confirmation of sequential assignments.

Figure 8.11 shows a ^{13}C-^{1}H SQC spectrum of a 10% $^{13}C/^{15}N$ labeled RNA. By using such signals, the structural change and/or interaction with other molecules can be monitored easily [30]. In the case of G- or U-labeled RNA, the imino ^{1}H-^{15}N

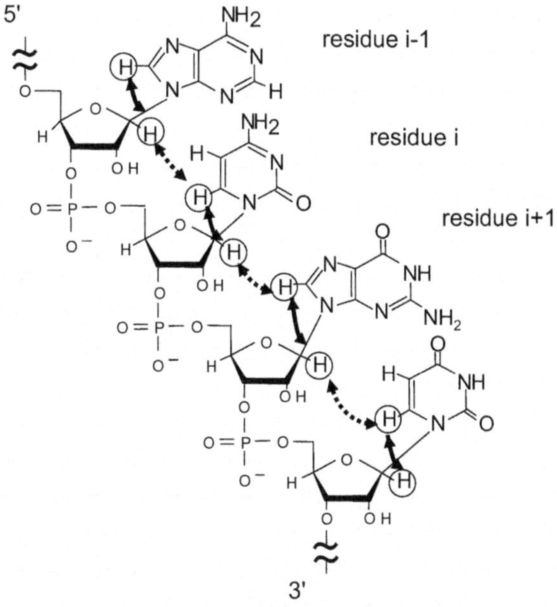

Fig. 8.8 Schematic representation of NOE connectivities. The H1' of residue i shows NOE cross peaks to the H8/H6 of residue i (intra-residue) and residue i + 1 (inter-residue). Intra-residue and inter-residue NOEs are shown by arrow and dotted arrow, respectively

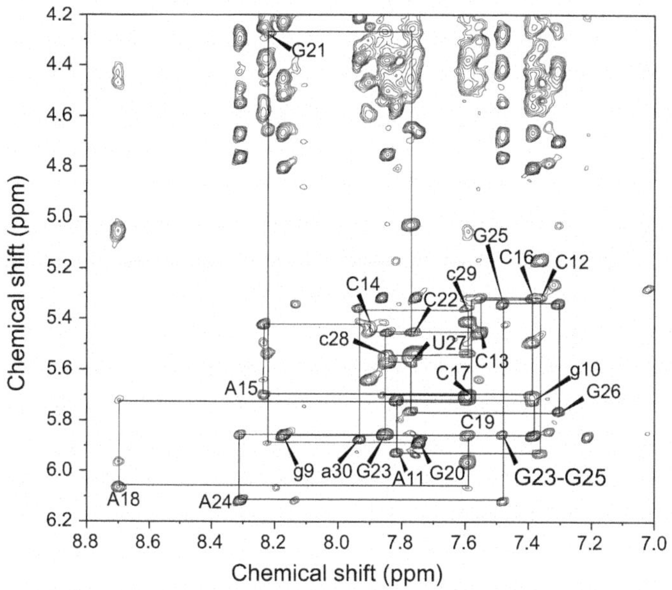

Fig. 8.9 A 2D NOESY spectrum of a 22-mer RNA in D_2O. The region for cross-peaks between aromatic H6 and H8 and ribose H1' is shown. Sequential connectivities are indicated by lines and intra-residue NOEs are labeled. [29] Reproduced from ref. by permission of Oxford University Press

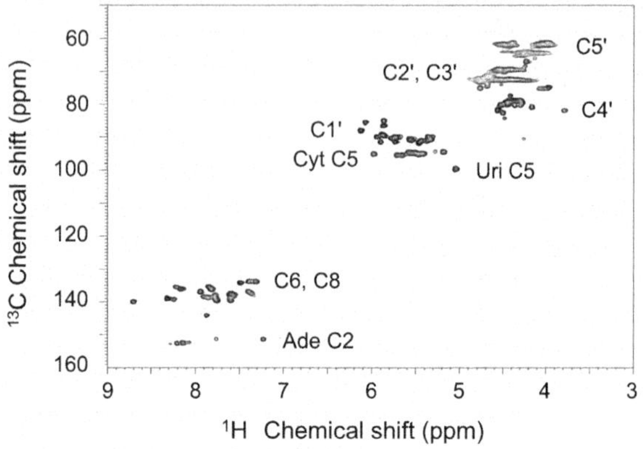

Fig. 8.10 ^{13}C-^{1}H SQC spectrum of a 22-mer RNA. C2 of adenosine (Ade), C5 of cytidine (Cyt), and uridine (Uri) can be assigned by their ^{13}C chemical shift

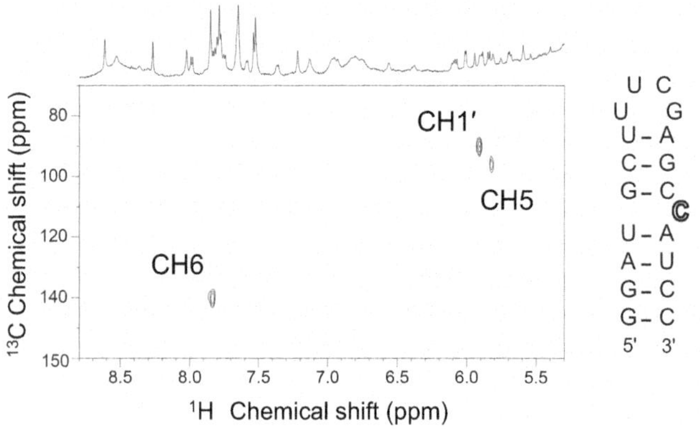

Fig. 8.11 ^{13}C-^{1}H SQC spectrum of a 19-mer RNA with a 10% $^{13}C/^{15}N$ labeled cytidine residue. The labeled residue is indicated in the secondary structure. The RNA sample concentration was 0.20 mM

signals can be observed in ^{15}N-^{1}H HMQC spectra, if stable base pairs are formed [30].

8.5 Structural Calculation

As a typical example of the RNA structure determination by NMR, the structure determination of a 22-mer RNA fragment is shown. The RNA obtained by SELEX against the AML1 Runt domain contains a DNA mimicry motif [31]. The non-labeled sample (0.6 mM) and uniformly $^{13}C/^{15}N$-labeled sample (0.5 mM) were synthesized using T7 RNA polymerase. NMR spectra were measured using Bruker DRX-500 and DRX-600 spectrometers at 10 °C. Imino proton resonances of G and U residues were distinguished by the ^{15}N-^{1}H SQC spectra and assigned by NOESY with mixing

Table 8.1 The calculation protocol used for the structure determination of a 22nt RNA [29]

Structure calculation without RDCs (CNS 1.1)		
Hot steps	20,000 K	60,000 steps (60 ps)
Cooling steps	20,000 K to 0 K	170,000 steps (170 ps)
Refinement with RDCs (Xplor-NIH 2.38)		
Refinement steps	2,000 K to 0 K	5,000 steps (10 ps)

times of 150 ms using the jump-and-return scheme for water suppression. Base pairing schemes were established by 2D HNN-COSY experiments. HCCH-COSY and HCCH-TOCSY were used to assign sugar spin systems. H2 protons of adenosines were assigned using HCCH-TOCSY and 2D SQC. NOE distance restraints from non-exchangeable protons were obtained using NOESY (mixing times of 100 and 400 ms) in D_2O. Dihedral restraints were obtained from TOCSY (mixing time of 50 ms). Partial alignment for the RDC measurements was achieved by the addition of approximately 16.9 mg/ml of Pf1 phage (ASLA Ltd.) to uniformly $^{13}C/^{15}N$-labeled samples. Single-bond ^{1}H-^{13}C RDC values for bases and ribose moieties were measured using non-decoupled ^{13}C-^{1}H SQC.

NOE from non-exchangeable protons were assigned using Felix and ARIA in combination with CNS (version 1.1). Restraints are summarized in Table 8.2. A set of 100 structures was calculated using a simulated annealing protocol with CNS 1.1 utilizing 354 distance restraints and 80 dihedral restraints (Table 8.3) starting from an extended structure. The protocol used for the structure calculation is summarized in Table 8.1. The alignment tensor was obtained by the grid search using the lowest-energy structures for the first calculation without RDCs. The final 10 structures with the lowest total energy were chosen (Fig. 8.12) and the structural statistics are shown in Table 8.4.

It is noted that an improved method for structure calculations was proposed [32] in which the van der Waals radius was modified from 0.9 to 1.10.

8.6 Interaction Analysis and Structure Screening

NMR is a useful tool for interaction analysis. The imino proton signals are often used for probes of interaction. This technique is quite important for RNA-targeted drug discovery [33]. Figure 8.13a shows the imino proton signals of a 19-mer RNA similar to that in Fig. 8.11 with a various amount of a fluoroquinolone derivative, KG022 [30]. Upon the addition of KG022, a new set of signals appeared, indicating a specific interaction between the RNA and KG022. The pyrimidine H5-H6 signals in TOCSY or HOHAHA spectra are also good probes of interaction. Figure 8.13b shows the drastic changes of the H5 and H6 chemical shifts of a 19-mer RNA upon binding of KG022, indicating the formation of a stable complex [30].

High-throughput analyses such as high-throughput sequencing give a large amount of information for novel functional RNA candidates. Thus, methods for

Table 8.2 Example of the constraints for structure determination [29]

NOE distance restraints		
between exchangeable protons from non-exchangeable protons	1.8 - 6.0 Å	
strong	1.8 - 3.0 Å	
medium	1.8 - 4.0 Å	
weak	1.8 - 5.0 Å	
very weak	1.8 - 7.0 Å	
A-helix restraints for stem region		
α	−62° ±30°	
β	−174° ±30°	
γ	49° ±30°	
δ	95° ±30°	
ε	−158° ±30°	
ζ	−73° ±30°	
Sugar packer restraints (δ angle) based on DQF-COSY H1'-H2' cross peak		
strong	C2'-*endo*	145° ±25°
no cross peak	C3'-*endo*	95° ±30°
Base pair restraints		
Hydrogen bond		
N1-H3 (AU)	1.8 - 2.2 Å	
H61-O4 (AU)	1.8 - 2.2 Å	
H1-N3 (GC)	1.8 - 2.2 Å	
O6-H41 (GC)	1.8 - 2.2 Å	
H21-O2 (GC)	1.8 - 2.2 Å	
Base pair planarity	N1 - N3 - C5	

high-throughput screening of RNA structure are required. Recently, two methods for RNA structure screening were proposed [34, 35]. The first method, In NMR tube transcription (INTT), is quite simple in which the transcription reaction by the T7 RNA polymerase is performed in the NMR tube and the NMR spectra of the transcript in the reaction mixture directly [34]. In the INTT method, no purification nor specific skills for RNA handling are required. The limitation of INTT is the condition of measurement is fixed by the transcription reaction. The second one accelerates the screening process by direct buffer exchange and concentration of the transcription mixture [35]. In this case, any buffer condition suitable for NMR measurements of the target RNA can be used. The structural profiling ability was applied to the monitoring of the concentration of specifically folded RNA during the process of the systematic evolution of ligands by the exponential enrichment (SELEX) method [36].

Table 8.3 Number of restraints used for the structure determination of a 22nt RNA [29]

Distance restraints	354
Total NOE	327
intra-residue	133
strong	44
medium	42
weak	3
very weak	44
inter-residue	194
strong	21
medium	36
weak	10
very weak	127
Hydrogen bonds	27
Dihedral angle restraints	72
Sugar pucker	19
C2'-*endo*	4
C3'-*endo*	15
Based on A-form geometry*	53
Base pair planarity	8
Residual Dipolar Coupling	10

*Restraints of angles γ, δ, ε, and ζ are used for the residue at 5'-terminal of helical A-form RNA, while those of angles α, β, γ, and δ are used at 3'-terminal.

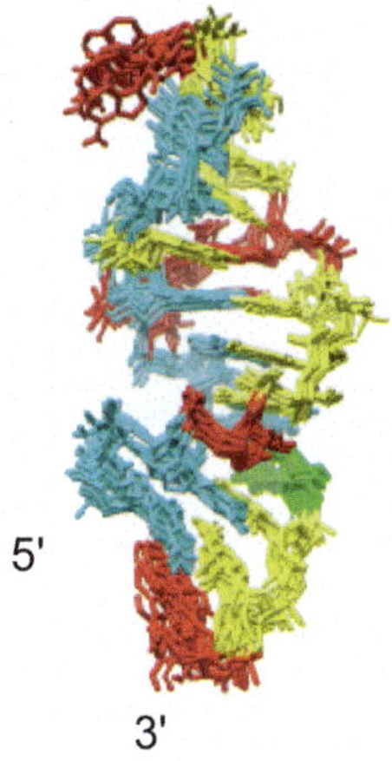

Fig. 8.12 Ensemble of 10 lowest energy structure of a 22-mer RNA. G, A, C, and U are indicated by blue, red, yellow, and green, respectively. [29] Reproduced from ref. by permission of Oxford University Press. The corresponding secondary structure is shown in Fig. 7

Table 8.4 Structural statistics for a 22nt RNA [29]

Distance constraints (Å)	0.0592± 0.0016
Dihedral angle constraints (°)	0.0934 ± 0.1826
r.m.s. deviations from idealized geometry	
Bond lengths (Å)	0.0065 ± 0.00019
Bond angles (°)	1.1898 ± 0.05405
Impropers (°)	1.0435 ± 0.14214
Averaged pairwise r.m.s. deviation (Å)	
All RNA heavy atoms	0.825

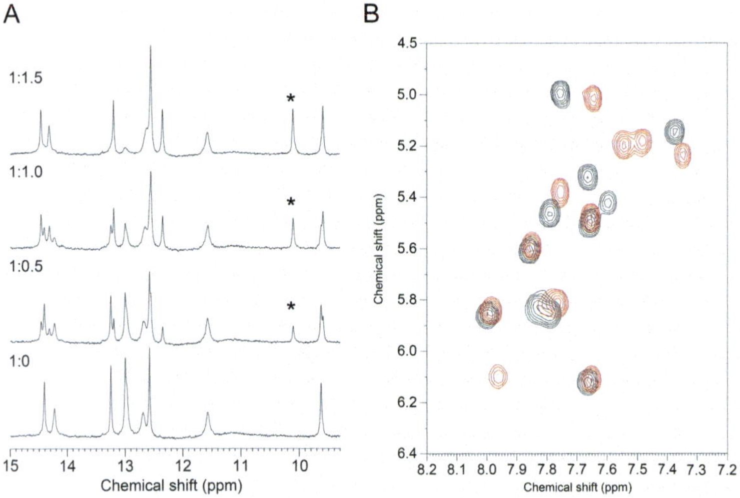

Fig. 8.13 NMR spectra of 19-mer RNAs. **a** Imino proton spectra in H_2O (5% D_2O) at 288 K. The molar ratio of RNA and a fluoroquinolone derivative, KG022 in the solution was indicated for each spectrum. The RNA sample concentration was 0.47 mM. The asterisks show the signal of KG022. **b** HOHAHA spectra showing the H5-H6 cross peaks of pyrimidine residues in D_2O at 288 K. Black: free RNA. Red: in the presence of KG022. The RNA sample concentration was 0.39 mM

8.7 Perspective

Some systems to support NMR signal analysis for RNA have been developed; for example, Chess2FLYA [37] and RNA-PAIRS [38]. Such systems will accelerate the NMR analysis of RNA.

Recently, functions of long RNAs such as untranslated regions of mRNAs, long non-coding RNAs (lncRNAs), and viral RNAs have attracted attention, and their NMR analyses have become important [3]. For example, the secondary structure of a 167nt long noncoding RNA was reported [39] by means of dividing the RNA into smaller units as described in the Sect. 8.1. The imino proton signals were used to confirm that the structure was maintained after dividing. Interestingly, it was suggested that the secondary structure determined by NMR is different from the computationally predicted structures. For larger RNA, methods to predict the structural domain in long sequences must be improved. GenoPoemics system for visualization of predicted secondary structures may be useful for such analysis [40].

In-cell NMR has become a powerful method used to directly characterize not only proteins but also nucleic acids in living cells [41]. In-cell NMR provided direct evidence for the formation of the hairpin structure of an RNA [42] and the interaction between an RNA and its ligand in living cells [43], It is expected that new knowledge about the structure and function of RNAs in living cells will be obtained by in-cell NMR.

References

1. Varani, G., Aboul-ela, F., Allain, F.H.-T.: NMR investigation of RNA structure. Prog. NMR Spectrosc. **29**, 51–127 (1996)
2. Fürtig, B., Richter, C., Wöhnert, J., Schwalbe, H.: NMR spectroscopy of RNA. Chem. Bio. Chem. **4**, 936–962 (2003)
3. Marušič, M., Toplishek, M., Plavec, J.: NMR of RNA - Structure and interactions. Curr. Opin. Struct. Biol. **79**, 102532 (2023)
4. Sato, K., Hamada, M., Asai, K., Mituyama, T.: CentroidFold: a web server for RNA secondary structure prediction. Nucleic. Acids Res. **37**, W277–W280 (2009)
5. Dawson, W., Fujiwara, K., Kawai, G.: Prediction of RNA pseudoknots using heuristic modeling with mapping and sequential folding. PLoS ONE **2**, 905 (2007)
6. Baba, S., Takahashi, K., Noguchi, S., Takaku, H., Koyanagi, Y., Yamamoto, N., Kawai, G.: Solution RNA structures of the HIV-1 dimerization initiation site in the kissing-loop and extended-duplex dimers. J. Biochem. **138**, 583–592 (2005)
7. Milligan, J.F., Groebe, D.R., Witherell, W.G., Uhlenbeck, O.C.: Oligoribonucleotide synthesis using T7 RNA polymerase and synthetic DNA templates. Nucleic. Acids Res. **15**, 8783–8798 (1987)
8. Ferré-D'Amaré, A.R., Doudna, J.A.: Use of cis- and trans-ribozymes to remove 5' and 3' heterogeneities from milligrams of in vitro transcribed RNA. Nucleic. Acids Res. **24**, 977–978 (1996)
9. Ohtsuki, T., Kawai, G., Watanabe, Y., Kita, K., Nishikawa, K., Watanabe, K.: Preparation of biologically active *Ascaris suum* mitochondrial $tRNA^{Met}$ with a TV-replacement loop by ligation of chemically synthesized RNA fragments. Nucleic. Acids Res. **24**, 662–667 (1996)
10. Ohtsuki, T., Kawai, G., Watanabe, Y.: The minimal tRNA: unique structure of *Ascaris suum* mitochondrial $tRNA^{Ser}{}_{UCU}$ having a short T arm and lacking the entire D arm. FEBS Let. **514**, 37–43 (2002)
11. Someya, T., Ando, A., Kimoto, M., Hirao, I.: Site-specific labeling of RNA by combining genetic alphabet expansion transcription and copper-free click chemistry. Nucleic Acids Res. **43**, 6665–6676 (2015)
12. Xu, J., Lapham, J., Crothers, D.M.: Determining RNA solution structure by segmental isotopic labeling and NMR: application to Caenorhabditis elegans spliced leader RNA 1. Proc. Natl. Acad. Sci. U.S.A. **93**, 44–48 (1996)
13. Kim, I., Lukavsky, P.J., Puglisi, J.D.: NMR study of 100 kDa HCV IRES RNA using segmental isotope labeling. J. Am. Chem. Soc. **124**, 9338–9339 (2002)
14. Tzakos, A.G., Easton, L.E., Lukavsky, P.J.: Complementary segmental labeling of large RNAs: economic preparation and simplified NMR spectra for measurement of more RDCs. J. Am. Chem. Soc. **128**, 13344–13345 (2006)

15. Kawahara, I., Haruta, K., Ashihara, Y., Yamanaka, D., Kuriyama, M., Toki, N., Kondo, Y., Teruya, K., Ishikawa, J., Furuta, H., Ikawa, Y., Kojima, C., Tanaka, Y.: Site-specific isotope labeling of long RNA for structural and mechanistic studies. Nucleic Acids Res. **40**, e7 (2012)
16. Liu, Y., Holmstrom, E., Zhang, J., Yu, P., Wang, J., Dyba, M.A., Chen, D., Ying, J., Lockett, S., Nesbitt, D.-D., A.R., Sousa R., Stagno J.R., Wang Y.X.: Synthesis and applications of RNAs with position-selective labelling and mosaic composition. Nature **522**, 368–372 (2015)
17. Longhini, A.P., LeBlanc, R.M., Becette, O., Salguero, C., Wunderlich, C.H., Johnson, B.A., D'Souza, V.M., Kreutz, C., Dayie, T.K.: Chemo-enzymatic synthesis of site-specific isotopically labeled nucleotides for use in NMR resonance assignment, dynamics and structural characterizations. Nucleic. Acids Res. **44**, e52 (2016)
18. Lukavsky, P.J., Puglisi, J.D.: Large-scale preparation and purification of polyacrylamide-free RNA oligonucleotides. RNA **10**, 889–893 (2004)
19. McKenna, S.A., Kim, I., Puglisi, E.V., Lindhout, D.A., Aitken, C.E., Marshall, R.A., Puglisi, J.D.: Purification and characterization of transcribed RNAs using gel filtration chromatography. Nat. Protoc. **2**, 3270–3277 (2007)
20. Plateau, P., Gueron, M.: Exchangeable proton NMR without base-line distorsion, using new strong-pulse sequences. J. Am. Chem. Soc. **104**, 7310–7311 (1982)
21. Piotto, M., Saudek, V., Sklenár, V.: Gradient-tailored excitation for single-quantum NMR spectroscopy of aqueous solutions. J. Biomol. NMR **2**, 661–665 (1992)
22. Dingley, A.J., Grzesiek, S.: Direct Observation of Hydrogen Bonds in Nucleic Acid Base Pairs by Internucleotide $^2J_{NN}$ Couplings. J. Am. Chem. Soc. **120**, 8293–8297 (1998)
23. Tjandra, N., Bax, A.: Direct measurement of distances and angles in biomolecules by NMR in a dilute liquid crystalline medium. Sci. **278**, 1111–1114 (1997)
24. Hansen, M.R., Mueller, L., Pardi, A.: Tunable alignment of macromolecules by filamentous phage yields dipolar coupling interactions. Nat. Struct. Biol. **5**, 1065–1074 (1998)
25. Hansen, M.R., Hanson, P., Pardi, A.: Filamentous bacteriophage for aligning RNA, DNA, and proteins for measurement of nuclear magnetic resonance dipolar coupling interactions. Methods Enzymol. **317**, 220–240 (2000)
26. Ottiger, M., Delaglio, F., Bax, A.: Measurement of J and dipolar couplings from simplified two-dimensional NMR spectra. J. Magn. Reson. **131**, 373–378 (1998)
27. Pervushin, K., Riek, R., Wider, G., Wüthrich, K.: Attenuated T2 relaxation by mutual cancellation of dipole-dipole coupling and chemical shift anisotropy indicates an avenue to NMR structures of very large biological macromolecules in solution. Proc. Natl. Acad. Sci. U.S.A. **94**, 12366–123671 (1997)
28. Bothe, J.R., Nikolova, E.N., Eichhorn, C.D., Chugh, J., Hansen, A.L., Al-Hashimi, H.M.: Characterizing RNA dynamics at atomic resolution using solution-state NMR spectroscopy. Nat. Methods **8**, 919–931 (2011)
29. Salmon, L., Yang, S., Al-Hashimi, H.M.: Advances in the determination of nucleic acid conformational ensembles. Annu. Rev. Phys. Chem. **65**, 293–316 (2014)
30. Ichijo, R., Kamimura, T., Kawai, G.: Interaction between a fluoroquinolone derivative KG022 and RNAs: Effect of base pairs 3′ adjacent to the bulged residues. Front. Mol. Biosci. **10**, 1145528 (2023)
31. Nomura, Y., Tanaka, Y., Fukunaga, J., Fujiwara, K., Chiba, M., Iibuchi, I., Tanaka, T., Nakamura, Y., Kawai, G., Kozu, T., Sakamoto, T.: Solution structure of a DNA mimicking motif of an RNA aptamer against transcription factor AML1 Runt domain. J. Biochem. **154**, 513–519 (2013)
32. Bermejo, G.A., Clore, M., Schwieters, C.D.: Improving NMR Structures of RNA. Struct. **24**, 806–815 (2016)
33. Morishita EC (2022) Discovery of RNA-targeted small molecules through the merging of experimental and computational technologies. Expert. Opin. Drug. Discov, https://doi.org/10.1080/17460441.2022.2134852
34. Okui, S., Kawai, G..: In NMR tube transcription for rapid screening of RNA conformation. Nucleosides Nucleotides Nucleic Acids **34**, 103–113 (2015)

35. Helmling, C., Keyhani, S., Sochor, F., Fürtig, B., Hengesbach, M., Schwalbe, H.: Rapid NMR screening of RNA secondary structure and binding. J. Biomol. NMR **63**, 67–76 (2015)
36. Amano, R., Aoki, K., Miyakawa, S., Nakamura, Y., Kozu, T., Kawai, G., Sakamoto, T.: NMR monitoring of the SELEX process to confirm enrichment of structured RNA. Sci. Rep. **7**, 283 (2017)
37. Aeschbacher, T., Schmidt, E., Blatter, M., Maris, C., Duss, O., Allain, F.H.-T., Güntert, P., Schubert, M.: Automated and assisted RNA resonance assignment using NMR chemical shift statistics. Nucleic Acids Res. **41**, e172 (2013)
38. Bahrami, A., Clos, L.J., II., Markley, J.L., Butcher, S.E., Eghbalnia, H.R.: RNA-PAIRS: RNA probabilistic assignment of imino resonance shifts. J. Biomol. NMR **52**, 289–302 (2012)
39. Ohyama, T., Takahashi, H., Sharma, H., Yamazaki, T., Gustincich, S., Ishii, Y., Carninci, P.: An NMR-based approach reveals the core structure of the functional domain of SINEUP lncRNAs. Nucleic Acids Res. **48**, 9346–9360 (2020)
40. Nakamura, S.: A Novel Virtual Spectrometry: Visualized Regulatory Motifs on *ADM, rPolB* and *CD83* mRNAs in Human-friendly Manners. J. Biochem. **146**, 251–261 (2009)
41. Yamaoki, Y., Nagata, T., Sakamoto, T., Katahira, M.: Recent progress of in-cell NMR of nucleic acids in living human cells. Biophys. Rev. **12**, 411–417 (2020)
42. Yamaoki, Y., Kiyoishi, A., Miyake, M., Kano, F., Murata, M., Nagata, T., Katahira, M.: The first successful observation of in-cell NMR signals of DNA and RNA in living human cells. Phys. Chem. Chem. Phys. **20**, 2982–2985 (2018)
43. Eladl, O., Yamaoki, Y., Kondo, K., Nagata, T., Katahira, M.: Detection of interaction between an RNA aptamer and its target compound in living human cells using 2D in-cell NMR. Chem. Commun. **59**, 102–105 (2022)

Chapter 9
NMR Data Science

Jun Kikuchi

Abstract NMR offers tremendous advantages in the analysis of molecularly complex compounds, such as crude biological extracts, supramacromolecular complexes, geochemical samples, and natural and artificial materials. Herein, I introduce the recent applications of several NMR approaches for evaluating homeostatic plasticity for human and environmental health using a data science approach. Further challenges in addressing the macromolecular complexity include determining the supramolecular structures, composition, and interactions of plant biomass, soil humic substances, aqueous particulate organic matter, and natural and artificial materials. Because solution and solid-state NMR can produce numerical matrix data (e.g., chemical shifts versus intensity) as well as time-domain NMR data (e.g., time versus relaxation decay) with high reproducibility and inter-institution convertibility, further data science approaches, such as multivariate analysis, machine learning, and optimization, are desired. Therefore, I also introduce informatics techniques for data pretreatment prior to solid-state NMR, feature extraction from heterogeneously measured spectroscopic data, and extraction of submerged information using data science approaches.

Keywords Molecular complexity · Machine learning · Optimization problem · Materials informatics

J. Kikuchi (✉)
RIKEN Center for Sustainable Resource Science, 1-7-22 Suehiro-Cho, Tsurumi-Ku, Yokohama 230-0045, Kanagawa, Japan
e-mail: jun.kikuchi@riken.jp

Graduate School of Medical Life Science, Yokohama City University, 1-7-29 Suehiro-Cho, Tsurumi-Ku, Yokohama 230-0045, Kanagawa, Japan

Graduate School of Bioagricultural Sciences, Nagoya University, 1 Furo-Cho, Chikusa-Ku, Nagoya 464-0810, Aichi, Japan

The NMR Society of Japan (ed.), *Experimental Approaches of NMR Spectroscopy II*,
https://doi.org/10.1007/978-981-97-6830-1_9

9.1 Introduction

Nuclear magnetic resonance (NMR) data science involves analytical cycles based on the concept of measurement informatics, which consists of sample preparation [1], NMR measurement [2], database construction [3], data pre-processing [4], and data analysis [5] (Fig. 9.1). Several NMR techniques facilitate the study of chemical structures [6] and dynamics [7]. Furthermore, NMR can be used to evaluate the components of mixtures from an exposome perspective [8]. In addition, solid-state NMR can be used to evaluate the highly ordered structures of solid samples [9]. NMR can be classified into two types based on applicability: high-field NMR for advanced analyses at the research level (such as multidimensional NMR [10] and solid-state NMR [11]), and low-field NMR for simple measurements in the production sites [12, 13]. For both approaches, the measured data are stored in databases. Consequently, the chemical shifts in the NMR spectra of samples can be assigned by referring to the spectral data of standard samples stored in databases [14]. For complex mixtures, the data are preprocessed to simplify the multicomponent signals [15], and the processed data are then evaluated using data-driven analysis [16]. Finally, based on evaluation and prediction, the sample preparation is optimized through trial and error experiments [17].

9.2 Metabolomics and Metabolic Profiling for Complex Small Molecules

9.2.1 Basics of Small Molecular Profiling

Metabolites are the final products of biological processes. Metabolic profiling, or metabolomics, is used to study organic metabolites in cells, tissues, and biological fluids [18–20]; it provides comprehensive datasets for target samples [21–23]. This section describes the range of samples, metabolites, and NMR experiments used in metabolomic studies.

Recently, solution NMR has been used for metabolic profiling, and the NMR spectra of mixtures of small biological molecules are subjected to further multivariate analyses to identify metabolite biomarkers [24–28] and evaluate nutrients in foods [29–32]. Innovations in nontargeted approaches for studying biological systems are important for increasing biomass production and improving the sustainability of such systems.

Homeostasis refers to the maintenance of the internal physiological environment within a certain range in response to changes in the external environment. However, pathological states collapse homeostasis and change the metabolite composition and/or profiles. Metabolomics exhibits great potential for the diagnosis and monitoring of diseases using samples such as blood, saliva, urine, and feces (Fig. 9.2). The NMR metabolomic analysis of human superorganisms has provided new insights

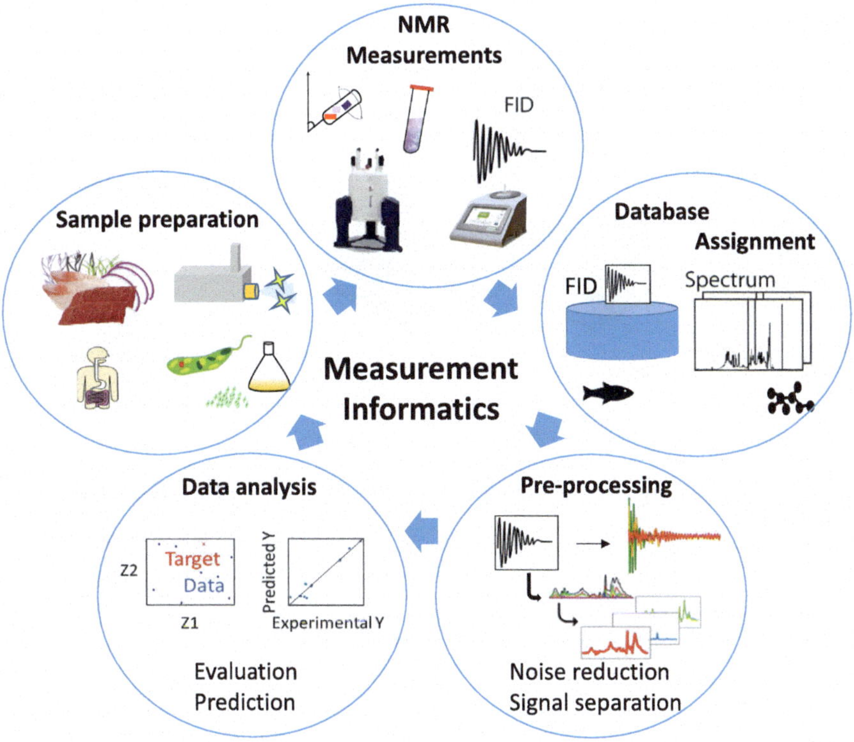

Fig. 9.1 Strategy for the analysis of complex molecular systems by NMR. The concept of measurement informatics is shown, which consists of a cycle of sample preparation, NMR measurement, database construction, data pre-processing, and data analysis

into the interactions between hosts and microorganisms during the establishment of immunological homeostasis in the intestines [33–36]. The evaluation of environmental homeostasis is becoming increasingly important with declining environmental health. Environmental homeostasis can be evaluated at multiple levels using NMR [16, 37–42].

9.2.2 *NMR Measurements and Data Processing*

9.2.2.1 Useful Pulse Sequences for Solution NMR

One-dimensional solution nuclear magnetic resonance (1D NMR) is often used for metabolic analysis. However, the analysis of mixtures is difficult when using traditional 1D NMR spectra because the signals overlap owing to signal splitting by spin coupling. As a solution, 2D *J*-resolved spectroscopy has been used to determine

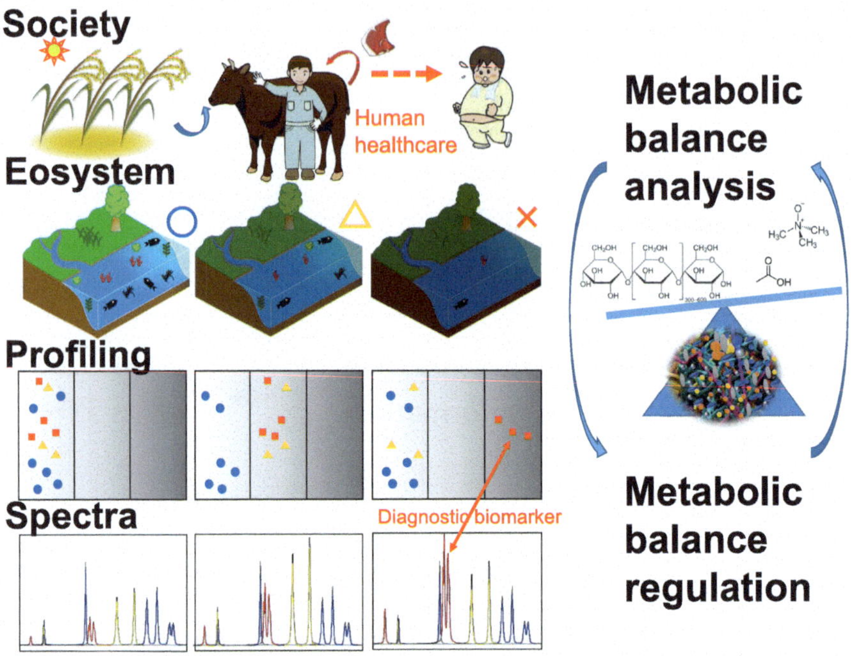

Fig. 9.2 Conceptual figure to elucidate metabolomics as indicators of homeostatic responses in plants, animals, and human superorganisms from input to output information analyzed by NMR. Metabolic compositional changes in biological samples potentially relay abundant information regarding homeostatic state; therefore, a data science approach can classify normal and abnormal states, following which diagnostic biomarkers can be identified from the complex NMR data

the spin–spin coupling constants, which were then used for structural analysis [43]. Furthermore, solution-state multidimensional NMR has been used to determine the correlation signals from homo- and heteronuclear experiments to identify metabolites. ^{13}C-heteronuclear single quantum coherence (HSQC) is a commonly used method in which metabolites can be identified by comparing the chemical shifts in the measured HSQC spectra with a chemical shift database. To increase the identification accuracy, ^{1}H–^{1}H correlation spectroscopy (COSY), ^{1}H–^{1}H totally correlated spectroscopy (TOCSY), and ^{1}H–^{1}H single quantum–double quantum correlation spectroscopy (SQ-DQ) have been used to confirm the correlation signals with adjacent atoms. Using HSQC-TOCSY and 3D HCCH-COSY, the adjacent correlation signals of the annotated metabolites in the HSQC spectra were easily confirmed [44].

9.2.2.2 Databases and Tools for NMR Analysis

Various methods have been used to measure metabolites owing to the diversity of their physical and scientific properties. A significant advantage of NMR is that the chemical shifts and coupling constants are absolute physical constants that can be readily and accurately reproduced in the spectra.

The following scheme was used for the analysis of the NMR data of metabolic mixtures: (1) spectral preprocessing, (2) statistical analysis and machine learning (ML) techniques for pattern recognition, (3) metabolite identification, and (4) pathway analysis (Fig. 9.3) [45, 46]. Many bioinformatics tools have been developed for data processing [47]. MetaboAnalyst3 [48–50] has a wide range of functions ranging from spectral processing to pathway analysis. For example, chemometric analysis, principal component analysis (PCA), and PLS discriminant analysis (PLS-DA) can be conducted, while ML approaches support hierarchical clustering with heatmaps, self-organizing maps, and k-means clustering. In addition, Random Forest and Support Vector Machine were used for supervised analysis.

Tools that support 2D-NMR and are useful for identifying metabolites include the Chenomx NMR Suite, COLMAR [51, 52], COLMARm [53], ECOMICS [54], MetaboLab [55], MetaboMiner [56], and rNMR [57]. Note that rNMR differs from existing software for NMR spectroscopy in that the analyses are based on regions of interest rather than peak lists.

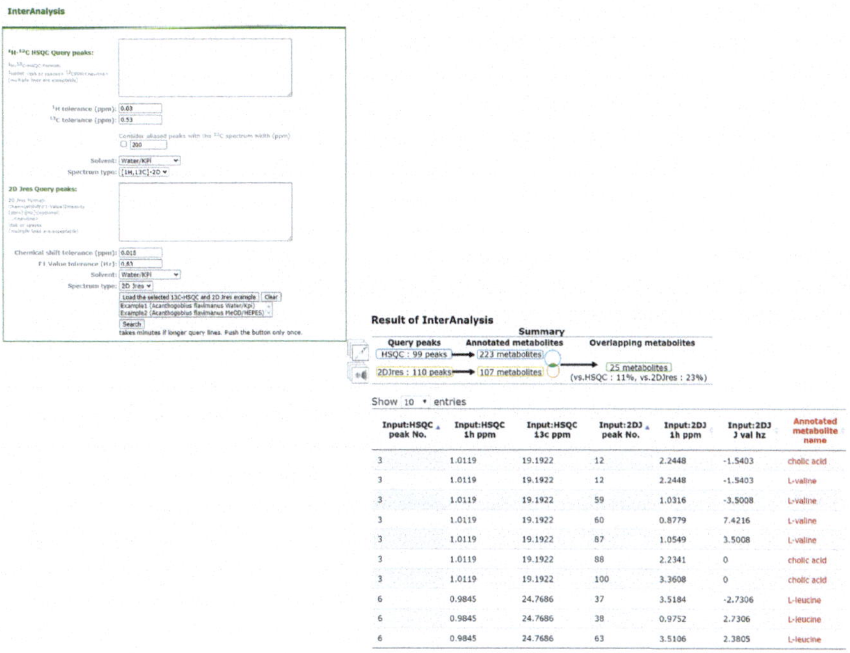

Fig. 9.3 Databases and web tools developed in the Kikuchi laboratory for metabolic analysis using NMR spectra. SpinAssign [117] and SpinCouple [120] are annotation tools for metabolites that use peaks from ^1H–^{13}C HSQC, 1D ^1H/^{13}C, and 2D-Jres; these web tools are integrated as InterAnalysis. The top and bottom of each section are the input and output concepts, respectively

The use of databases of compounds and pathways enables the identification of metabolites from complex spectra and the evaluation of their functions [58]. We have integrated all our original NMR databases (SpinAssign, SpinCouple, SpinMacro (previously ECOMICS), and FoodPro) as the InterSpin database [3] (Fig. 9.3).

9.2.2.3 Signal Assignments and Structure Elucidation

Evaluation of the physical properties of compounds using theoretical values from quantum chemical calculations has been a topic of computer science research in recent years [59]. The theoretical values of NMR chemical shifts and spin coupling constants and the relationships between the molecular structure and electronic states can be determined by solving equations such as the Schrödinger equation, or by approximating the behavior of atoms and electrons. Density functional theory (DFT) is used to ensure that the calculation time and accuracy are guaranteed to some extent. Other calculation methods, basis sets, and solvents can also be selected [60, 61].

Recently, improvements in the accuracy of assigning NMR spectra to metabolic mixtures using theoretical chemical shifts have been reported. Misawa et al. proposed a new annotation method using SENsitivity improvement by Signal Integration (SENSI) and quantum chemical calculations [62]. Komatsu et al. identified small biological molecules using 3D NMR and quantum chemical calculations [10]. In this study, the correlation signals of the methanol-soluble fractions of ^{13}C-labeled land plants were detected using 3D-(H)CCH-TOCSY and ^{13}C–^{13}C-COSY. Ito et al. designed a new assignment method for the NMR signals of metabolites using a fragment assembly approach and quantum chemical calculations [63]. In the study, metabolite structures were estimated by automatic chain assignments and fragment assembly (Fig. 9.4), and then evaluated by quantum chemical calculations. The initial conformation should be carefully set owing to the conformational dependence of the theoretical calculations [64]. Such "de novo" structure elucidation has been further developed by using chemoinformatics descriptors [14, 15, 65], ML, and tree search-type reinforcement learning [6]. The results suggest that quantum chemical calculations can be used to assign NMR signals to biological metabolic mixtures.

9.2.2.4 Spectral Pretreatment for Data Science

As experimentally obtained NMR spectra may include useful and valuable information, analyses such as multivariate statistical analysis and ML should be incorporated into the analytical flow of metabolomic studies. To enable an appropriate analytical flow, data pretreatment and preprocessing, such as the standardization of datasets, are very important.

Peak alignment, a computational approach for accurately aligning peaks in NMR spectra, is an important data pretreatment method because NMR chemical shifts are altered by various factors including pH, temperature, physicochemical interactions, the background matrix, and ion intensity. Spectral buckling and binning are the most

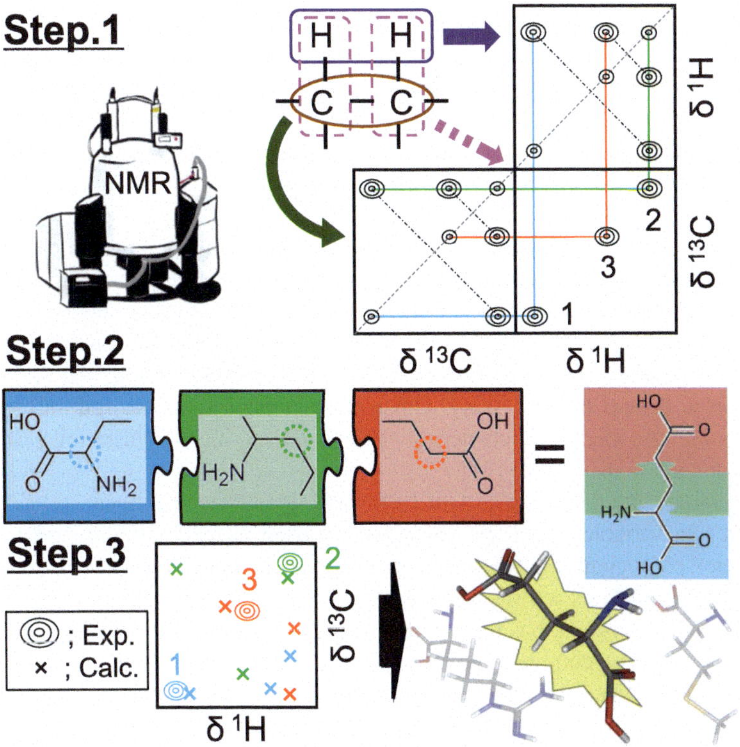

Fig. 9.4 Strategy for assigning metabolites via fragment assembly and quantum chemical calculations. Correlation signals are detected with high sensitivity using labeled samples. Step 1: automatic chain assignment derives the chemical backbone of metabolic mixtures. Step 2: partial structures of each component are derived by analyzing the most common substructure, and the entire structure can be estimated by connecting each fragment. Step 3: theoretical NMR properties of the estimated structure are calculated using the quantum mechanics (QM) method. The assignment allows comparison between experimental and theoretical chemical shifts. Reproduced from Ref. [102] with permission from the American Chemical Society

commonly employed methods. Binning splits a small bucket containing a range of variations for a specific peak shift, and the intensity of each bucket is calculated as the area under the curve. One drawback of the binning method is the considerable reduction in spectral resolution [66].

Scaling and normalization are also important for data pretreatment and preprocessing. Extreme differences in scale between multiple parameters can obscure the data, leading to inadequate interpretation. Typically, an internal standard such as 4,4-dimethyl-4-silapentane-1-sulfonic acid or trimethylsilane is used for normalization between datasets. Z-scoring is an alternative method that uses a standard normal distribution with an average of zero and a standard deviation of one [67]. However, using this method for analyzing an entire spectrum results in the noise also being normalized. Meanwhile, probabilistic quotient normalization (PQN) is not susceptible to outliers because the median of numerous estimated values is used [68].

9.2.2.5 Multi-Variate Analysis for Classification and Regression for a Large Number of NMR Samples

In metabolomic studies, the inclusion of multivariate analyses (such as chemometrics, pattern recognition, and data mining) is imperative. Methods for multivariate analysis fall into two categories: unsupervised (e.g., PCA) and supervised (e.g., partial least squares, PLS). PCA is the most widely used metabolomics method. It is typically used for data overview, trend (or cluster) identification, and dimension reduction in dataset preprocessing. Other representative unsupervised multivariate analyses include hierarchal clustering (HCA) [69], K-means clustering [70], and self-organizing maps (SOM) [71].

Correlation-based analyses, which involve identifying a linear relationship between two variables, have also been widely used in metabolomic studies [72]. Correlation coefficients are typically visualized as heat maps and network diagrams. Market basket analysis is a statistical approach for identifying the co-occurrence of variables in datasets, and has recently been applied to metabolomic studies [73].

PLS is the most commonly used supervised multivariate analysis method for metabolomic studies. PLS and related analyses are the most common classification and regression tools used to analyze metabolomic data. PLS-related analyses are powerful data-mining methods because they enable the analysis of large and highly complex datasets with collinearity and noise. However, alternative ML methods that are suitable for use in the field of metabolomics have been recently introduced [74–79].

9.2.3 Applications to Plant, Animal, and Microbial Systems

9.2.3.1 Application to Plant Systems

The various plants and algae in natural environments serve as primary producers and help maintain ecosystems. Therefore, studies on compounds in plants and algae are expected to be useful not only for agriculture, but also for other fields [80, 81]. Metabolic profiling using ^{1}H NMR concurrently tracks the changes in metabolites in plants and algae [82, 83]. Plants and algae are autotrophic organisms that are unable to move independently. Numerous organisms (e.g., grains, fruits, and seaweeds) have been studied using this method, and a variety of their parts (e.g., roots, leaves, and stems) have been analyzed [84]. Characterization based on differences in geographical origin and plant variety is possible using multivariate analyses, such as PCA. Such techniques can detect differences in the presence and contents of low-molecular-weight molecules, such as amino acids, organic acids, and sugars. Both low-molecular-weight molecules and polymers can be evaluated using dimethylsulfoxide (DMSO) as the solvent [85].

9.2.3.2 Application to Animal Systems

Living organisms that consume plant biomass, such as termites, are among the most attractive for this type of examination because of their considerable global impact on the decomposition of dead plant materials. The primary constituent of plant biomass is cellulose, which comprises glucose linked by β-1,4-glucosidic bonds. Although most animals cannot use this polysaccharide, termites are thought to be capable of hydrolyzing 74–99% of the cellulose in woody materials. This efficient cellulolytic system in termites has attracted the attention of many researchers and has been studied extensively in recent decades. An emerging technique for clarifying complex metabolic processes such as cellulose digestion involves the generation of comprehensive metabolite profiles. We recently demonstrated the utility of this technique for animal systems by feeding ^{13}C-glucose and amino acids to *Bombyx mori* and monitoring the time-dependent metabolic dynamics during insect morphogenesis [44].

In the context of symbiotic interactions, we successfully demonstrated the use of ^{13}C-labeled compounds to elucidate the critical role of acetate [24] and reuterin [35] production played by resident gut bacteria in mice. The advantages of using ^{13}C-labeling methodologies in combination with multidimensional NMR analyses include the separation of crowded signals from metabolite mixtures and the opportunity to perform quantitative multivariate analyses of time series in the acquired data at an atomic level.

The growth and activity of beneficial microbial symbionts in the human gastrointestinal tract are enhanced by functional foods, such as fructooligosaccharide (FOS) and galactooligosaccharides [36, 86]. Therefore, evaluating the effects of functional foods on the dietary interactive modulation between the host and microbial symbionts is important for human health [87–90]. In the initial validation stage, evaluations are usually performed using animal models such as mice and rats [91, 92]. However, animal experiments are time-consuming and involve ethical issues; thus, a simple and rapid in vitro method that eliminates the need for animal experiments has been sought as a better means for evaluating the effects of functional foods [93].

NMR-based metabolomic approaches are widely used for the metabolic characterization of animals and their intestinal symbiotic systems [79, 94, 95]. In human studies, the metabolic characterization of noninvasive samples, such as urine, saliva, and feces, are well known to provide abundant vital and beneficial information regarding human health and disease.

We described an advanced analytical approach for evaluating metabolic, mineral, and microbiota fluctuations in all fecal, urinary, and salivary samples from humans. These fluctuations are synchronously related to variations in dietary intake when the subjects are allowed to consume unlimited and uncontrolled diets, as found using data-driven approaches [62, 73, 96].

Aquaculture also provides increasingly important sources of protein for human consumption. In recent years, aquaculture technology has developed, and culture methods have become more intensive; however, several issues remain, such as determining the aquaculture environment, appropriate feeding, and disease control and prevention. As an example of the application of the exposome paradigm to aquaculture, the muscle quality of natural fish varies depending on the water temperature and nutrients in the growth environment (Fig. 9.5), as well as the plankton and small fish that are consumed as feed in this environment; this determines the market value of fish meats, which can vary greatly [97, 98]. Thus, analysis data on the muscle quality of natural fishes can be used to determine their origin and contribute to maintaining and evaluating their market value. To this end, we focused on the recent progress in NMR instruments, not only for high-end laboratory analysis but also for benchtop analysis, and developed a peak-separation method from multi-variate analysis [99]. If a high-end instrument is used, it is possible to more clearly discriminate changes in the chemical composition of the muscle caused by differences in the living environment and food. Therefore, by introducing cutting-edge ML calculations, such as deep learning and ensemble deep learning, into NMR analysis to evaluate the exposome during a fish lifetime [100], we can either extract important factors (i.e., NMR peaks of metabolites) to determine the origin of natural fish or predict important factors involved in fish growth [101] (Fig. 9.5). In addition, such analytical methods can make other important contributions to the fishery industry, such as for determining conditions such as water temperature and oxygen concentration for the management of aquaculture environments that affect muscle quality and for pinpointing important components in feed that are related to growth potential.

9.2.3.3 Application to Microbial Systems

Primordial microbial communities that perform anaerobic digestion are considered the most ancient metabolic communities. These communities became segregated from aerobic environments upon the emergence of photosynthetic organisms, which increased the initial oxygen concentration on Earth. Many microbial species in anaerobic environments, such as in soil and aquatic sediments, degrade the biomass produced by photosynthetic organisms. Thus, anaerobic digestion is essential for the biogeochemical cycling of organic and inorganic matter. Among these, anaerobic microbial communities that are responsible for the digestion of plant saccharides into short-chain fatty acids (SCFAs) have been investigated in several systems, including animal guts, waste management systems [102, 103] and soil from paddy fields [104].

Although ^{1}H-NMR is easy to perform and versatile, its limited spectral resolution is insufficient for addressing the overlapping signals of saccharides. The introduction of ^{13}C-nuclei into NMR analysis is a useful technological advancement for the resolution of overlapping signals [105]. The advantages of ^{13}C-labeling methods for NMR include the characterization of complex components, determination of compound structures, and tracking of microbial metabolic pathways [106].

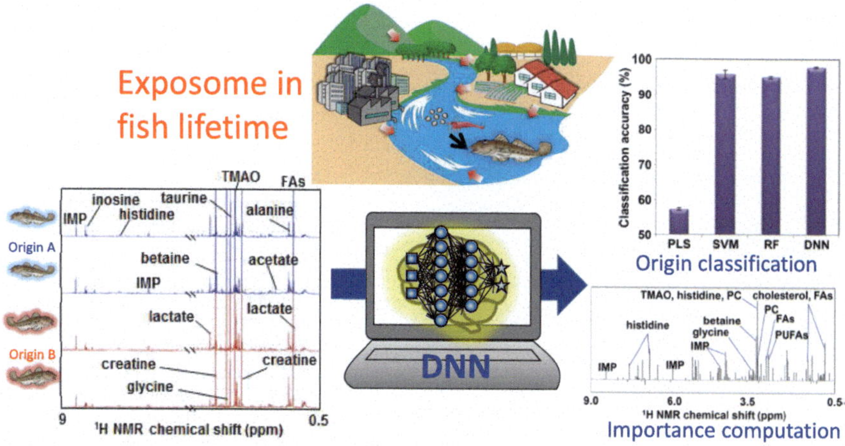

Fig. 9.5 Data-driven approach for evaluating the exposome during the fish lifetime. The muscle quality of fishes (determined based on their metabolic and relaxometic profiling) can vary depending on their environment and feed. Therefore, feed optimization and wastewater recycling should work closely as a "pair of wheels" in sustainable aquaculture. For establishing land-based aquaculture to move toward a sustainable society, advanced water treatment technology will be required, which in turn requires research on wastewater recycling systems. Optimized feed should be well metabolized into fish muscle, and not excreted as feces into the aquatic environment

9.3 Biomass Profiling for Complex Macromolecules

9.3.1 Basics of Macromolecular Profiling

Plant biomass is the most abundant and important component of the terrestrial biosphere. Its major components are cellulose, hemicellulose, and lignin (Fig. 9.6), which are complex molecules that are abundantly produced in plant cell walls. Cellulose is a linear condensation polymer that consists of β-1,4-linked *D*-glucose units; the degree of polymerization ranges from 100 to 20 000. Hemicelluloses are branched, have a lower molecular mass than cellulose, and are primarily composed of glucose, mannose, galactose, xylose, and arabinose. Lignins are reticulated, cross-linked macromolecules composed of phenylpropanoid units, including *p*-hydroxyphenyl, guaiacyl, and syringyl groups. These persistent substances have highly formed supramolecular structures that include cellulose and hemicellulose likely cemented by lignin—such a supramolecular structure is known as lignocellulose.

Comprehensive analysis of biological macromolecules without requiring purification is the next frontier in the advancement of research for gaining insights into living systems. Breakthrough technologies in basic biological research toward polysaccharide complexity can be developed by using the profiling of biomass

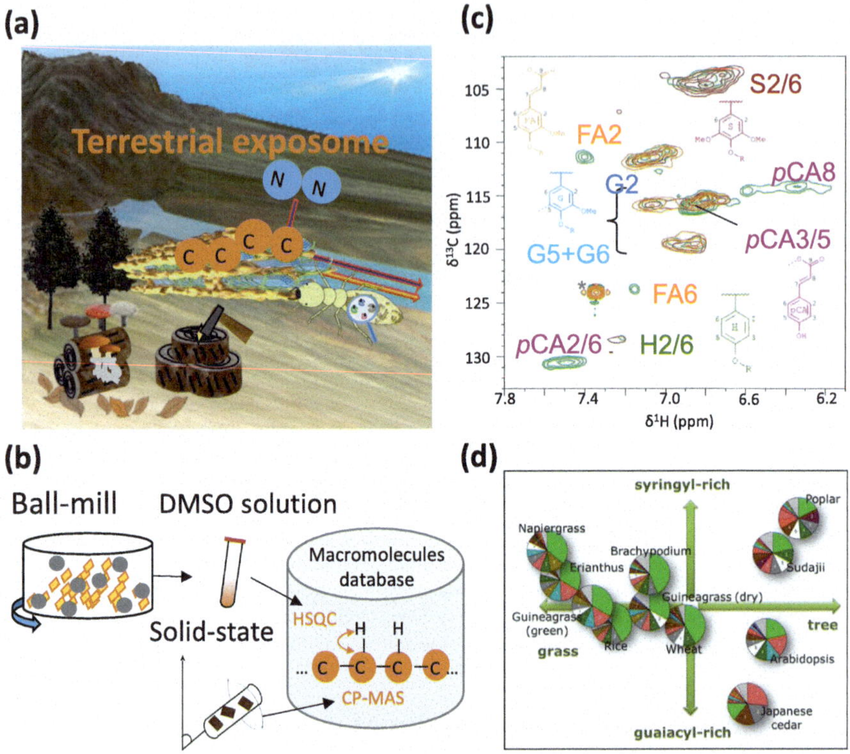

Fig. 9.6 Carbon and nitrogen circulation and tools for characterizing the terrestrial exposome and biomass. Coupled with nitrogen fixation, carbon from plants is decomposed and metabolized into amino acids by the gut microbes of termites. Circulation of carbon and nitrogen via their feces and remains represents the terrestrial exposome, which enriches and affects both forest and aquatic ecosystems. Web tools and NMR databases for biomass macromolecules are useful for evaluating the terrestrial exposome; for example, ball-milled samples dissolved in dimethyl sulfoxide (DMSO) can be compositionally evaluated by solution NMR, and mushroom cultivation and wood quality can be evaluated by solid-state NMR

synthesis and degradation as intermediate-monitoring systems. Innovation in nontargeted approaches for a wide range of ecosystem biological systems is important for increasing biomass production and improving the sustainability of such systems.

9.3.2 *NMR Measurements and Data Processing*

9.3.2.1 Useful Pulse Sequences

Signals in the ^{1}H NMR spectra of polysaccharides in vascular plants, such as cell-wall polysaccharides, are often broad lines because of the high viscosities of the samples.

Thus, ^{1}H–^{13}C HSQC was used for cell wall composition analyses. Multidimensional NMR with ^{13}C-labeled plants has provided detailed resonance assignments for plant cell walls, particularly the polysaccharide components [107]. 3D HCCH-Cozy experiments are powerful tools for the resonance assignment of polysaccharides because the crowded peaks from polysaccharides are separated into a 3D space. Notably, ^{1}H–^{13}C HSQC-NOESY spectra provide inter-sugar correlations for polysaccharides; this information cannot be obtained from ^{1}H–^{1}H COSY, ^{1}H–^{1}H TOCSY, ^{1}H–^{13}C HSQC, or 3D HCCH-Cozy spectra [108].

Solid-state NMR detects nonpolar polymers, and many pulse sequences have been developed for this technique. However, the application of multidimensional NMR pulse sequences has been delayed because the anisotropic effects of dipole interactions are strong, and measuring the ^{1}H detection type is very difficult. To overcome these problems, 1D cross-polarization magic-angle spinning (CP-MAS) uses a greater number of scans to detect low-sensitivity ^{13}C, and 1D non-quaternary suppression (NQS)-MAS emphasizes quaternary carbons [109, 110].

There are several relaxation time constants, including the longitudinal relaxation of ^{1}H (^{1}H T_1) and ^{13}C (^{13}C T_1), rotating-frame ^{1}H (^{1}H $T_{1\rho}$), and transverse relaxation of ^{1}H (^{1}H T_2) and ^{13}C (^{13}C T_2). The sensitive frequencies of molecular motion are different in each relaxation process. The relaxation time constants represent the types of molecular motion caused by rapid ^{1}H–^{1}H spin diffusion within the domains. In contrast, ^{13}C relaxation time constants represent the local molecular motion of each nucleus.

In general, ^{1}H T_1, ^{13}C T_1, and ^{1}H $T_{1\rho}$ were longer and ^{1}H T_2 and ^{13}C T_2 were shorter in cellulose than in the matrix polysaccharides in hydrated plant cell walls. Hong et al. investigated the differences in the physicochemical properties of cell walls via ^{13}C T_1 and ^{1}H local $T_{1\rho}$ (without spin diffusion by using ^{1}H–^{1}H homonuclear decoupling) using the primary cell walls of various *Arabidopsis* plants, including wild-type, chemically pectin-depleted, and homozygous mutants [111, 112].

The difference in relaxation times of the components can be used as a basis for spectral filtering. Komatsu and Kikuchi developed ^{13}C–^{13}C correlation spectroscopy coupled with spectral filtering techniques to analyze matrix polysaccharides by removing the signals derived from cellulose [113]. Dipolar dephasing is used as a spectral filter, as coherence can be de-phased by ^{1}H–^{13}C dipolar interactions; the amplitude strongly depends on molecular motion.

Another option is to label the sample completely with ^{13}C, which allows for a measurement rate that is 100 times greater than the ^{1}H–^{13}C coupling detection rate and 10,000 times greater than the ^{13}C–^{13}C coupling detection rate. ^{1}H–^{13}C coupling has been used to evaluate crystalline polysaccharides, lipids, and low-molecular-weight molecules using double CP, solid-state HSQC, and high-resolution (HR)-MAS. ^{13}C–^{13}C coupling has been used to detect the correlation signal between adjacent carbons using dipolar-assisted rotational resonance (DARR), a natural abundance double quantum transfer experiment (INADEQUATE), and 3D ^{13}C–^{13}C–^{13}C (CCC)-MAS [11].

9.3.2.2 Peak Separation for Broad Macromolecular Spectra

The overlapping of peaks in NMR spectra is one of the biggest challenges encountered in the analysis of metabolic mixtures. Solid-state NMR typically has very low resolution; thus, what is observed as a single peak is more likely a combination of multiple peaks derived from several chemical components. Peak-separation techniques can be applied to address this problem. Multivariate curve resolution-alternating least squares (MCR-ALS) is a peak separation method based on optimization by the iterative least squares method to calculate components [114]. Non-negative matrix factorization (NMF) is an algorithm used to analyze matrices consisting of zero and/or positive values; it can decompose multivariate and spectral data into a small number of basic functions and extract local features. NMF can be used for peak separation in solution- and solid-state NMR spectra [4, 115].

9.3.3 Applications to Material, Biological, and Geochemical Systems

9.3.3.1 Application to Cellulosic Material Systems

Cellulose is the most common chemical polymer on land and is an alternative energy and chemical resource with the greatest potential on Earth. PCA has been used to extract the components from the solid-state NMR spectra of bacterial cellulose [116]. Polymers such as cellulose have several domain structures, and these structures and their dynamics can be reflected using a variety of solid-state spectra derived from different parameters [116].

Further statistical approaches have been used to characterize the heterogeneous structures of bacterial cellulose samples pretreated with four types of ionic liquids (ILs). The structural heterogeneity of these samples has been measured using Fourier-transform infrared spectroscopy and solid-state NMR methods such as cross-polarization magic-angle spinning and dipolar-assisted rotational resonance. The obtained data matrices are then evaluated by PCA. The measured one-dimensional data clearly revealed the modification of crystalline cellulose, and the statistical approach revealed subtle structural changes that occurred upon pretreatment with different ILs [117].

Although crystalline cellulose has been intensively investigated, the structure of amorphous cellulose, including its local structures, has not been clarified. Thus, we prepared ^{13}C-labeled amorphous cellulose and fully assigned ^{13}C NMR chemical shifts by 2D ^{13}C–^{13}C correlation spectroscopy, conducted both in IL solution and in regenerated solid states. Based on the assigned experimental ^{13}C NMR chemical shifts, a candidate was proposed for the local structures of amorphous cellulose via molecular dynamics and quantum chemistry computer simulations; the suggested structure was twisted, exposing more hydrophilic surfaces than extended crystalline cellulose [118].

The anaerobic digestion of biomacromolecules in various microbial ecosystems is influenced by variations in the type, quality, and quantity of chemical components. Herein, we describe a characterization strategy using NMR spectroscopy to target solid-insoluble biomass, catabolized soluble metabolites, and product gases. ^{13}C-labeled cellulose produced by *Gluconacetobacter xylinus* was added as a substrate to stirred tank reactors and gradually degraded for 120 h. Time-course variations in the structural heterogeneity of the cellulose during catabolism were determined using solid-state NMR, and the soluble metabolites produced by the cellulose degradation were monitored using solution NMR [102, 103].

9.3.3.2 Application to Plant Biomass Systems

Recycling in ecosystems occurs via symbiosis between termite gut microbial communities, specific protists and fungi, and soil microbial communities. Plant-derived saccharides are degraded by soil microbial communities and eventually return to the soil [105]. Lignocellulose, a highly ordered and sparingly soluble polymer mixture formed from the polysaccharides cellulose and hemicellulose and the polymer lignin, is the most abundant biomass on land (Fig. 9.6) [107, 119–122]. This carbon source is produced by trees, and is decomposed and metabolized to amino acids via nitrogen fixation by the intestinal microbial community of termites, eventually becoming the proteins that constitute the termite body [123]. The decomposition of macromolecular complexes such as lignocellulose is also an important process in the circulation of various elements. Carbon sources from plants and atmospheric nitrogen are decomposed and metabolized into amino acids via nitrogen fixation by the gut microbes of termites. Carbon and nitrogen circulation via their feces and remains is a terrestrial exposome that enriches and affects forest and aquatic ecosystems. Web tools and NMR databases for biomass macromolecules are useful for evaluating terrestrial exposomes; for example, ball-milled samples dissolved in dimethyl sulfoxide (DMSO) solvent can be compositionally analyzed by solution NMR, mushroom cultivation can be analyzed by solid-state NMR, and wood quality can be evaluated. For example, humus that accumulates in the soil after biomass decomposition can adsorb metal elements, which are then taken up by detritus-eating animals that ingest the humus; these animals are responsible for the bioaccumulation of various metals. Plankton and algae such as *E. gracilis* [119] are deeply involved in the cycle of materials and the food chain of aquatic ecosystems. The chemical components of seaweeds are used in foods, feeds, fertilizers, and fine chemicals, and the adsorption of metals by polysaccharides is attracting attention in water treatment processes. However, seaweed composition fluctuates seasonally and cannot be artificially controlled. Therefore, to capture the complex seasonal changes that determine the components of naturally growing seaweeds, it is necessary to comprehensively measure these components over time and evaluate them in an integrated manner [114].

9.3.3.3 Application to Biodegradable Polymers

Plastics are important components of plastic bottles and synthetic fibers; however, the discharge of plastic wastes into the environment and their subsequent outflow from rivers to oceans has resulted in a worldwide problem of marine pollution by highly stable microplastics, [124] which greatly affects marine exposure (Fig. 9.7, step 1) [125–128]. The degradation of plastics can be investigated using in vitro and in vivo experiments conducted in the ocean or within fish intestinal environments. Biodegradable plastics, which can decompose in the environment, are a promising solution to environmental problems caused by plastic disposal; [129–132] therefore, the environmental degradation rates and pathways for major types of thermoplastic polymers have been investigated [133]. Meanwhile, as alternatives to petroleum resources, microbial products and plant biomass can be used to produce macromolecular materials such as plastics and feedstock [134]. For example, polymers such as polylactic acid (PLA) [135], poly-ε-caprolactone [136], and cellulose [102, 103, 107, 113, 116–118, 137] have been prepared as high-performance materials with various properties. To develop such materials, it is necessary to analyze the microbial and plant biomass with multiple macromolecular components in complex biochemical systems. To this end, solution- and solid-state NMR spectroscopy can be used to characterize the native structure, components, and dynamics of these samples at the atomic level (Fig. 9.7, Step 2) [138, 139]. In addition, time-series NMR measurements can produce ^{1}H anisotropic and ^{13}C spectra as descriptors of structure and properties during the biodegradation process.

The higher-order structure of a material significantly affects its macroscopic properties [140]. Traditional trial-and-error design approaches for new materials face significant challenges owing to their vast design space. In addition, computational technologies such as density functional theory and molecular dynamics [118] are usually computationally expensive; thus, it is difficult to calculate molecular structures from material properties. In order to address these problems, ML-assisted material design has emerged as a breakthrough tool in several scientific areas [141]. Furthermore, a material informatics approach has recently been considered for material design [142]. In this approach, big data deposited in databases, as well as structural data of highly ordered structures obtained during the materials production process [143, 144], are used to predict the physicochemical properties of new materials. Molding conditions are also important for developing materials with the desired physical properties. In this regard, NMR, particularly low-magnetic-field NMR, which is used for routine material evaluation, produces vast datasets [3] that can be used in the prediction of NMR signals to aid in the cycle of developing materials and identifying structures with the desired properties. Coupled with NMR measurements, inverse design using ML can be used to design polymers with the desired phase behavior. For example, ML has been used for the cloud-point engineering of polymers whose compositions have been determined by ^{1}H NMR spectroscopy [145]. NMR chemical shifts are rich in information and encode the structural features of molecules contributing to the physical, chemical, and biological properties of a material; thus, they have also been used as descriptors in a previous study that

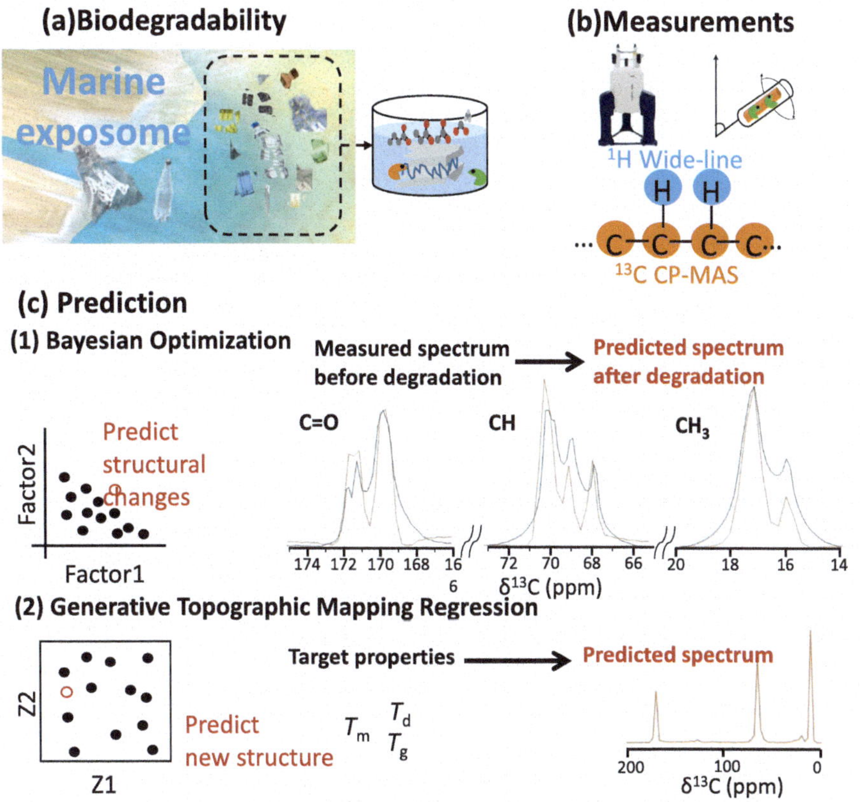

Fig. 9.7 Data-driven cycle for the development of biodegradable plastics. **a** Plastic pollution of the ocean owing to human society negatively affects the marine exposome. The degradation process of plastics can be investigated by in vitro and in vivo experiments conducted in either the ocean or fish intestinal environments. **b** Time-series NMR measurements produce ^{1}H anisotropic and ^{13}C spectra as descriptors of structures and properties during the biodegradation process. **c** In the development of biodegradable plastics, it is difficult to control physical properties related to biodegradability and durability. Data-driven scientific methods such as Bayesian optimization and generative topographic mapping regression (GTMR) allow for efficient material design based on the iterative cycle of sample preparation, sample measurement, and analysis of measured data

modeled the quantitative structure–activity/property relationship using data-driven generative topographic mapping regression (GTMR) for data analysis [146]. When designing biodegradable plastics with highly ordered structures, physical properties such as the glass-transition, melting, and degradation temperatures (Tg, Tm, and Td), which are related to biodegradability and durability, must be controlled (Fig. 9.7, step 3). For an efficient material design based on the iterative cycle of preparation, sample measurement, and analysis of measured data, GTMR was applied to the inverse analysis of CP-MAS spectra to determine the structure of the material in its native solid state based on its thermophysical properties [5]. Furthermore, Bayesian optimization has been proposed for both real and virtual degrading experiments for bioplastics

[147]. From the results of the degradation experiments, the average molecular weight was obtained and used as the objective variable, whereas the initial crystallinity, moisture content in the compost environment, degradation period, and NMR spectra obtained via PCA were used as explanatory variables. These values were subjected to a decomposition degree predictive model comprising ML. The optimum decomposition degree, various analytical values, and experimental conditions were explored using a combination of the decomposition-degree predictive model and Bayesian optimization (Fig. 9.8).

Microalgae (e.g., *Chlamydomonas reinhardtii* and *Phaeodactylum tricornutum*) are also suitable for environmental and biotechnological applications based on the exploitation of solar light [148]. Polystyrene is generally considered durable and biodegradable, but recently, the biodegradation of polystyrene by enterobacteria of mealworms has been reported as a novel method to deal with plastic waste [149]. Bioderived polyesters such as PLA are interesting alternatives to petrochemical-based plastics [150]. PLA is unlikely to hydrolyze at low temperatures in marine environments; however, its carbonyl group absorbs UV radiation below 280 nm, making it susceptible to photodegradation in marine environments. Lastly, biodegradation of polyethylene microplastics by the marine fungus *Zalerion maritimum* was reported based on Fourier transform infrared (FT-IR) spectroscopy, NMR, and scanning electron microscopy (SEM) data [151].

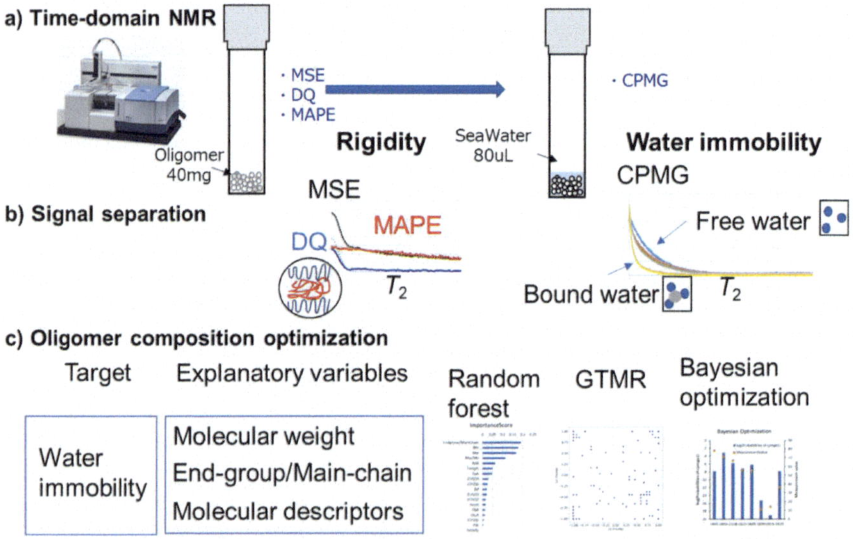

Fig. 9.8 Oligomer composition optimization for desired water-immobility using time-domain NMR. **a** MSE, DQ, MAPE, and CPMG data were acquired by time-domain NMR. **b** T2 relaxation times were separated based on the mobility and water-immobility of oligomers. **c** Random forest and GTMR were used to search for important factors from the molecular weights, terminal ratios, and molecular descriptors for water immobility, and Bayesian optimization was used to optimize the compositions

9.3.3.4 Application to Geochemical Samples

A wide variety of microorganisms, including archaea, bacteria, fungi, plankton, and protists, are present throughout the biosphere in soil, sediment, underground areas, wetlands, rivers [152], oceans, and so on. In soil environments, both solution and solid-state NMR measurements have been used to characterize cellulosic supramolecular structures in rice straws and their degradation profiles caused by a microbial community in paddy fields [104]. NMR-based metabolomic studies can also be used to analyze the decomposition of biochar [153] and fish [71] by soil microbial communities.

The chemical components and structures of soil organic matter have been evaluated in numerous studies using NMR-based approaches [154, 155]. For instance, solution NMR has revealed that biomacromolecules derived from microorganisms and plants are the major components of soil and aqueous humic substances [156, 157]. Using a DMSO-solubilization protocol for humic substances followed by alkaline fractionation, we also revealed that eutrophied sediments from metropolitan sites exhibit high amounts of peptides and inorganic nitrogen [158]. Solid-state NMR is also a major tool for the characterization of organic matter. A variety of solid-state NMR spectroscopies, including DP- and CP-MAS, DP- and CP-TOSS, and ^{1}H–^{13}C HETCOR, have been used to characterize the chemical structure and composition of organic matter [159–161].

9.4 Future Perspectives

Because NMR produces vast amounts of data and exhibits inter-institutional compatibility [162], it is highly suitable for data science applications combined with informatics techniques such as multivariate analysis, ML, and database analysis. Many targeted analyses focus on an individual compound or a type of compound; this requires pre-processing procedures for sample preparation, pre-treatment, and extraction. In contrast, owing to its high reproducibility, ability for quantification, and high throughput, the NMR-based non-targeted approach is superior for analyzing both large-scale data from time-series samples in environmental evaluation and multiple samples generated during polymer development and manufacturing. With advances in informatics technology for the pre-processing of low-resolution NMR data, cost-effective benchtop NMR is expected to become more applicable for on-site assessments in the natural environment and in the fields of human health, agriculture, aquaculture, and industrial processing [163–165]. Furthermore, artificial intelligence and robotics for the automation of research and development are expected to become key technologies in NMR data science (Fig. 9.9).

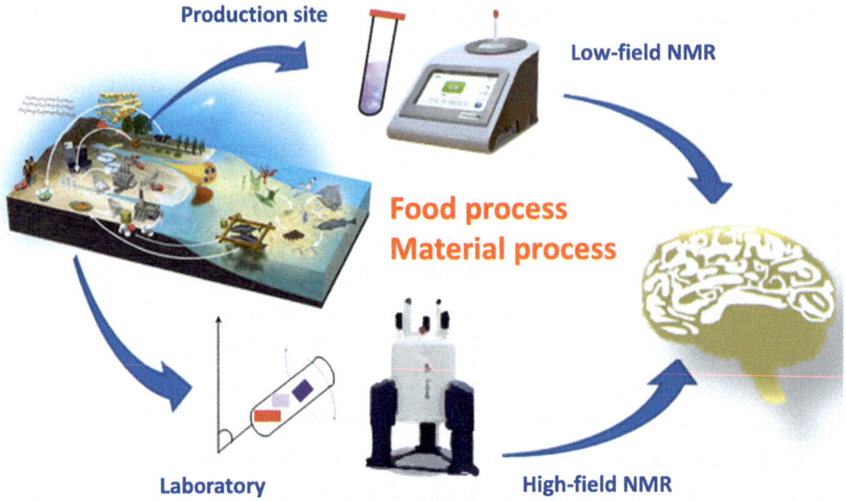

Fig. 9.9 NMR data science evaluating the homeostatic states of molecular complexity in the natural environment as well as human society from laboratory-based high-field NMR data and low-field NMR data from the production sitess

References

1. Deborde, C. et al.: Optimizing 1D H-1-NMR profiling of plant samples for high throughput analysis: extract preparation, standardization, automation and spectra processing. Metabolomics. **15** (2019)
2. Kruk, J., et al.: NMR techniques in metabolomic studies: a quick overview on examples of utilization. Appl. Magn. Reson. **48**, 1–21 (2017)
3. Yamada, S., et al.: InterSpin: integrated supportive webtools for low-and high-field NMR analyses toward molecular complexity. ACS Omega. **4**, 3361–3369 (2019)
4. Wei, F., Ito, K., Sakata, K., Date, Y., Kikuchi, J.: Pretreatment and integrated analysis of spectral data reveal seaweed similarities based on chemical diversity. Anal. Chem. **87**, 2819–2826 (2015)
5. Yamada, S., Chikayama, E., Kikuchi, J.: Signal deconvolution and generative topographic mapping regression for solid-state NMR of multi-component materials. Int. J. Mol. Sci. **22** (2021)
6. Zhang, J., et al.: NMR-TS: de novo molecule identification from NMR spectra. Sci. Technol. Adv. Mater. **21**, 552–561 (2020)
7. Yamada, S., Tsuboi, Y., Yokoyama, D., Kikuchi, J.: Polymer composition optimization approach based on feature extraction of bound and free water using time-domain nuclear magnetic resonance. J. Magn. Reson. **351**, 107438 (2023)
8. Kikuchi, J., Yamada, S.: The exposome paradigm to predict environmental health in terms of systemic homeostasis and resource balance based on NMR data science. RSC Adv. **11**, 30426–30447 (2021)
9. Hara, K., Yamada, S., Kurotani, A., Chikayama, E., Kikuchi, J.: Materials informatics approach using domain modelling for exploring structure-property relationships of polymers. Sci. Rep. **12**, 10558 (2022)
10. Komatsu, T., Ohishi, R., Shino, A., Kikuchi, J.: Structure and metabolic-flow analysis of molecular complexity in a 13C-labeled tree by 2D and 3D NMR. Angew. Chem. Int. Ed. **55**, 6000–6003 (2016)

11. Komatsu, T., Kobayashi, T., Hatanaka, M., Kikuchi, J.: Profiling planktonic biomass using element-specific, multicomponent nuclear magnetic resonance spectroscopy. Environ. Sci. Technol. (2015)
12. Chikayama, E., et al.: FoodPro: a web-based tool for evaluating covariance and correlation NMR spectra associated with food processes. Metabolites. **6** (2016)
13. Hara, K., Yamada, S., Chikayama, E., Kikuchi, J.: Parameter visualization of benchtop nuclear magnetic resonance spectra toward food process monitoring. Processes. **10** (2022)
14. Ito, K., Obuchi, Y., Chikayama, E., Date, Y., Kikuchi, J.: Exploratory machine-learned theoretical chemical shifts can closely predict metabolic mixture signals. Chem. Sci. **9**, 8213–8220 (2018)
15. Takamura, A., Tsukamoto, K., Sakata, K., Kikuchi, J.: Integrative measurement analysis via machine learning descriptor selection for investigating physical properties of biopolymers in hairs. Sci. Rep. **11**, 24359 (2021)
16. Kikuchi, J., Yamada, S.: NMR window of molecular complexity showing homeostasis in superorganisms. Analyst. **142**, 4161–4172 (2017)
17. Yamawaki, R., Tei, A., Ito, K., Kikuchi, J.: Decomposition factor analysis based on virtual experiments throughout bayesian optimization for compost-degradable polymers. Appl. Sci-Basel. **11** (2021)
18. Clayton, T.A., et al.: Pharmaco-metabonomic phenotyping and personalized drug treatment. Nature. **440**, 1073–1077 (2006)
19. Schlipalius, D.I., et al.: A core metabolic enzyme mediates resistance to phosphine gas. Science (New York, N.Y.) **338**, 807–810 (2012)
20. Auro, K., et al.: A metabolic view on menopause and ageing. Nat. Commun. **5**, 4708 (2014)
21. Suhre, K., et al.: A genome-wide association study of metabolic traits in human urine. Nat. Genet. **43**, 565–569 (2011)
22. Kettunen, J., et al.: Genome-wide association study identifies multiple loci influencing human serum metabolite levels. Nat. Genet. **44**, 269–276 (2012)
23. Kettunen, J., et al.: Genome-wide study for circulating metabolites identifies 62 loci and reveals novel systemic effects of LPA. Nat. Commun. **7**, 11122 (2016)
24. Fukuda, S., et al.: Bifidobacteria can protect from enteropathogenic infection through production of acetate. Nature. **469**, 543–547 (2011)
25. Furusawa, Y., et al.: Commensal microbe-derived butyrate induces the differentiation of colonic regulatory T cells. Nature. **504**, 446–450 (2013)
26. Brindle, J.T., et al.: Rapid and noninvasive diagnosis of the presence and severity of coronary heart disease using 1H-NMR-based metabonomics. Nat. Med. **8**, 1439–1444 (2002)
27. Holmes, E., et al.: Human metabolic phenotype diversity and its association with diet and blood pressure. Nature. **453**, 396–400 (2008)
28. Smith, M.I., et al.: Gut microbiomes of Malawian twin pairs discordant for kwashiorkor. Science (New York, N.Y.). **339**, 548–554 (2013)
29. Watanabe, M., Ohta, Y., Licang, S., Motoyama, N., Kikuchi, J.: Profiling contents of water-soluble metabolites and mineral nutrients to evaluate the effects of pesticides and organic and chemical fertilizers on tomato fruit quality. Food Chem. **169**, 387–395 (2015)
30. Tomita, S., et al.: A NMR-based, non-targeted multistep metabolic profiling revealed l-rhamnitol as a metabolite that characterised apples from different geographic origins. Food Chem. **174**, 163–172 (2015)
31. Tomita, S., et al.: A survey of metabolic changes in potato leaves by NMR-based metabolic profiling in relation to resistance to late blight disease under field conditions. Magn. Reson. Chem. **55**, 120–127 (2017)
32. Sekiyama, Y., Okazaki, K., Kikuchi, J., Ikeda, S.: NMR-based metabolic profiling of field-grown leaves from sugar beet plants harbouring different levels of resistance to cercospora leaf spot disease. Metabolites. **7** (2017)
33. Li, M., et al.: Symbiotic gut microbes modulate human metabolic phenotypes. Proc. Natl. Acad. Sci. U. S. A. **105**, 2117–2122 (2008)

34. Sugahara, H., et al.: Probiotic bifidobacterium longum alters gut luminal metabolism through modification of the gut microbial community. Sci. Rep. **5**, 13548 (2015)
35. Morita, H., et al.: Comparative genome analysis of lactobacillus reuteri and lactobacillus fermentum reveal a genomic island for reuterin and cobalamin production. DNA Res. **15**, 151–161 (2008)
36. Kato, T., et al.: Multiple omics uncovers host-gut microbial mutualism during prebiotic fructooligosaccharide supplementation. DNA Res. **21**, 469–480 (2014)
37. Bundy, J.G., et al.: Metabolic profile biomarkers of metal contamination in a sentinel terrestrial species are applicable across multiple sites. Environ. Sci. Technol. **41**, 4458–4464 (2007)
38. Collette, T.W., et al.: Impacts of an anti-androgen and an androgen/anti-androgen mixture on the metabolite profile of male fathead minnow urine. Environ. Sci. Technol. **44**, 6881–6886 (2010)
39. Ellis, R.P., et al.: (1)H NMR metabolomics reveals contrasting response by male and female mussels exposed to reduced seawater pH, increased temperature, and a pathogen. Environ. Sci. Technol. **48**, 7044–7052 (2014)
40. Samuelsson, L.M., Bjorlenius, B., Forlin, L., Larsson, D.G.: Reproducible (1)H NMR-based metabolomic responses in fish exposed to different sewage effluents in two separate studies. Environ. Sci. Technol. **45**, 1703–1710 (2011)
41. Whitfield Aslund, M.L., et al.: Earthworm sublethal responses to titanium dioxide nanomaterial in soil detected by (1)H NMR metabolomics. Environ. Sci. Technol. **46**, 1111–1118 (2012)
42. Yoshida, S., Date, Y., Akama, M., Kikuchi, J.: Comparative metabolomic and ionomic approach for abundant fishes in estuarine environments of Japan. Sci. Rep. **4** (2014)
43. Fonville, J.M., et al.: Evaluation of full-resolution J-resolved 1H NMR projections of biofluids for metabonomics information retrieval and biomarker identification. Anal. Chem. **82**, 1811–1821 (2010)
44. Chikayama, E., Suto, M., Nishihara, T., Shinozaki, K., Kikuchi, J.: Systematic NMR analysis of stable isotope labeled metabolite mixtures in plant and animal systems: coarse grained views of metabolic pathways. PLoS ONE. **3**, e3805 (2008)
45. Aguiar-Pulido, V., et al.: Metagenomics, metatranscriptomics, and metabolomics approaches for microbiome analysis. Evol. Bioinform. Online. **12**, 5–16 (2016)
46. Johnson, S.R., Lange, B.M.: Open-access metabolomics databases for natural product research: present capabilities and future potential. Front. Bioeng. Biotechnol. **3**, 22 (2015)
47. Izquierdo-Garcia, J., et al.: Descriptive review of current NMR-based metabolomic data analysis packages. Prog. Nucl. Magn. Reson. Spectrosc. **59**, 263–270 (2011)
48. Xia, J., Psychogios, N., Young, N., Wishart, D.S.: MetaboAnalyst: a web server for metabolomic data analysis and interpretation. Nucleic Acids Res. **37**, W652-660 (2009)
49. Xia, J., Mandal, R., Sinelnikov, I.V., Broadhurst, D., Wishart, D.S.: MetaboAnalyst 2.0-a comprehensive server for metabolomic data analysis. Nucleic. Acids. Res. **40**, W127–133 (2012)
50. Xia, J., Sinelnikov, I.V., Han, B., Wishart, D.S.: MetaboAnalyst 3.0-making metabolomics more meaningful. Nucleic. Acids. Res. **43**, W251–257 (2015)
51. Zhang, F., Robinette, S.L., Bruschweiler-Li, L., Brüschweiler, R.: Web server suite for complex mixture analysis by covariance NMR. Magn. Reson. Chem. **47**(Suppl 1), S118-122 (2009)
52. Bingol, K., et al.: Unified and isomer-specific NMR metabolomics database for the accurate analysis of (13)C-(1)H HSQC spectra. ACS Chem. Biol. **10**, 452–459 (2015)
53. Bingol, K., Li, D., Zhang, B., Bruschweiler, R.: Comprehensive metabolite identification strategy using multiple two-dimensional NMR spectra of a complex mixture implemented in the COLMARm web server. Anal. Chem. **88**, 12411–12418 (2016)
54. Ogata, Y., et al.: ECOMICS: a web-based toolkit for investigating the biomolecular web in ecosystems using a trans-omics approach. Plos One. **7** (2012)
55. Ludwig, C., Günther, U.L.: MetaboLab–advanced NMR data processing and analysis for metabolomics. BMC Bioinform. **12**, 366 (2011)

56. Xia, J., Bjorndahl, T.C., Tang, P., Wishart, D.S.: MetaboMiner–semi-automated identification of metabolites from 2D NMR spectra of complex biofluids. BMC Bioinform. **9**, 507 (2008)
57. Lewis, I.A., Schommer, S.C., Markley, J.L.: RNMR: open source software for identifying and quantifying metabolites in NMR spectra. Magn. Reson. Chem. **47**(Suppl 1), S123-126 (2009)
58. Alonso, A., Marsal, S., Julià, A.: Analytical methods in untargeted metabolomics: state of the art in 2015. Front. Bioeng. Biotechnol. **3**, 23 (2015)
59. Willoughby, P.H., Jansma, M.J., Hoye, T.R.: A guide to small-molecule structure assignment through computation of ((1)H and (1)(3)C) NMR chemical shifts. Nat. Protoc. **9**, 643–660 (2014)
60. Lodewyk, M.W., Siebert, M.R., Tantillo, D.J.: Computational prediction of 1H and 13C chemical shifts: a useful tool for natural product, mechanistic, and synthetic organic chemistry. Chem. Rev. **112**, 1839–1862 (2012)
61. Odoh, S.O., Cramer, C.J., Truhlar, D.G., Gagliardi, L.: Quantum-chemical characterization of the properties and reactivities of metal-organic frameworks. Chem. Rev. **115**, 6051–6111 (2015)
62. Misawa, T., Komatsu, T., Date, Y., Kikuchi, J.: SENSI: signal enhancement by spectral integration for the analysis of metabolic mixtures. Chem. Commun. **52**, 2964–2967 (2016)
63. Ito, K., Tsutsumi, Y., Date, Y., Kikuchi, J.: Fragment assembly approach based on graph/network theory with quantum chemistry verifications for assigning multidimensional NMR signals in metabolite mixtures. ACS Chem. Biol. **11**, 1030–1038 (2016)
64. Chikayama, E., Shimbo, Y., Komatsu, K., Kikuchi, J.: The effect of molecular conformation on the accuracy of theoretical 1H and 13C chemical shifts calculated by ab initio methods for metabolic mixture analysis. J. Phys. Chem. B **120**, 3479–3487 (2016)
65. Ito, K., Xu, X., Kikuchi, J.: Improved prediction of carbonless NMR spectra by the machine learning of theoretical and fragment descriptors for environmental mixture analysis. Anal. Chem. **93**, 6901–6906 (2021)
66. Anderson, P.E., Reo, N.V., DelRaso, N.J., Doom, T.E., Raymer, M.L.: Gaussian binning: a new kernel-based method for processing NMR spectroscopic data for metabolomics. Metabolomics. **4**, 261–272 (2008)
67. Fukuda, S., et al.: Evaluation and characterization of bacterial metabolic dynamics with a novel profiling technique, real-time metabolotyping. PLoS ONE. **4**, e4893 (2009)
68. Dieterle, F., Ross, A., Schlotterbeck, G., Senn, H.: Probabilistic quotient normalization as robust method to account for dilution of complex biological mixtures. Application in 1H NMR metabonomics. Anal. Chem. **78**, 4281–4290 (2006)
69. Date, Y., Sakata, K., Kikuchi, J.: Chemical profiling of complex biochemical mixtures from various seaweeds. Polym. J. **44**, 888–894 (2012)
70. Ogawa, D.M.O., et al.: Biogeochemical typing of paddy field by a data-driven approach revealing sub-systems within a complex environment–a pipeline to filtrate, organize and frame massive dataset from multi-omics analyses. PLoS ONE. **9**, e110723 (2014)
71. Ogura, T., Hoshino, R., Date, Y., Kikuchi, J.: Visualization of microfloral metabolism for marine waste recycling. Metabolites. **6** (2016)
72. Date, Y., Iikura, T., Yamazawa, A., Moriya, S., Kikuchi, J.: Metabolic sequences of anaerobic fermentation on glucose-based feeding substrates based on correlation analyses of microbial and metabolite profiling. J. Proteome Res. **11**, 5602–5610 (2012)
73. Shiokawa, Y., Misawa, T., Date, Y., Kikuchi, J.: Application of market basket analysis for the visualization of transaction data based on human lifestyle and spectroscopic measurements. Anal. Chem. **88**, 2714–2719 (2016)
74. Chatzimichali, E.A., Bessant, C.: Novel application of heuristic optimisation enables the creation and thorough evaluation of robust support vector machine ensembles for machine learning applications. Metabolomics. **12** (2016)
75. Yamazaki, K., et al.: Oral pathobiont-induced changes in Gut microbiota aggravate the pathology of nonalcoholic fatty liver disease in mice. Front. Immunol. **12**, 766170 (2021)
76. Shima, H., et al.: Identification of salmoniformes aquaculture conditions to increase creatine and anserine levels using multiomics dataset and nonnumerical information. Front. Microbiol. **13**, 991819 (2022)

77. Yokoyama, D., Suzuki, S., Asakura, T., Kikuchi, J.: Microbiome and metabolome analyses in different closed-circulation aquarium systems and their network visualization. ACS Omega. **7**, 30399–30404 (2022)
78. Yokoyama, D., Suzuki, S., Asakura, T., Kikuchi, J.: Chemometric analysis of NMR spectra and machine learning to investigate membrane fouling. ACS Omega. **7**, 12654–12660 (2022)
79. Shima, H., Sakata, K., Kikuchi, J.: Prediction of influence transmission by water temperature of fish intramuscular metabolites and intestinal microbiota factor cascade using Bayesian networks. Appl. Sci.-Basel. **13** (2023)
80. Ward, J.L., Baker, J.M., Llewellyn, A.M., Hawkins, N.D., Beale, M.H.: Metabolomic analysis of arabidopsis reveals hemiterpenoid glycosides as products of a nitrate ion-regulated, carbon flux overflow. Proc. Natl. Acad. Sci. U. S. A. **108**, 10762–10767 (2011)
81. Poulson-Ellestad, K.L., et al.: Metabolomics and proteomics reveal impacts of chemically mediated competition on marine plankton. Proc. Natl. Acad. Sci. U. S. A. **111**, 9009–9014 (2014)
82. Vergara, F., Kikuchi, J., Breuer, C.: Artificial autopolyploidization modifies the tricarboxylic acid cycle and GABA shunt in arabidopsis thaliana Col-0. Sci. Rep. **6**, 26515 (2016)
83. Okamoto, M., Tsuboi, Y., Chikayama, E., Kikuchi, J., Hirayama, T.: Metabolic movement upon abscisic acid and salicylic acid combined treatments. Plant. Biotechnol.-Nar. **26**, 551–560 (2009)
84. Mochida, K., Furuta, T., Ebana, K., Shinozaki, K., Kikuchi, J.: Correlation exploration of metabolic and genomic diversity in rice. BMC Genomics. **10**, 568 (2009)
85. Mansfield, S.D., Kim, H., Lu, F.C., Ralph, J.: Whole plant cell wall characterization using solution-state 2D NMR. Nat. Protoc. **7**, 1579–1589 (2012)
86. Kato, T., et al.: Integrated multi-omics analysis reveals differential effects of fructo-oligosaccharides (FOS) supplementation on the human gut ecosystem. Int. J. Mol. Sci. **23** (2022)
87. van Duynhoven, J., et al.: Metabolic fate of polyphenols in the human superorganism. Proc. Natl. Acad. Sci. U. S. A. **108**(Suppl 1), 4531–4538 (2011)
88. Kitami, T., et al.: Deep phenotyping of myalgic encephalomyelitis/chronic fatigue syndrome in Japanese population. Sci. Rep. **10**, 19933 (2020)
89. Miyauchi, E., et al.: Dysbiosis of gut microbiome is relevant to oral microbiome profiles in patients with periodontitis. Microbiome (2023)
90. Shiokawa, Y., Date, Y., Kikuchi, J.: Application of kernel principal component analysis and computational machine learning to exploration of metabolites strongly associated with diet. Sci. Rep. **8**, 3426 (2018)
91. Hsieh, C.Y., et al.: Strengthening of the intestinal epithelial tight junction by Bifidobacterium bifidum. Physiol. Rep. **3** (2015)
92. Kato, T., et al.: Oral Administration of porphyromonas gingivalis alters the Gut microbiome and serum metabolome. mSphere. **3** (2018)
93. Date, Y., et al.: In vitro evaluation method for screening of candidate prebiotic foods. Food Chem. **152**, 251–260 (2014)
94. Vergara, F., Shino, A., Kikuchi, J.: Cannibalism affects core metabolic processes in helicoverpa armigera larvae-A 2D NMR metabolomics study. Int. J. Mol. Sci. **17** (2016)
95. Asakura, T., Sakata, K., Yoshida, S., Date, Y., Kikuchi, J.: Noninvasive analysis of metabolic changes following nutrient input into diverse fish species, as investigated by metabolic and microbial profiling approaches. PeerJ **2**, e550 (2014)
96. Misawa, T., Date, Y., Kikuchi, J.: Human metabolic, mineral, and microbiota fluctuations across daily nutritional intake visualized by a data-driven approach. J. Proteome Res. **14**, 1526–1534 (2015)
97. Wei, F., et al.: Fish ecotyping based on machine learning and inferred network analysis of chemical and physical properties. Sci. Rep. **11**, 3766 (2021)
98. Date, Y., et al.: Relaxometric learning: a pattern recognition method for T2 relaxation curves based on machine learning supported by an analytical framework. BMC chemistry **15**, 13 (2021)

99. Yamada, S., Kurotani, A., Chikayama, E., Kikuchi, J.: Signal deconvolution and noise factor analysis based on a combination of time-frequency analysis and probabilistic sparse matrix factorization. Int. J. Mol. Sci. **21** (2020)
100. Date, Y., Kikuchi, J.: Application of a deep neural network to metabolomics studies and its performance in determining important variables. Anal. Chem. **90**, 1805–1810 (2018)
101. Asakura, T., Date, Y., Kikuchi, J.: Application of ensemble deep neural network to metabolomics studies. Anal. Chim. Acta **1037**, 230–236 (2018)
102. Yamazawa, A., Iikura, T., Shino, A., Date, Y., Kikuchi, J.: Solid-, solution-, and gas-state NMR monitoring of C-13-cellulose degradation in an anaerobic microbial ecosystem. Molecules **18**, 9021–9033 (2013)
103. Yamazawa, A., et al.: Cellulose digestion and metabolism induced biocatalytic transitions in anaerobic microbial ecosystems. Metabolites. **4**, 36–52 (2013)
104. Ogura, T., Date, Y., Kikuchi, J.: Differences in cellulosic supramolecular structure of compositionally similar rice straw affect biomass metabolism by paddy soil microbiota. Plos One. **8** (2013)
105. Ogura, T., Date, Y., Tsuboi, Y., Kikuchi, J.: Metabolic dynamics analysis by massive data integration: application to tsunami-affected field soils in Japan. ACS Chem. Biol. **10**, 1908–1915 (2015)
106. Date, Y., et al.: New monitoring approach for metabolic dynamics in microbial ecosystems using stable-isotope-labeling technologies. J. Biosci. Bioeng. **110**, 87–93 (2010)
107. Komatsu, T., Kikuchi, J.: Comprehensive signal assignment of 13C-labeled lignocellulose using multidimensional solution NMR and 13C chemical shift comparison with solid-state NMR. Anal. Chem. **85**, 8857–8865 (2013)
108. Mori, T., et al.: Multidimensional high-resolution magic angle spinning and solution-state NMR characterization of 13C-labeled plant metabolites and lignocellulose. Sci. Rep. **5**, 11848 (2015)
109. Mao, J.D., Schmidt-Rohr, K.: Accurate quantification of aromaticity and nonprotonated aromatic carbon fraction in natural organic matter by 13C solid-state nuclear magnetic resonance. Environ. Sci. Technol. **38**, 2680–2684 (2004)
110. Mao, J., Kong, X., Schmidt-Rohr, K., Pignatello, J.J., Perdue, E.M.: Advanced solid-state NMR characterization of marine dissolved organic matter isolated using the coupled reverse osmosis/electrodialysis method. Environ. Sci. Technol. **46**, 5806–5814 (2012)
111. Wang, T., Salazar, A., Zabotina, O.A., Hong, M.: Structure and dynamics of brachypodium primary cell wall polysaccharides from two-dimensional 13C solid-state nuclear magnetic resonance spectroscopy. Biochem. **53**, 2840–2854 (2014)
112. Wang, T., Park, Y.B., Cosgrove, D.J., Hong, M.: Cellulose-pectin spatial contacts are inherent to never-dried arabidopsis primary cell walls: evidence from solid-state nuclear magnetic resonance. Plant Physiol. **168**, 871–884 (2015)
113. Komatsu, T., Kikuchi, J.: Selective signal detection in solid-state NMR using rotor-synchronized dipolar dephasing for the analysis of hemicellulose in lignocellulosic biomass. J. Phys. Chem. Lett. **4**, 2279–2283 (2013)
114. Ito, K., Sakata, K., Date, Y., Kikuchi, J.: Integrated analysis of seaweed components during seasonal fluctuation by data mining across heterogeneous chemical measurements with network visualization. Anal. Chem. **86**, 1098–1105 (2014)
115. Oita, A., et al.: Profiling physicochemical and planktonic features from discretely/continuously sampled surface water. Sci. Total. Environ. **636**, 12–19 (2018)
116. Okushita, K., Komatsu, T., Chikayama, E., Kikuchi, J.: Statistical approach for solid-state NMR spectra of cellulose derived from a series of variable parameters. Polym. J. **44**, 895–900 (2012)
117. Okushita, K., Chikayama, E., Kikuchi, J.: Solubilization mechanism and characterization of the structural change of bacterial cellulose in regenerated states through ionic liquid treatment. Biomacromol. **13**, 1323–1330 (2012)
118. Mori, T., et al.: Exploring the conformational space of amorphous cellulose using NMR chemical shifts. Carbohydr. Polym. **90**, 1197–1203 (2012)

119. Komatsu, T., Ohishi, R., Shino, A., Kikuchi, J.: Structure and Metabolic-Flow Analysis of Molecular Complexity in a (13) C-Labeled Tree by 2D and 3D NMR. Angew. Chem. Int. Ed. Engl. **55**, 6000–6003 (2016)
120. Akiyoshi, N., et al.: Functional analysis of poplar SOMBRERO-type NAC transcription factors proposed novel strategy to modify woody cell wall property. Plant. Cell. Physiol. (2021) In press
121. Umeda-Hara, C., et al.: Tetraploidization promotes radial stem growth in poplars. Plant. Biotechnol. (Tokyo) **39**, 215–220 (2022)
122. Nakano, Y., et al.: Enhancement of secondary cell wall formation in poplar xylem using a self-reinforced system of secondary cell wall-related transcription factors. Front. Plant Sci. **13**, 819360 (2022)
123. Tokuda, G., et al.: Metabolomic profiling of 13C-labelled cellulose digestion in a lower termite: insights into gut symbiont function. Proc. Biol. Sci. **281**, 20140990 (2014)
124. Wang, G.X., Huang, D., Ji, J.H., Volker, C., Wurm, F.R.: Seawater-degradable polymers-fighting the marine plastic pollution. Adv. Sci. **8** (2021)
125. Borrelle, S.B., et al.: Predicted growth in plastic waste exceeds efforts to mitigate plastic pollution. Sci. **369**, 1515-+ (2020)
126. Kubowicz, S., Booth, A.M.: Biodegradability of plastics: challenges and misconceptions. Environ. Sci. Technol. **51**, 12058–12060 (2017)
127. Rhodes, C.: Plastic pollution and potential solutions. Sci. Prog. **101**, 207–260 (2018)
128. Amelia, T., et al.: Marine microplastics as vectors of major ocean pollutants and its hazards to the marine ecosystem and humans. Prog. Earth Planet. Sci. **8** (2021)
129. Mohanty, A.K., Vivekanandhan, S., Pin, J.M., Misra, M.: Composites from renewable and sustainable resources: challenges and innovations. Science **362**, 536–542 (2018)
130. Filiciotto, L., Rothenberg, G.: Biodegradable plastics: standards, policies, and impacts. Chemsuschem **14**, 56–72 (2021)
131. Haider, T., Volker, C., Kramm, J., Landfester, K., Wurm, F.: Plastics of the future? the impact of biodegradable polymers on the environment and on society. Angew. Chem.-Int. Ed. **58**, 50–62 (2019)
132. Shah, A.A., Hasan, F., Hameed, A., Ahmed, S.: Biological degradation of plastics: a comprehensive review. Biotechnol. Adv. **26**, 246–265 (2008)
133. Chamas, A., et al.: Degradation rates of plastics in the environment. Acs Sustain. Chem. & Eng. **8**, 3494–3511 (2020)
134. Zhu, Y., Romain, C., Williams, C.K.: Sustainable polymers from renewable resources. Nature. **540**, 354–362 (2016)
135. Inkinen, S., Hakkarainen, M., Albertsson, A., Sodergard, A.: From lactic acid to poly(lactic acid) (PLA): characterization and analysis of PLA and its precursors. Biomacromol. **12**, 523–532 (2011)
136. Schaler, K., Achilles, A., Barenwald, R., Hackel, C., Saalwachter, K.: Dynamics in crystallites of poly(epsilon-caprolactone) as investigated by solid-state NMR. Macromolecules. **46**, 7818–7825 (2013)
137. Foston, M.: Advances in solid-state NMR of cellulose. Curr. Opin. Biotechnol. **27**, 176–184 (2014)
138. Eden, M.: Editorial for the special issue on solid-state NMR spectroscopy in materials chemistry. Molecules. **25** (2020)
139. Kikuchi, J., Ito, K., Date, Y.: Environmental metabolomics with data science for investigating ecosystem homeostasis. Prog. Nucl. Mag. Res. Sp. **104**, 56–88 (2018)
140. Schlagnitweit, J., et al.: A solid-state NMR method to determine domain sizes in multi-component polymer formulations. J. Magn. Reson. **261**, 43–48 (2015)
141. Chen, G., et al.: Machine-learning-assisted de novo design of organic molecules and polymers: opportunities and challenges. Polymers. **12** (2020)
142. Himanen, L., Geurts, A., Foster, A., Rinke, P.: Data-driven materials science: status, challenges, and perspectives (vol 6, 1900808, 2019). Adv. Sci. **7** (2020)

143. Ma, R., Luo, T.: PI1M: a benchmark database for polymer informatics. J. Chem. Inf. Model. **60**, 4684–4690 (2020)
144. Granda, J.M., Donina, L., Dragone, V., Long, D.L., Cronin, L.: Controlling an organic synthesis robot with machine learning to search for new reactivity. Nature. **559**, 377–381 (2018)
145. Kumar, J., et al.: Machine learning enables polymer cloud-point engineering via inverse design. Npj Comput. Mater. **5** (2019)
146. Verma, R.P., Hansch, C.: Use of 13C NMR chemical shift as QSAR/QSPR descriptor. Chem. Rev. **111**, 2865–2899 (2011)
147. Yamawaki, R., Tei, A., Ito, K., Kikuchi, J.: Decomposition factor analysis based on virtual experiments throughout Bayesian optimization for compost-degradable polymers. Appl. Sci. **11**, 2820 (2021)
148. Barone, G., Ferizovic, D., Biundo, A., Lindblad, P.: Hints at the applicability of microalgae and cyanobacteria for the biodegradation of plastics. Sustain. **12** (2020)
149. Yang, Y., et al.: Biodegradation and mineralization of polystyrene by plastic-eating mealworms: part 1. Chemical and physical characterization and isotopic tests. Environ. Sci. & Technol. **49**, 12080–12086 (2015)
150. Castro-Aguirre, E., Iniguez-Franco, F., Samsudin, H., Fang, X., Auras, R.: Poly(lactic acid)-mass production, processing, industrial applications, and end of life. Adv. Drug Deliv. Rev. **107**, 333–366 (2016)
151. Paco, A., et al.: Biodegradation of polyethylene microplastics by the marine fungus Zalerion maritimum. Sci. Total. Environ. **586**, 10–15 (2017)
152. Yokoyama, D., Kikuchi, J.: Inferring microbial community assembly in an urban river basin through geo-multi-omics and phylogenetic bin-based null-model analysis of surface water. Environ. Res. **231**, 116202 (2023)
153. Ogura, T., et al.: Improvement of physical, chemical, and biological properties of aridisol from Botswana by the incorporation of torrefied biomass. Sci. Rep. **6**, 28011 (2016)
154. Kudo, H., et al.: Bayesian network highlights the contributing factors for efficient arsenic phytoextraction by Pteris vittata in a contaminated field. Sci. Total. Environ. **899**, 165654 (2023)
155. Ichihashi, Y., et al.: Multi-omics analysis on an agroecosystem reveals the significant role of organic nitrogen to increase agricultural crop yield. Proc. Natl. Acad. Sci. U.S.A. **117**, 14552–14560 (2020)
156. Goldberg, S.J., et al.: Refractory dissolved organic nitrogen accumulation in high-elevation lakes. Nat. Commun. **6**, 6347 (2015)
157. Mladenov, N., et al.: Dissolved organic matter quality in a shallow aquifer of Bangladesh: implications for arsenic mobility. Environ. Sci. Technol. **49**, 10815–10824 (2015)
158. Asakura, T., Date, Y., Kikuchi, J.: Comparative analysis of chemical and microbial profiles in estuarine sediments sampled from Kanto and Tohoku regions in Japan. Anal. Chem. **86**, 5425–5432 (2014)
159. Cao, X., Lattao, C., Pignatello, J.J., Mao, J., Schmidt-Rohr, K.: Sorption selectivity in natural organic matter probed with fully deuterium-exchanged and carbonyl-13C-labeled benzophenone and 1H–13C NMR spectroscopy. Environ. Sci. Technol. **48**, 8645–8652 (2014)
160. Kaiser, E., Simpson, A.J., Dria, K.J., Sulzberger, B., Hatcher, P.G.: Solid-state and multidimensional solution-state NMR of solid phase extracted and ultrafiltered riverine dissolved organic matter. Environ. Sci. Technol. **37**, 2929–2935 (2003)
161. Lattao, C., et al.: Sorption selectivity in natural organic matter studied with nitroxyl paramagnetic relaxation probes. Environ. Sci. Technol. **46**, 12814–12822 (2012)
162. Dona, A.C., et al.: Precision high-throughput proton NMR spectroscopy of human urine, serum, and plasma for large-scale metabolic phenotyping. Anal. Chem. **86**, 9887–9894 (2014)
163. Nagata, A., Kose, K., Terada, Y.: Development of an outdoor MRI system for measuring flow in a living tree. J. Magn. Reson. **265**, 129–138 (2016)

164. Parker, T., et al.: 60 MHz 1H NMR spectroscopy for the analysis of edible oils. Trends Analyt Chem **57**, 147–158 (2014)
165. Kikuchi, J.: Practical aspects of the analysis of low- and high-field NMR data from environmental samples. Methods Mol. Biol. **2037**, 315–331 (2019)

Chapter 10
NMR of Paramagnetic Compounds

Yasuhiko Yamamoto and Tomokazu Shibata

Abstract Paramagnetic ions and molecules have been exploited quite extensively as extrinsic shift and relaxation probes for investigating the structure and dynamics of biological molecules. The prodigious growth of related research areas is easily discernible as the remarkably widening scope of application in diverse fields in life and material sciences. Sperm whale myoglobin (Mb) is well-known as the first protein to have its three-dimensional structure revealed by X-ray crystallographic study and is also known as one of the first paramagnetic proteins studied by NMR. The heme Fe atom in Mb can exhibit a variety of oxidation, ligation, and spin states. In this chapter, Mb is selected as a reference paramagnetic compound to provide an overview of the relationship between the spectral features and the number of unpaired electrons, because the effects of a change in the spin quantum number *S*, i.e., the number of unpaired electrons, on NMR spectral parameters of a single compound can be readily understood. Field-dependent broadening of signals of proteins with a series of *S* values is also described.

Keywords Heme electronic structure · Myoglobin · Paramagnetic effect · Spin state · Unpaired electron

10.1 Introduction

The use of paramagnetic ions or molecules as extrinsic shift and relaxation probes for investigating the structures of biological molecules not only in solution but also in solid states has been exploited much more than originally expected [1–45]. In addition, dynamic nuclear polarization (DNP) ascribing to the magnetization transfer

Y. Yamamoto (✉)
Department of Chemistry, University of Tsukuba, Tsukuba 305-8571, Japan
e-mail: yamamoto@chem.tsukuba.ac.jp

T. Shibata
Department of Bioscience and Bioinformatics, Faculty of Computer Science and Systems Engineering, Kyushu Institute of Technology, Iizuka 820-8502, Japan
e-mail: shiba535@bio.kyutech.ac.jp

The NMR Society of Japan (ed.), *Experimental Approaches of NMR Spectroscopy II*,
https://doi.org/10.1007/978-981-97-6830-1_10

occurring from unpaired electrons to nuclei through stochastic modulation of the hyperfine interaction between electron and nuclear spins is gaining increasing attention for a variety of applications, showing its potential in solution and solid-state NMR as well as in MRI [38, 46–49]. NMR study of paramagnetic compounds originated in solid-state physics in the middle of the 1950s [50, 51] and, as is well known, the technique was soon introduced to chemistry and biochemistry. By 1970, the physical principles underlying the technique were well-understood and the basic concepts of the technique, together with the theoretical framework for interpreting NMR parameters of paramagnetic compounds, had been well-established [1, 2, 10, 52–61]. The prodigious growth of related research areas is easily witnessed by the remarkably widening scope of application in diverse fields in life and material sciences. A large number of reviews on NMR study of paramagnetic compounds have already appeared, some [10, 19, 62–64] comprehensive and others [3–9, 11–18, 20, 21, 23, 26–29, 38, 40, 65] more focused. Considering the aim of the present book, this chapter should serve as a reference source for active researchers and as an introduction to this subject for a novice, with particular emphasis on the experimental aspects. In view of the multiple purposes of the chapter, we selected hemoproteins, particularly sperm whale myoglobin (Mb (Fig. 10.1a)), as reference compounds. As is well known, Mb was the first protein to have had its three-dimensional structure revealed by X-ray crystallographic study [66]. Mb is also known as one of the first paramagnetic proteins studied by NMR [67], and paramagnetic hemoproteins such as Mb, cytochrome *c*, and hemoglobin were extensively investigated by Shulman [68], Wüthrich [69], Ogawa [70], La Mar [71], and Morishima [72] during the late 1960s to the late 1970s. In addition, paramagnetic Mb has been used for one of the landmark NMR measurements, which demonstrated that, in a magnetic field-oriented molecule, a magnetic dipole–dipole interaction does not average to zero [73] and yields a measurable dipolar coupling useful for the refinement of structural and dynamic properties of molecules [74].

Mb is an oxygen (O_2) storage hemoprotein with a molecular weight of about 17 kDa [75]. The heme cofactor (Fig. 10.1b) is buried inside the protein matrix composed of 153 amino acid residues, and its binding to the protein is stabilized by the coordination bond between the heme iron (Fe) atom and the nitrogen one (N_ε) of the proximal His (His93) [66, 76–78] (Fig. 10.1c), together with hydrophobic interaction of the heme cofactor with the surrounding amino acid residues in the heme pocket, and the formation of salt bridges between the heme propionate groups and nearby polar amino acid side chains [76–78]. The heme Fe atom in Mb is generally either in the ferrous (Fe^{2+}) or ferric (Fe^{3+}) state. The numbers of electrons in the 3d orbitals of Fe^{2+} and Fe^{3+} are six and five, respectively (Inset of Fig. 10.2). Hence, the total spin quantum number S is the integer and half-integer for Fe^{2+} and Fe^{3+}, respectively. Depending upon the degree of spin pairing of electrons in the 3d orbitals, Fe^{2+} can have 4, 2, or 0 unpaired electrons, corresponding to $S = 2$, 1, or 0, respectively, and Fe^{3+} corresponding to 5, 3, or 1 unpaired electron, $S = 5/2$, 3/2, or 1/2, respectively. Based on an octahedral ligand field, the energy levels of the five 3d orbitals are split into two groups in such a way that the levels of the d_{z^2} and $d_{x^2-y^2}$ orbitals are higher than those of the other three orbitals, d_{xy}, d_{yz}, and d_{xz}. The spin state of Mb depends

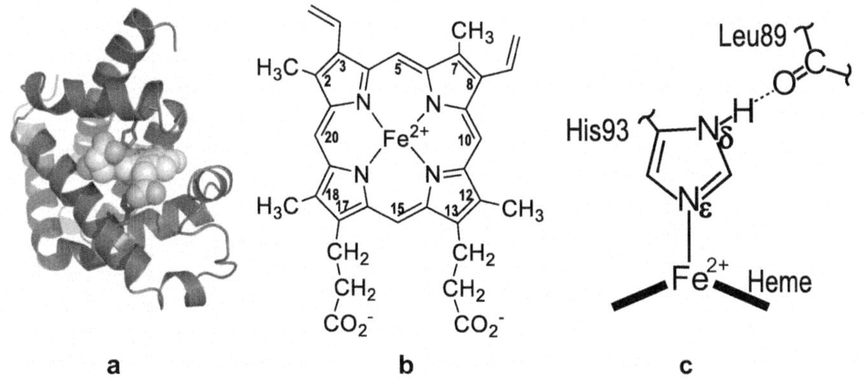

Fig. 10.1 **(a)** Schematic representation of the structure of sperm whale myoglobin (Mb; protein data bank ID 1A6M). The polypeptide chain is illustrated as a ribbon model and the heme is drawn as a space-filling model. **(b)** Structure and numbering system of heme. **(c)** Schematic drawing of the heme coordination structure in deoxy Mb [76]. The hydrogen bonding interaction of His93 $N_{\delta}H$ with the Leu89 carbonyl oxygen atom is represented by a broken line

on the chemical nature of thc ligand. For heme Fe^{2+}, the deoxy form (deoxy Mb) is penta-coordinated with a high-spin configuration, $S = 2$, and the oxy form ($Mb(O_2)$) or carbonmonoxy form (Mb(CO)) possesses a low-spin configuration, $S = 0$. On the other hand, the binding of ligands of relatively weak field strength such as H_2O to heme Fe^{3+} gives a high-spin state $S = 5/2$ ($metMb(H_2O)$), and low-spin state $S = 1/2$ is obtained with a strong ligand such as CN^- ($metMb(CN^-)$). In addition, the binding of ligands of intermediate field strength such as N_3^- ($metMb(N_3^-)$) exhibits a thermal equilibrium between $S = 1/2$ and $S = 5/2$ states. Complexes with $S = 1$ and 3/2 are not so common and have been obtained in some particular systems such as Fe^{4+}–peroxo species ($S = 1$) and Fe^{3+} species coupled with O_2^- ($S = 3/2$) [79].

Aside from the importance of Mb in the early history of the NMR study of paramagnetic compounds, the significance of Mb in NMR studies is mainly two-fold, as follows. Firstly, as described above, the heme Fe atom in Mb can exhibit a variety of oxidation, ligation, and spin states, and hence the effects of a change in the S value, i.e., the number of unpaired electrons, on NMR spectral parameters of a single compound can be readily understood. Secondly, the relationship between the contact shift (δ_c), due to a delocalized unpaired electron, and a pseudo-contact one (δ_{pc}), due to a through-space dipolar interaction with the unpaired electron spin(s), as to paramagnetic shift (δ_{para}) is clearly reflected in the spectra of paramagnetic Mbs, and thus a comprehensive description of the δ_c value can be given in terms of delocalization of unpaired electron density from the paramagnetic center. We intend to portray a range of techniques applied to paramagnetic Mbs, which exhibit widespread utility in life and material sciences.

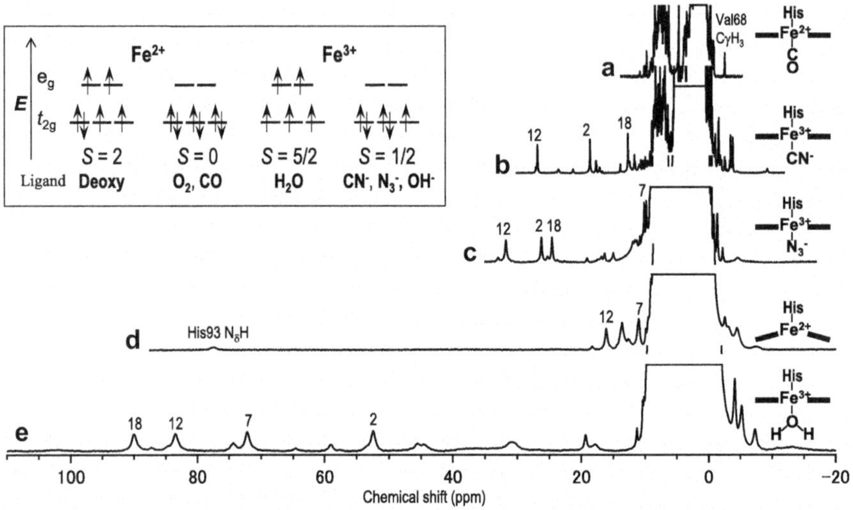

Fig. 10.2 400 MHz ^{1}H NMR spectra of (**a**) carbonmonoxy form (Mb(CO)), (**b**) met-cyano form (metMb(CN^-)), (**c**) met-azido form (metMb($N_3{}^-$)), and (**d**) deoxy form (deoxy Mb) of Mb in 90% ^1H_2O/10% ^2H_2O, pH 7.4, at 25 °C, and **e** met-aquo form (metMb(H_2O)) of the protein in 90% ^1H_2O/10% ^2H_2O, pH 6.5, at 25 °C. The heme Fe oxidation and ligation states of the proteins are schematically illustrated on the right-hand side of the spectra. The assignments of heme methyl proton signals, i.e., 2-, 7-, 12-, and 18-CH_3, are indicated by the corresponding numbers in the spectra. In **a**, four signals are observed at ~3 ppm. In **b**, the 7-CH_3 signal at ~5 ppm is buried in the diamagnetic envelope, and in **d**, 2- and 18-CH_3 signals are not assigned yet. Assignments of the Val68 CγH_3 signal of Mb(CO) and the His93 N_δH proton signal of deoxy Mb are also indicated. In the inset, typical heme Fe oxidation, spin, and ligation states of the protein are illustrated. The intermediate spin complexes, i.e., Fe^{2+} $S = 1$ and Fe^{3+} $S = 3/2$ complexes, are obtained in some particular systems [79]

10.2 Paramagnetic Effects

The theory behind NMR parameters in paramagnetic systems has been treated thoroughly elsewhere [2, 10, 13, 16]. Only a qualitative description of paramagnetic shifts and relaxation, which will suffice to allow an appreciation of the spectra presented, is given below.

10.2.1 Paramagnetic Shifts

Analysis of δ_{para} in terms of the interaction between nuclear and electron spins provides a wealth of information about electronic and molecular structures [2, 8–21]. The observed shift (δ_{obs}) of a paramagnetic compound is given as in Eq. (10.1), where δ_{dia} and δ_{para} are the diamagnetic and paramagnetic contributions, respectively.

$$\delta_{obs} = \delta_{dia} + \delta_{para} \tag{10.1}$$

δ_{ia} is the shift that would have been observed if the molecule contained no unpaired electron, and δ_{para} is expressed as the sum of δ_c and δ_{pc},

$$\delta_{para} = \delta_c + \delta_{pc} \tag{10.2}$$

δ_c reflects the electronic structure of the molecule, and the metal-centered δ_{pc}, due to the magnetic dipolar field arising from delocalized unpaired electron(s) at the paramagnetic center, has been used extensively to refine the molecular structure.

In addition to δ_{para} itself, its temperature dependence also provides valuable information about the molecule. Since both δ_c and δ_{pc} are proportional to the reciprocal of absolute temperature (T), plots of δ_{obs} against $1/T$, often called Curie plots, exhibit a straight line with the intercept at $1/T \rightarrow 0$, which is equal to δ_{dia}. However, anomalous Curie plots are occasionally observed, and the thermodynamic nature of the thermal spin equilibrium in metMb(N_3^-) [80], and the thermal equilibrium between ^5E, $(d_{xy})(d_{xz})^2(d_{yz})(d_{z^2})(d_{x^2-y^2})$, and $^5B_2$, $(d_{xy})^2(d_{xz})(d_{yz})(d_{z^2})(d_{x^2-y^2})$, states of high-spin heme Fe^{2+} in deoxy Mb [81] (see Sect. 10.3.3.2) have been quantitatively characterized through analysis of such anomalous temperature-dependent shift changes of the signals.

10.2.2 Paramagnetic Relaxation

The analysis of nuclear relaxation has provided a wealth of information about the structural and dynamic properties of molecules. In particular, in the case of paramagnetic metalloproteins, the dynamic nature of a molecule is sharply manifested in paramagnetic relaxation observed for paramagnetically shifted NMR signals [2, 10, 13, 14, 16].

The nuclear relaxation rate (R_{obs}) in a paramagnetic system is expressed as the sum of diamagnetic (R_{dia}) and paramagnetic (R_{para}) terms

$$R_{obs} = R_{dia} + R_{para} \tag{10.3}$$

R_{dia} is the sum of the contributions of the dipole–dipole interaction (R_{dia}^{DD}), chemical shift anisotropy (R_{dia}^{CSA}), and others (R_{dia}^{other})

$$R_{dia} = R_{dia}^{DD} + R_{dia}^{CSA} + R_{dia}^{other} \tag{10.4}$$

The contributions of R_{dia}^{CSA} to the nuclear spin–lattice (R_{1dia}) and spin–spin (R_{2dia}) relaxation rates (R_{1dia}^{CSA} and R_{2dia}^{CSA}, respectively) depend on the magnetic field strength and can be written as follows [56]:

$$R_{1dia}^{CSA} = \frac{6}{40}\omega_I^2\delta_z^2\left(1+\frac{\eta^2}{3}\right)J(\omega_I) \tag{10.5}$$

$$R_{2dia}^{CSA} = \frac{1}{40}\omega_I^2\delta_z^2(3J(\omega_I)+4J(0)) \tag{10.6}$$

$$J(\omega_I) = \frac{2\tau_r}{1+\omega_I^2\tau_r^2} \tag{10.7}$$

where δ_z is associated with the principal components of the chemical shift tensor, η is the asymmetric parameter of the molecule under consideration, and $J(\omega_I)$ represents the spectral density function.

On the other hand, R_{para} is expressed as the sum of the contributions of metal-centered (R_{para}^{MC}) and ligand-centered (R_{para}^{LC}) dipolar terms, the contact hyperfine interaction (R_{para}^{C}), and Curie spin relaxation (R_{para}^{Curie}) [2], as illustrated in Fig. 10.3.

$$R_{para} = R_{para}^{MC} + R_{para}^{LC} + R_{para}^{C} + R_{para}^{Curie} \tag{10.8}$$

Using the Solomon-Bloembergen equations [52, 82], together with the expression of R_{para}^{Curie} [83, 84], paramagnetic contributions to the nuclear spin–lattice and spin–spin relaxation rates ($R_{1\text{para}}$ and $R_{2\text{para}}$, respectively) in a paramagnetic system are expressed by

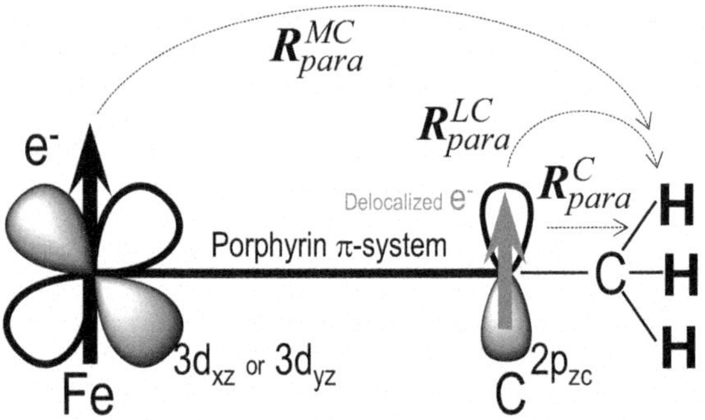

Fig. 10.3 Schematic representation of paramagnetic relaxation mechanisms for heme CH_3 protons. Paramagnetic metal-centered relaxation (R_{para}^{MC}) is due to the dipole–dipole interaction between the nuclear spin and electron spin localized at heme Fe. Paramagnetic ligand-centered relaxation (R_{para}^{LC}) is due to the dipole–dipole interaction between the nuclear spin and electron spins delocalized from heme Fe into the ligand atoms, and, in the case of the heme CH_3 proton, R_{para}^{LC} due to the electron spin delocalized into the p_z orbital of the carbon atom to which CH_3 is covalently attached is predominant. Paramagnetic contact relaxation (R_{para}^{C}) arises from the contact hyperfine interaction between the nuclear spin and electron spin delocalized into the 1s orbital of the CH_3 hydrogen atoms. Curie spin relaxation (R_{para}^{Curie}) arises from the dipole–dipole interaction between the nuclear spin and the time-average of the electron magnetic moment (not shown)

$$R_{1para} = \frac{2}{15}\left(\frac{\mu_0}{4\pi}\right)^2 \gamma_I^2 \beta_e^2 \mu_B^2 S(S+1)\left(\gamma_M^{-6} + \Sigma\rho^2\gamma_L^{-6}\right) \times \left[\frac{\tau_{c2}}{1+(\omega_I-\omega_s)^2\tau_{c2}^2} + \frac{3\tau_{c1}}{1+\omega_I^2\tau_{c1}^2} + \frac{6\tau_{c2}}{1+(\omega_I+\omega_s)^2\tau_{c2}^2}\right] + \frac{2}{3}S(S+1)\left(\frac{A}{\hbar}\right)^2\left[\frac{\tau_{e2}}{1+(\omega_I+\omega_s)^2\tau_{e2}^2}\right] + \frac{2}{5}\left(\frac{\mu_0}{4\pi}\right)^2 \frac{\omega_I^2\beta_e^4\mu_B^4 S^2(S+1)^2}{(3kT)^2 r_M^6}\left(\frac{3\tau_r}{1+\omega_I^2\tau_r^2}\right) \quad (10.9)$$

$$R_{2para} = \frac{1}{15}\left(\frac{\mu_0}{4\pi}\right)^2 \gamma_I^2 \beta_e^2 \mu_B^2 S(S+1)\left(\gamma_M^{-6} + \Sigma\rho^2\gamma_L^{-6}\right) \times \left[4\tau_{c1} + \frac{3\tau_{c2}}{1+(\omega_I-\omega_S)^2\tau_{c2}^2} + \frac{3\tau_{c1}}{1+\omega_I^2\tau_{c1}^2} + \frac{6\tau_{c2}}{1+\omega_S^2\tau_{c2}^2} + \frac{6\tau_{c2}}{1+(\omega_I+\omega_s)^2\tau_{c2}^2}\right] + \frac{1}{3}S(S+1)\left(\frac{A}{\hbar}\right)^2\left[\tau_{e1} + \frac{\tau_{e2}}{1+(\omega_I-\omega_S)^2\tau_{e2}^2}\right] + \frac{1}{5}\left(\frac{\mu_0}{4\pi}\right)^2 \frac{\omega_I^2\beta_e^4\mu_B^4 S^2(S+1)^2}{(3kT)^2 r_M^6}\left(4\tau_{c2} + \frac{3\tau_r}{1+\omega_I^2\tau_r^2}\right) \quad (10.10)$$

$$\tau_{c1}^{-1} = T_{1e}^{-1} + \tau_r^{-1} + \tau_{ex}^{-1} \quad (10.11)$$

$$\tau_{c2}^{-1} = T_{2e}^{-1} + \tau_r^{-1} + \tau_{ex}^{-1} \quad (10.12)$$

$$\tau_{e1}^{-1} = T_{1e}^{-1} + \tau_{ex}^{-1} \quad (10.13)$$

$$\tau_{e2}^{-1} = T_{2e}^{-1} + \tau_{ex}^{-1} \quad (10.14)$$

where in the present case, r_M and r_L are the distances between the proton of interest and the Fe atom, and between the proton and the pyrrole carbon atom, respectively, ρ is the unpaired electron density at the pyrrole carbon atoms (Fig. 10.1b), $\left(\frac{A}{\hbar}\right)$ is the apparent hyperfine coupling constant, ω_I and ω_S are the Larmor frequencies of the nucleus and electron, respectively, T_{1e} and T_{2e} are the electron longitudinal and transverse relaxation times, respectively, and τ_{ex} is the electron exchange time. The other parameters are as usual. In large molecules with highly resolved NMR spectra, $T_{1e}, T_{2e} \ll \tau_r$, and at high magnetic field, $\omega_I^2 T_{1e}^2 \ll 1$, $1 < \omega_S^2 T_{1e}^2$, and $1 < \omega_S^2 T_{2e}^2$, Eqs. (10.9) and (10.10) are reduced to the following equations [85, 86]:

$$R_{1para} = \frac{2}{15}\left(\frac{\mu_0}{4\pi}\right)^2 \gamma_I^2 \beta_e^2 \mu_B^2 S(S+1)\left(\gamma_M^{-6} + \Sigma\rho^2\gamma_L^{-6}\right)T_{1e} \quad (10.15)$$

$$R_{2para} = \frac{7}{15}\left(\frac{\mu_0}{4\pi}\right)^2 \gamma_I^2 \beta_e^2 \mu_B^2 S(S+1)\left(\gamma_M^{-6} + \Sigma\rho^2\gamma_L^{-6}\right)T_{1e} + \frac{1}{3}S(S+1)\left(\frac{A}{\hbar}\right)^2 T_{1e} + \frac{4}{5}\left(\frac{\mu_0}{4\pi}\right)^2 \frac{\omega_I^2 \beta_e^4 \mu_B^4 S^2(S+1)^2}{(3kT)^2 r_M^6}\tau_r \quad (10.16)$$

Equation (10.16) dictates that $R_{2\text{para}}$ depends on the field strength due to the contribution of R_{para}^{Curie}, the third term on the right-hand side of the equation (see Sect. 10.3).

10.3 ^{1}H NMR Spectra of Myoglobin with 0, 1, 4, or 5 Unpaired Electrons

In this section, ^{1}H NMR spectra of Mbs in a variety of oxidation, ligation, and spin states are overviewed to understand the relationship between the number of unpaired electrons and the spectral features. In addition, 400, 600, and 800 MHz ^{1}H spectra of each form of the protein are compared with each other to observe the field dependence of the spectral parameters.

10.3.1 Overview

400 MHz ^{1}H spectra of Mb(CO), metMb(CN^-), metMb(N_3^-), deoxy Mb, and metMb(H_2O) are shown in Fig. 10.2. Heme methyl (CH_3) proton signals are often used as sensitive probes for the heme electronic structures of hemoproteins [6, 14]. In addition to ready observation of the signals with three-proton intensity, the δ_{para} values of the hemoprotein provide a measure of the nonequivalence of the four pyrrole environment of the porphyrin ring, i.e., the in-plane asymmetry of the heme electronic structure, because there is a CH_3 group on each pyrrole (Fig. 10.1b), and delocalization of unpaired electron(s) from the heme π-system to the s orbital of the CH_3 protons necessary for the Fermi contact interaction occurs effectively through hyperconjugation [2, 10]. A comparison of the δ_{para} values of the heme CH_3 proton signals between heme model compounds and hemoproteins in every oxidation/spin state indicated that the spread of the four signals is always much larger in the protein environment [6]. The spread of the four heme CH_3 proton signals in the model compound is due to the intrinsic in-plane asymmetry of the heme electronic structure, and hence the increase in the spread of the signals in the protein reflects the asymmetric nature of the heme-protein linkage, which is relevant to regulation of heme Fe reactivity, and hence the protein function [2, 6, 14].

In the presence of a large spread of the four heme CH_3 proton signals, the average shift of the signals is used as a measure of the electronic nature of the heme Fe atom in the protein. The average shift of the signals of diamagnetic Mb(CO) is ~3 ppm

[87] (Fig. 10.2a), and those of paramagnetic metMb(CN^-) and metMb(H_2O) are ~14 [88] and ~75 ppm [89, 90], respectively (Fig. 10.2b and e, respectively). Thus, the average shift of the paramagnetically-shifted heme CH_3 proton signals is roughly proportional to the number of unpaired electrons, i.e., the S value. The average shift of the signals of metMb(N_3^-), i.e., ~23 ppm [91] (Fig. 10.2c), is somehow between those of metMb(CN^-) and metMb(H_2O), because of the intermediate ligand field strength of N_3^-, which renders the energy difference between the Fe^{3+} low-spin ($S = 1/2$) and high-spin ($S = 5/2$) states comparable to the thermal energy [91–94]. Finally, among the physiologically active forms of the protein, deoxy Mb is the least understood in terms of the molecular properties of its heme-active site [81, 95–97]. Despite extensive efforts, the ground state electronic structure of the Fe^{2+} high-spin ($S = 2$) state of deoxy Mb remains to be elucidated [81]. Due to out-of-plane displacement of the Fe atom from the porphyrin plane in the penta-coordinated heme complex in the protein [77] (Fig. 10.1c), the average shift of the four heme CH_3 proton signals of deoxy Mb, i.e., ~15 ppm [96], is rather small, while the signal of the axial His93 $N_\delta H$ proton is observed at ~77 ppm [98] (Fig. 10.2d).

10.3.2 Diamagnetic Carbonmonoxy Form (Mb(CO))

Mb(CO) is often considered a model of Mb(O_2) that tends to be spontaneously oxidized to metMb(H_2O) through a reaction known as autoxidation. In the absence of an unpaired electron, Mb(CO) is just like an ordinary diamagnetic protein. So the benefit of the higher resolution in the spectrum recorded at higher magnetic-field strength can be clearly witnessed in Fig. 10.4. The ring current effect of the porphyrin moiety of heme is considerably greater than that of the phenyl group such as phenylalanine. As a result, the Val68 $C_\gamma H_3$ proton signal is resolved at ~ −2.5 ppm [99], and signals due to heme 5, 10, 15, and 20-H protons (Fig. 10.1b) are resolved in a downfield-shifted region, i.e., > ~9.5 ppm [87]. Hence, the porphyrin π-system of heme itself can be considered as a sort of intrinsic shift reagent. Furthermore, some NH protons involved in intramolecular hydrogen bonding are also resolved below ~9.5 ppm [99, 100].

10.3.3 Paramagnetic Deoxy Form (Deoxy Mb) with S = 2

10.3.3.1 Paramagnetic Shift

The His93 $N_\delta H$ proton signal is resolved at ~77 ppm [98], and the heme CH_3 proton ones in the range of < ~20 ppm [96] (Fig. 10.2d). The line widths of signals due to the His93 $N_\delta H$ and heme CH_3 protons of the protein in a 400 MHz spectrum at 25 °C (Fig. 10.2d) are ~450 and ~120 Hz, respectively, and their relaxation times are < ~10 ms. Consequently, as far as the acquisition of only the paramagnetically shifted

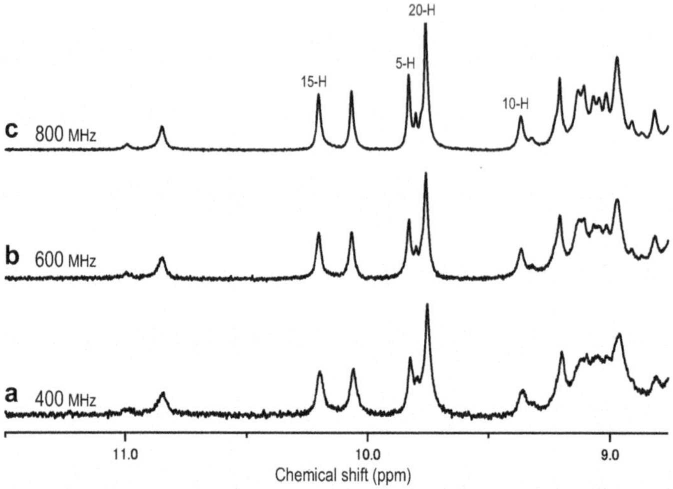

Fig. 10.4 Portions, 8.8–11.5 ppm, of ^1H NMR spectra of Mb(CO) in 90% ^1H_2O/10% ^2H_2O, pH 7.4, at 25 °C, recorded at ^1H frequencies of 400 (**a**), 600 (**b**), and 800 MHz (**c**). Assignments of *meso* 5-, 10-, 15-, and 20-H proton (see Fig. 10.1b) signals are indicated in **c**. The signals resolved below ~9.5 ppm are due to NH protons involved in intramolecular hydrogen bonding [99, 100]

signals of the protein is concerned, the repetition time can be set considerably short, provided that the experimental setting is properly optimized. A drawback to such a measurement is that it will distort the spectral features of signals due to protein protons far away from the paramagnetic center because of the long relaxation times. A relatively large number of dummy scans prior to data acquisition will help to reduce the spectral distortion caused by short repetition times.

The unpaired electrons of the heme Fe atom of deoxy Mb are delocalized into the His93 imidazole through the Fe-His σ and π bonds, i.e., σ- and π-delocalization, respectively [2, 10, 101]. The delocalization of the positive spin of an unpaired electron through the σ- and π-delocalization results in a net spin density of the opposite sign, that is, negative, and a positive spin on the His93 N_ε atom, respectively, and the spin on the His93 N_ε atom is further delocalized, through the π system of the imidazole ring, into the N_δ atom, and then finally into the 1 *s* orbital of the His93 $N_\delta H$ hydrogen, through σ spin polarization [102]. As a result, the positive spin is delocalized into the 1s orbital of the His93 $N_\delta H$ hydrogen through σ-delocalization, whereas the negative spin is delocalized through π-delocalization. Thus, the δ_c value for the His93$N_\delta H$ shift is determined as the difference between the σ- and π-delocalization, and hence large positive δ_c values indicate that the σ -delocalization dominates over the π-delocalization [101]. The strength of the Fe-His bond is possibly affected by steric and electronic factors. La Mar and de Ropp [103] demonstrated, in a study involving horse radish ferrous peroxidase, that the proximal His $N_\delta H$ proton shift is greatly affected by the hydrogen bond between the $N_\delta H$ proton and a protein acceptor residue through its effect on the delocalization of the unpaired electron along the N_δ-H bond, and that simultaneous consideration of changes of the stretching frequency of the Fe-His bond [104] and the axial His N_δH proton shift enables differentiation of the steric and electronic influences on the Fe-His bond [103]. In the case of deoxy Mb, the delocalization of the unpaired electron from the N_δ atom to the 1s orbital of the N_δH hydrogen decreases with increasing strength of the His93-Leu89 H-bond (Fig. 10.1c) because the interaction between the N_δH hydrogen and the unpaired electron density on the His93 imidozole ring is hampered by enhancement of polarization of the N_δ-H bond. Furthermore, a change in the strength of the His93-Leu89 hydrogen bond (Fig. 10.1c) also influences the Fe-His bond strength in such a manner that the Fe-His bond becomes stronger with increasing His93-Leu89 hydrogen bond strength because of enhancement of the Fe $\leftarrow$ His σ donation. Consequently, the strengthening of the Fe-His bond through an increase in the His93-Leu89 H-bond strength results in an increase in the stretching frequency of the Fe-His bond and, inversely, a decrease in the His93N_δH shift [103].

In contrast to effective delocalization of unpaired electrons from the heme Fe atom to the His93 imidazole π system through the Fe-His bond, unpaired electron density delocalized from the Fe atom into the porphyrin π system in the penta-coordinated heme complex is considerably smaller than that in a hexa-coordinated one such as metMb(H_2O) (see Sect. 10.3.6) due to the relatively weak electronic interaction between the Fe orbitals and the porphyrin π system in the former complex [102]. As a result, the heme CH_3 proton signals of deoxy Mb exhibit relatively small δ_{para} values. In addition, the Val68 $C_\gamma H_3$ proton signal is resolved in the upfield-shifted region, i.e., $< \sim -7$ ppm [100], in the ^1H spectrum of deoxy Mb, due to magnetic anisotropy of the Fe^{2+} high-spin state [95].

The signals of deoxy Mb increased dramatically with increasing the field strength (Fig. 10.5), and the field-dependent broadening was enhanced for signals of larger paramagnetic molecules [105]. According to the theory proposed by Gueron [83], there will be net polarization of the Fe electron spin magnetic moment that will be oriented along the direction of the external magnetic field. Modulation of this dipolar field due to the spin polarization (known as the "Curie spin") through rotational diffusion will then introduce an extra field-dependent term into the expression for transverse relaxation (Eq. (10.16)). Since the Curie spin contribution is proportional to the square of the magnetic field strength [83, 84], the benefit of higher resolution in a spectrum recorded at higher magnetic-field strength is no longer expected for deoxy Mb.

10.3.3.2 Temperature-Dependent Shift Changes

O_2 is reversibly bound to heme Fe^{2+} in deoxy Mb (Fig. 10.1c). The reaction between O_2 and heme Fe^{2+} is not as simple as it appears [106]. Upon the binding of O_2 to heme Fe^{2+} in the protein, ground state triplet O_2, i.e., $S = 1$, is bound to high-spin quintet heme Fe^{2+}, i.e., $S = 2$, in deoxy Mb to yield low-spin singlet heme Fe^{2+}, i.e., $S = 0$, in Mb(O_2), which is considered as a mixture of Weiss [107], Pauling [108,

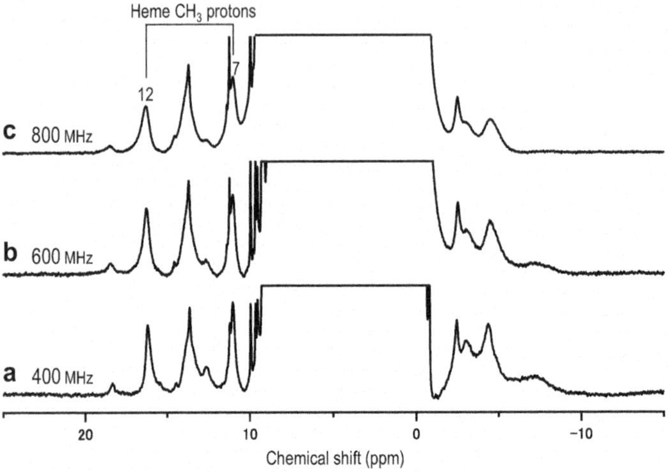

Fig. 10.5 Portions, −15–25 ppm, of ^{1}H NMR spectra of deoxy Mb in 90% ^1H_2O/10% ^2H_2O, pH 7.4, at 25 °C, recorded at ^{1}H frequencies of 400 (**a**), 600 (**b**), and 800 MHz (**c**). Assignments of heme 7-and 12-CH_3 signals are indicated in (c) [96]

109], and McClure forms [110]. Hence, the reaction involves inversion of at least one electron spin. Since chemical reactions do not normally change the spin-states of electrons, the O_2 binding to heme Fe^{2+} in the protein is formally spin-forbidden, which means that it is slow.

Heme Fe is thought to play a crucial role by enhancing the spin-transition probability for accelerating the spin-forbidden reaction between O_2 and heme Fe^{2+} in deoxy Mb. The high-spin heme Fe^{2+} of deoxy Mb has been suggested to have excited states with unpaired electrons close in energy to the ground state, enhancing crossover probability through the spin–orbit coupling of Fe to allow reactions with triplet O_2. Theoretical studies indicated that the low-lying electronic terms of the high-spin heme Fe^{2+} in deoxy Mb are $^5B_2$, ^5E, ^3E, and ^1A, with $^5B_2$ or one of the ^5E components as the ground state [111] (Fig. 10.6). The d-orbital splittings derived from the ^5E, $(d_{xy})(d_{xz})^2(d_{yz})(d_{z^2})(d_{x^2-y^2})$, or $(d_{xy})(d_{xz})(d_{yz})^2(d_{z^2})(d_{x^2-y^2})$, and $^5B_2$, $(d_{xy})^2(d_{xz})(d_{yz})(d_{z^2})(d_{x^2-y^2})$ states for the high-spin heme Fe^{2+} are thought to be close in terms of energy [111].

The mixing of the d-electron configurations of the ^5E and $^5B_2$ states of heme Fe^{2+} of deoxy Mb at ambient temperature could be characterized through analysis of the temperature-dependent shift changes of paramagnetically-shifted ^{19}F NMR signals due to ^{19}F-labeled heme cofactors incorporated into the proteins [112]. Trifluoromethyl group(CF_3)-substituted heme cofactors, i.e., 2,8-DPF [113] and 7-PF [114] (Fig. 10.6a), were used to prepare reconstituted Mbs, i.e., Mb(2,8-DPF) and Mb(7-PF), respectively. ^{19}F NMR spectra of deoxy and carbonmonoxy forms of the reconstituted proteins are compared to each other, i.e., Mb(2,8-DPF) versus Mb(2,8-DPF)(CO) and Mb(7-PF) versus Mb(7-PF)(CO), in Fig. 10.7 [81]. The appearance of two signals in the spectra of the Mb(2,8-DPF) proteins indicated that the chemical

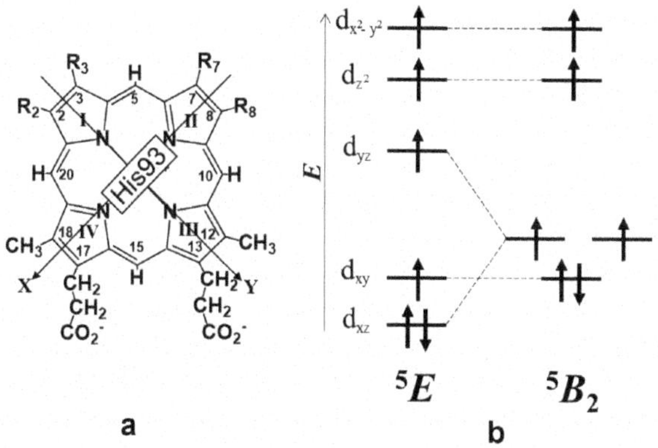

Fig. 10.6 (**a**) The structures and numbering system for 13,17-bis(2-carboxylatoethyl)-3,7,12,18-tetramethyl-2,8-bis(trifluoromethyl)-porphyrinatoiron(II) (2,8-DPF) ($R_2 = R_8 = CF_3$, $R_3 = R_7 = CH_3$) [113], and 13,17-bis(2-carboxylatoethyl)-3,8-diethyl-2,12,18-trimethyl-7-trifluoromethylporphyrinatoiron(II) (7-PF) ($R_2 = CH_3$, $R_3 = R_8 = C_2H_5$, $R_7 = CF_3$) [114], and orientation of the axial His93 imidazole, with respect to the heme cofactor, in native Mb, with hydrogen bonding between the His93 $N_\delta H$ hydrogen and Leu89 carbonyl oxygen atoms (Fig. 10.1c) [77]. (**b**) d-electron configurations of the ^5E, $(d_{xy})(d_{xz})^2(d_{yz})(d_{z^2})(d_{x^2-y^2})$, and $^5B_2$, $(d_{xy})^2(d_{xz})(d_{yz})(d_{z^2})(d_{x^2-y^2})$ states of the high-spin heme Fe(II) in deoxy Mb

equivalence of the two CF_3 groups of 2,8-DPF is removed through the asymmetric heme-protein interaction in the protein [115, 116]. On the other hand, the observation of the two signals, i.e., signals *m* and *M*, in the spectra of the Mb(7-PF) proteins indicated that 7-PF is incorporated into the heme pocket with the two orientations that differ by 180° rotation about the 5,15-*meso* axis [71] (see inset of Fig. 10.7).

^{19}F NMR spectra of deoxy Mb(2,8-DPF) and Mb(7-PF) recorded at various temperatures are shown in Fig. 10.8. The effect of Curie spin relaxation [83, 105] on the appearance of the signals, i.e., broadening of the signals with decreasing temperature, is observed in Fig. 10.8. The contributions of the multiple electronic states to the electronic nature of deoxy Mb are reflected in anomalous temperature-dependent shift changed in the signals, such as curvature and anti-Curie slope, in the Curie plots (Fig. 10.9).

As described in Sect. 10.2.1, δ_{obs} of a signal converges to δ_{dia} as $T \to \infty$. Hence, the δ_{para}s of CF_3 ^{19}F signals of the deoxy proteins in the ^5E and $^5B_2$ states, $\delta(^5E)$ and $\delta(^5B_2)$, respectively, are given by

$$\delta(^5E) = E/T \tag{10.17}$$

$$\delta(^5B_2) = B/T \tag{10.18}$$

where *E* and *B* are constants. Since the interconversion between the two states is faster than the NMR timescale, the δ_{para} value observed for a given CF_3 signal is expressed as,

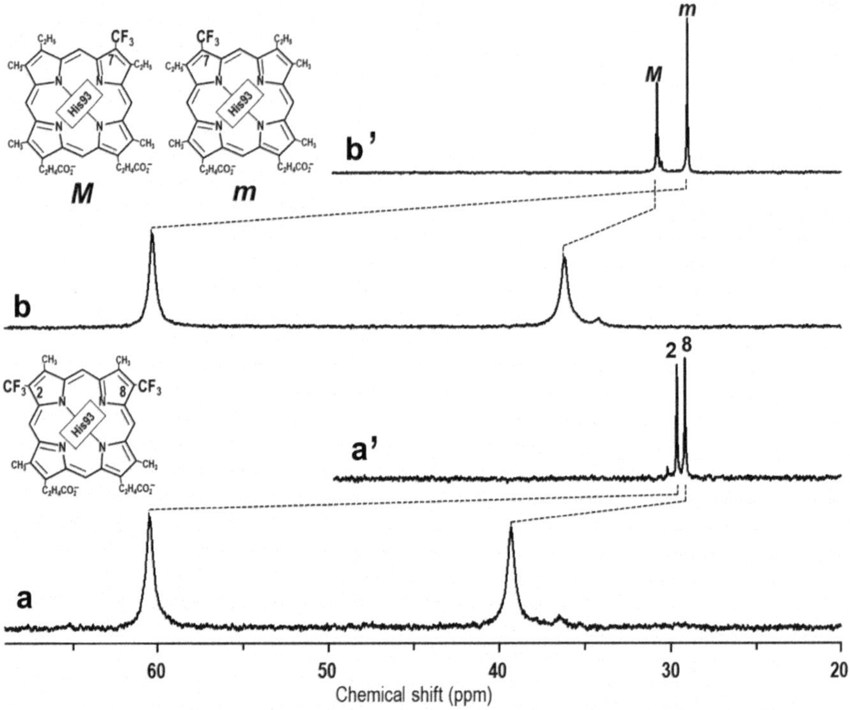

Fig. 10.7 ^{19}F NMR spectra of **(a)** deoxy Mb(2,8-DPF), (a') Mb(2,8-DPF)(CO), **(b)** deoxy Mb(7-PF), and (b') Mb(7-PF)(CO) at 25 °C in 90% H_2O/10% ^2H_2O at pH 6.5. The orientations of 2,8-DPF and 7-PF relative to His93 are illustrated in the insets. Signal assignments are indicated in the spectra. 2-CF_3 and 8-CF_3 signals of Mb(2,8-DPF)(CO) are observed at 29.68 and 29.21 ppm, respectively, [81] and signals *M* and *m* of Mb(7-PF)(CO) are detected at 30.81 and 29.03 ppm, respectively [117]. The shifts of the signals of the carbonmonoxy form can be used as δ_{dia} values for the corresponding signals of the deoxy form

$$\delta_{para} = \delta(^5E)\alpha + \delta(^5B_2)(1-\alpha) \tag{10.19}$$

where α represents the fraction of the ^5E state. The equilibrium constant K (= [^5E]/[$^5B_2$]) is expressed by α as follows:

$$K = \alpha/(1-\alpha) \tag{10.20}$$

Using Eqs. (10.1) and (10.17)–(10.20), the van't Hoff equation is written as

$$\ln[\{B - T(\delta_{\text{obs}} - \delta_{\text{dia}})\}/\{(T(\delta_{\text{obs}} - \delta_{\text{dia}}) - E)\}] = -(\Delta H/R)(1/T) + \Delta S/R \tag{10.21}$$

where ΔH, ΔS, and R on the right side of the equation are as usual.

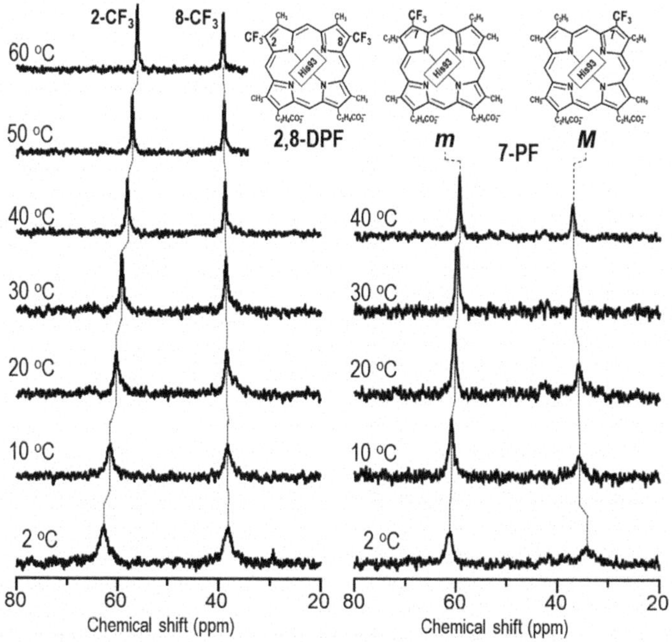

Fig. 10.8 471 MHz ^{19}F NMR spectra of deoxy Mb(2,8-DPF) (left) and Mb(7-PF) (right) in 90% H_2O/10% ^2H_2O at pH 6.5 and the indicated temperatures. The molecular structures of 2,8-DPF and 7-PF, together with two different orientations of 7-PF relative to His93, i.e., the *m* and *M* forms, are illustrated in the insets. Signal assignments are shown with the spectra

The Curie plots of the signals were fitted to Eq. (10.21) to yield the optimal *E*, *B*, ΔH, and ΔS values, using a curve-fitting program reported by Neya and Funasaki [80]. Using $\delta_{\text{dia}} = 29.21$ ppm [117], the fitting of the Curie plots for the 8-CF_3 signal of deoxy Mb(2,8-DPF) with Eq. (10.21) yielded values of -20 ± 10 kJmol^{-1} and -57 ± 38 JK^{-1} mol^{-1} for ΔH and ΔS, respectively, [81] and those for the other signals were similarly analyzed (Fig. 10.9). The obtained ΔH and ΔS values indicated that the energy difference between the ^5E and $^5B_2$ states of the protein at ambient temperature is comparable to thermal energy, suggesting that the presence of the low-lying excited state of a high-spin quintet heme Fe^{2+} in deoxy Mb accelerates the spin-forbidden O_2 binding to heme Fe^{2+} in the protein through the interconversion between the states.

10.3.4 Paramagnetic Met-Cyano Form (metMb(CN^-)) with S = 1/2

Owing to the short spin–lattice relaxation time of electrons (T_{1e}), i.e., ~2×10^{-12} s [4, 20], and the large magnetic anisotropy of hexa-coordinated Fe^{3+} heme complexes

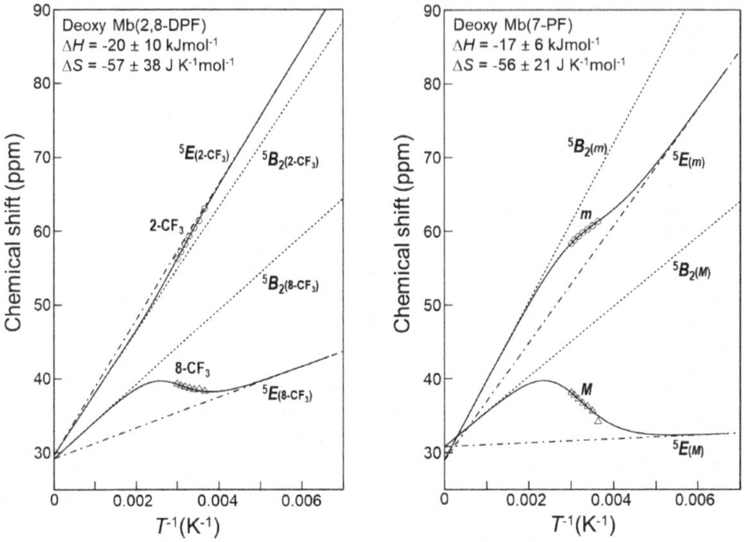

Fig. 10.9 Curie plots of the ^{19}F NMR signals of deoxy Mb(2,8-DPF) (left) and Mb(7-PF) (right) in 90% H_2O/10% ^2H_2O at pH 6.5, together with fitting the data to Eq. (10.21) using a curve-fitting program reported by Neya and Funasaki [80]. The curie plots of the signals exhibited not only apparent positive (2-CF_3 and *m*) and negative slopes (*M*), but also curvature (8-CF_3). The plots for the 2-CF_3 signal of deoxy Mb(2,8-DPF) and signal *M* of deoxy Mb(7-PF) were fitted with the parameters yielded on analysis of the plots for 8-CF_3 and signal *m*, respectively. The ΔH and ΔS values obtained on the analysis are shown with the plots. The Curie plots for the signals of the proteins in the ^5E and $^5B_2$ states, i.e., ^5E(i) and $^5B_2$(i), where i = 2-CF_3, 8-CF_3, *m*, or *M*, determined using the *E* and *B* values in Eqs. (10.17) and (10.18) yielded on curve-fitting, are shown as dot-and-dash and broken lines, respectively

with $S = 1/2$ [118], metMb(CN^-) exhibits well-resolved and relatively narrow signals useful for characterizing the electronic/molecular structure of its heme active site [14, 118]. The asymmetric nature of the heme-protein linkage can be readily characterized through analysis of the δ_{para} values of the four heme CH_3 proton signals of metMb(CN^-). The energy levels for the d_{xz} and d_{yz} orbitals are affected by the interaction with axial ligands and a single unpaired electron resides in either the d_{xz} or d_{yz} orbital, whichever possesses the highest energy. Consequently, depending upon the relative energy of the d_{xz} and d_{yz} orbitals, π spin delocalization occurs into either pyrroles I, III or II, IV (Fig. 10.10). Since a diatomic cyanide ion is coordinated to the heme Fe atom with its orientation nearly normal as to the heme plane, its coordination does not affect the degree and pattern of the in-plane asymmetry of the heme electronic structure. Consequently, the energy levels for the d_{xz} and d_{yz} orbitals are predominantly affected by the orientation of the His93 imidazole ring, relative to the heme, through a repulsive interaction between the p_z orbital of the Fe-bound nitrogen atom of His93 imidazole and the d_{xz} or d_{yz} one of the heme Fe atom (Fig. 10.10a). The Fe-His coordination bond is roughly along the heme normal, and hence, with the orientation of His93 imidazole, as shown in Fig. 10.10a [119], i.e., the projection of the His93 imidazole ring onto the heme plane is along the nitrogen atoms of pyrroles II and IV, and the p_z orbital of the Fe-bound nitrogen atom in the yz plane interact repulsively with the d_{yz} one, raising the energy level of d_{yz} orbital relative to that of the d_{xz} one (Fig. 10.10b). Consequently, the unpaired electron spin density delocalized from the heme Fe atom into the pyrroles I and III π-system is much greater than that into pyrroles II and IV one, due to the unpaired electron delocalized from the singly occupied d_{yz} orbital directly to the former pyrroles. In contrast, the energy level of the d_{xz} orbital is higher than that of the d_{yz} one with the orientation of the His93 imidazole ring, of which the projection onto the heme plane is along the nitrogen atoms of pyrroles I and III, and the unpaired electron spin density delocalized from the heme Fe atom into the pyrroles II and IV π-system is greater than that into the pyrroles I and III one (Fig. 10.10c). Since the orientation of the His93 imidazole ring in Mb is similar to that illustrated in Fig. 10.10a [76, 77], the δ_c values of the signals due to the heme CH_3 groups at positions 2 and 12 are larger than those at positions 7 and 18.

The large magnetic anisotropy of metMb(CN^-) induces large δ_{pc} values for signals arising from protons located close to the heme Fe atom, as can be witnessed by the observation of many resolved signals due to amino acid protons (Fig. 10.11). The combined use of NMR and X-ray studies allows the determination of principal components and orientation, with respect to the heme, of the magnetic susceptibility

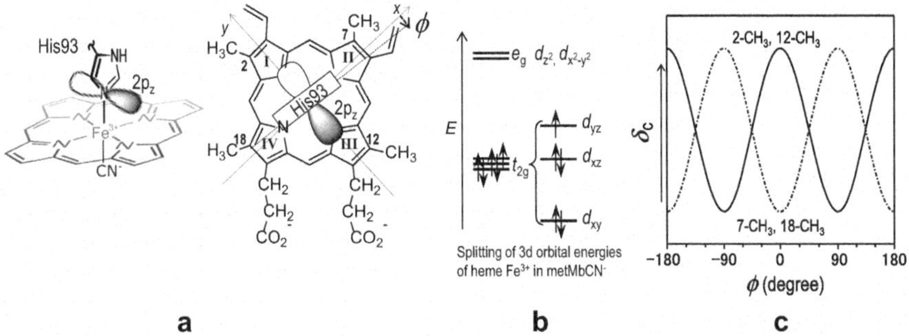

Fig. 10.10 **(a)** Schematic representation of the orientation of the His93 imidazole ring and $2p_z$ orbital of the His93 N_ε atom, with respect to the heme in Mb [76, 77]. Heme peripheral side chains are eliminated for clarity. Angle ϕ is defined as the angle between the projection of the His93 imidazole plane onto the heme one and the N_{II}-Fe-N_{IV} axis. **(b)** Splitting of 3d orbital energies of heme Fe^{3+} in metMb(CN^-). With $\phi = \sim 0°$, the energy level of the $3d_{yz}$ orbital is higher than that of the $3d_{xz}$ one. Hence, electron spin density delocalized from the heme Fe atom into the pyrroles I and III π-system is much greater than that into the pyrroles II and IV one, due to unpaired electron delocalized from the singly occupied $3d_{yz}$ orbital directly to the former pyrroles. Coordination of CN^- with its orientation nearly normal to the heme plane does not significantly affect the energy levels of the $3d_{yz}$ and $3d_{xz}$ orbitals. **(c)** Semi-quantitative representation of ϕ dependence of the contact shift (δ_c) of heme 2-and 12-CH_3 (solid line), and 7-and 18-CH_3 (alternate long and short dash line) signals of metMb(CN^-). The δ_c value was estimated with the assumption of four-fold symmetry in the electronic structure of the porphyrin moiety of heme

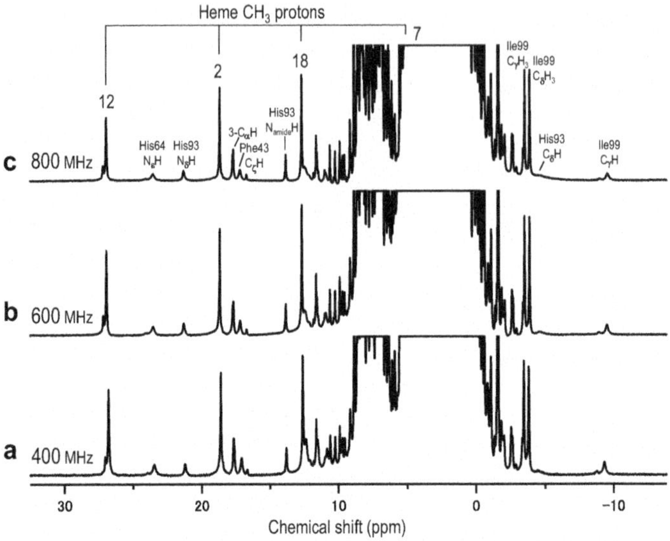

Fig. 10.11 Portions, −13.9–32.5 ppm, of ^{1}H NMR spectra of metMb(CN^-) in 90% $^{1}H_2O$/10% $^{2}H_2O$, pH 7.4, at 25 °C, recorded at ^{1}H frequencies of 400 (**a**), 600 (**b**), and 800 MHz (**c**). Assignments of heme CH_3 and some amino acid proton signals are indicated in (c)

tensor in the protein [118]. Formulation of the δ_{pc} value is useful for detailed characterization of the local structure. In addition, due to the small Curie spin relaxation expected from the small S value, the signals of metMb(CN^-) exhibit small field-dependent broadening, i.e., the line widths of heme 12-CH_3 signals recorded at 400, 600, and 800 MHz were 39, 44, and 66 Hz, respectively (Fig. 10.11).

10.3.5 Paramagnetic Met-Azido Form (MetMb(N_3^-)) with Mainly S = 1/2

The met-azido derivatives of hemoproteins in general share the property that they exhibit a thermal equilibrium between low-spin, $S = 1/2$, and high-spin, $S = 5/2$, states, due to the intermediate field strength of N_3^- [92, 93]. Hence, changes in the population of the two spin states in the equilibrium provide a sensitive probe of the effective axial field strength. Among various techniques used to characterize the thermal spin equilibria in proteins, NMR is probably the most sensitive one and provides two separate probes of the average spin magnetization of the heme Fe atom, i.e., the heme CH_3 proton and axial His93 N_δ H proton δ_{para} values [80, 91]. The thermal spin equilibrium in metMb(N_3^-) is manifested in the curvature or anti-Curie behavior of Curie plots, i.e., the plots of the δ_{obs} values against the reciprocal of absolute temperature ($1/T$), of these signals [80, 91]. Fitting of the Curie plots using

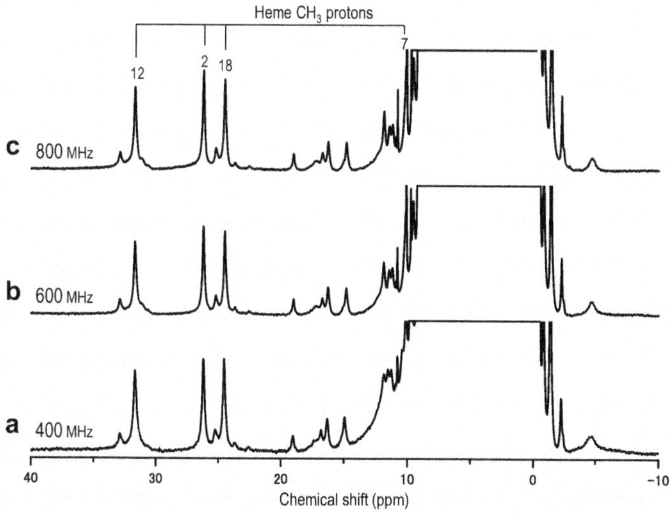

Fig. 10.12 Portions, −10–40 ppm, of ^{1}H NMR spectra of metMb(N_3^-) in 90% ^1H_2O/10% ^2H_2O, pH 7.4, at 25 °C, recorded at ^{1}H frequencies of 400 (**a**), 600 (**b**), and 800 MHz (**c**). Assignments of heme CH_3 signals are indicated in (c)

the van't Hoff equation provides thermodynamic parameters of the equilibrium [80]. In the case of metMb(N_3^-), the high-spin contents at ambient temperature are ~20%, and the values of ~−13 kJ mol^{-1} and ~−30 J T^{-1} mol^{-1} have been reported for the enthalpy change (ΔH) and molar entropy (ΔS), respectively [105]. ΔH is simply determined by the difference in thermodynamic energy between the low-spin and high-spin states, and ΔS is determined by the structural and dynamic natures of the heme environment. The appreciable contribution of the high-spin state to the ground state electron configuration of heme Fe^{3+} in metMb(N_3^-) leads to sizable Curie spin relaxation, which considerably broadens paramagnetically-shifted signals with increasing magnetic field strength, e.g., the line width of the heme 12-CH_3 signal at 25 °C was ~140 Hz at 400 MHz, ~160 Hz at 600 MHz, and ~200 Hz at 800 MHz (Fig. 10.12).

10.3.6 Paramagnetic Met-Aquo Form (MetMb(H_2O)) with S = 5/2

In the presence of five unpaired electrons in the system, the signals of metMb(H_2O) are observed over the range of ~110 to ~−30 ppm [89, 90]. Although the signals are considerably broad due to their relatively long T_{1e}, i.e., < ~1 × 10^{-10}s [4, 20], owing to the large chemical shift range, the signals are fairly well-separated (Fig. 10.13). In Fe^{3+} high-spin complexes, the relaxation process for the heme proton signals is

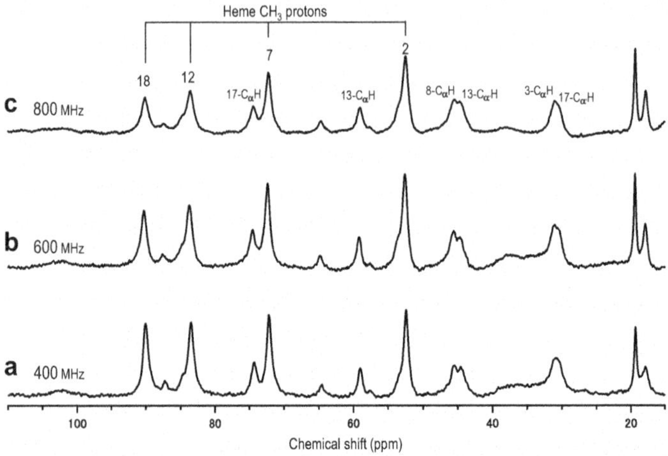

Fig. 10.13 Portions, 15–110 ppm, of ^{1}H NMR spectra of metMb(H_2O) in 90% ^1H_2O/10% ^2H_2O, pH 7.0, at 25 °C, recorded at ^{1}H frequencies of 400 (**a**), 600 (**b**), and 800 MHz (**c**). Assignments of heme side chain proton signals are indicated in (**c**)

modulated mainly by T_{1e}, which is in turn determined by modulation of the zero-field level of the complex [120, 121]. Some Mbs possess penta-coordinated hemes in their Fe^{3+} high-spin complexes because of the absence of Fe-bound H_2O [122]. The heme proton signals of the Mbs with penta-coordinated Fe^{3+} high-spin hemes are narrower than those of the proteins with hexa-coordinated ones, because of the larger zero-field splitting in the former hemes, due to the lower symmetry in the ligand field around the heme Fe, which leads to reduction of T_{1e} [122].

As indicated in Fig. 10.2e, signals resolved outside of the so-called "diamagnetic envelope", i.e., ~10–~0 ppm, are mostly due to heme peripheral side chain protons, and only a few signals due to amino acid protons are resolved. These spectral features indicated that δ_c is large, while δ_{pc} is relatively small. The δ_{pc} contribution of metMb(H_2O) is due to magnetic anisotropy caused by zero-field splitting at the heme Fe, and is in practice axially symmetric [123]. Since the Curie spin relaxation contribution is significant in high-spin metMb(H_2O), the line widths of paramagnetically shifted signals increased dramatically with increasing magnetic field strength, e.g., the line width of the heme 12-CH_3 signal at 25 °C was ~380 Hz at 400 MHz, ~560 Hz at 600 MHz, and ~950 Hz at 800 MHz (Fig. 10.13).

10.4 NMR Measurements

10.4.1 NOEs in Paramagnetic Compounds

Despite the initial prejudice that the presence of unpaired electron(s) in paramagnetic molecules seriously diminishes the magnitude of NOEs through "paramagnetic leakage", and hence the observation of NOEs in paramagnetic systems is practically unrealistic, nowadays NOE measurements have been extensively applied for signal assignments and structural characterization in various paramagnetic systems [17]. The observation of relatively large NOEs in paramagnetic systems [88, 90, 119, 124, 125] could be rationalized on the basis of the nature of relaxation in the systems. The steady-state NOE in an isolated two-spin is given by Eq. (10.22)

$$NOE_{i \to j} = \frac{\sigma_{ij}}{\rho_j} \tag{10.22}$$

where σ_{ij} and ρ_j are the cross-relaxation rate between the interacting protons, i and j, and the intrinsic spin–lattice relaxation rate of spin j, respectively. For paramagnetic system, $\rho_j = \rho_{dia} + \rho_{para}$ (ρ_{dia} and ρ_{para} are the diamagnetic and paramagnetic contributions to ρ_j, respectively), and for a dominant paramagnetic relaxation ($\rho_{dia} \ll \rho_{para}$), the steady-state NOEs are considerably reduced through paramagnetic leakage. Although the magnitude is relatively small, the observation of NOEs has been reported for the most accessible oxidation and spin states of proteins.

Since σ_{ij} is given by

$$\sigma_{ij} = \frac{\hbar^2 \gamma_H^2}{10r^6}\left[\frac{6\tau_c}{1 + 4\omega_H^2 \tau_c^2} - \tau_c\right] \tag{10.23}$$

where γ_H is the gyromagnetic ratio of proton, r is the distance between the protons, τ_c is the correlation time that modulates the dipolar interaction between the two spins, and ω_H. is the Larmor frequency of proton. In the case of macromolecules, at a high magnetic field and a long τ_c, the molecular motion occurs in the slow motion limit ($\omega_H^2 \tau_c^2 \gg 1$), and Eq. (10.23) is reduced to

$$\sigma_{ij} = -\frac{\hbar^2 \gamma_H^2 \tau_c}{10r^6} \tag{10.24}$$

Thus the magnitude of the NOEs is simply proportional to τ_c, and hence, as far as σ_{ij} is concerned, the observation of NOEs could be more promising for a larger molecule than for a small one. In practice, however, for a diamagnetic system with a large τ_c, the detrimental effects of spin diffusion preclude the collection of any useful data in NOE studies. However, as demonstrated by La Mar and co-workers [90], paramagnetism not only undermines spin diffusion, and hence allows observation of specific NOEs for larger systems, but also improves the magnitude of NOEs with increasing τ_c. Consequently, a large molecular size is likely to be an advantage rather than a disadvantage for observing useful NOEs in a paramagnetic system [126] (Fig. 10.14).

The optimization of experimental parameters in 1D and 2D NOE measurements on rapidly relaxing systems, where σ_{ij} is much smaller than ρ_j, has been fully discussed

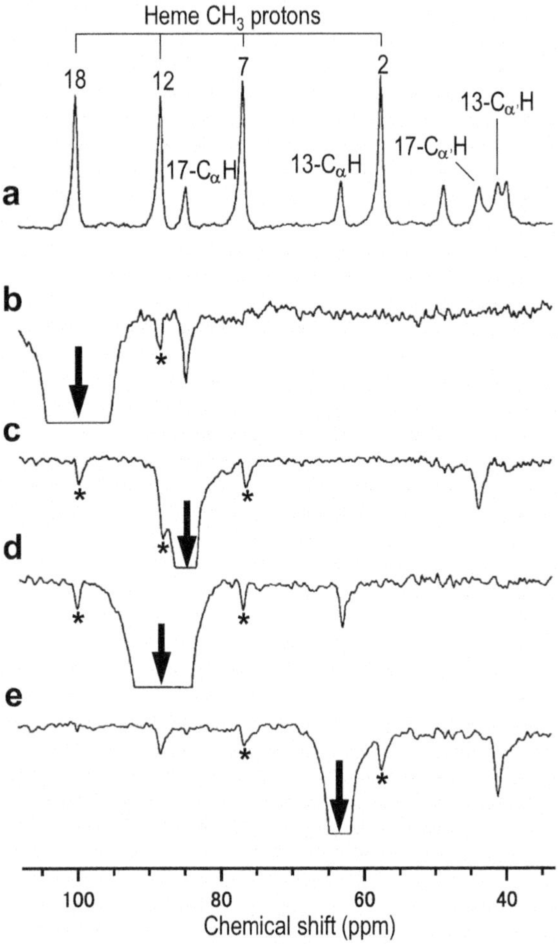

Fig. 10.14 Downfield-shifted portions, 33–108 ppm, of the 500 MHz ^{1}H NMR spectrum and NOE difference spectra of shark *Galeorhinus japonicas* metMb in ^2H_2O, pH 7.0, at 35 °C. **(a)** Reference spectrum and **(b)–(e)** NOE difference spectra generated by subtracting the reference spectrum (a) from the spectrum recorded with the saturation of the desired signal. Arrows indicate signals that are saturated and peaks due to the decoupler pulse power spillage are indicated by *. (b) Saturation of the 18-CH_3 signal exhibiting an NOE to 17–C_αH one. **(c)** Saturation of 17–C_αH signal exhibiting an NOE to 17–$C_{\alpha'}$H one. **(d)** Saturation of the 12-CH_3 signal exhibiting an NOE to a 13–C_αH one. **(e)** Saturation of the 13–C_αH signal exhibiting an NOE to a 13–$C_{\alpha'}$H one. The spectra were taken from [126]

[17, 90]. Considering $\sigma_{ij} \ll \rho_j$ in a paramagnetic system, the magnitude of NOEs to be detected is quite small. In the spectrum of a paramagnetic molecule, fast relaxing signals exhibit large δ_{para} and hence are resolved outside of the diamagnetic envelope. The observation of NOEs between a rapidly relaxing signal and a slowly relaxing one, resonating in the diamagnetic envelope, can be much more effectively performed by conventional 1D NOE difference spectroscopy, provided that the signal-to-noise ratio of spectra is high enough to detect small NOEs. The accumulation of a large number of transients is a simple and easy approach to achieving a satisfactory signal-to-noise ratio. A short acquisition time due to a large spectral width and a relatively low digital resolution generally used for the observation of the spectra of a paramagnetic system, together with rapid repetition owing to fast paramagnetic relaxation, significantly contributes to effective data collection. In general, the detection of a small intensity change becomes more difficult as the signal broadens. Hence, T_2 is often the determining factor for observing the effect; since $T_2 \leq T_1$, this is particularly true for signals with short T_1s.

Practically, there are also other limitations to the NOE method applied to paramagnetic systems. First, when saturating a fast relaxing signal by decoupler irradiation, the necessarily strong decoupler RF field irradiation can induce interfering off-resonance perturbations over a wide region of the spectrum [90]. In such a case, true NOEs may be difficult to distinguish from an off-resonance effect, often called "decoupler power spillage". Comparison of the relative magnitudes of the effects under conditions of a reduced off-resonance effect and the determination of the actual power profile, as a function of offset, of the decoupler RF field irradiation can usually alleviate the problem. Second, observation of a small intensity difference in a broad signal demands a high dynamic range for the receiver and the digitizer. Selective excitation pulse sequences such as a simple 1–1 jump and return one [127] are helpful for suppressing the diamagnetic signals and allowing the full utilization of the dynamic range.

10.4.2 Hydrogen Exchange Rates

Biological macromolecules exhibit freedom of various structural changes, which are often relevant to their functions. Consequently, in order to elucidate the functions of the molecules, their static structures described by X-ray crystallographic coordinates must be augmented by their dynamic properties manifested in various spectroscopic data. Among a variety of spectroscopic techniques available for studies of their dynamic nature, NMR is quite unique in terms of not only its sensitivity to dynamic structures exhibiting a wide range of time scales but also its ability to provide information from which detailed descriptions of the dynamic structures of molecules in solution can be made.

Upon such dynamic studies involving NMR, it is important to consider the "NMR time scale". For example, in the case of a molecule with interconvertible conformations, i.e., conformers I and II, which possibly exhibit two distinct NMR signals, whether or not we can observe two separate signals due to conformers I and II depends on the time scale of the interconversion. If the interconversion is sufficiently slow compared with the NMR time scale, signals I and II can be separately observed. On the other hand, if the interconversion is faster than the NMR time scale, a single signal representing the time-averaged conformation of conformers I and II is observed. The NMR time scale of the interconversion between conformers I and II is given by the difference in resonance frequency between signals I and II ($\Delta\delta_{I/II}$), and hence depends on magnetic field strength (B_0) in such a manner that the NMR time scale increases with increasing B_0 because $\Delta\delta_{I/II}$ is proportional to B_0. Taking advantage of the large δ_{para}, a paramagnetic counterpart possibly exhibits the $\Delta\delta_{I/II}$ value that is large enough to allow observation of separate signals I and II. Thus the NMR time scale can be controlled through the paramagnetic effect.

The exchange rates (k_{ex} s) of labile protons provide a wealth of information about the structural and functional features of proteins. For example, the X-ray structures of Mb [76, 77] suggested that the equilibrium structure does not possess channels large enough to permit ligand entry to and exit from the heme cavity. Theoretical calculations [128] have shown that the protein backbone and side chain motions of Mb modulate the energy barrier for O_2 entry. Thus the dynamics, as well as the structure of a protein, are relevant to its function [129, 130]. One of the most powerful methods for studying protein dynamics is the observation of the exchange behavior of labile protons, which are buried in the protein interior [131–133]. The time scale of the exchange of labile protons can be considered as a measure of the local dynamic stability [130–133].

k_{ex} is derived from the saturation factor (F) given by [134]

$$F = \frac{I}{I_0} = T_{1(int)}^{-1} \times \left(T_{1(int)}^{-1} + \tau_i^{-1}\right)^{-1} \tag{10.25}$$

where I and I_0 are the intensities of a labile proton signal with and without saturation of the ^1H_2O signal, respectively. $T_{1(int)}$ is the intrinsic spin–lattice relaxation time, with which F is evaluated. τ_i is the lifetime of the labile proton in the protein environment, and hence $\tau_i^{-1} = k_{ex}$. Consequently, independent measurements of $T_{1(int)}$ and F allow the determination of k_{ex}.

If $T_{1(int)}$ is independently known and remains constant over the pH range used, k_{ex} is obtained through Eq. (10.25′).

$$k_{ex} = \tau_i^{-1} = T_{1(int)}^{-1} \times (1 - F)F^{-1} \tag{10.25′}$$

With the occurrence of exchange, and when $T_{1(int)}$ cannot be determined directly, k_{ex} is obtained through Eq. (10.26).

$$k_{ex} = T'^{-1}_{1(int)} \times (1 - F) \tag{10.26}$$

where $T'^{-1}_{1(int)} = T_{1(int)}^{-1} + \tau_i^{-1}$. In such a case, $T_{1(int)}$ can be expressed as

$$T^{-1}_{1(int)} = F \times T'^{-1}_{1(int)} \tag{10.27}$$

$T_{1(int)}$ can be measured as follows. At pH values where τ_i of the labile proton is long compared with $T_{1(int)}$, i.e., $T_{1(int)} \ll \tau_i$, a selective saturation recovery experiment with a pulse sequence for solvent suppression is performed. At pH values where $\tau_i \approx T_{1(int)}$, $T'^{-1}_{1(int)}$ is obtained by selective inversion recovery with saturation of the H_2O signal at all times except during acquisition.

In general, k_{ex} is expressed as [131]

$$k_{ex} = k_A[H^+]^A + k_B[OH^-]^B + k_W[H_2O] \tag{10.28}$$

where k_A, k_B, and k_W are the rate constants for H^+, OH^-, and H_2O catalysis, respectively. Exponents A and B are at unity with k_A and k_B close to those observed for a model compound, and are less than unity for protons buried in the protein interior or/and involved in hydrogen bonds [131]. When k_A and k_B are comparable in magnitude to each other, the pH value at the minimum k_{ex} is close to ~7. A change in either k_A or k_B alters the pH value at the minimum k_{ex} [130].

For example, the His93 $N_\delta H$ proton signal of deoxy Mb is observed at ~77 ppm [96] (Fig. 10.2d). The His93 $N_\delta H$ proton of the protein is hydrogen-bonded to the carbonyl oxygen atom of Leu89, as illustrated in Fig. 10.1c, and exhibits $T_{1(int)} =$ ~15 ms and $k_{ex} =$ ~30 s^{-1} at pH = 10.6 and 40 °C. Interestingly, the k_{ex} value of the His93 $N_\delta H$ protons of the subunits of the deoxy form of tetrameric human adult hemoglobin, i.e., $k_{ex} =$ ~10^{-4}–10^{-5} s^{-1}, are several orders of magnitude smaller than that of deoxy Mb [135].

A 400 MHz ^1H NMR spectrum of bovine metMbCN$^-$ at 25 °C is shown in Fig. 10.15. In the spectrum, His64 $N_\varepsilon H$ and His93 $N_\delta H$ proton signals were observed at 23.31 and 20.96 ppm, respectively, and exhibited $T_{1(int)}$ values of 8.4 and 22 ms, respectively [136]. The F values of the His64 $N_\varepsilon H$ and His93 $N_\delta H$ protons of the protein at various pH values were measured (Fig. 10.16a), and then the pH-profiles of the k_{ex} values of the protons were calculated from the F and $T_{1(int)}$ values [136] (Fig. 10.16b). The k_{ex} values of the His64 $N_\varepsilon H$ and His93 $N_\delta H$ protons exhibited distinctly different pH-profiles. The pH-profile of the k_{ex} value of the His64 $N_\varepsilon H$ proton reflected both acid- and base-catalyzed NH hydrogen exchange, with the minimum k_{ex} value at pH ~9.5. On the other hand, only the base-catalyzed exchange was manifested in the pH-profile of the k_{ex} value of His93 $N_\delta H$ hydrogen atom. Thus the k_{ex} values of the NH protons provide valuable information about the dynamic nature of the heme cavity of the protein.

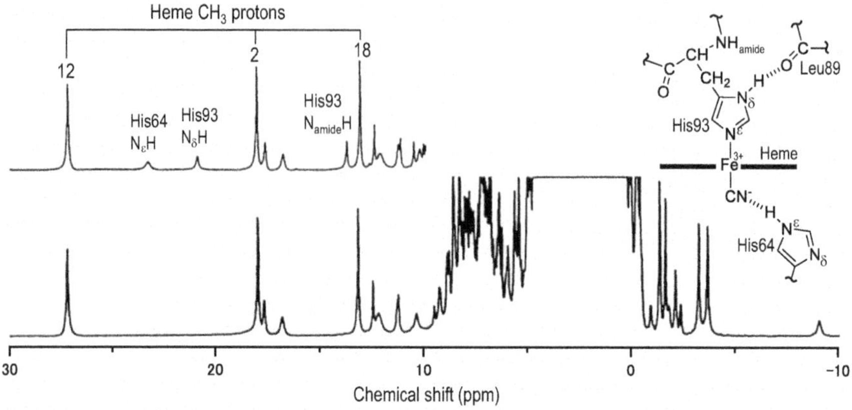

Fig. 10.15 400 MHz ^{1}H NMR spectrum of bovine metMb(CN$^-$) in ^2H_2O, pH 7.0, at 25 °C. The downfield-shifted portion of the spectrum of the protein in 90% ^1H_2O/10% ^2H_2O, pH 7.0, at 25 °C is shown in the inset. Assignments of 2-, 12-, and 18-CH_3 signals and His64 N_εH, His93 N_δH and His93 N_{amide}H ones are shown in the spectrum. The location of these NH protons in the protein is shown in the inset. The spectra were taken from [135]

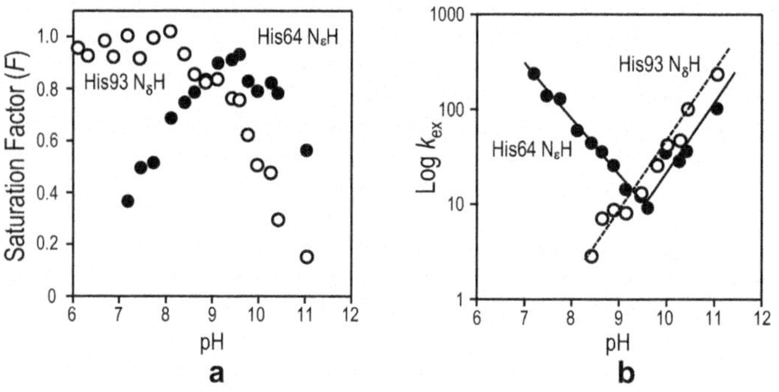

Fig. 10.16 pH-profiles of (**a**) saturation factor (F) and (**b**) logarithm of k_{ex} (s^{-1}) for the His64 N_εH (•) and His93 N_δH protons (○) in bovine metMb(CN$^-$) in 90% ^1H_2O/10% ^2H_2O at 25 °C [135]. The exchange behavior of histidyl imidazole ring NH hydrogen atoms sharply reflects their local environment. The H^+ and OH^- catalyses involve an attack at different sites of the neutral imidazole side chain, and both acid-and base-catalyzed exchange were observed for the His64 N_εH proton. On the other hand, only base-catalyzed exchange was observed for the His93 N_δH proton because of the coordination of His93 to heme Fe, which precludes acid-catalyzed exchange

10.5 Concluding Remarks

This chapter has described the fundamentals of ^{1}H NMR spectra of paramagnetic compounds. Using Mb as a reference compound, the spectral characteristics of paramagnetic compounds possessing a series of S values have been described in order to provide an overview of the effects of changes in the number of unpaired electrons on the spectral features. A few techniques applied to paramagnetic compounds, which provide widespread utility in life and material sciences, have also been described. As described in this chapter, the use of electron spin is opening up new horizons for NMR studies in life and material sciences.

Acknowledgements The authors are grateful to Prof. Toshiyuki Tanaka (Graduate School of Life and Environmental Sciences, University of Tsukuba) for the use of a Bruker AVANCE III 800 spectrometer at TARA Center, University of Tsukuba.

References

1. Wüthrich, K.: Structural studies of hemes and hemoproteins by nuclear magnetic resonance spectroscopy. Struct. Bond. **8**, 53–121 (1970)
2. La Mar, G.N., Horrocks, Jr. DeW, W., Holm, R.H. (eds.): NMR of Paramagnetic Molecules, Principles and Applications. Academic Press, New York (1973)
3. Morrow, J.S., Gurd, F.R.N.: Nuclear magnetic resonance studies of hemoglobin: functional state correlations and isotopic enrichment strategies. CRC Crit. Rev. Biochem. **3**, 221–287 (1975)
4. Wüthrich, K.: NMR in Biological Research: Peptides and Proteins, Chap. 6, North-Holland Publishing Co., Amsterdam (1976)
5. Morishima, I., Ogawa, S., Inubushi, T., Iizuka, T.: Nuclear magnetic resonance studies of high-spin ferric hemoproteins. Adv. Biophys. **11**, 217–245 (1978)
6. La Mar, G.N.: Model compound as aids in interpreting NMR spectra of hemoproteins. In: Shulman, R.G. (ed.) Biological Application of Magnetic Resonance, pp. 305–343. Academic Press, New York (1979)
7. Inagaki, F., Miyazawa, T.: NMR analyses of molecular conformations and conformational equilibria with the lanthanide probe method. Prog. NMR Spectrosc. **14**, 67–111 (1981)
8. Keller, R.M., Wüthrich, K.: Multiple irradiation ^{1}H NMR experiments with hemopoteins. Biol. Magn. Reson. **3**, 1–51 (1981)
9. Jardetzky, O., Roberts, G.C.K.: NMR in Molecular Biology, Chap. III, pp 69–114. Academic Press, New York (1981)
10. Bertini, I., Luchinat, C.: NMR of Paramagnetic Molecules in Biological Systems. Benjamin/ Cummings Publishing, Menlo Park, CA (1986)
11. Satterlee, J.D.: MNR spectroscopy of paramagnetic haem proteins. Annu. Rep. NMR Spectrosc. **17**, 79–178 (1986)
12. Satterlee, J.D.: Proton NMR studies of biological problems involving paramagnetic heme proteins. Met. Ions Biol. Syst. **21**, 121–185 (1986)
13. Bertini, I., Turano, P., Vila, A.J.: Nuclear magnetic resonance of paramagnetic metalloproteins. Chem. Rev. **93**, 2833–2932 (1993)
14. Yamamoto, Y.: NMR study of active sites in paramagnetic haemoproteins. Annu. Rep. NMR Spectrosc. **36**, 1–77 (1998)
15. La Mar, G.N., Satterlee, J.D., de Ropp, J.S.: Nuclear Magnetic Resonance of Hemoproteins. In: Kadish, K.M., Smith, K.M., Guilard, R. (eds.) The Porphyrin Handbook, pp. 185–298. Academic Press, San Diego (2000)
16. Bertini, I., Luchinat, C., Parigi, G.: Solution NMR of Paramagnetic Molecules. Elsevier, Amsterdam (2001)
17. de Ropp, J.S., Yu, L.P., La Mar, G.N.: 2D NMR of paramagnetic metalloenzymes: cyanide-inhibited horseradish peroxidase. J. Biomol. NMR **1**, 175–190 (1991)
18. Banci, L., Piccioli, M., Scozzafava, A.: Advances in the NMR investigation of paramagnetic molecules in solution. Coord. Chem. Rev. **120**, 1–28 (1992)

19. Ubbink, M., Worrall, J.A.R., Canters, G.W., Groenen, E.J.J., Huber, M.: Paramagnetic resonance of biological metal centers. Annu. Rev. Biophys. Biomol. Struct. **31**, 393–422 (2002)
20. Amesano, F., Banci, L., Piccioli, M.: NMR structures of paramagnetic metalloproteins. Quart. Rev. Biophys. **38**, 167–219 (2005)
21. Bertini, I., Luchinat, C., Parigi, G., Pierattelli, R.: NMR spectroscopy of paramagnetic metalloproteins. ChemBioChem **6**, 1536–1549 (2005)
22. Tang, C., Iwahara, J., Clore, G.M.: Visualization of transient encounter complexes in protein-protein association. Nat. **444**, 383–386 (2006)
23. Clore, G.M., Tang, C., Iwahara, J.: Elucidating transient macromolecular interactions using paramagnetic relaxation enhancement. Curr. Opin. Struct. Biol. **17**, 603–616 (2007)
24. Tang, C., Schwieters, C.D., Clore, G.M.: Open-to-closed transition in apo maltose-binding protein observed by paramagnetic NMR. Nat. **449**, 1078–1082 (2007)
25. John, M., Schmitz, C., Park, A.Y., Dixon, N.E., Huber, T., Otting, G.: Sequence-specific and stereospecific assignment of methyl groups using paramagnetic lanthanides. J. Am. Chem. Soc. **129**, 13749–13757 (2007)
26. Pintacuda, G., John, M., Su, X.C., Otting, G.: NMR structure determination of protein-ligand complexes by lanthanide labeling. Acc. Chem. Res. **40**, 206–212 (2007)
27. Bertini, I., Luchinat, C., Parigi, G., Pierattelli, R.: Perspectives in paramagnetic NMR of metalloproteins. Dalton. Trans. 3782–3790 (2008)
28. Clore, G.M., Iwahara, J.: Theory, practice, and applications of paramagnetic relaxation enhancement for the characterization of transient low-population states of biological macromolecules and their complexes. Chem. Rev. **109**, 4108–4139 (2009)
29. Otting, G.: Protein NMR using paramagnetic ions. Annu. Rev. Biophys. **39**, 387–405 (2010)
30. Bertini, I., Emsley, L., Lelli, M., Luchinat, C., Mao, J., Pintacuda, G.: Ultrafast MAS solid-state NMR permits extensive ^{13}C and ^{1}H detection in paramagnetic metalloproteins. J. Am. Chem. Soc. **132**, 5558–5559 (2010)
31. Iwahara, J., Clore, G.M.: Structure-independent analysis of the breadth of the positional distribution of disordered groups in macromolecules from order parameters for long, variable-length vectors using NMR paramagnetic relaxation enhancement. J. Am. Chem. Soc. **132**, 13346–13356 (2010)
32. Bertini, I., Giachetti, A., Luchinat, C., Parigi, G., Petoukhov, M.V., Pierattelli, R., Ravera, E., Svergun, D.I.: Conformational space of flexible biological macromolecules from average data. J. Am. Chem. Soc. **132**, 13553–13558 (2010)
33. Ravera, E., Salmon, L., Fragai, M., Parigi, G., Al-Hashimi, H., Luchinat, C.: Insights into domain-domain motions in proteins and RNA from solution NMR. Acc. Chem. Res. **47**, 3118–3126 (2014)
34. Yamaguchi, T., Sakae, Y., Zhang, Y., Yamamoto, S., Okamoto, Y., Kato, K.: Exploration of conformational spaces of high-mannose-type oligosaccharides by an NMR-validated simulation. Angew. Chem. Int. Ed. Engl. **53**, 10941–10944 (2014)
35. Brath, U., Swamy, S.I., Veiga, A.X., Tung, C.-C., Van Petegem, F., Erdélyi, M.: Paramagnetic ligand tagging to identify protein binding sites. J. Am. Chem. Soc. **137**, 11391 – 11398 (2015)
36. Schlagnitweit, J., Tang, M., Baias, M., Richardson, S., Schantz, S., Emsley, L.: Nanostructure of materials determined by relayed paramagnetic relaxation enhancement. J. Am. Chem. Soc. **137**, 12482–12485 (2015)
37. Matei, E., Gronenborn, A.M.: ^{19}F paramagnetic relaxation enhancement: a valuable tool for distance measurements in proteins. Angew. Chem. Int. Ed. Engl. **55**, 150–154 (2016)
38. Ravera, E., Luchinat, C., Parigi, G.: Basic facts and perspectives of overhauser DNP NMR. J. Magn. Reson. **264**, 78–87 (2016)
39. Kato, K., Yamaguchi, T.: Paramagnetic NMR probes for characterization of the dynamic conformations and interactions of oligosaccharides. Glycoconj. J. **32**, 505–513 (2015)
40. Feintuch, A., Otting, G., Goldfarb, D.: Gd^{3+} spin labeling for measuring distances in biomacromolecules: why and how? Methods Enzymol. **563**, 415–457 (2015)

41. Bertmer, M.: Paramagnetic solid-state NMR of materials. Solid State Nucl. Magn. Reson. **81**, 1–7 (2017)
42. Chen, J.L., Wang, X., Yang, F., Cao, C., Otting, G., Su, X.C.: 3D structure determination of an unstable transient enzyme intermediate by paramagnetic NMR spectroscopy. Angew. Chem. Int. Ed. Engl. **55**, 13744–13748 (2016)
43. Pilla, K.B., Otting, G., Huber, T.: Pseudocontact shift-driven iterative resampling for 3D structure determinations of large proteins. J. Mol. Biol. **428**, 522–532 (2016)
44. Orton, H.W., Kuprov, I., Loh, C.-T., Otting, G.: Using paramagnetism to slow down nuclear relaxation in protein NMR. J. Phys. Chem. Lett. **7**, 4815–4818 (2016)
45. Pilla, K.B., Otting, G., Huber, T.: 3D computational modeling of proteins using sparse paramagnetic NMR data. Methods Mol. Biol. **1526**, 3–21 (2017)
46. Can, T.V., Ni, Q.Z., Griffin, R.G.: Mechanisms of dynamic nuclear polarization in insulating solids. J. Magn. Reson. **253**, 23–35 (2015)
47. Su, Y., Andreas, L., Griffin, R.G.: Magic angle spinning NMR of proteins: high-frequency dynamic nuclear polarization and ^{1}H detection. Annu. Rev. Biochem. **84**, 465–497 (2015)
48. Matsuki, Y., Idehara, T., Fukazawa, J., Fujiwara, T.: Advanced instrumentation for DNP-enhanced MAS NMR for higher magnetic fields and lower temperatures. J. Magn. Reson. **264**, 107–115 (2016)
49. Comment, A.: Dissolution DNP for in vivo preclinical studies. J. Magn. Reson. **264**, 39–48 (2016)
50. McConnell, H.M., Holm, C.H.: Proton resonance shifts in nickelocene. J. Chem. Phys. **27**, 314–315 (1957)
51. McConnell, H.M., Holm, C.H.: Proton resonance shifts in paramagnetic metal aromatic complexes. J. Chem. Phys. **28**, 749–750 (1958)
52. Bloembergen, N.: Proton relaxation times in paramagnetic solutions. J. Chem. Phys. **27**, 572–573 (1957)
53. McConnell, H.M., Chesnut, D.B.: Theory of isotropic hyperfine interactions in π-electron radicals. J. Chem. Phys. **28**, 107–117 (1958)
54. McConnell, H.M., Robertson, R.E.: Isotropic nuclear resonance shifts. J. Chem. Phys. **29**, 1361–1365 (1958)
55. Karplus, M.: Contact electron-spin coupling of nuclear magnetic moments. J. Chem. Phys. **30**, 11–15 (1959)
56. Abragam, A.: The Principles of Nuclear Magnetism. Oxford University Press, Oxford (1961)
57. Karplus, M., Fraenkel, G.K.: Theoretical interpretation of carbon-13 hyperfine interactions in electron spin resonance spectra. J. Chem. Phys. **35**, 1312–1323 (1961)
58. Fitzgerald, R.J., Drago, R.S.: Contact-shift studies and delocalization mechanisms of Nickel (II)-benzylamine complexes. J. Am. Chem. Soc. **89**, 2879–2883 (1967)
59. Fitzgerald, R.J., Drago, R.S.: Contact-shift studies, delocalization mechanisms, and extended Hückel calculations of nickel(II)-alkylamine complexes. J. Am. Chem. Soc. **90**, 2523–2527 (1968)
60. Kurland, R., McGarvey, B.R.: Isotropic NMR shifts in transition metal complexes: the calculation of the Fermi contact and pseudocontact terms. J. Magn. Reson. **2**, 286–301 (1970)
61. Abragam, A., Bleaney, B.: Electron paramagnetic resonance of transition metal ions. Oxford University Press, Oxford (1970)
62. Schwarzhans, K.E.: NMR spectroscopy of paramagnetic complexes. Angew. Chem. Int. Ed. Engl. **9**, 946–953 (1970)
63. Crans, D.C., Yang, L., Gaidamauskas, E., Khan, R., Jin, W., Simonis, U.: Applications of paramagnetic NMR spectroscopy for monitoring transition metal complex stoichiometry and speciation. In: Telser, J. (ed.) Paramagnetic Resonance of Metallobiomolecules, pp. 304–326. ACS Book Series 858, American Chemical Society (2003)
64. Bertini, I., Luchinat, C., Parigi, G., Ravera, E.: NMR of paramagnetic molecules, applications to metallobiomolecules and models, 2nd edn. Elsevier Science (2016)
65. Walker, F.A.: Magnetic spectroscopic (EPR, ESEEM, Mössbauer, MCD and NMR) studies of low-spin ferriheme centers and their corresponding heme proteins. Coord. Chem. Rev. **185–186**, 471–534 (1999)

66. Kendrew, J.C., Bodo, G., Dintzis, H.M., Parrish, R.G., Wyckoff, H., Phillips, D.C.: A three-dimensional model of the myoglobin molecule obtained by x-ray analysis. Nat. **181**, 662–666 (1958)
67. Kowalsky, A.: Nuclear magnetic resonance studies of proteins. J. Biol. Chem. **237**, 1807–1819 (1962)
68. Shulman, R.G., Wüthrich, K., Yamane, T., Antonini, E., Brunori, M.: Nuclear magnetic resonances of reconstituted myoglobins. Proc. Natl. Acad. Sci. U.S.A. **63**, 623–628 (1969)
69. Wüthrich, K., Shulman, R.G., Peisach, J.: High-resolution proton magnetic resonance spectra of sperm whale cyanometmyoglobin. Proc. Natl. Acad. Sci. U.S.A. **60**, 373–380 (1968)
70. Ogawa, S., Shulman, R.G., Yamane, T.: High resolution nuclear magnetic resonance spectra of hemoglobin. I. The cyanide complexes of α and β chains. J. Mol. Biol. **70**, 291–300 (1972)
71. La Mar, G.N., Budd, D.L., Viscio, D.B., Smith, K.M., Langry, K.C.: Proton nuclear magnetic resonance characterization of heme disorder in hemoproteins. Proc. Natl. Acad. Sci. U.S.A. **75**, 5755–5759 (1978)
72. Morishima, I., Iizuka, T.: Nuclear magnetic resonance studies of hemoproteins. Unusual temperature dependence of hyperfine shifts and spin equilibrium in ferric myoglobin and hemoglobin derivatives. J. Am. Chem. Soc. **96**, 5279–5283 (1974)
73. Tolman, J.R., Flanagan, J.M., Kennedy, M.A., Prestegard, J.H.: Nuclear magnetic dipole interactions in field-oriented proteins: information for structure determination in solution. Proc. Natl. Acad. Sci. U.S.A. **92**, 9279–9283 (1995)
74. Tjandra, N., Omichinski, J.G., Gronenborn, A.M., Clore, G.M., Bax, A.: Use of dipolar ^{1}H–^{15}N and ^{1}H–^{13}C couplings in the structure determination of magnetically oriented macromolecules in solution. Nat. Struct. Biol. **4**, 732–738 (1997)
75. Antonini, E., Brunori, M.: Hemoglobin and myoglobin in their reactions with ligands. North Holland Publishing, Amsterdam (1971)
76. Takano, T.: Structure of myoglobin refined at 2.0 Å resolution. I. Crystallographic refinement of metmyoglobin from sperm whale. J. Mol. Biol. **110**, 537–568 (1977)
77. Takano, T.: Structure of myoglobin refined at 2.0 Å resolution. II. Structure of deoxy myoglobin from sperm whale. J. Mol. Biol. **110**, 569–584 (1977)
78. Vojtechovský, J., Chu, K., Berendzen, J., Sweet, R.M., Schlichting, I.: Crystal structures of myoglobin-ligand complexes at near-atomic resolution. Biophys. J. **77**, 2153–2174 (1999)
79. Hong, S., Sutherlin, K.D., Park, J., Kwon, E., Siegler, M.A., Solomon, E.I., Nam, W.: Crystallographic and spectroscopic characterization and reactivities of a mononuclear non-haem iron(III)-superoxo complex. Nat. Commun. **5**, 5440–5446 (2014)
80. Neya, S., Funasaki, N.: Proton nuclear magnetic resonance investigation of the spin-state equilibrium of the α and β subunits in intact azidomethemoglobin. Biochem. **25**, 1221–1226 (1986)
81. Shibata, T., Kanai, Y., Nishimura, R., Xu, L., Moritaka, Y., Suzuki, A., Neya, S., Nakamura, M., Yamamoto, Y.: Characterization of ground state electron configurations of high-spin quintet ferrous heme iron in deoxy myoglobin reconstituted with trifluoromethyl group-substituted heme cofactors. Inorg. Chem. **55**, 12128–12136 (2016)
82. Solomon, I.: Relaxation processes in a system of two spins. Phys. Rev. **99**, 559–565 (1955)
83. Gueron, M.: Nuclear relaxation in macromolecules by paramagnetic ions: a novel mechanism. J. Magn. Reson. **19**, 58–66 (1975)
84. Vega, A.J., Fiat, D.: Nuclear relaxation processes of paramagnetic complexes the slow-motion case. Mol. Phys. **31**, 347–355 (1976)
85. Unger, S.W., Jue, T., La Mar, G.N.: Proton NMR dipolar relaxation by delocalized spin density in low-spin ferric porphyrin complexes. J. Magn. Reson. **61**, 448–456 (1985)
86. Yamamoto, Y.: Analysis of 13C relaxation of heme peripheral methyl group in ferric low-spin myoglobin. J. Magn. Reson. **B103**, 72–76 (1994)
87. Mabbutt, B.C., Wright, P.E.: Assignment of heme and distal amino acid resonances in the ^{1}H-NMR spectra of the carbon monoxide and oxygen complexes of sperm whale myoglobin. Biochim. Biophys. Acta **832**, 175–185 (1985)

88. Emerson, S.D., La Mar, G.N.: Solution structural characterization of cyanometmyoglobin: resonance assignment of heme cavity residues by two-dimensional NMR. Biochem. **29**, 1545–1556 (1990)
89. La Mar, G.N., Budd, D.L., Smith, K.M., Langry, K.C.: Nuclear magnetic resonance of high-spin ferric hemoproteins. Assignment of proton resonances in met-aquo myoglobins using deuterium-labeled hemes. J. Am. Chem. Soc. **102**, 1822–1827 (1980)
90. Unger, S.W., Lecomte, J.T.J., La Mar, G.N.: The utility of the nuclear Overhauser effect for peak assignment and structure elucidation in paramagnetic proteins. J. Magn. Reson. **64**, 521–526 (1985)
91. Yamamoto, Y., Chujo, R., Suzuki, T.: NMR study of *Galeorhinus japonicus* myoglobin. ^{1}H-NMR evidence for a structural alteration on the active site of *G. japonicus* myoglobin upon azide ion binding. Eur. J. Biochem. **198**, 285–291 (1991)
92. Beetlestone, J., George, P.: A magnetochemical study of equilibria between high and low spin states of metmyoglobin complexes. Biochem. **3**, 707–714 (1964)
93. Iizuka, T., Kotani, M.: Analysis of a thermal equilibrium phenomenon between high-spin and low-spin states of ferrimyoglobin azide. Biochim. Biophys. Acta **154**, 417–419 (1968)
94. Neya, S., Morishima, I.: Interaction of methemoglobin with inositol hexaphosphate. Presence of the T state in human adult methemoglobin in the low spin state. J. Biol. Chem. **256**, 793–798 (1981)
95. Yamamoto, Y., Iwafune, K., Chujo, R., Inoue, Y., Imai, K., Suzuki, T.: ^{1}H-NMR comparative study of the active site in shark (*Galeorhinus japonicus*), horse, and sperm whale deoxy myoglobins. J. Biochem. **112**, 414–420 (1992)
96. La Mar, G.N., Davis, L., Johnson, R.D., Smith, W.S., Hauksson, J.B., Budd, D.L., Dalichow, F., Langry, K.C., Morris, I.K., Smith, K.M.: Nuclear magnetic resonance investigation of the electronic structure of deoxymyoglobin. J. Am. Chem. Soc. **115**, 3869–3876 (1993)
97. Bertin, I., Luchinat, C., Turano, P., Battaini, G., Casella, L.: The magnetic properties of myoglobin as studied by NMR spectroscopy. Chem. Eur. J. **9**, 2316–2322 (2003)
98. La Mar, G.N., Budd, D.L., Goff, H.: Assignment of proximal histidine proton nmr peaks in myoglobin and hemoglobin. Biochem. Biophys. Res. Commun. **77**, 104–110 (1977)
99. Thériault, Y., Pochapsky, T.C., Dalvit, C., Chiu, M.L., Sligar, S.G., Wright, P.E.: ^{1}H and ^{15}N resonance assignments and secondary structure of the carbon monoxide complex of sperm whale myoglobin. J. Biomol. NMR **4**, 491–504 (1994)
100. Yamamoto, Y.: ^{1}H NMR probes for inter-segmental hydrogen bonds in myoglobins. J. Biochem. **120**, 126–132 (1996)
101. Bougault, C.M., Dou, Y., Ikeda-Saito, M., Langry, K.C., Smith, K.M., La Mar, G.N.: Solution ^{1}H NMR study of the electronic structure and magnetic properties of high-spin ferrous or deoxy myoglobins. J. Am. Chem. Soc. **120**, 2113–2123 (1998)
102. Nishimura, R., Shibata, T., Tai, H., Ishigami, I., Yanagisawa, S., Ogura, T., Neya, S., Suzuki, A., Yamamoto, Y.: Effect of the electron density of the heme Fe atom on the Fe-histidine coordination bond in deoxy myoglobin. Bull. Chem. Soc. Jpn **87**, 905–911 (2014)
103. La Mar, G.N., de Ropp, J.S.: Proton NMR characterization of the state of protonation of the axial imidazole in reduced horseradish peroxidase. J. Am. Chem. Soc. **104**, 5203–5206 (1982)
104. Matsukawa, S., Mawatari, K., Yoneyama, Y., Kitagawa, T.: Correlation between the iron-histidine stretching frequencies and oxygen affinity of hemoglobins. A continuous strain model. J. Am. Chem. Soc. **107**, 1108–1113 (1985)
105. Johnson, M.E., Fung, L.W.-M., Ho, C.: Magnetic field and temperature Induced line broadening in the hyperfine-shifted proton resonances of myoglobin and hemoglobin. J. Am. Chem. Soc. **99**, 1245–1250 (1977)
106. Jensen, K.P., Ryde, U.: How O_2 binds to heme. J. Biol. Chem. **279**, 14561–14569 (2004)
107. Weiss, J.J.: Nature of the iron-oxygen bond in oxyhaemoglobin. Nature **202**, 83–84 (1964)
108. Pauling, L., Coryell, C.D.: The magnetic properties and structure of hemoglobin, oxyhemoglobin and carbonmonoxyhemoglobin. Proc. Natl. Acad. Sci. U.S.A. **22**, 210–216 (1936)
109. Pauling, L.: Nature of the iron-oxygen bond in oxyhaemoglobin. Nat. **203**, 182–183 (1964)

110. McClure, D.S.: Electronic structure of transition-metal complex ions. Radiat. Res. Suppl. **2**, 218–242 (1960)
111. Eicher, H., Bade, D., Parak, F.: Theoretical determination of the electronic structure and the spatial arrangement of ferrous iron in deoxygenated sperm whale myoglobin and human hemoglobin from Mössbauer experiments. J. Chem. Phys. **64**, 1446–1455 (1976)
112. Yamamoto, Y.: NMR studies of *b*-type haemoproteins reconstituted with a ring-fluorinated haem. Annu. Rep. NMR Spectrosc. **57**, 51–98 (2006)
113. Shibata, T., Nagao, S., Fukaya, M., Tai, H., Nagatomo, S., Morihashi, K., Matsuo, T., Hirota, S., Suzuki, A., Imai, K., Yamamoto, Y.: Effect of heme modification on oxygen affinity of myoglobin and equilibrium of the acid-alkaline transition in metmyoglobin. J. Am. Chem. Soc. **132**, 6091–6098 (2010)
114. Toi, H., Homma, M., Suzuki, A., Ogoshi, H.: Paramagnetic ^{19}F n.m.r. spectra of iron(III) porphyrins substituted with CF_3 groups and reconstituted myoglobin. J. Chem. Soc. Chem. Commun. 1791–1792 (1985)
115. Yamamoto, Y., Hirai, Y., Suzuki, A.: ^{19}F NMR study of protein-induced rhombic perturbations on the electronic structure of the active site of myoglobin. J. Biol. Inorg. Chem. **5**, 455–462 (2000)
116. Hirai, Y., Nagao, S., Mita, H., Suzuki, A., Yamamoto, Y.: ^{19}F NMR study on the heme electronic structure in oxy and carbonmonoxy reconstituted myoglobins. Bull. Chem. Soc. Jpn **77**, 1485–1486 (2004)
117. Yamamoto, Y., Nagao, S., Hirai, Y., Inose, T., Terui, N., Mita, H., Suzuki, A.: NMR investigation of the heme electronic structure in deoxymyoglobin possessing a fluorinated heme. J. Biol. Inorg. Chem. **9**, 152–160 (2004)
118. Emerson, S.D., La Mar, G.N.: NMR determination of the orientation of the magnetic susceptibility tensor in Cyanometmyoglobin: a new probe of steric tilt of bound ligand. Biochem. **29**, 1556–1566 (1990)
119. Yamamoto, Y., Nanai, N., Chujo, R., Suzuki, T.: Heme methyl hyperfine shift pattern as a probe for determining the orientation of the functionally relevant proximal histidyl imidazole with respect to the heme in hemoproteins. FEBS Lett. **264**, 113–116 (1990)
120. La Mar, G.N., Walker, F.A.: Proton nuclear magnetic resonance line widths and spin relaxation in paramagnetic metalloporphyrins of chromium(III), manganese(III), and iron(III). J. Am. Chem. Soc. **95**, 6950–6956 (1973)
121. La Mar, G.N., de Ropp, J.S., Smith, K.M., Langry, K.C.: Proton nuclear magnetic resonance investigation of the electronic structure of compound I of horseradish peroxidase. J. Biol. Chem. **256**, 237–243 (1981)
122. Yamamoto, Y., Osawa, A., Inoue, Y., Chujo, R.: A ^{1}H-NMR study of electronic structure of the active site of *Galeorhinus japonicus*. Eur. J. Biochem. **192**, 225–229 (1990)
123. Kao, Y.-H., Lecomte, J.T.J.: Determination of the zero-field splitting constant for proton NMR chemical shift analysis in metaquomyoglobin. The dipolar shift as a structural probe. J. Am. Chem. Soc. **115**, 9754–9762 (1993)
124. Trewhella, J., Wright, P.E., Appleby, C.A.: Molecular basis for proton-dependent anion binding by soybean leghaemoglobin *a*. Nat. **280**, 87–88 (1979)
125. Johnson, R.D., Ramaprasad, S., La Mar, G.N.: A method of assigning functionally relevant amino acid residue resonances in paramagnetic hemoproteins using proton NOE measurements. J. Am. Chem. Soc. **105**, 7205–7206 (1983)
126. Yamamoto, Y., Inoue, Y., Chujo, C., Suzuki, T.: ^{1}H-NMR study of heme propanoate mobility in the active site of myoglobin from *Galeorhinus japonicus*. Eur. J. Biochem. **189**, 567–573 (1990)
127. Plateau, P., Guéron, M.: Exchangeable proton NMR without base-line distortion, using new strong-pulse sequences. J. Am. Chem. Soc. **104**, 7310–7311 (1982)
128. Case, D.A., Karplus, M.: Dynamics of ligand binding to heme proteins. J. Mol. Biol. **132**, 343–368 (1979)
129. Karplus, M., McCammon, J.A.: The internal dynamics of globular proteins. CRC Crit. Rev. Biochem. **9**, 293–349 (1981)

130. McCammon, J.A., Karplus, M.: The dynamic picture of protein structure. Acc. Chem. Res. **16**, 187–193 (1983)
131. Woodward, C.K., Hilton, B.D.: Hydrogen exchange kinetics and internal motions in proteins and nucleic acids. Annu. Rev. Biophys. Bioeng. **8**, 99–127 (1979)
132. Krishna, M.M.G., Hoang, L., Lin, Y., Englander, S.W.: Hydrogen exchange methods to study protein folding. Methods **34**, 51–64 (2004)
133. Anthis, N.J., Clore, G.M.: Visualizing transient dark states by NMR spectroscopy. Quart. Rev. Biophys. **48**, 35–116 (2015)
134. Lecomte, J.T.J., La Mar, G.N.: Proton NMR study of labile proton exchange in the heme cavity as a probe for the potential ligand entry channel in myoglobin. Biochem. **24**, 7388–7395 (1985)
135. Han, K.-H., La Mar, G.N.: Nuclear magnetic resonance study of the isotope exchange of the proximal histidyl ring labile protons in hemoglobin A. The exchange rates and mechanisms of individual subunits in deoxy and oxy-hemoglobin. J. Mol. Biol. **189**, 541–552 (1986)
136. Yamamoto, Y., Kurihara, N., Egawa, T., Shimada, H., Ishimura, Y.: Hydrogen bonding interaction of the amide group of Asn and Gln at distal E7 of bovine myoglobin with bound-ligand and its functional consequences. Biochim. Biophys. Acta **1433**, 27–44 (1999)

Chapter 11
NMR of Quadrupole Nuclei in Organic Compounds

Kazuhiko Yamada

Abstract General aspects of solid-state nuclear magnetic resonance (NMR) of quadrupole nuclei in organic solids, including theoretical background on quadrupole interactions and analysis of the characteristic line shapes that arise from quadrupole and chemical shift interactions, have been described. Two theoretical approaches for spectral simulations, the perturbation and the direct diagonalization methods, are discussed with examples of solid-state ^{17}O ($I = 5/2$), ^{33}S ($I = 3/2$), and $^{79/81}$Br ($I = 3/2$) NMR of organic compounds, as well as those of inorganic compounds with larger quadrupole interactions. When the magnitude of the quadrupole interactions is smaller than that of the Zeeman interactions, the perturbation method, in which a mathematical formula can be obtained to express the first and second-order quadrupole interactions under static or magic-angle spinning (MAS) conditions, is applicable. Otherwise, the direct diagonalization method, in which the combined Zeeman and quadrupole Hamiltonian are numerically calculated to provide probabilities for each transition, is applied for spectral simulations. A few experimental techniques to obtain NMR spectra broadened by large quadrupole interactions are briefly described.

Keywords Solid-state NMR · Quadrupole nuclei · Field-swept NMR · Frequency-swept NMR

11.1 Introduction

Almost all the atoms in the periodic table exhibit non-zero spin quantum number (I). It is therefore theoretically possible to obtain those NMR spectra. Accessible NMR nuclei, however, are limited to those with $I = ½$, such as ^{1}H, ^{13}C, and ^{15}N. It is important to point out that most of the atoms in the periodic table have quadrupole nuclei ($I > 1/2$), which may give rise to additional nuclear spin interactions, known

K. Yamada (✉)
Interdisciplinary Science Unit, Multidisciplinary Sciences Cluster, Research and Education Faculty, Kochi University, Oko Campus, Nankoku 783-8505, Kochi, Japan
e-mail: kyamada@kochi-u.ac.jp

The NMR Society of Japan (ed.), *Experimental Approaches of NMR Spectroscopy II*,
https://doi.org/10.1007/978-981-97-6830-1_11

as quadrupole interactions [1, 2]. In general, the solid-state NMR experiments of quadrupole nuclei in organic compounds are much more difficult than those in inorganic compounds, since the magnitude of quadrupolar interactions of the former is much larger. Nevertheless, thanks to recent developments in solid-state NMR technology, the number of accessible quadrupole nuclei of organic compounds has gradually increased, and the research areas, to which solid-state NMR of quadrupole nuclei can be applied, have undoubtedly expanded in the past decade.

Solid-state NMR spectra of quadrupole nuclei in organic compounds generally exhibit characteristic line shapes [3–5]. Unlike in cases of $I = 1/2$, sharp NMR signals may not be obtained, even if a standard magic-angle spinning (MAS) method is used. A spectral simulation is commonly applied for such complicated NMR spectra to obtain NMR parameters. In a typical spectral simulation, a theoretical NMR spectrum is calculated, based on Hamiltonians that express the suitable nuclear spin interactions under the associated experimental conditions. The relevant NMR parameters are input and varied until a theoretical line shape reproduces the experimental one. Although this is one of the most basic approaches for the analysis of NMR spectra of quadrupole nuclei, the author feels that there have been very little literatures or textbooks to explain the fundamental knowledge for graduated students and young scientists. In this chapter, the experimental and theoretical aspects of solid-state NMR analysis of quadrupole nuclei in organic compounds, with particular emphasis on spectral simulations, are presented.

11.2 Theoretical Background of Quadrupole Interactions

In cases of $I > 1/2$, i.e., quadrupole nuclei, there is a nuclear spin interaction between the quadrupole moment and the electric-field-gradient (EFG) that a molecule forms [1, 2]. This is called a quadrupole interaction and can be expressed by a second-rank tensor since it has anisotropy (in the same way as other nuclear spin interactions, such as chemical shielding (CS) interactions, dipole–dipole interactions, and indirect spin–spin interactions). Two independent NMR parameters, namely, the quadrupole coupling constant, denoted as e^2qQh^{-1} or C_Q [Hz], and the asymmetry parameter, denoted as η_Q, are used to describe a nuclear quadrupole interaction, since the EFG tensor, **V**, is traceless. The EFG tensor in the principal axis system (the PAS frame) is diagonal:

$$\boldsymbol{V} = \begin{bmatrix} V_{XX} & 0 & 0 \\ 0 & V_{YY} & 0 \\ 0 & 0 & V_{ZZ} \end{bmatrix} \tag{11.1}$$

where $\mathrm{V_{XX}} + \mathrm{V_{YY}} + \mathrm{V_{ZZ}} = 0$ with $|V_{XX}| \leq |V_{YY}| \leq |V_{ZZ}|$. According to the definition,

$$e^2qQh^{-1} = eQh^{-1} \times V_{ZZ} \tag{11.2}$$

$$\eta_Q = \frac{V_{XX} - V_{YY}}{V_{ZZ}} \tag{11.3}$$

where $0 \leq \eta_Q \leq 1$. The former and the latter describe the magnitude and symmetry of the EFG tensors, respectively.

The Hamiltonian representing a nuclear quadrupole interaction is defined by

$$\mathcal{H}_Q = \frac{C_Q}{4I(2I-1)}\left[\frac{\sqrt{6}}{3}\left(3I_Z^2 - I^2\right)V_0 + (I_Z I_+ + I_+ I_Z)V_{-1} - (I_Z I_- + I_- I_Z)V_{+1} + I_+^2 V_{-2} + I_-^2 V_{+2}\right] \tag{11.4}$$

where $I_\pm$ are the raising and lowering spin operators, and V is the EFG principal tensor component in the spherical tensor representation [2, 3]. Equation (11.4) can be further simplified when the magnitude of the Zeeman interactions is much larger than that of the quadrupole interactions, which is recognized as the high-field approximation:

$$\mathcal{H}_Q^{(1)} = \frac{\sqrt{6}C_Q}{12I(2I-1)}\left[3I_Z^2 - I(I+1)\right]V_0 \tag{11.5}$$

$$\mathcal{H}_Q^{(2)} = -\frac{1}{\nu_0}\left[\frac{C_Q}{4I(2I-1)}\right]^2\left\{2V_{-1}V_1 I_Z\left[4I(I+1) - 8I_Z^2 - 1\right] + 2V_{-2}V_2 I_Z\left[2I(I+1) - 2I_Z^2 - 1\right]\right\} \tag{11.6}$$

where ν_0 is the Larmor frequency. The former and the latter correspond to the Hamiltonians of the first-order quadrupole interaction and the second-order quadrupole interaction, respectively. For a spectral simulation, a suitable Hamiltonian must be chosen, depending on the magnitude of the quadrupole interactions. For example, the first-order quadrupole Hamiltonian is used to describe the experimental NMR spectra broadened by much smaller quadrupole interaction, compared to the Zeeman interaction. It should be noted that the Larmor frequency is included in the denominator of Eq. (11.6), indicating that the magnitude of the second-order quadrupole interaction is inversely proportional to the strength of the external magnetic field. Therefore, a higher magnetic field is advantageous for the observation of NMR spectra broadened by the second-order quadrupole interactions.

The perturbation theory is valid only when one interaction is much larger than the other. Equations (11.5) and (11.6) are not suitable, when the magnitude of the quadrupole interactions is comparable to or larger than that of the Zeeman interaction. In such cases, it is necessary to numerically solve the combined Zeeman and quadrupole Hamiltonians to obtain transition probabilities. The full quadrupole Hamiltonian, i.e., Eq. (11.4), combined with the Zeeman Hamiltonian, $\nu_0 I_z$, is diagonalized to find the eigenvalues, from which the transition amplitudes can be calculated for each transition. In this chapter, this procedure is called the direct diagonalization method [6, 7].

To express the magnitude of quadrupole interactions, the quadrupole frequency, ν_Q, is often used as an alternative to a C_Q value:

$$\nu_Q = \frac{3C_Q\sqrt{1+\eta_Q^2/3}}{2I(2I-1)} \tag{11.7}$$

This frequency corresponds to the resonant frequency, observed in NQR or zero-field NMR measurements, and can be served as a chemical shift, which will provide structural information.

In a short summary, a suitable theoretical model, i.e., the choice of the first- and/or second-order quadrupole Hamiltonians, or the direct diagonalization method, should be chosen carefully in a spectral simulation to reproduce the experimental NMR spectrum, depending on whether the Larmor frequency is larger or smaller than ν_Q or C_Q values. In the next two sections, basic approaches for the analyses of one-dimensional NMR spectra of quadrupole nuclei by the perturbation and the direct diagonaliznsistent with the experimend with experimental examples.

11.3 Simulating Solid-State NMR of Quadrupole Nuclei Based on the Perturbation Method

Solid-state oxygen-17 (^{17}O, I, = 5/2, natural abundance = 0.038%, $\gamma = 3.62808 \times 10^7$ rad T^{-1} s^{-1}, and quadrupole moment (Q) = 2.558 fm^2) NMR, which becomes one of the most powerful tools for investigating physical and chemical properties in organic and biological chemistry, is reviewed with spectral simulations by the perturbation method [5]. Figure 11.1 shows the (upper) experimental and (lower) calculated ^{17}O MAS spectra of ^{17}O-enriched Fmoc-amino acid, measured at 11.7 T with a sample spinning frequency of 19.9 ± 0.2 kHz [8]. The peak at approximately 380 ppm marked by an asterisk arises from the MAS rotor material, ZrO_2. The use of MAS rotors made of silicon nitride is highly recommended, if available. Figure 11.2 shows the (upper) experimental and (lower) calculated ^{17}O static spectra of [^{17}O]-Fmoc-amino acid, observed at 21.8 T [8]. The calculated total spectrum (site A + site B) is the sum of site A and site B sub-spectra with a 1:1 intensity ratio. Both the MAS and static ^{17}O NMR spectra exhibit characteristic line shapes, which are typically observed in solid-state NMR of half-integer quadrupole nuclei in organic compounds. To obtain experimental NMR parameters, a theoretical NMR spectrum is calculated by entering NMR parameters, which are then varied until the calculated NMR spectrum is consistent with the experimental one. The magnitude of a quadrupole interaction is expressed as a product of the quadrupole moment and the EFG, which is closely related to the molecular symmetry around the nuclei. For solid-state ^{17}O NMR, the molecular symmetries of organic and biological compounds are generally low, which results in larger C_Q values (~10 MHz), compared with those of inorganic compounds (~1 MHz). The second-order quadrupole interaction, i.e., Eq. (11.6), with the central transition $m = 1/2 \Leftrightarrow m = 1/2$, where m is a quantum

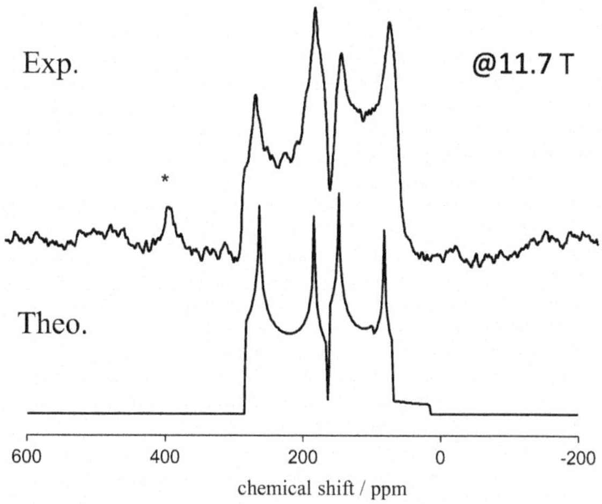

Fig. 11.1 Experimental and calculated ^{17}O MAS spectra of [^{17}O]-Fmoc-amino acid, measured at 11.7 T. The peak marked by * arises from the MAS rotor material, ZrO_2

magnetic number, is often considered for spectral simulations of organic and biological solids. Although they can provide the same results, spectral simulations for the satellite transitions are rarely used due to the weak signals and the wide width of the line, and the details are omitted here due to space limitations. For organic and biological solids, the relevant ^{17}O NMR parameters are the EFG and CS tensors. The following three cases of spectral simulations for solid-state ^{17}O NMR spectra will be given: The analyses of (1) static NMR spectra under the influence of second-order quadrupole interactions, (2) fast MAS NMR conditions under the influence of second-order quadrupole and CS interactions, and (3) static NMR spectra under the influence of second-order quadrupole and CS interactions.

11.3.1 Static NMR Spectra Under the Influence of Second-Order Quadrupole Interactions

In the cases of ^{17}O ($I = 5/2$) NMR, the Zeeman effect splits an energy level into six components in the presence of a static magnetic field [9]. The central transition, which is our main interest, is not affected by the first-order quadrupole interaction so no change is made in the resonant frequency. This implies that the contribution of the first-order one only produces a sharp NMR signal observed at the Larmor frequency. On the other hand, the second-order quadrupole interaction affects the energy levels involved in the central transition as well as the satellite transitions, indicating that,

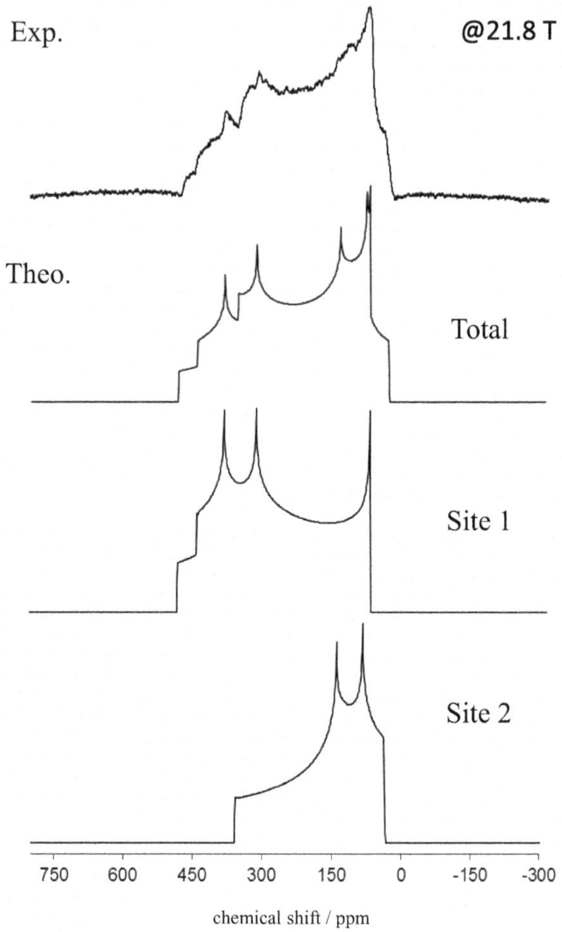

Fig. 11.2 Experimental and calculated static ^{17}O spectra of [^{17}O]-Fmoc-amino acid, observed at 21.8 T. The calculated total spectrum (site A + site B) is the sum of site A and site B sub-spectra with a 1:1 intensity ratio

compared to the Larmor frequency, the resonant frequency, $\nu_Q^{(2)}$, is shifted according to

$$\begin{aligned}\nu_Q^{(2)} &= \langle -\frac{1}{2}|\mathcal{H}_Q^{(2)}| - \frac{1}{2}\rangle - \langle \frac{1}{2}|\mathcal{H}_Q^{(2)}|\frac{1}{2}\rangle \\ &= -\frac{1}{\nu_0}\left[\frac{C_Q}{4I(2I-1)}\right]^2 \{2V_{-1}V_1[3-4I(I+1)] + V_{-2}V_2[3-4I(I+1)]\}\end{aligned} \tag{11.8}$$

where the bra-ket means the energy level at m, and the V components are expressed in terms of the laboratory frame. In order to simulate static ^{17}O NMR spectra, the V components in Eq. (11.8) must be converted into ones expressed in terms of the PAS frame. For this purpose, it is convenient to use the following relationship:

$$V_i = \sum_{j=-2}^{2} \mathcal{D}^{(2)}_{j,i}(\phi, \theta, 0) V^{PAS}_j \tag{11.9}$$

where $\mathcal{D}^{(2)}_{j,i}$ is the Wigner rotation matrix, which is defined in the literature [2]. The orientation between the PAS and the laboratory frames used in this chapter is given in Fig. 11.3a (In the initial state of a spectral simulation, orientations of EFG and CS tensors are assumed to be coincided, although CS and EFG tensors very often exhibit different orientations, as shown in Fig. 11.3b. The orientation of a CS tensor (δ_{11}, δ_{22}, δ_{33}) will be discussed later.) For example, V_0 can be rewritten as

$$\begin{aligned} V_0 &= \mathcal{D}^{(2)}_{-2,0}(\phi,\theta,0)V^{PAS}_{-2} + \mathcal{D}^{(2)}_{-1,0}(\phi,\theta,0)V^{PAS}_{-1} + \mathcal{D}^{(2)}_{0,0}(\phi,\theta,0)V^{PAS}_{0} + \mathcal{D}^{(2)}_{1,0}(\phi,\theta,0)V^{PAS}_{1} + \mathcal{D}^{(2)}_{2,0}(\phi,\theta,0)V^{PAS}_{2} \\ &= \sqrt{3/2}eQ\left[1/2\left(3\cos^2\theta - 1\right) + 1/2\eta_Q \sin^2\theta\cos 2\phi\right] \end{aligned} \tag{11.10}$$

where $V^{PAS}_0 = \sqrt{3/2}eq$, $V^{PAS}_{\pm1} = 0$, and $V^{PAS}_{\pm2} = 1/2eq$. Similarly, $V_{\pm1}$ and $V_{\pm2}$ in the laboratory frame can be converted into the PAS frame, which is then inserted into Eq. (11.8). The NMR frequency arising from the second-order quadrupolar interaction under static conditions, $\nu^{(2)}_{Q,STATIC}$, is given by

$$\begin{aligned} \nu^{(2)}_{Q,STATIC} &= -1/6\nu_0\left[3C_Q/2I(2I-1)\right]^2[I(I+1) - 3/4]\left[A\cos^4\theta + B\cos^2\theta + C\right] \\ A &= -27/8 + 9/4\eta_Q\cos 2\phi - 3/8\left(\eta_Q\cos 2\phi\right)^2 \\ B &= 30/8 - 1/2\eta_Q^2 - 2\eta_Q\cos 2\phi + 3/4\left(\eta_Q\cos 2\phi\right)^2 \\ C &= -3/8 + 1/3\eta_Q^2 - 1/4\eta_Q\cos 2\phi - 3/8\left(\eta_Q\cos 2\phi\right)^2 \end{aligned} \tag{11.11}$$

If the sample is a single crystal, the observed NMR frequency will be calculated with a single pair of ϕ and θ from the above equation. Polycrystalline or polymeric materials are very often used for solid-state NMR experiments so space-averaging in Eq. (11.11) is required to produce the theoretical powder pattern. Figure 11.4a shows the calculated static ^{17}O NMR spectra as a function of η_Q from 0 to 1.0 in steps of 0.2 [10]. These calculations are carried out using a C_Q value of 10 MHz and ν_0 of 100 MHz. The powder patterns exhibit complicated line shapes, and the line shape varies greatly, depending on η_Q. In addition, the spectral width is highly dependent on C_Q and ν_0. Larger C_Q values give wider spectral widths, while higher ν_0 gives smaller widths. Again, note that ν_0 is included in the denominator of Eqs. (11.6) and (11.11). A computer simulation is practically required for extracting C_Q and η_Q from experimental NMR spectra. As discussed above, the powder pattern of Eq. (11.11) is depicted by inputting C_Q and η_Q, and the parameters are changed until they conform to the experimental one. The error bars should always be considered.

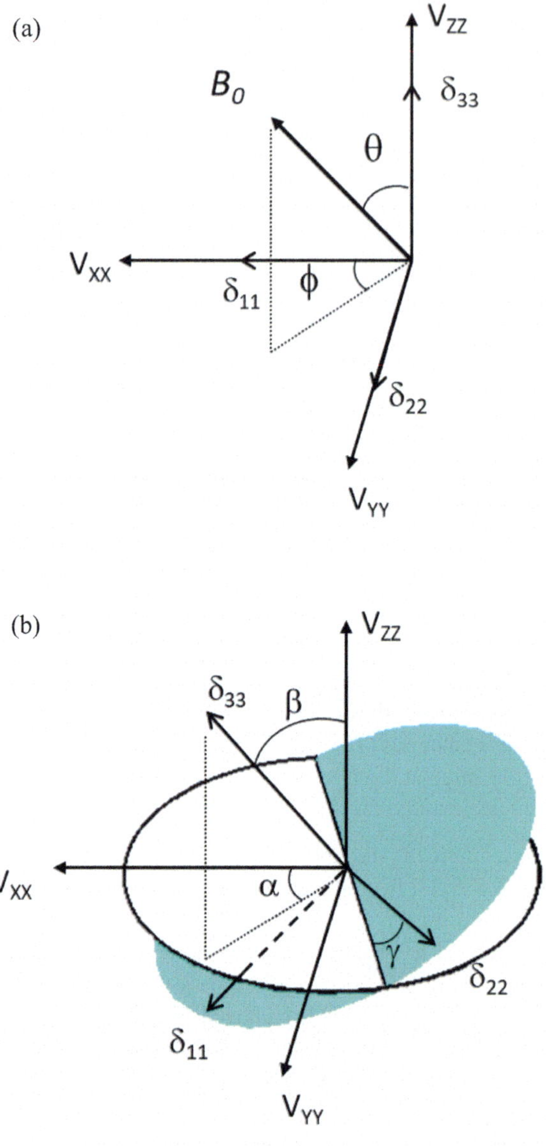

Fig. 11.3 **a** Initial alignment of the ^{17}O EFG and CS tensors and direction of the external magnetic field, B_0. EFG and CS tensors are assumed to be collinear. **b** Euler angles relating EFG and CS tensors for the case that the orientations of EFG and CS tensors are different

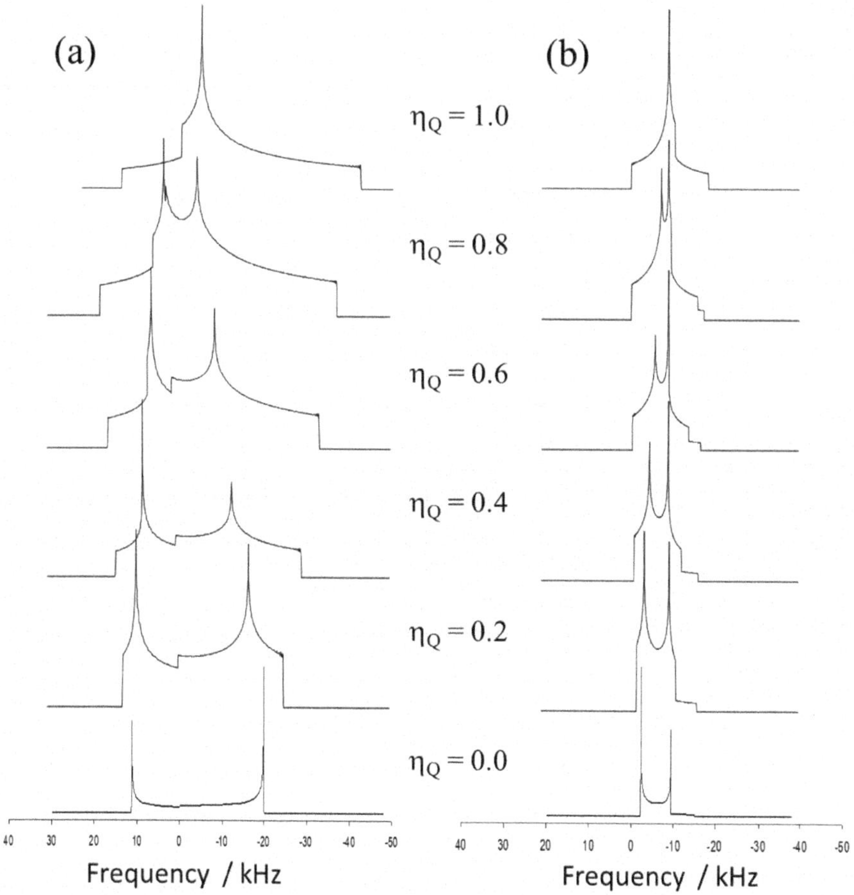

Fig. 11.4 Dependence of η_Q on **a** static and **b** MAS NMR spectra of half-integer quadrupole nuclei, calculated by Eqs. (11.11) and (11.14)

11.3.2 Fast MAS NMR Conditions Under the Influence of Second-Order Quadrupole and CS Interactions

In the presence of second-order quadrupole interaction, high-resolution NMR spectra are not achievable by conventional MAS methods, even if the sample spinning frequency is sufficiently high compared to the spectral width [2, 3]. This is because the second-order quadrupole interactions under MAS conditions are not simply proportional to the term of $3\cos^2\theta - 1$, which is related to the Hamiltonians of the first-order nuclear spin interactions such as dipole–dipole, CS, and first-order quadrupolar ones [2] As a result, the effect of a MAS method for the second-order quadrupolar interactions is to partially narrow it, and a characteristic line shape still appears in the MAS NMR spectrum.

To calculate the MAS NMR spectrum of a half-integer quadrupole nucleus, a coordinate system of the V component in Eq. (11.8) must be changed stepwise from the laboratory frame to the MAS rotor frame, and the PAS frame. The following two relationships are used:

$$V_i = \sum_{j=-2}^{2} \mathcal{D}_{j,i}^{(2)}(\omega_R t, \theta_M, 0) V_j^{MAS} \tag{11.12}$$

$$V_j^{MAS} = \sum_{k=-2}^{2} \mathcal{D}_{k,j}^{(2)}(\phi, \theta, 0) V_k^{PAS} \tag{11.13}$$

where ω_R and θ_M are the sample spinning frequency in angular velocity and the magic angle of 54.7, respectively. In Eq. (11.13), θ and ϕ describe the direction of the MAS rotor with respect to the PAS frame in a similar manner to the orientation in Fig. 11.3a, and the third angle is set to zero, since the experimental conditions are assumed to be in the fast MAS range, which means that the sample spinning frequency is high enough compared to the spectral width. In cases of slow or intermediate spinning ranges, however, the third angle in Eq. (11.13) becomes non-zero, and spinning side bands appear in the MAS NMR spectra, which makes it more complicated to analyze the line shape [11].

By inserting Eqs. (11.12) and (11.13) into Eq. (11.8), the frequency contribution from the second-order quadrupole interaction under the fast MAS condition, $\nu_{Q,MAS}^{(2)}$, is expressed as

$$\nu_{Q,MAS}^{(2)} = -1/6\nu_0\left[3C_Q/2I(2I-1)\right]^2[I(I+1)-3/4]\left[Dcos^4\theta + Ecos^2\theta + F\right]$$

$$\mathrm{D} = 21/16 - 7/8\eta_Q cos2\phi + 7/48\left(\eta_Q cos2\phi\right)^2$$

$$\mathrm{E} = -9/8 + 1/12\eta_Q^2 + \eta_Q cos2\phi + 7/24\left(\eta_Q cos2\phi\right)^2$$

$$\mathrm{F} = 5/16 - 1/8\eta_Q cos2\phi + 7/48\left(\eta_Q cos2\phi\right)^2 \tag{11.14}$$

Figure 11.4b shows the simulated ^{17}O MAS spectra as a function of η_Q from 0 to 1 in steps of 0.2 [10]. These calculations are carried out using the same conditions as those used for the static ^{17}O NMR spectra in Fig. 11.4a. The spectral width of a MAS NMR spectrum is approximately one-third narrower than that of the static NMR spectrum. A MAS spectrum also exhibits characteristic line shapes, from which information on C_Q and η_Q can be extracted. If the sample spinning frequency is much faster than the magnitude, the anisotropy of the CS tensors is assumed to be removed by MAS, and the scalar value of δ_{iso} should be added to Eq. (11.14) for the spectral simulation. As an example, the one-dimensional ^{17}O MAS NMR spectrum of two oxygen sites of Fmoc-amino acid, C = O and COH, are shown in Fig. 11.1, and the spectral simulation yields the following parameters: C = O:δ_{iso} = 305 ± 2 ppm, C_Q = 7.90 ± 0.08 MHz, η_Q = 0.18 ± 0.04; COH, δ_{iso} = 180 ± 2 ppm, C_Q = 7.05 ± 0.08 MHz, η_Q = 0.15 ± 0.04. The frequency contribution from the second-order quadrupole interactions, combined with the anisotropy of the CS interactions under slow/intermediate MAS conditions, is given in the literature [10], and a complicated line shape is expected to appear in the MAS NMR spectrum (a computer spectral simulation is not trivial).

Overlapping of multiple sites in MAS NMR spectra may result in complicated line shapes, and it is very difficult or nearly impossible to analyze such NMR spectra. To separate each site in MAS NMR spectra of quadrupole nuclei, the following experimental approaches are commonly used; dynamic-angle spinning (DAS) [12], double rotation (DOR) [13], multiple-quantum MAS (MQMAS) [14], and satellite-transition MAS (STMAS) [15], all of which are designed to remove the second-order quadrupole interactions, i.e., the elimination of the anisotropy in Eq. (11.14). Probably, MQMAS is the most widely used, since it can be performed with a normal MAS probe. In other words, special NMR probes are needed for the other methods, so that only a limited number of NMR users can use them. Figure 11.5 shows the contour plot of the two-dimensional (2D) ^{17}O triple-quantum z-filtered MQMAS spectrum of [^{17}O]-L-valine obtained by rotor synchronization for t_1 increments with a MAS rate of 15.00.1 kHz [16]. The Y- and X-axes projections of the 2D MQMAS spectrum are shown on the side and at the top of the figure, respectively. As indicated in Fig. 11.5, four isotropic peaks, which arise from a pair of magnetically inequivalent carboxylate oxygen sites, are observed in the 2D MQMAS spectrum. The slice along the t_2 direction (X-axis direction) of each peak corresponds to the 1D ^{17}O MAS spectrum for each oxygen site, and spectral analysis can be performed easily. However, MQMAS spectra do not generally have quantitative information on peak intensities. Moreover, such cross-sectional spectra are often distorted by multiple-quantum effects. Therefore, spectral simulation of the 1D MAS spectrum is always required to confirm the final ^{17}O NMR parameters. The NMR parameters, estimated from the analysis of MQMAS spectra, are useful as the initial values for the spectral simulations of the 1D MAS spectrum. The results of the 1D MAS NMR spectrum of [^{17}O]-L-valine, assisted by those of the MQMAS, are given in Fig. 11.6, where the calculated spectrum is the sum of the O1, O2, O3, and O4 sub-spectra with a 1:1:1:1 intensity ratio: O1, δ_{iso} = 287 ± 1 ppm, C_Q = 7.10 ± 0.02 MHz, η_Q = 0.55 ± 0.02, O2, δ_{iso} = 280 ± 1 ppm, C_Q = 7.45 ± 0.02 MHz, η_Q = 0.67 ± 0.02, O3, δ_{iso} = 313 ± 1 ppm, C_Q = 7.85 ± 0.02 MHz, η_Q = 0.31 ± 0.02, O4, δ_{iso} = 264 ± 1 ppm, C_Q = 7.20 ± 0.02 MHz, η_Q = 0.60 ± 0.02.

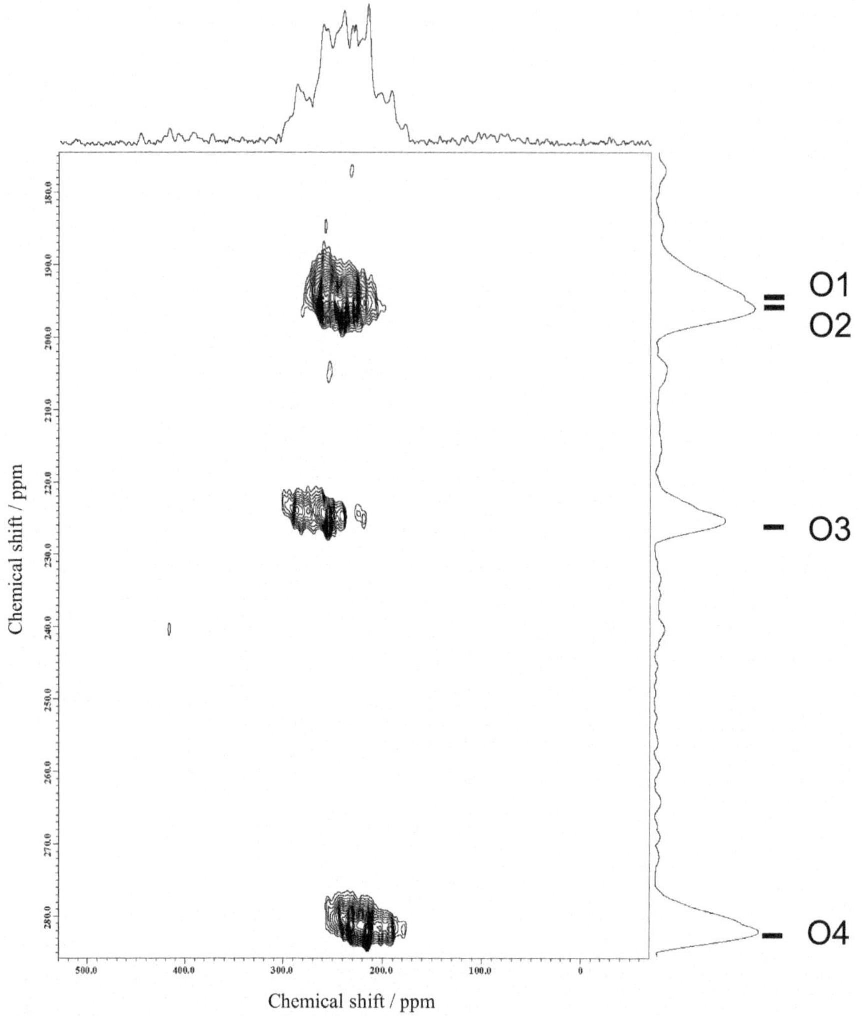

Fig. 11.5 Contour plot of the ^{17}O z-filter 3QMQMAS spectrum of $[^{17}O]$-L-valine recorded at 16.4 T

11.3.3 Static NMR Spectra Under the Influence of Second-Order Quadrupole and CS Interactions

One of the advantages of solid-state NMR is that tensor information on nuclear spin interaction can be obtained [2]. In a typical analysis of a static ^{17}O NMR spectrum of organic and biological compounds, the following parameters will be obtained; CS tensors (δ_{11}, δ_{22}, δ_{33}), EFG tensors (C_Q and η_Q), and the Euler angles (α, β, γ), which exhibits the relative orientation between the two tensors. To calculate a static ^{17}O

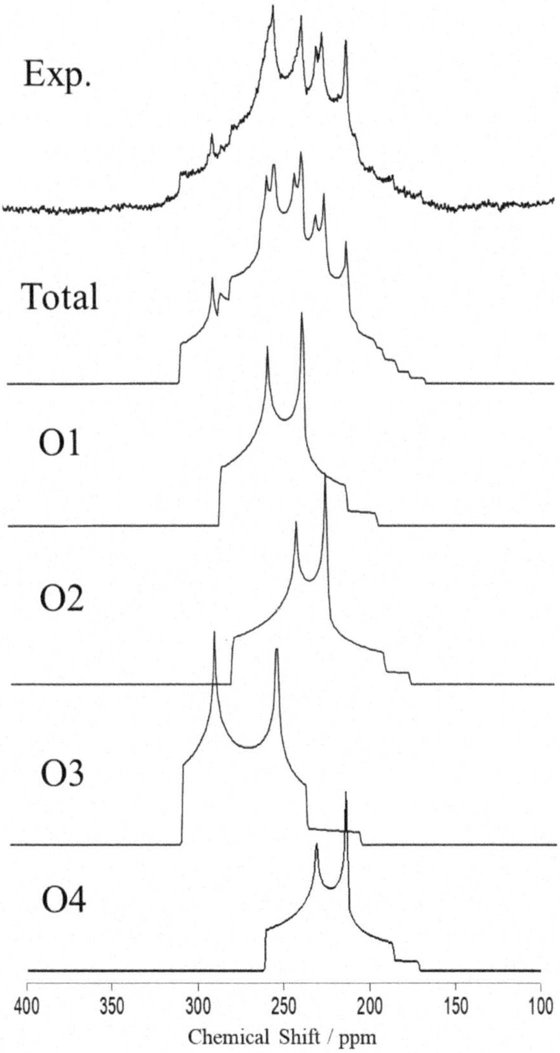

Fig. 11.6 Experimental and calculated 1D ^{17}O MAS spectra of [^{17}O]-L-valine recorded at 16.4 T. The calculated total spectrum is the sum of O1, O2, O3, and O4 sub-spectra with a 1:1:1:1 intensity ratio

NMR spectrum, the following equation to express the anisotropy of CS interactions, ν_{CS}, is added to Eq. (11.11):

$$\nu_{CS} = \delta_{11} sin^2\theta cos^2\phi + \delta_{22} sin^2\theta sin^2\phi + \delta_{33} cos^2\theta \tag{11.15}$$

where θ and ϕ describe the orientation of the magnetic field in the principal axis system of the CS tensor illustrated in Fig. 11.3a [1]. Very often, however, the CS and EFG tensors exhibit different orientations in the molecular frame of reference. This simply means that θ and ϕ in Eq. (11.15) are not identical to those in Eq. (11.11). It is necessary to consider and introduce a relative orientational term between the two tensors, using the three Euler angles (α, β, γ), as shown in Fig. 11.3b. It is convenient to rotate the PAS of a CS tensor, based on that of an EFG tensor. Unfortunately, there have been several different definitions of the Euler angles, describing the relative orientation of the two tensors; the direction of rotation or initial coordinates may differ, and the two Euler rotations of Z-X-Z or Z-Y-Z exist. Therefore, it is sometimes difficult or misleading to obtain information on tensor orientations from the literature. In this section, the approach of Eichele et al. is described [17], since the author believes that this procedure is the most straightforward. The alignment of the two tensors in the magnetic field is illustrated in Fig. 11.3a. The transformation matrix of R(α, β, γ) can be written as

$$R(\alpha, \beta, \gamma) = \begin{pmatrix} \cos\gamma\ \cos\beta\ \cos\alpha - \sin\gamma\ \sin\alpha & \cos\gamma\ \cos\beta\ \sin\alpha - \sin\gamma\ \cos\alpha & -\cos\gamma\ \sin\beta \\ -\sin\gamma\ \cos\beta\ \cos\alpha - \cos\gamma\ \sin\alpha & -\sin\gamma\ \cos\beta\ \sin\alpha - \cos\gamma\ \cos\alpha & -\sin\gamma\ \sin\beta \\ \sin\beta\ \cos\alpha & \sin\beta\ \sin\alpha & \cos\beta \end{pmatrix} \tag{11.16}$$

The directional cosines in Eq. (11.15) are transformed according to the above transformation [3]. The final mathematical formula is given in the literature [4]. When (α, β, γ) = (0, 0, 0), the orientations of the CS tensor become coincident with those of the EFG tensor, i.e., a simple addition of Eqs. (11.11)–(11.15) [18]. It is worth noting that the sign of an EFG tensor cannot be distinguished by a conventional NMR measurement. Therefore, the addition of 180° to one of the Euler angles has changed no NMR line shapes. For example, the sets of Euler angles of (α, β, γ), (α, β, γ + 180°), (α, β +180°, −γ + 180°), (−α, β + 180°, γ) and (α+180°, β+180°, −γ) will reproduce identical NMR line shapes.

As an example, the experimental and the best-fitted theoretical static solid-state ^{17}O NMR spectrum of Fmoc-amino acid has been shown in Fig. 11.2. For a sub-NMR spectrum, a static NMR line shape, arising from the sum of the CS and the second-order quadrupolar contributions (taking the relative orientations into consideration), is calculated. The spectral analysis yields the following NMR parameters: C = O, $\delta_{11} = 478 \pm 9$ ppm, $\delta_{22} = 409 \pm 9$ ppm, $\delta_{33} = 28 \pm 9$ ppm, $\alpha = 14 \pm 10°$, $\beta = 90 \pm 8°$, $\gamma = 134 \pm 10°$; COH, $\delta_{11} = 325 \pm 9$ ppm, $\delta_{22} = 108 \pm 9$ ppm, $\delta_{33} = 107 \pm 9$ ppm, $\alpha = 10 \pm 15°$, $\beta = 12 \pm 8°$, $\gamma = 16 \pm 10°$. There are eight independent NMR parameters (δ_{11}, δ_{22}, δ_{33}, C_Q, η_Q, α, β, γ) for each oxygen site. In the present case, a total of 16 independent NMR parameters are changed for the spectral simulation. Using the results from the analysis of the ^{17}O MAS experiments (see Fig. 11.1), which yields δ_{iso}, C_Q, and η_Q for each site, the number of independent NMR parameters can be reduced to 10. Unfortunately, it is still not straightforward to analyze a static ^{17}O NMR spectrum, because a slight change in NMR parameters, particularly the Euler angles, has a significant effect on theoretical line shape. Prior to a computational spectral simulation, it is advantageous to understand the possible set of Euler angles for each functional group. There have been a few literatures on the summary of ^{17}O NMR parameters, including tensor orientations [5]. Quantum chemical calculations are one of the most useful tools for predicting the Euler angles. One should be kept in mind that the current *ab initio* calculations produce highly accurate results for ^{17}O EFG tensors, but not for ^{17}O CS tensors. It is also effective to use multiple magnetic fields for solid-state ^{17}O NMR measurements, by which different line shapes of static ^{17}O NMR spectra are obtained, and the accuracy of the spectral fitting is dramatically improved. It takes advantage of the fact that a line shape, arising from a combination of CS and EFG tensors, varies greatly with the magnetic field strength. The spectral analyses of static solid-state ^{17}O NMR spectra of polycrystalline amino acid (two inequivalent oxygen sites of a carboxylate group), observed in various magnetic fields, are shown in Fig. 11.7 [19]. In the three different magnetic fields, the theoretical NMR spectra are calculated with the same parameters, although they exhibit different line shapes.

The three Euler angles exhibit the relative orientation between two NMR tensors, but not a physical property. One may need the tensor orientation in terms of the molecular frame. One of the best techniques for obtaining orientational information on NMR tensors is a single-crystal NMR experiment [20]. However, a large single crystal suitable for solid-state NMR experiments is often difficult to obtain. In the case of ^{17}O NMR, the need to introduce ^{17}O-labels makes it even more difficult to grow large single crystals because of the small quantity of a labeled sample. An alternative method to establish the absolute tensor orientation in the molecular frame is to use quantum chemical calculations. It is assumed that the orientations of the ^{17}O EFG tensors obtained from high-level quantum chemical calculations are correct. With the Euler angles combined results of the absolute orientations, it is possible to determine the absolute orientations of the ^{17}O CS tensor with respect to the molecular frame. The schematic representation of this approach is given in Fig. 11.8 [4]. The absolute orientations of a CS tensor are reflected by three-dimensional electronic information and thus provide a more detailed molecular structure.

11.4 Solid-State NMR of Quadrupole Nuclei Based on Direct Calculation Methods

The perturbation method, i.e., Eqs. (11.5) and (11.6), cannot be used for spectral simulations, when the magnitude of the quadrupole interactions is comparable to or larger than that of the Zeeman interactions. In such cases, the direct diagonalization method is a promising approach to analyzing an NMR spectrum broadened by a quadrupole interaction. For simplicity, this section focuses on the cases in which a sample is static and a quadrupolar interaction is dominant. Figure 11.9 shows the static NMR spectra of half-integer quadrupole nuclei, calculated by (a) the direct diagonalization method and (b) the perturbation method [7]. These calculations are performed using $C_Q = 150$ MHz, $\eta_Q = 0.1$, $\nu_0 = 100$ MHz, and $I = 3/2$. In the later approach, only the central transition was considered, so that the baseline on both sides is slightly less raised. Similar line shapes are observed for both NMR spectra in the range of the central transitions, but the line width calculated by the perturbation method is slightly expanded to the lower frequency. This difference is thought to be the main limit of the perturbation theory. It is potentially erroneous to

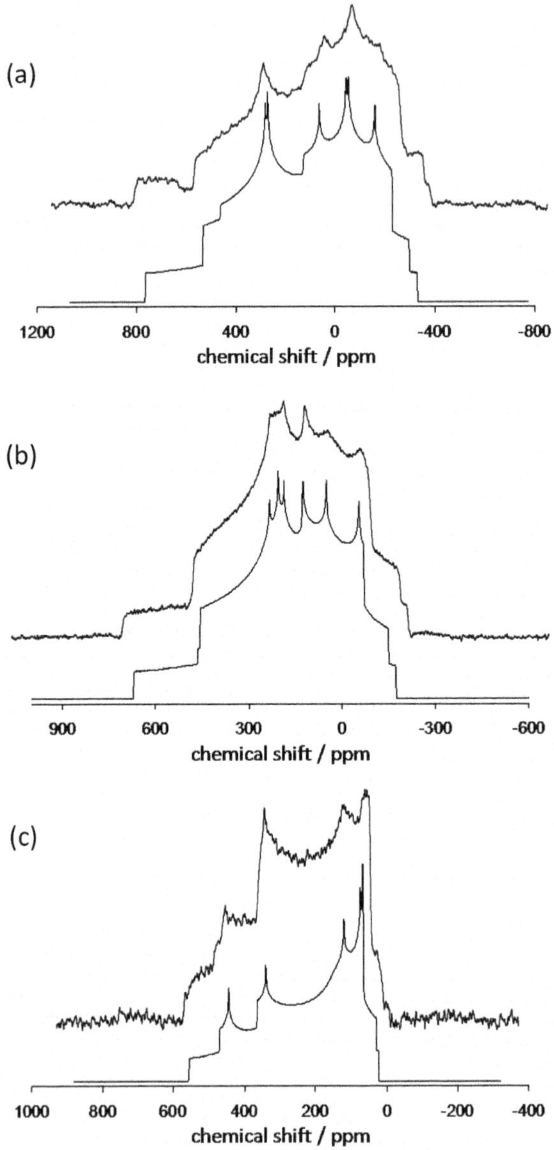

Fig. 11.7 Experimental and calculated static ^{17}O NMR spectra for [^{17}O]-L-alanine hydrochloride, recorded at **a** 9.4, **b** 11.7, and **c** 21.6 T

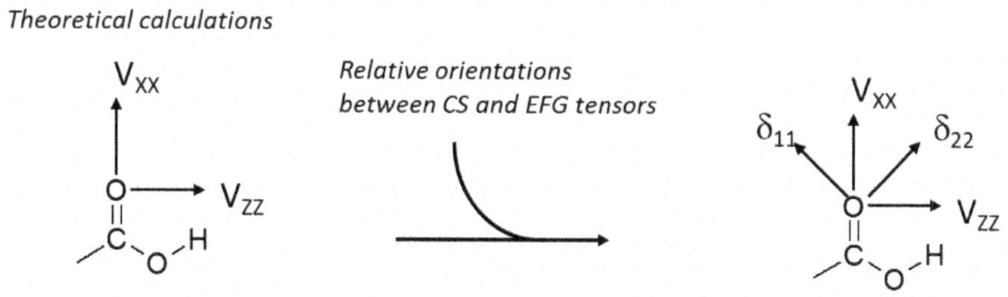

Fig. 11.8 Schematic representation to obtain information on CS tensor orientation in terms of molecular frames, combined with the Euler angles determined experimentally

apply the perturbation method when the magnitude of the quadrupole interactions is larger than that of the Zeeman interactions, or when the strength of the external magnetic field is lower even if the magnitude of a quadrupole interaction is small.

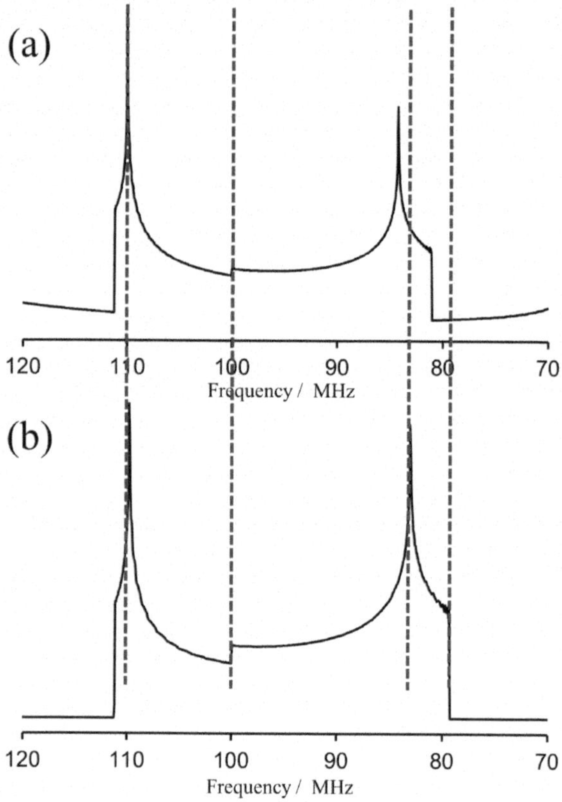

Fig. 11.9 Theoretical NMR spectra of half-integer quadrupole nuclei, based on **a** the direct diagonalization method **b** the perturbation method. See the text in the details

Traditionally, there have been two experimental techniques to observe solid-state NMR spectra broadened by large quadrupole interactions, i.e., frequency-swept NMR and field-swept NMR. A schematic representation of the two methods is given in Fig. 11.10. Both methods may have advantages and disadvantages. For frequency-swept NMR, spectral analysis is straightforward, and experiments can be performed with a conventional NMR instrument under a static magnetic field. However, much effort is needed to manually adjust a probe-tuning, unless a fully automated system is used. Moreover, there may be a limitation in hardware, i.e., the capacity of the condensers of an NMR probe, which determines the limits of the observable frequency range. For field-swept NMR, on the other hand, a spectral analysis is complicated and time-consuming. A field-variable superconducting magnet, which is not commonly used in the research fields of chemistry and biochemistry, is also required for the NMR measurements. However, there is no limitation of observable frequency range, caused by the hardware. Although both methods may suffer from a low sensitivity, a suitable method should be selected, depending on sample conditions. In the rest of this section, bromine and sulfur solid-state NMR are presented as examples of frequency-swept NMR and field-swept NMR, respectively.

Bromine is the third-lightest halogen and plays important roles in a variety of materials including flame retardants, dye compounds, and gasoline additives. There are two NMR-accessible stable bromine isotopes: ^{79}Br ($I = 3/2$, natural abundance $= 50.54\%$, $\gamma = 6.7228 \times 10^7$ rad T^{-1} s^{-1}, and $Q = 0.37 \times 10^{28}$ Qm^{-2}) and ^{81}Br ($I = 3/2$, natural abundance $= 49.46\%$, $\gamma = 7.2468 \times 10^7$ rad T^{-1} s^{-1}, and $Q =$

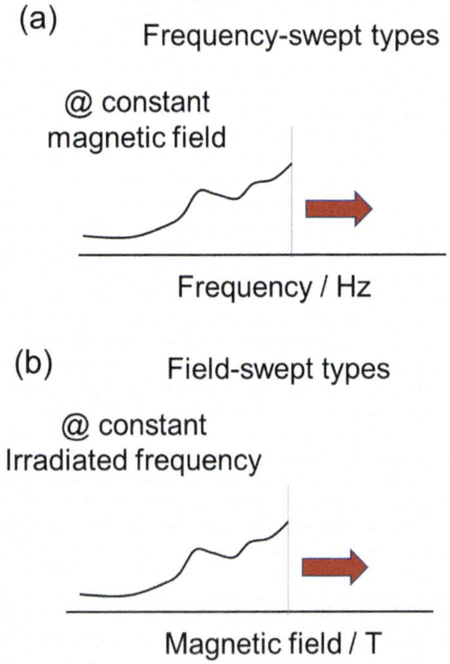

Fig. 11.10 Schematic representation of the two experimental techniques to record solid-state NMR spectra broadened by large quadrupole interactions: **a** frequency-swept types and **b** field-swept types of NMR experiments. Both methods, based on Fourier Transformation NMR, are not a classical CW NMR technique

0.31×10^{28} Qm^{-2}). ^{79}Br may be commonly used since its resonant frequency is very close to that of ^{13}C, and KBr is used for adjusting the magic angle prior to MAS experiments. The magnitude of quadrupole interactions for $^{79/81}Br$ depends on the bonding mode. Inorganic compounds or those forming ionic bonds, in which molecular symmetry is high, generally exhibit smaller quadrupole interactions, while organic compounds, or those forming covalent bonds larger. The latter increases its spectral width. For example, the spectral width in ^{79}Br NMR, the C_Q value of which is more than 30 MHz, is broadened to the order of megahertz.

Figure 11.11 shows the (a) calculated and (b) experimental static ^{79}Br frequency-swept NMR spectra of SrBr $6H_2O$ measured at 11.7 T [7]. The experimental spectrum is the sum of (c) five sub-NMR spectra, each of which is observed by the Oldfield echo [21] with steps of 300 kHz in the irradiation frequency. The spectral simulation based on the direct diagonalization method provides the following NMR parameters: $C_Q = 33.2 \pm 0.5$ MHz and $\eta_Q = 0.01 \pm 0.01$. The line shape is insensitive to changes in the CS interactions, so the effect of the CS interactions (Eq. (11.15)) is safely ignored in the spectral simulation. If the NMR probe can cover the frequency band, frequency-swept NMR is one of the best experimental methods to observe an NMR spectrum broadened by large quadrupole interactions.

Organosulfur compounds are essential for a variety of biological activities. For example, sulfur-containing L-methionine and L-cysteine are two of the twenty common amino acids in the human body, and many organosulfur compounds such as penicillin have been marketed as drugs or used in drug development [22, 23]. Solid-state sulfur NMR (^{33}S, $I = 3/2$, natural abundance = 0.75%, $\gamma = 2.055 \times 10^7$ $radT^{-1}\ s^{-1}$, $Q = 5.5 \times 10^{26}$ Qm^{-2}) has a long history, but still presents difficulties depending on the magnitude of quadrupole interactions, i.e., the bonding mode. It is safe to say that solid-state ^{33}S NMR studies of organic compounds that form covalent bonds have been undeveloped.

Figure 11.12 shows a calculated 2D solid-state ^{33}S NMR spectrum of a typical organic compound [7]. The calculations are performed by the direct diagonalization method with $C_Q = 43.68$ MHz and $\eta_Q = 0.57$, which are obtained from the analysis of the ^{33}S NMR spectra of α-S_8 [6, 24]. It is convenient to use this figure to discuss the relationship between frequency-swept and field-swept NMR. The slices in the Y-axis direction correspond to frequency-swept NMR spectra in a static magnetic field, while those in the X-axis direction correspond to field-swept NMR spectra at a fixed irradiation frequency. Figure 11.13 shows the calculated ^{33}S NMR spectrum sliced at 10 T in Fig. 11.12, which corresponds to a static ^{33}S NMR spectrum at 10 T. It is observed that the spectral width spread over 30 MHz, suggesting that it is very difficult or nearly impossible to observe a ^{33}S NMR spectrum under a moderate magnetic field. Figure 11.14a shows the experimental field-swept ^{33}S NMR spectrum of α-S_8 at the Larmor frequency of 16.2 MHz, acquired at 260 K [6]. Two sharp peaks are observed at approximately 1.3 and 1.4 T, which are assigned to the NMR signals of ^{65}Cu and ^{63}Cu, respectively, in the solenoid coil of the NMR probe. The best-fitted calculated field-swept ^{33}S NMR spectrum of α-S_8 is shown in Fig. 11.14b, which is the slice in the X-axis direction at 16.2 MHz in Fig. 11.12. Indeed, such spectral simulations can be time-consuming. Ichijo et al. [25] have demonstrated that C_Q and

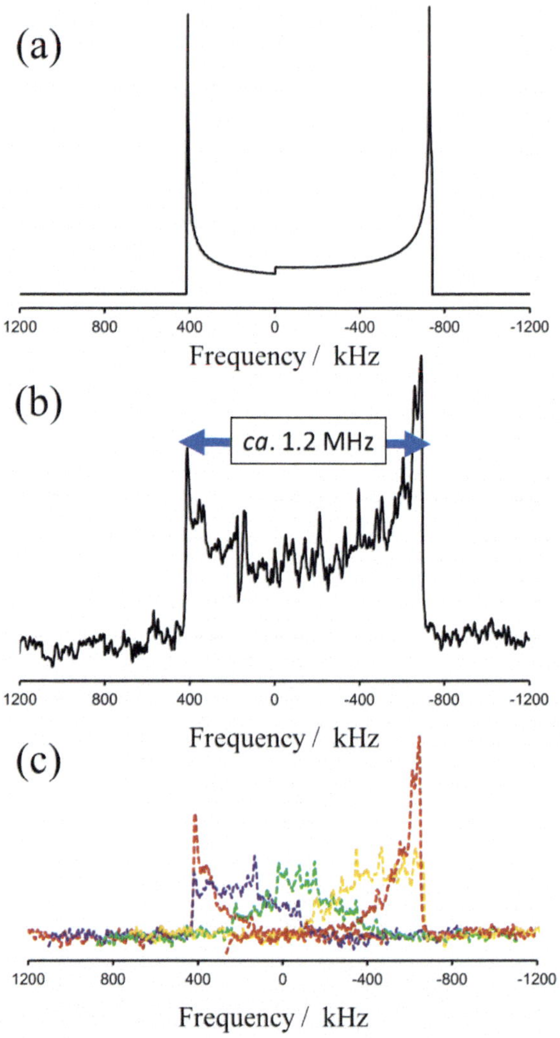

Fig. 11.11 **a** Calculated and **b** experimental static ^{79}Br ($I = 3/2$) frequency-swept NMR spectra. The experimental spectrum is the sum of **c** five sub-NMR spectra

η_Q can be simply determined from the positions of singularities in a field-swept NMR spectrum. This approach is based on the fact that the positions of singularities arise from the 4 points of (θ,ϕ) in spatial averaging; ($\pi/2$, $\pi/2$), ($\pi/2$, $3\pi/2$), ($\pi/2$, 0), and ($\pi/2$, π), where θ and ϕ are polar angles between the PAS and the laboratory frames. In Fig. 11.14a, for example, a shoulder and a peak are observed at 2.11 and 3.96 T, respectively. Using the above correlation, a possible combination of C_Q and η_Q can be calculated for each singularity. Figure 11.15 shows the correlation between C_Q and η_Q for the singularities in Fig. 11.14a. From the intersection of the curves, EFG parameters can be determined. This approach dramatically reduces the time required for the spectral analysis of field-swept NMR spectra. It is also useful for separating multiple sites. Figure 11.16 shows (upper) experimental and (lower) theoretical field-swept solid-state ^{115}In NMR spectra of indium oxide, In_2O_3, observed at 58.18 MHz. There are two crystallographically nonequivalent indium atoms in the cubic form of indium oxide, In(1) and In(2) [26]. The two indium sites can be easily separated using this singularity-analysis since theoretical peaks are simply aligned to the positions of each singularity. $C_Q = 178 \pm 6$ MHz and $\eta_Q = 0.04 \pm 0.04$, and $C_Q = 121 \pm 6$ MHz and $\eta_Q = 0.9 \pm 0.1$ are obtained for In(1) and In(2), respectively.

Recently, frequency-swept solid-state ^{33}S NMR that uses an extremely low magnetic field (~0.10 T) has been reported as an alternative to the above two methods, which corresponds to the slice in the Y-axis direction at 0.10 T or lower in Fig. 11.12. The advantage of this method is that spectral width becomes much narrower than that at higher magnetic fields, i.e., the experimental conditions are much closer to those

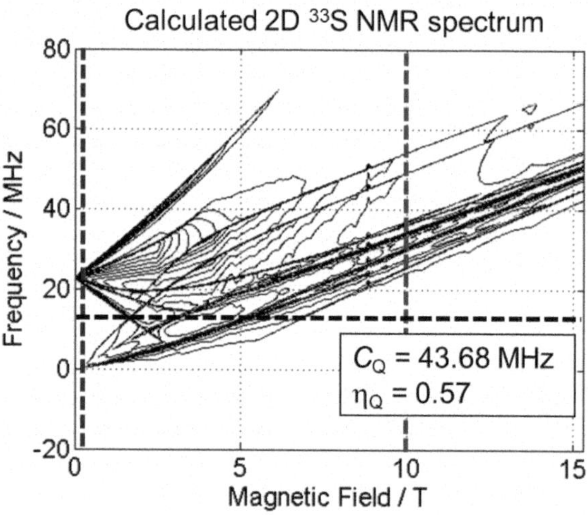

Fig. 11.12 Calculated 2D solid-state ^{33}S NMR spectrum using C_Q = 43.68 MHz and $\eta_Q = 0.57$. The slices to the Y-axis direction correspond to frequency-swept NMR spectra at constant magnetic fields, while those to the X-axis direction to field-swept NMR spectra at irradiated frequencies

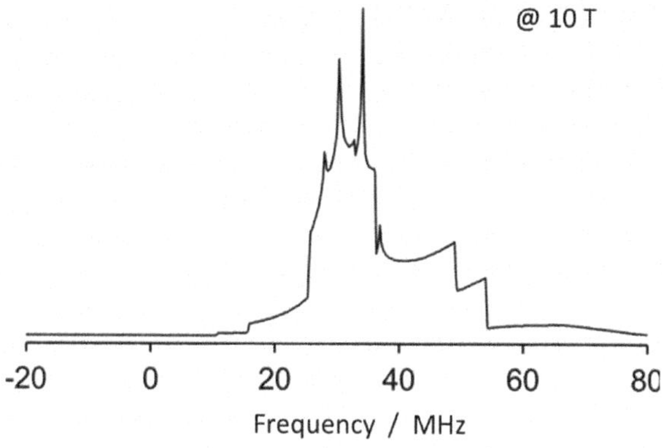

Fig. 11.13 Calculated frequency-swept ^{33}S solid-state NMR spectrum at 10 T, which corresponds to the slice in the vertical red line in the NMR spectrum in Fig. 11.12

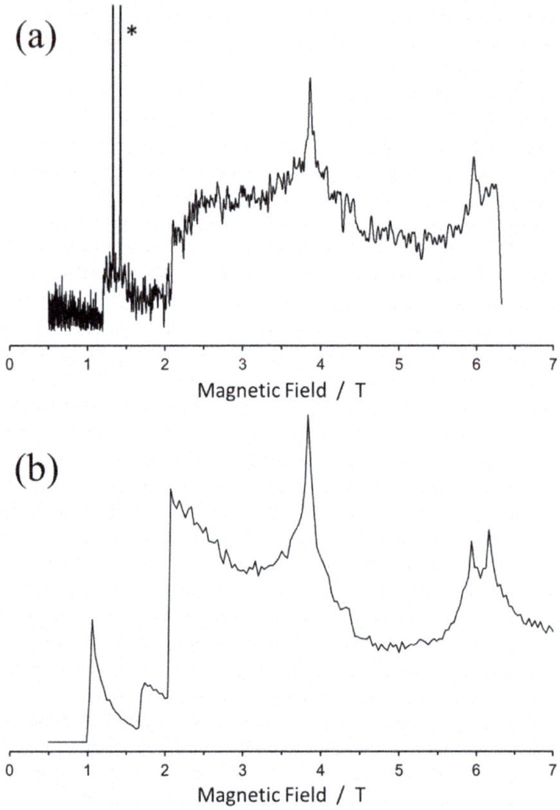

Fig. 11.14 **a** Experimental and **b** best-fitted calculated field-swept ^{33}S NMR spectra of α-S_8, acquired at 16.2 MHz. The spectral analysis, assuming that there is a single sulfur site, gives $C_Q = 43.68$ MHz and $\eta_Q = 0.57$

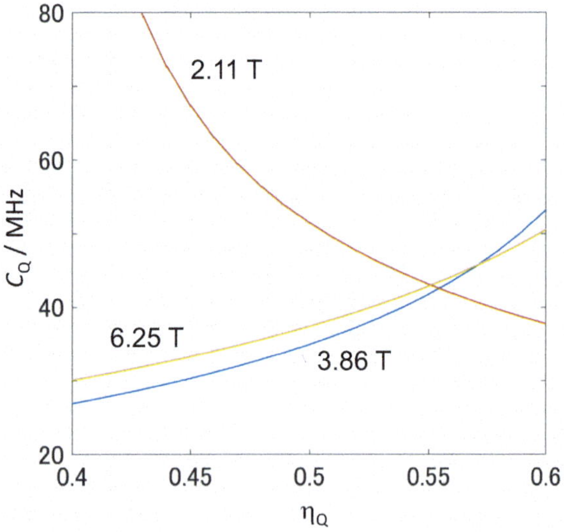

Fig. 11.15 An example of singularity-analysis for field-swept solid-state NMR; the correlation between C_Q and η_Q for the singularities in Fig. 11.14. From the intersection, the two NMR parameters can be obtained

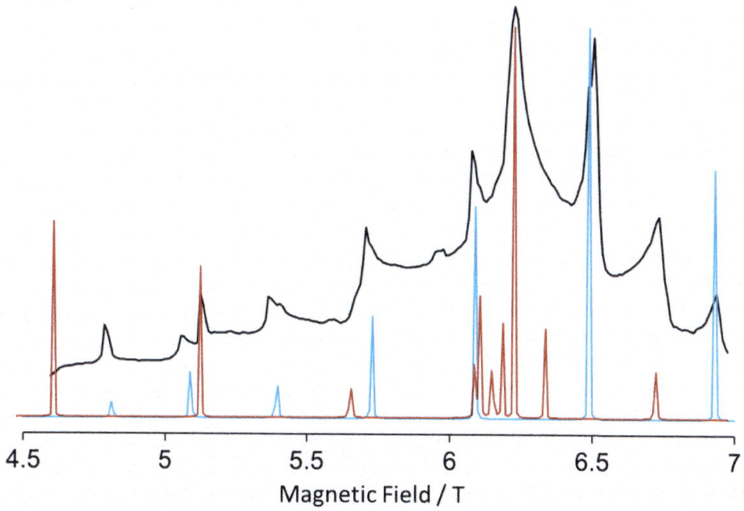

Fig. 11.16 The experimental field-swept solid-state ^{115}In NMR spectrum of indium oxide, acquired at 58.19 MHz and $T = 100$ K (black line). Singularity analysis is applied, and the blue and red lines are assigned to In(1) and In(2), respectively

of zero-field NMR or NQR experiments where a sharp signal is generally observed at ν_Q. It should be noted that NQR data alone cannot yield EFG information, since the spin number of ^{33}S is 3/2 (see Eq. (11.7)). The experimental and best-fit calculated ^{33}S NMR spectra of [^{33}S]-diphenyl disulfide observed at (a) 0.06 and (b) 0.10 T, which correspond to the Larmor frequencies of 0.1959 and 0.3265 MHz, respectively, are shown in Fig. 11.17 [27]. Unfortunately, the two sulfur sites in the disulfide bridge cannot be distinguished in the experimental NMR spectra due to poor spectral resolution. Assuming that there is a single sulfur site, the analysis of the NMR spectra yielded the following parameters: $C_Q = 43.3 \pm 1.9$ MHz and $\eta_Q = 0.70 \pm 0.20$.

In a short summary, the direct diagonalization method can be used to analyze NMR spectra of organic compounds broadened by large quadrupole interactions, for which the perturbation method is no longer valid. Frequency-swept and field-swept NMR are useful techniques for recording ultra-wide line NMR spectra. From the view of the instrumental limitations, the author feels that field-swept NMR has many advantages for potentially increasing signal intensity. The resonant frequency is fixed so that, for example, an NMR probe containing a high-temperature superconducting (HTS) coil with an extremely high Q value [28], or a new technology called an optomechanics system [29], in which an NMR signal can be transformed into laser radiation via a micro-condenser, is applied to field-swept NMR.

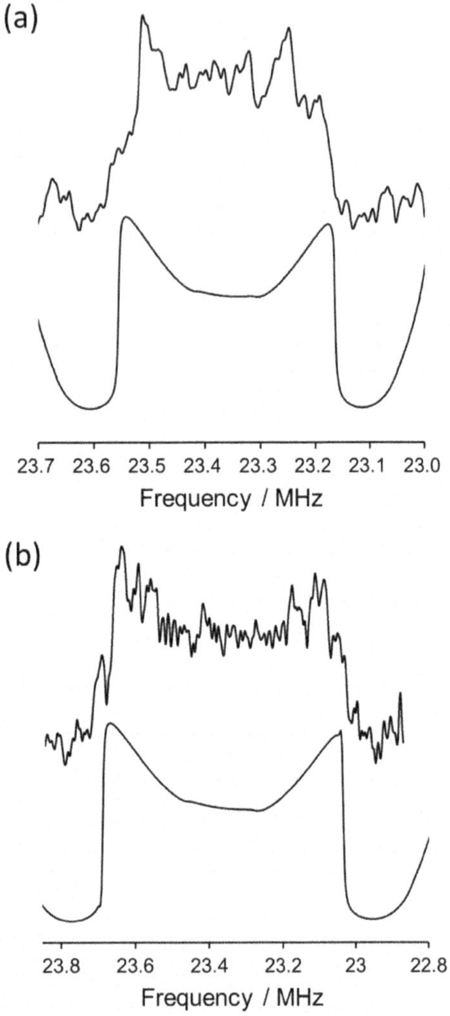

Fig. 11.17 Experimental (upper trace) and best-fitted calculated (lower trace) frequency-swept ^{33}S NMR spectra of [^{33}S]-diphenyl disulfide, acquired at **a** 0.50 and **b** 0.10 T

11.5 Conclusions

A general review of quadrupole solid-state NMR of organic compounds has been presented. When the magnitude of the quadrupole interactions is much smaller than that of the Zeeman interactions, the perturbation theory should be applied, and numerical formulas to express the first- and second-order Hamiltonians of quadrupole interactions under static or MAS conditions are obtained. As examples, analyses of the solid-state ^{17}O NMR spectra of organic compounds are presented. On the other hand, when the magnitude of the quadrupole interactions is comparable to or larger than that of the Zeeman interactions, the direct diagonalization method, in which a combined Zeeman and quadrupole Hamiltonian is numerically calculated to obtain the eigenvalues, should be used for the spectral simulation, because the perturbation theory, which is no longer valid, may lead to the incorrect conclusion. To record NMR spectra broadened by large quadrupole interactions, there have been traditionally two experimental techniques: frequency-swept NMR and field-swept NMR. Examples of both methods are presented, using experimental results for solid-state ^{79}Br and ^{33}S NMR of organic compounds and those forming covalent bonds. Frequency-swept solid-state NMR that uses lower magnetic fields is also briefly explained.

Recently, ultra-high field magnets made of HTS materials have been developed by NMR manufacturers. Unlike in the cases of low-temperature superconducting (LTS) materials, HTS magnets are expected to make it possible to dramatically increase magnetic field strength and have already allowed ^{1}H Larmor frequencies of more than 1.0 GHz. Currently, HTS magnets of 1.2 GHz, which corresponds to 28.18 T, are commercially available, and 1.3 GHz magnet is being developed in Japan. Ultra-high field magnets will be helpful for recording NMR spectra broadened by large quadrupole interactions because the magnitude of second-order quadrupole interactions is inversely proportional to the strength of the external magnetic field.

There are still many undeveloped research areas in solid-state NMR of quadrupolar nuclei, simply because quadrupole interactions are too large to be handled by the currently available magnetic field strengths. For further development, it is necessary to consider alternative strategies other than higher magnetic fields. This is because of the following two reasons: (1) For NMR users, magnetic field strengths of 1.2 GHz are still not sufficient to observe NMR spectra of the quadrupole nuclei, which current NMR instruments cannot access. For instance, solid-state ^{33}S NMR spectra of organic compounds are expected to exhibit spectral widths in excess of several megahertz, even if 1.2 GHz magnets are applied. An increase of about ~20% of magnetic strength would make little difference in the Larmor frequencies. The Larmor frequencies for low γ increase only by a few megahertz, as the strength increases from 1.0 to 1.2 GHz. (2) The effects of CS interactions have been ignored. The magnitude of CS interactions is proportional to the strength of the external magnetic field. It has been reported that the spectral width of solid-state ^{17}O NMR spectra for amino acids increases as the magnetic field strength increases. This is because a dominant factor for the determination of spectral width is no longer a quadrupole interaction, but a CS interaction. The author believes that field-swept NMR, in which there is principally no limitation on the observable frequency range, is one of the best candidates for this purpose. Although it may suffer from low sensitivities, which leads to long-time measurements or no signals received, the latest technology promises to achieve highly sensitive NMR signals.

References

1. Slichter, C.P.: Principles of Magnetic Resonance. Springer-Verlag, Berlin (1990)
2. Mehring, M.: Principles of High-Resolution NMR in Solids. Springer-Verlag, New York (1983)

3. Vega, A.J.: Quadrupolar nuclei in solids. In: Grant, D.M., Harris, R.K. (eds.) Encyclopedia of Nuclear Magnetic Resonance, pp. 3869–3889. Wiley, Chichester (1996)
4. Yamada, K., Dong, S., Wu, G.: Solid-state ^{17}O NMR investigation of the carbonyl oxygen electric-field-gradient tensor and chemical shielding tensor in amides. J. Am. Chem. Soc. **122**, 11602–11608 (2000)
5. Yamada, K.: Recent applications of solid-state ^{17}O NMR. In: Webb, G.A. (ed.) Annual Reports on NMR Spectroscopy, Chapter 3, pp. 115–158. Academic Press (2010)
6. Yamada, K., Kitagawa, K., Takahashi, M.: Field-swept ^{33}S NMR study of elemental sulfur. Chem. Phys. Lett. **618**, 20–23 (2015)
7. Yamada, K.: Development of sulfur NMR in chemistry. J. Cryo. Super. Soc. Jpn. **51**, 64–70 (2016)
8. Yamada, K., Hashizume, D., Shimizu, T., Ohki, S., Yokoyama, S.: A solid-state ^{17}O NMR, X-ray, and quantum chemical study of N-α-Fmoc-protected amino acids. J. Mol. Struct. **888**, 187–196 (2008)
9. Abragam, A.: The Principles of Nuclear Magnetism. Oxford University Press, Oxford (1961)
10. Yamada, K.: Solid-state NMR in half-integer quadrupolar spins: oxygen-17 solid-state NMR in organic compounds. J. Spectrosc. Soc. Jpn. **55**, 99–109 (2006)
11. Skibsted, J., Nielsen, N.C., Bildsoe, H., Jakobsen, H.J.: Satellite transitions in MAS NMR spectra of quadrupolar nuclei. J. Magn. Reson. **95**, 88 (1991)
12. Mueller, K.T., Sun, B.Q., Chingas, G.C., Zwanziger, J.W., Terao, T., Pines, A.: Dynamic-angle spinning of quadrupolar nuclei. J. Magn. Reson. **86**, 470 (1990)
13. Samoson, A., Kundla, E., Lippmaa, E.: High resolution MAS-NMR of quadrupolar nuclei in powders. J. Magn. Reson. **49**, 350–357 (1982)
14. Frydman, L., Harwood, J.S.: Isotropic spectra of half-integer quadrupolar spins from bidimensional magic-angle spinning NMR. J. Am. Chem. Soc. **117**, 5367–5368 (1995)
15. Gan, Z.: Isotropic NMR spectra of half-integer quadrupolar nuclei using satellite transitions and magic-angle spinning. J. Am. Chem. Soc. **122**, 3242–3243 (2000)
16. Yamada, K., Nemoto, T., Asanuma, M., Honda, H., Yamazaki, T., Hirota, H.: Both experimental and theoretical investigations of solid-state ^{17}O NMR for L-valine and L-isoleucine. Solid State Nucl. Magn. Reson. **30**, 182–191 (2006)
17. Eichele, K., Wasylishen, R.E.: WSOLIDS NMR Simulation Package, Ver. 1.17.30 (2001)
18. Baugher, J.F., Taylor, P.C., Oja, T., Bray, P.J.: Nuclear magnetic resonance powder patterns in the presence of completely asymmetric quadrupole and chemical shift effects: application to metavanadates. J. Chem. Phys. **50**, 4914–4925 (1969)
19. Yamada, K., Shimizu, T., Yamazaki, T., Ohki, S.: Determination of the orientations for the ^{17}O NMR tensors in a polycrystalline L-alanine hydrochloride. Solid State Nucl. Magn. Reson. **33**, 88 (2008)
20. Scheubel, W., Zimmerman, H., Haeberlen, U.: Oxygen-17 quadrupole coupling and nuclear magnetic shielding tensors in benzophenone. J. Magn. Reson. **63**, 544–555 (1985)
21. Kunwar, A.C., Turner, G.L., Oldfield, E.: Solid-state spin-echo Fourier transform NMR of potassium-39 and zinc-67 salts at high field. J. Magn. Reson. **69**, 124–127 (1986)
22. Barbarella, G.: Sulfur-33 NMR. Prog. Nucl. Magn. Reson. Spectrosc. **25**, 317–343 (1993)
23. Wagler, T.A., Daunch, W.A., Panzner, M., Youngs, W.J., Rinaldi, P.L.: Solid-state ^{33}S MAS NMR of inorganic sulfates. J. Magn. Reson. **170**, 336–344 (2004)
24. O'Dell, L.A., Moudrakovski, I.L.: Testing the sensitivity limits of ^{33}S NMR: An ultra-wideline study of elemental sulfur. J. Magn. Reson. **207**, 345–347 (2010)
25. Ichijo, N., Takeda, K., Yamada, K., Takegoshi, K.: Determination of nuclear quadrupolar parameters using singularities in field-swept NMR patterns. J. Chem. Phys. **145**, 134201–134208 (2016)
26. Marezio, M.: Refinement of the crystal structure of In_2O_3 at two wavelengths. Acta Cryst. **20**, 723–728 (1966)
27. Yamada, K., Aoki, D., Kitagawa, K., Takata, T.: Frequency-swept solid-state ^{33}S NMR of an organosulfur compound in an extremely low magnetic field. Chem. Phys. Lett. **630**, 86–90 (2015)

28. Yamada, K., Takahashi, M., Tritrakarn, T., Kato, S., Okamura, T., Irie, K., Hoshi, H., Saito, A.: Development of high-temperature superconducting coil for solid-state NMR experiments. Bull. Chem. Soc. Jpn **96**, 550–554 (2023)
29. Bagci, T., Simonsen, A., Schmid, S., Villanueva, L.G., Zeuthen, E., Appel, J., Taylor, J.M., Sørensen, A., Usami, K., Schliesser, A., Polzik, E.S.: Optical detection of radio waves through a nanomechanical transducer. Nat. **507**, 81–85 (2014)

Chapter 12
Quadrupole Nuclei in Inorganic Materials

Toshikazu Takahashi

Abstract The solid-state NMR of quadrupolar nuclei is becoming more important to date. The anisotropy of quadrupolar coupling cannot be eliminated by magic-angle spinning. Although large quadrupolar couplings often prevent us from observing NMR signals of quadrupolar nuclei as sufficiently separated signals, the recent development of the equipment and methods including DNP, are spreading out opportunities for NMR observation to a wider range of materials and in deeper levels of information. The quadrupolar parameters, which can be taken from NMR spectra of quadrupolar nuclei, have a potential usefulness to distinguish atoms depending on their local electrostatic environment. MQMAS or STMAS enables us not only to observe quadrupolar nuclei in separated signals but also to get quadrupolar parameters and isotopic chemical shifts of each separated signal component in a 2D spectrum. J- or D-HMQC, CP-HETCOR and related techniques provide us with information about inter-atomic chemical bonding or geometrical correlations. When the sensitivity is insufficient, besides isotope enrichment, for quadrupolar nuclei specifically, enhancement through intra-spin population transfer methods, RAPT, FAM, DFS, WURST, and HS can be applied. WURST, when combined with the QCPMG method, can provide us a method for obtaining NMR spectra for those nuclei having huge quadrupolar couplings, which make it difficult for us to observe the eintire signal shape. This article deals with the NMR methods being standard at present, or becoming standard in the near future, especially for practical samples, for example, mixed samples or amorphous samples. A number of experimental hints and a couple of experimental examples of ^{27}Al and ^{17}O are described.

Keywords Quadrupolar interactions · MQMAS · STMAS · HMQC · HETCOR

T. Takahashi (✉)
Interdisciplinary Research Center for Catalytic Chemistry, Institute of Advanced Industrial Science and Technology, 1-1-1 Higashi, Tsukuba 305-8565, Ibaraki, Japan
e-mail: takahashi.toshikazu@aist.go.jp

The NMR Society of Japan (ed.), *Experimental Approaches of NMR Spectroscopy II*,
https://doi.org/10.1007/978-981-97-6830-1_12

12.1 Introduction

Quadrupolar nuclei, having spin quantum number $I > 1/2$, are consisting the majority of the periodic table of elements. Above 75% of naturally occurring elements have at least one isotope that has $I > 1/2$. Above 65% of the naturally occurring elements have only quadrupolar nuclei in their naturally occurring isotope lists [1]. Although the NMR spectra of quadrupolar nuclei have, relatively, not frequently appeared in the literature compared to those of the spin 1/2 nuclei such as ^{1}H, ^{13}C, ^{15}N, ^{19}F, ^{31}P, and ^{29}Si, now quadrupolar nuclei, such as ^{7}Li, ^{11}B, ^{14}N, ^{17}O, ^{23}Na, ^{25} Mg, ^{27}Al, ^{39} K, ^{41}Ca, ^{95}Mo, ^{133}Cs, and ^{139}La, are becoming more frequently investigated than before. Needless to say, quadrupolar nuclei play indispensable roles in organic, inorganic, or hybrid materials, those being used in our daily lives. The NMR for quadrupolar nuclei is, in its principle, already established in very early times of NMR history [2–4]. Although, compared to the spin 1/2 nuclei, there still remain some difficulties in the observations of the quadrupolar nuclei by NMR, the development of NMR equipment, especially in the magnetic field strength, which is still under development using high-temperature superconducting wire [5], and the increasing detection capabilities by the development of high-speed MAS [6], low noise circuits, and cryogenic circuits including probes [7] are expanding the range of observation opportunities for the quadrupolar nuclei in various materials. Software development including NUS [8] and related time-reducing technologies will also be important in the current development for the same purpose. Now the opportunity is spread to the materials of electric power storages, catalysts, intelligent components of the construction materials or glasses, shieling components of barrier films, and surface-modified or surface-anchored materials used for providing functional surfaces such as separation or detection of particular molecules, high-performance adsorption materials, rustproofed, anti-microbial or mold-prevention functions and so on.

Before the closing remarks in this section, the author will briefly comment on the recently developed dynamic nuclear polarization (DNP) technologies, because this new technology will be opening another step of opportunity for solid-state NMR of quadrupolar nuclei [9]. DNP is the name of the methods for observing dynamically polarized nuclei, usually being conducted under low or very low temperatures, through achieving a larger magnetic polarization on nuclear spins compared to those of room temperature even in the highest magnetic fields ordinarily used in the current NMR apparatus. The magnetization transfer from long-lived triplet molecules, electron spins of organic radicals, or para-hydrogen are used to promote a large content of magnetic polarization on the nuclear spins. Among them, methods using organic radical polarizers (the source of nuclear polarization formation) are recently developed, and the specific equipment for solid samples [10] is commercialized.[1] For the liquid samples, the target sample mixed with an organic polarizer is frozen within a glassy spin-delivering solvent (having protons in a suitable concentration and distribution) and stored in a medium magnetic field where the sample

[1] Bruker and JEOL are still developing DNP MAS NMR systems.

is irradiated microwaves to promote magnetization transfer from the much more polarized electron spins to the ^1H nuclear spin. Then after a time during which the nuclear spin polarization is sufficiently developed, the frozen solid sample is rapidly dissolved by a hot solvent, and then the solution is instantaneously transferred into the observation vessel set in a standard high-field NMR magnet through a transfer tube equipped between the two magnets. For the solid samples, high-field (currently 400–800 MHz) DNP MAS NMR equipment is developed and commercialized. This equipment also uses organic radicals as the polarizer, and solvents frozen in a glassy solid containing rich protons allowing the spin diffusion from the polarizer to the sample (In some cases, the objective nuclei are magnetized directly by the polarizer radicals without any proton mediators). The sample and the polarizer and the solvent are mixed and packed in the MAS rotor. Then, the sample in a rotor is loaded into the low-temperature MAS probe, the rotor rapidly frozen under a magic angle spinning in the probe around 100 K. In the standard cases, the frozen sample is continuously irradiated by high-power microwaves to promote magnetization transfer from the electrons to nuclei. Although the sample should be mixed with organic radicals, and the enhancement coefficient is usually smaller than 100, which is much lower than the theoretical limit (660), the highly enhanced sensitivity will open a new era to the solid-state NMR. Although sometimes we still need to prepare samples with some isotopes enriched, the development of NMR techniques is thus promising for the future development of material sciences and technologies as one of the key technologies of observing and analysis tools for materials.[2]

In this contribution, we will describe about solid-state NMR of quadrupolar nuclei. The purpose of this contribution is not to provide comprehensive knowledge about the NMR or that of quadrupolar nuclei. It is almost concentrated on helping readers who will think and learn deeply about NMR and will do experiments to observe solid-state NMR of their objective materials and extract useful information from them and develop knowledge of their own or, in some appropriate opportunity, sharing knowledge with somebody else around the readers.

12.2 Anisotropy of Magnetic Interactions

NMR is an observation method of magnetically active nuclei electromagnetically under the influence of Zeeman interaction. When a nuclear spin is put in a uniform magnetic field, the energy levels of a spin having spin quantum number I is quantized by that interaction energy and split into $2I + 1$ states identified by the magnetic quantum number $I_z(orm) = I, I - 1, \cdots, -I$. The energy difference between each of the neighboring two levels is $\gamma\hbar H = \hbar\omega = h\nu$, if there are no other interactions of significance. This interaction energy is called Zeeman energy, and the states are Zeeman states. Each the $\hbar$ and h (=$2\pi\hbar$) is Planck constant used in combination with angular velocity or frequency, respectively. γ is the gyromagnetic (or magnetogyric)

[2] Readers will be able to get some information of DNP in the Chap.17 of this book.

ratio of the nucleus that is a specific constant depending on the particular nucleus. H is the magnetic field that the nucleus is feeling at a point during a time that the signal coherence continuing. The energy of Zeeman states of a quadrupolar nucleus can be described as

$$E = \gamma \hbar I_z H (I_z = I, I-1, \cdots, -I) \tag{12.1}$$

We can detect nuclear spins through coherently observed electromagnetic waves emitted from the sample, which have been stimulated electromagnetically immediately before. These emitted waves can be transferred into the peaks in the frequency domain through the Fourier transformation. In the spectrum of spins in a single magnetic environment, we can see a signal having a frequency that corresponds to the energy difference between the two neighboring states [3].

$$\Delta E = \gamma \hbar H = h\nu \tag{12.2}$$

The NMR observation is based on the detection of this differential energy through an electromagnetic wave detection circuit including the sample. Here, the magnetic field H is not exactly the same as the main magnetic field we will apply to the sample using the magnet, however, H includes the secondary (and higher ordered) magnetic fields within the material induced by the main field (Fig. 12.1). H may include static and any fluctuated components, which means it depends not only on the molecular structure but also on the orientation of the molecule relative to the field. Furthermore, it will be influenced by the relative distance and orientation of the neighboring molecules (Fig. 12.2).

Among all the interactions shown in Fig. 12.1, the most important one is thought to be the dipole–dipole interaction. Because every magnetic source can be represented by the sum of micro magnetic dipole moments. The magnetic fields, as like as electric fields, are constructed by every small magnetic source with the principle of superposition of electromagnetic field at a time. The Hamiltonian of dipole–dipole interaction between two spins $\hat{I}_1$ and $\hat{I}_2$ aligned parallel or anti-parallel orientation to the magnetic field is expressed in Eqs. 12.3–12.4, which consists of 6 terms named "dipole alphabet", A to F, respectively [11]. All of them are described using the polar coordinate representation of the vector connecting one dipole at the origin and the other dipole center; the length (r) and the angle with the magnetic field (θ) and the azimuthal phase (ϕ) of it. Every term of Eq. 12.3 is constituted by the product of the spins and space-dependent terms. And the interacting two spins' gyromagnetic ratio and the spin quantum operators.

$$\hat{H}_d = \frac{\gamma_1 \gamma_2 \hbar^2}{r^3} \left(\hat{A} + \hat{B} + \hat{C} + \hat{D} + \hat{E} + \hat{F} \right) \tag{12.3}$$

$$\hat{A} = \hat{I}_{1Z} \hat{I}_{2Z} \left(1 - 3\cos^2 \theta \right)$$

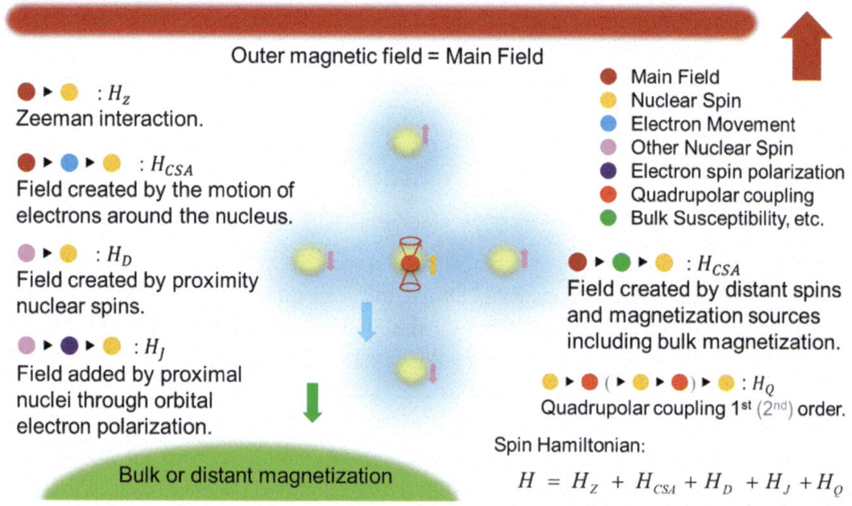

Fig. 12.1 Various origin of primary and secondary magnetic fields that will affect the magnetic field on the location of the spin

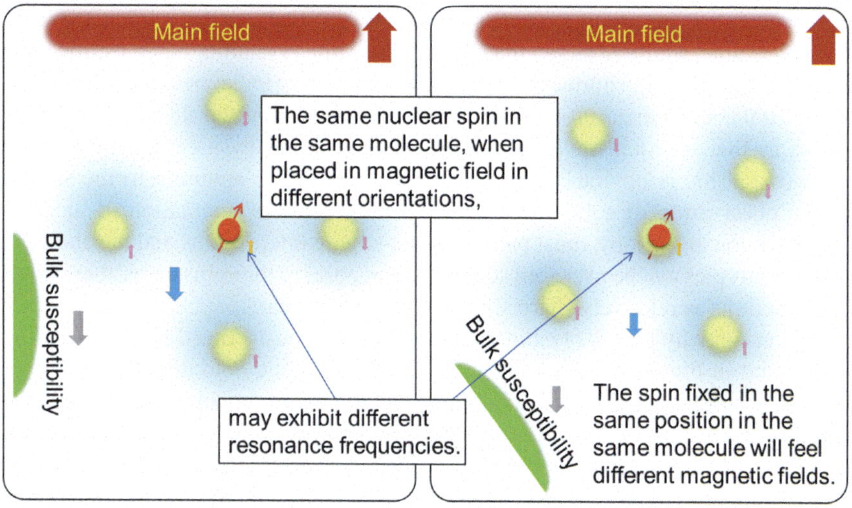

Fig. 12.2 A spin within the same site of the same molecule will feel a different magnetic field depending on the orientation of the molecular axis relative to the field

$$\hat{B} = -\frac{1}{4}\left(\hat{I}_{1+}\hat{I}_{2-} + \hat{I}_{1-}\hat{I}_{2+}\right)\left(1 - 3\cos^2\theta\right)$$

$$\hat{C} = -\frac{3}{2}\left(\hat{I}_{1+}\hat{I}_{2Z} + \hat{I}_{1Z}\hat{I}_{2+}\right)\sin\theta\cos\theta e^{-i\varphi}$$

$$\hat{D} = -\frac{3}{2}(\hat{I}_{1-}\hat{I}_{2Z} + \hat{I}_{1Z}\hat{I}_{2-}) \sin\theta \cos\theta e^{+i\varphi}$$

$$\hat{E} = -\frac{3}{4}\hat{I}_{1+}\hat{I}_{2+} \sin^2\theta e^{-2i\varphi}$$

$$\hat{F} = -\frac{3}{4}\hat{I}_{1-}\hat{I}_{2-} \sin^2\theta e^{+2i\varphi} \tag{12.4}$$

Among all the terms demonstrated above, the zero-quantum shift term $(\widehat{A})$, which will affect to the resonance frequency of the nuclei and the zero-quantum flip-flop term $(\widehat{B})$ are particularly important for the NMR spectrum. Both of them become zero where the $\theta =$ magic angle, 54.736°.

If these influences are directly added to the energy observed in the spectrum, NMR will be far from a method through which we can extract useful information because each molecule is under a different magnetic environment at a time that will be observed at different frequencies. However, which is particularly characteristic of this method,[3] the observed energy is well averaged on each spin during a period of its coherence duration. In fact, this averaging duration may sustain several seconds yields a very narrow signal with a width of less than 1 Hz, meaning that the due time of the coherence sustaining is longer than 1 s (=1/(1 Hz)). In the solid-state NMR, the typical line width is larger than 10 Hz even for crystallite samples, meaning the coherence time is shorter than 0.1 s (=1/(10 Hz)). However, although there is no molecular movement around the spin that will average the field automatically as like as that in liquid-state NMR, the field averaging by a certain mechanism still works. This is obvious when we apply magic angle spinning to the sample for the observation of solid-state NMR. In this case, the orientation of the bulk magnetization anisotropy, most part of dipole–dipole interactions, and the chemical shift anisotropies can be averaged so that we can observe the spins under these kinds of interactions as relatively sharp signals in the NMR charts.

The reason why the dipole–dipole interaction will be averaged by the MAS can be depicted in Fig. 12.3. In a sufficiently strong magnetic field directing $+ z$ axis, every spin will be precessing around the main field vector in parallel or antiparallel directions in most of the time. In this situation, the average magnetic field produced by a spin will have a particular orientation dependence having a shape as like as that of $d_z{}^2$ orbital. Under MAS conditions, from a fixed point of view on the spin A, the spin B interacting this spin will be seen as rotating along a circle centered on the magic angle axis passing the spin A. As a result, the magnetic field effect caused by spin B on the point spin A will be the same to those as if that of spin C may cause spin A. Because a dipole moment does not have any z-axis component at the cone surface having a magic angle relative to the z-axis (main field). Zero quantum

[3] ESR, Mössbauer spectrum and other related methods are having similar characteristics with respective different averaging times.

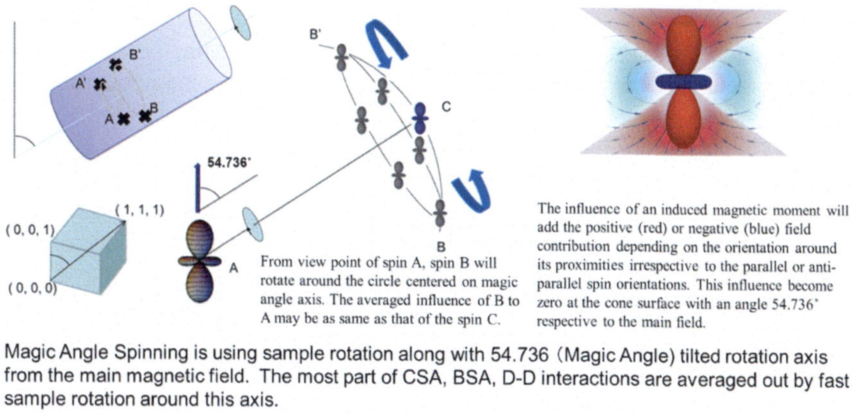

Fig. 12.3 Magic angle spinning under a strong field will cancel out the dipole–dipole interactions

contribution (chemical shift) of the dipole–dipole interaction between the two spins will be averaged out in this situation.

There are several numbers of anisotropic interactions that cannot be averaged completely by MAS. These interactions comprise the factors of 'inhomogeneous broadening'. One of the origins of inhomogeneous broadening comes from an anisotropy in the magnetic susceptibility of a local magnetization source within the sample. The secondary magnetic field caused by a molecular residue will normally be canceled out by orientation averaging caused by free molecular motions (typically in liquid states) or magic angle spinning (in solid samples). However, when the molecular residue has a relatively strong anisotropy in its magnetic susceptibility, a residual magnetic moment will appear which will not be averaged completely even after orientation averaging between the molecular axes and the magnetic field. These phenomena are called residual dipolar coupling (RDC) in liquid state NMR [12], which is responsible for NOE interaction in the liquid state. In the solid-state NMR particularly, in some cases, this type of interaction causes splitting of the signals. These phenomena are obvious especially the relatively sharp signals of spin 1/2 nucleus, which has a strong dipole coupling to a quadrupolar nucleus [13–15]. These phenomena are called residual dipolar shifts or splitting (RDS) in solid-state NMR [16, 17], however, the difference between them may exist only on the level of field averaging relative to the field difference. A large anisotropy on magnetic susceptibility produces a small anisotropic field that causes small chemical shifts to the nuclei located on the close proximities of that residue depending on the relative direction between the orientation vector connecting to the spin from the functional group (more exactly, the center of the source of the anisotropy) and the orientation of the group's residual magnetization orientation. In the liquid-state NMR, this interaction will affect only small chemical shifts on its close proximities which doesn't cause any broadening to the signals, because the orientation averaging of a molecule within a liquid is perfect owing to the molecular motions, which are faster than 10^{10}

Hz. However, in some cases of larger molecular systems, the anisotropy of this interaction becomes obvious. Thus, RDC potentially causes both the line-broadening and field-dependent shifts to the spins in the close proximities. A quadrupolar nucleus may, as like a permanent RDC, have anisotropy that cannot be averaged by MAS. The main term being responsible for this line-broadening is the second-order quadrupolar interaction. In solid-state NMR including quadrupolar nuclei, we often encounter the RDC caused by pure dipole–dipole, and quadrupole-dipole couplings (sometimes called quadrupole-dipole cross-term, especially the two interactions cannot easily be separated [16, 18]). Although RDC becomes relatively smaller in the higher magnetic fields, this kind of interaction is one of the main reasons for line-broadening in solid-state NMR. Every kind of magnetic susceptibility anisotropies may cause such kind of inhomogeneous (not caring whether or not this term is suitable) broadenings on its surface by similar mechanisms [19].

Different from those nuclei having spin number 0 or 1/2, the charge distributions on a quadrupolar nucleus are not in perfect spherical symmetry. This situation leads to an interaction between a quadrupolar nucleus and an electrostatic field surrounding it mainly through the field's second-order gradient in the 3D space at the point where the nucleus is located. Since the 'spin' is one of the substantial properties of nuclei that interact with the magnetic fields exclusively, the nuclear spin may not interact with the electrostatic field directly. However, because of the strong binding between the nucleon spins inside a nucleus, the magnetic moment of a nucleus would scarcely allow a single nucleon spin to interact with the exterior magnetic field as an independent spin. The nuclear spins do not indicate any sub-structures other than I_z dependence when it is investigated by the photons of their energy around Larmor-frequencies (typically several 10's of MHz to around 1 GHz), which are used in contemporary high-resolution NMR apparatus, and have extremely low energy even compared to the environmental photons, i.e., environmental noise.[4] Consequently, a spin behaves as like a fixed magnetic moment in any electro-photonic experiments except for those using very high-energy photons of ranges of γ-ray frequencies as the incident light sources. In literature [2], the authors described that within one nucleus, consisting of a number of sub-particles such as quarks or gluons, the angular motion of the charges within the nucleus is strongly restricted by the nuclear spin that will restrict strongly the moving charges distributed within the nuclei being under a higher spatial symmetry. This conjecture is reasonable considering the electric energy within a nucleus may be around 10^{17} Hz which is much larger than the energy of magnetic resonance, ~10^9 Hz, or quadrupolar interactions, ~10^7 Hz. The deviation of the charge distribution within a nucleus will be minimized by taking a higher rotational symmetry.[5]

[4] It is at least two orders lower even compared to the peak frequency of the Cosmic 2.73 K background radiation, having its peak frequency at 160.2 GHz, which is around 2 orders lower to that of the earth's environmental temperatures.

[5] According to Cohen 1954, the order of quadrupolar interaction is as large as $Q'_{jk}\left(\partial^2 V/\partial x_i \partial x_j\right) \sim er_n^2\left(e/r_e^3\right) = eV_0(r_n/r_e)^2$. The quadrupolar interaction is around 10^{-8} compared to the electrostatic force between the charges within the nucleus.

A question will possibly be argued, why the non-spherical shape of the nuclei is possible, when the angular momentum state of that nuclei is zero, which is observed exclusively in the ground state of nuclei found in nature. Since, when a nucleus is in its ground state, it stands in the smallest angular momentum condition $L = 0$, in that state, the uncertainty of the orientation vector of this complex particle will be perfectly undeterminable compared to any of the exterior world frame, i.e., it exists in perfect spherical symmetry. However, a small deviation from the spherical symmetry will be possible due to the perturbation of the magnetic interactions between the spin and the orbital within the nucleus, which causes the ground state nucleus to mix with the higher rotational quantum states with even angular momentum quantum numbers, for example, $L = 2, 4$. This will enable the shape of the nucleus to be subject to a very small deviation from a perfect spherical symmetry observable from the exterior frames. The quadrupolar coupling that is observed in NMR spectra can be described as the small contribution of the higher orders of rotational states introduced by perturbation of magnetic interactions within a nucleus that interact with the second-order gradient of the external electric field.

The $2I + 1$ energy levels of the quadrupolar nucleus with a spin quantum number I can be confirmed by observation of quadrupolar splitting in the NMR signals. Figure 12.4 demonstrates a static NMR spectrum, taken under a stable magnetic field, of a virtual single crystal sample containing a single site spin 5/2 nuclei. In a static field, each spin will be observed as $2I + 1$ signals separated by $2\nu_Q$ (Quadrupolar frequency) approximately each signal corresponds to the transition between the magnetic quantum number (also corresponding to the stationary states of observable $\hat{I}_z$, with z-axis taken to be parallel to the magnetic field) m and $m - 1$ states. This separation frequency is corresponding to the first-order quadrupolar interaction energy. When the quadrupolar interaction is sufficiently small, the second-order shift can be negligible and the signals separated only by the influence of first-order quadrupolar interaction will be observed. The signal intensity of them is proportional to $\zeta^2(m, m - 1) = I(I + 1) - m(m - 1)$. For the case of $I = 5/2$, the ratio is 5:8:9:8:5, depending on m. This ratio is confirmed experimentally for some gel samples [20] under special conditions, liquid–crystal and solid single-crystal samples containing quadrupolar nuclei [21] (Fig. 12.4). If the single crystal in the magnetic field is rotated around an axis for an angle, then the frequency splitting of the spin will be changed, however, that of the central band changes only a small amount (by CSA or dipolar coupling or higher order of quadrupolar couplings). For the sample consisting of many numbers of crystallites, the signal will be observed in a different shape. The signal will be distributed along with the frequency range in which orientation dependence of the interaction sweeps spherically. The angular dependence of this frequency change is similar to those observed in a spin 1/2 nuclei having a chemical shift anisotropy (CSA). Indeed, as with CSA, the first-order quadrupolar interaction can be canceled by magic angle spinning. The signal position will be influenced by not only the first-order quadrupolar interaction but also the second-order quadrupolar interaction. This is obvious when large quadrupolar interactions act. The interaction anisotropy is much different from that of the chemical shift anisotropy and becomes

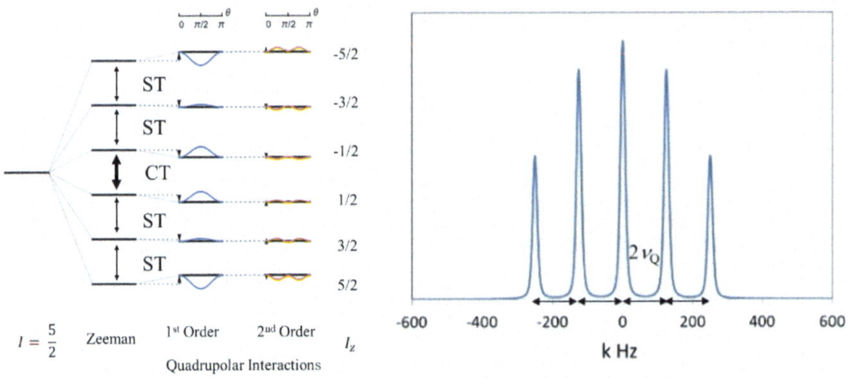

Fig. 12.4 The energy level diagram of a virtual ^{27}Al nucleus ($I = 5/2$) under a static field (~9.4 MHz). The energy levels of Zeemn states are perturbed by 1st and 2nd quadrupolar interactions. Polar angle (θ) dependence of anisotropies of quadrupolar interactions for each of the Zeeman states ($\eta = 0$) are illustrated with amplitudes expanded with ratios being around 500, and 5000 times, for the 1st and 2nd interactions, respectively. A virtual spectrum of a ^{27}Al sample with $2\nu_Q = 125$kHz, where the influence of the second-order interaction is too small to be seen. Gaussian (20 kHz) and Lorentzian (10 kHz) broadenings are applied

one that cannot be averaged perfectly by magic angle spinning. Using this anisotropy positively, crystal structure analysis using NMR is possible and several numbers of groups are developing this method [22]. For single-crystal ensembles, crystal orientation control techniques can be introduced [23]. In this method, the orientation of every single crystal is controlled by exerting a magnetic field in some viscous media.

We would like to make a reminder regarding the frequency distribution made by a square pulse. Square-shaped sine waves are most frequently used in ordinary NMR experiments. The Fourier transform of a square pulse (Fig. 12.5a) provides a sinc function ($\text{sinc}x \equiv (\sin x)/x$) (Fig. 12.5b), which represents the frequency distribution of the photons contained in this pulse. It is a symmetrical function with respect to the sign of ω, having a large amplitude centered at $\omega = 0$, and decaying with a $1/\omega$ dependence, showing oscillation crossing $y = 0$ line at $\omega = \pm 2n\pi/\tau\,(n = 1, 2, \cdots)$. In the case of a sine wave with a square envelope, the frequency distribution by Fourier transform, is centered at the frequency of the original sine wave, and like a simple square wave, it oscillates and decays with $2\pi/\tau$ and $1/\Delta\omega$ dependences, respectively, but when we see only the real part of frequency distribution, it is slightly more complicated depending on the phase difference between the square wave and the sine wave (Fig. 12.5c). However, in every case, the absolute values of these distributions do not change compared to that of the first one (Fig. 12.5d). When we square the absolute amplitude of these distribution functions, we can get the energy distribution of the square-enveloped pulse along with the frequency domain (Fig. 12.5e). Compared with that of amplitude intensity, the photon energy distribution shows a further concentration in the center peak, clearly showing that the zero-energy crossing points at $\omega = \pm 2n\pi/\tau\,(n = 1, 2, \cdots)$. Because the irradiation time of the pulse is lengthening during the pulse continuing, the frequency at which

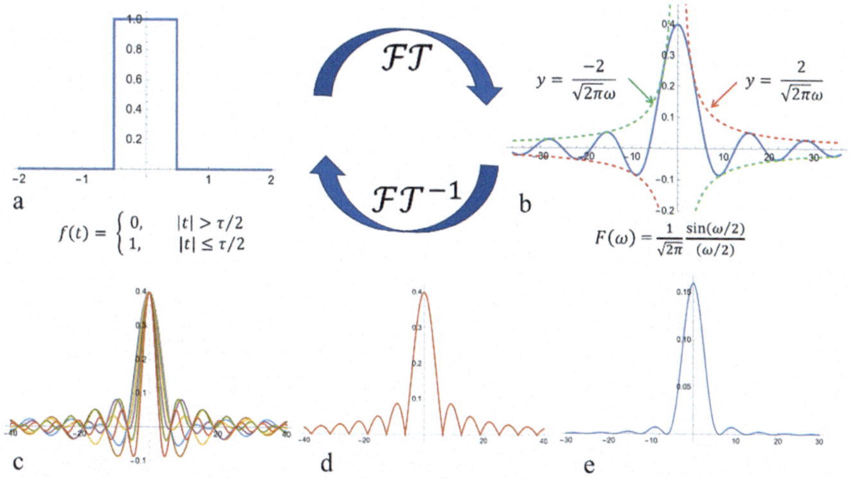

Fig. 12.5 Fourier transform of a square pulse (**a**). The frequency distribution of the Fourier transform of (**a**), (**b**). The real part of the Fourier transforms of square enveloped sine waves with 10 different phase mismatches (**c**). The absolute values of every frequency distribution function in (**c**), (**d**). The square of **d** showing the frequency distribution of photon energy (**e**)

a certain amount of energy is irradiated later becomes zero, this seems strange. One thing can be said is that the Fourier component has both +/− signs of amplitudes in the frequency domain, and at the zero-energy point, the plus and minus components of the amplitude of that frequency are summed up to become zero. The resulting irradiation energy distribution (photon number distribution) can thus be regarded as representing an amplitude excess in one direction to the other.

It should be noted that the longer the square-enveloped pulse irradiated, the narrower the irradiation frequency range becomes. In most NMR experiments, a short pulse with sufficiently strong power is ordinarily used for irradiating a full range of the frequency of which every type of nuclei of that element may be included. However, this is not always possible, being limited by the specification of the spectrometer and the probe, so compromised conditions must be selected.

Figure 12.6 shows a graph of the Fourier transform of a sine wave with a rectangular envelope given by a probe as the sum of two complex-valued functions, each of which is expressed in polar coordinates representing complex values with respect to frequency. It can be seen that the two complex-valued waves are centered at the same absolute frequency with +/− signs, respectively. In this case, one peak shows the amplitude of the opposite sign of the other. It can also be seen that the continuous phase changes with small amplitude changes around the irradiation center.

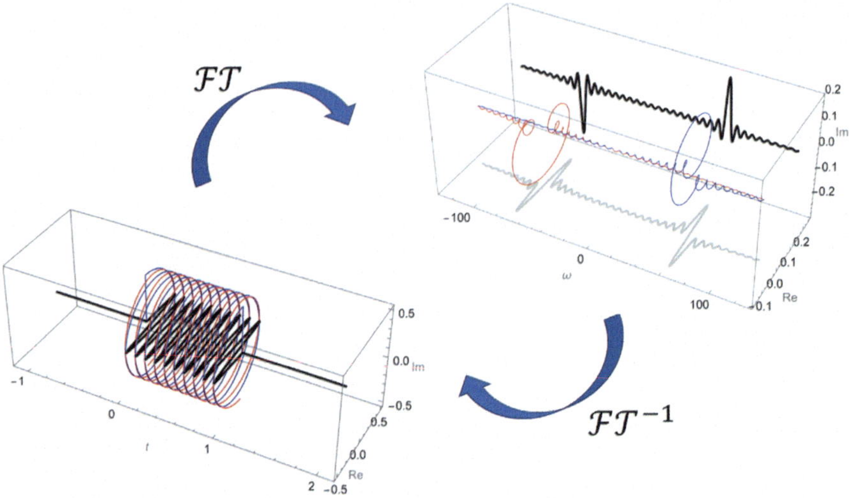

Fig. 12.6 Fourier transform of square-enveloped sine wave represented as a sum of two complex waves with opposite polarities

12.3 Orientation Dependence of Quadrupolar Interaction

In this part, we will follow the reference [24]. The orientation dependency of the quadrupolar interaction can be described below. For simplicity, it is ordinarily described on its principal axis system (PAS) of the quadrupolar interaction. The quadrupolar interaction Hamiltonian (H_Q) is written as

$$\hat{H}_Q = \frac{eQ}{6I(2I-1)\hbar}\left\{\sum_{\alpha,\beta=x,y,z} V_{\alpha\beta}\left[\frac{3}{2}(\hat{I}_\alpha\hat{I}_\beta + \hat{I}_\beta\hat{I}_\alpha) - \delta_{\alpha\beta}I(I+1)\right]\right\},$$

$$V_{\alpha\beta} = \left.\frac{\partial^2 U}{\partial\alpha\partial\beta}\right|_{r=0} \tag{12.5}$$

where $\delta_{\alpha\beta}$ is Kronecker delta. U is the electric potential formed by the charge distributions outside of the nucleus, $V_{\alpha\beta}$ are the Cartesian components of EFG at the origin. Because the matrix $\boldsymbol{V} = \{V_{\alpha\beta}\}$ should be real and symmetry from its definition, it will have three real eigenvalues and can be diagonalized by applying an orthogonal transformation T

$$\boldsymbol{V} = T\begin{bmatrix} V_{xx} & 0 & 0 \\ 0 & V_{yy} & 0 \\ 0 & 0 & V_{zz} \end{bmatrix}T^{-1} \tag{12.6}$$

can be obtained in the newly defined coordination system (x, y, z), determined also with the convention of $|V_{zz}| \geq |V_{yy}| \geq |V_{xx}|$. This new coordination system, of which the z-axis is set parallel orientation to the maximum electric field gradient, is called the principal axis system (PAS) of the quadrupolar interaction. For simplicity, we will discuss the quadrupolar interaction based on this PAS, exclusively here-after. The electrostatic force at the origin should be neutral because we are considering only the static charge distribution that should be balanced at the origin, so we obtain that the Laplacian of the field at the origin should be zero ($V_{xx} + V_{yy} + V_{zz} = 0$). Thus, the independent components of this matrix will be reduced to two, quadrupolar coupling constant, eq, the largest component of the gradient tensor, and the asymmetric parameter, η, defined by

$$eq = V_{zz}, \eta = \frac{V_{xx} - V_{yy}}{V_{zz}} \tag{12.7}$$

If the sign of V_{zz} is positive, then those of V_{xx} and V_{yy} both are negative, or the other case of fully inverse signs is possible. In any case, η varies from 0 to 1. If η is zero, then $V_{xx} = V_{yy} = (-1/2)V_{zz}$. If $\eta = 1$, then $V_{xx} = 0$ and $V_{yy} = -V_{zz}$. Consequently, V_{xx} varies between $-(1/2)V_{zz}$ and zero, and V_{yy} varies between $-V_{zz}$ and $-(1/2)V_{zz}$. If the charge distribution around the nuclei is in the axial symmetry, and if the polar orientations along with the symmetry axis are positively charged, then eq becomes positive and if negatively charged, eq becomes negative. If the charge distribution around the nuclei is having a perfect tetrahedral or octahedral symmetry, then $eq = 0$. Thus, the orientation of the PAS is closely related to the coordination environment of the nuclei. Only the two parameters, eq and η can be obtained in NMR experiments as the independent parameters of the electric field gradient.

The electronic field gradient can be re-written by the spherical tensor expressions on the PAS in order to improve the visibility of the expressions in the intermediate calculations. In those expressions,

$$V_0^{PAS} = \sqrt{\frac{3}{2}}eq, V_1^{PAS} = V_{-1}^{PAS} = 0, V_2^{PAS} = V_{-2}^{PAS} = \frac{1}{2}eq\eta \tag{12.8}$$

From now on, we will attempt to explain the anisotropy of quadrupolar interaction quantitatively to a certain level. As for the more generalized derivations of formula, we recommend the readers refer to the documents [24] or books [25, 26]. The author has given great numbers of help from these two books everywhere in this contribution.

When the quadrupolar interaction is relatively small compared to the Zeeman interaction, the quadrupolar interaction can be expressed correctly by using perturbation theory. Here we will express the Hamiltonian of the quadrupolar interaction by the first and the second order of perturbation to the Zeeman interaction as

$$\hat{\mathcal{H}}_Q^{[1]} = \frac{eQ}{4I(2I-1)\hbar}\frac{\sqrt{6}}{3}\left[3\hat{I}_z^2 - I(I+1)\right]V_0 \tag{12.9}$$

and

$$\hat{\mathcal{H}}_Q^{[2]} = -\frac{1}{\omega_0}\left[\frac{eQ}{4I(2I-1)\hbar}\right]^2 \times \left\{ \begin{array}{l} 2V_{-1}V_1\hat{I}_z\left[4I(I+1) - 8\hat{I}_z^2 - 1\right] \\ +2V_{-2}V_2\hat{I}_z\left[2I(I+1) - 2\hat{I}_z^2 - 1\right] \end{array} \right\} \quad (12.10)$$

The first-order quadrupolar interaction is independent to ω_0, whereas the second-order interaction is proportional to C_Q^2/ω_0. This is the reason why a strong magnetic field is suitable for the observation of quadrupolar nuclei.

The shift of the first-order quadrupolar interaction for the transition energy from the m to $m-1$ states, being characterized by their magnetic quantum number is

$$\begin{aligned} \omega_{m,\,m-1}^{(1)} &= \left\langle m-1\middle|\mathcal{H}_Q^{[1]}\middle|m-1\right\rangle - \left\langle m\middle|\mathcal{H}_Q^{[1]}\middle|m\right\rangle \\ &= \frac{3eQ}{4I(2I-1)}\frac{\sqrt{6}}{3}(1-2m)V_0 \end{aligned} \quad (12.11)$$

Thus, the resonance frequency will split into $2I+1$ lines separated by a constant

$$2\nu_Q^{PAS} = \frac{3e^2qQ}{2I(2I-1)h} = \frac{3C_Q}{2I(2I-1)},\ 2\omega_Q^{PAS} = \frac{3\pi C_Q}{2I(2I-1)}. \quad (12.12)$$

The parameter $2\nu_Q{}^{PAS}$, (or in angular velocity, $2\omega_Q^{PAS}$) often called quadrupolar splitting frequency and the C_Q, defined by $\chi = e^2qQ/\hbar$, or$C_Q = e^2qQ/h$ is called as quadrupolar coupling constant, which is also denoted as NQCC or QCC or C_q or C_{qcc} in the literature. Thus, the frequency of the central line is not shifted by the first-order interaction.

The second-order quadrupolar shift can be calculated as

$$\omega_{m,\,m-1}^{(2)} = \left\langle m-1\middle|\mathcal{H}_Q^{[2]}\middle|m-1\right\rangle - \left\langle m\middle|\mathcal{H}_Q^{[2]}\middle|m\right\rangle$$

$$= -\frac{2}{\omega_0}\left[\frac{eQ}{4I(2I-1)\hbar}\right]^2 \times \left\{ \begin{array}{l} V_{-1}V_1[24m(m-1) - 4I(I+1) + 9] \\ +V_{-2}V_2[6m(m-1) - 2I(I-1) + 3] \end{array} \right\} \quad (12.13)$$

In Eq. 12.11 and 12.13, the anisotropic part of the interactions is included in terms correlated with V_0, V_1, V_{-1}, V_2 and V_{-2}. These terms can be transferred to those in another coordination frame using Wigner rotation matrix elements [21] defined by the Euler angles α, β and γ.

$$V_i = \sum_{j=-2}^{2} \mathcal{D}_{j,i}^{(2)}(\alpha, \beta, \gamma)V_j^{PAS} \quad (12.14)$$

In a single crystal, the orientation-dependent quadrupolar splitting frequency can be written using the Euler angles as

$$\omega_Q = \frac{3\chi}{4I(2I-1)}\left[\frac{1}{2}(3\cos^2\beta - 1) + \frac{1}{2}\eta\sin^2\beta\cos 2\alpha\right] \tag{12.15}$$

where χ is another quadrupolar coupling constant used with angular velocity, $\chi = 2\pi C_Q$. The first-order quadrupolar shift under static conditions is

$$\omega_{m-1,m}^{(1)static} = (1-2m)\omega_Q \tag{12.16}$$

The shift of the center band will become zero, and the other lines are equally separated. The separation frequency is dependent on the orientation. For the second-order shifts, the orientation dependency is expressed as

$$\omega_{-1/2,1/2}^{(2)static} = -\frac{1}{6\omega_0}\left[\frac{3\chi}{2I(I-1)}\right]^2\left[I(I+1) - \frac{3}{4}\right]$$
$$\times\left[A(\alpha,\eta)\cos^4\beta + B(\alpha,\eta)\cos^2\beta + C(\alpha,\eta)\right] \tag{12.17}$$

with

$$A(\alpha,\eta) = -\frac{27}{8} + \frac{9}{4}\eta\cos 2\alpha - \frac{3}{8}(\eta\cos 2\alpha)^2$$

$$B(\alpha,\eta) = \frac{30}{8} - \frac{1}{2}\eta^2 - 2\eta\cos 2\alpha + \frac{3}{4}(\eta\cos 2\alpha)^2$$

$$C(\alpha,\eta) = -\frac{1}{8} + \frac{1}{3}\eta^2 - \frac{1}{4}\eta\cos 2\alpha - \frac{3}{8}(\eta\cos 2\alpha)^2 \tag{12.18}$$

Under MAS conditions, the first-orr quadrupolar interaction becomes zero. The second-order quadrupolar interaction in the center band is

$$\omega_{-1/2,1/2}^{(2)MAS} = -\frac{1}{6\omega_0}\left[\frac{3\chi}{2I(2I-1)}\right]^2\left[I(I+1) - \frac{3}{4}\right]$$
$$\times\left[D(\alpha,\eta)\cos^4\beta + E(\alpha,\eta)\cos^2\beta + F(\alpha,\eta)\right] \tag{12.19}$$

with

$$D(\alpha,\eta) = \frac{21}{16} - \frac{7}{8}\eta\cos 2\alpha + \frac{7}{48}(\eta\cos 2\alpha)^2$$

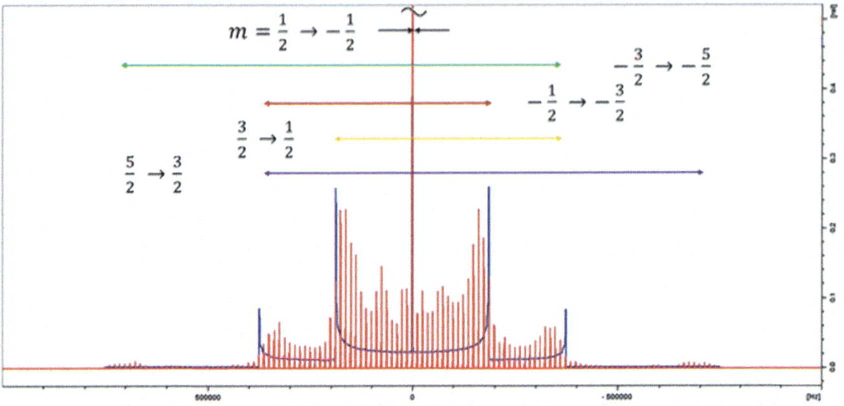

Fig. 12.7 The powder pattern NMR spectrum for a virtual ^{27}Al ($I = 5/2$) nucleus with $\nu_Q = 450$ kHz, $\eta_Q = 0$. A static spectrum of an assembly of crystallites randomly oriented (blue line) with a MAS spectrum of the same sample ($\nu_R = 12.5$ kHz) is superposed with the intensity calibrated to 1/10 (red line). The line broadening factor is 10 Hz in both spectra. These spectra are simulated by 'sola', the solid-state NMR line-shape analyzing application implemented in TOPSPIN4®

$$E(\alpha,\eta) = -\frac{9}{8} + \frac{1}{12}\eta^2 + \eta\cos 2\alpha - \frac{7}{24}(\eta\ \cos 2\alpha)^2$$

$$F(\alpha.\eta) = \frac{5}{16} - \frac{1}{8}\eta\cos 2\alpha + \frac{7}{48}(\eta\cos 2\alpha)^2 \tag{12.20}$$

Here, the orientation of the magnetic field vector is described on the basis of the Cartesian coordinates of the PAS of the quadrupolar interaction.

Figure 12.7 shows spectra of a virtual ^{27}Al under MAS and static conditions. We should note that each of the signals from the ST bands overlaps with the CT band region. This fact makes the nutation pattern of a ^{27}Al signal complex. The second-order interaction effect is still hidden within every peak in this chart.

Using MAS, the second-order effect on the line width is reduced to about 30% [27] that enables us to observe Al[O_4] and Al[O_6] in γ-Al_2O_3 as separate signals in 400 MHz magnets (Fig. 12.8). One important thing should be pointed out. The second-order quadrupolar interactions consist of not only orientation-dependent terms but also an orientation-independent part that will result in a field-dependent permanent shift relative to the δ_{iso} in the spectrum. For this reason, the peak position of NMR signals of quadrupolar nuclei are field-dependent. For a signal in which the environment of quadrupolar nuclei has a certain deviation when assuming a symmetrical Gaussian distribution to the C_Q, the apparent signal will be distributed unsymmetrically in the frequency domain as can be seen in Fig. 12.8 [28].

We want to notify another challenging feature of the NMR experiment of quadrupolar nuclei. There is a difficulty evoked by the signal mixing with different neighboring Zeeman states pairs (we call them as 'different bands' hereafter) mutually overlapping in the frequency domain. Because one band's ceiling is another

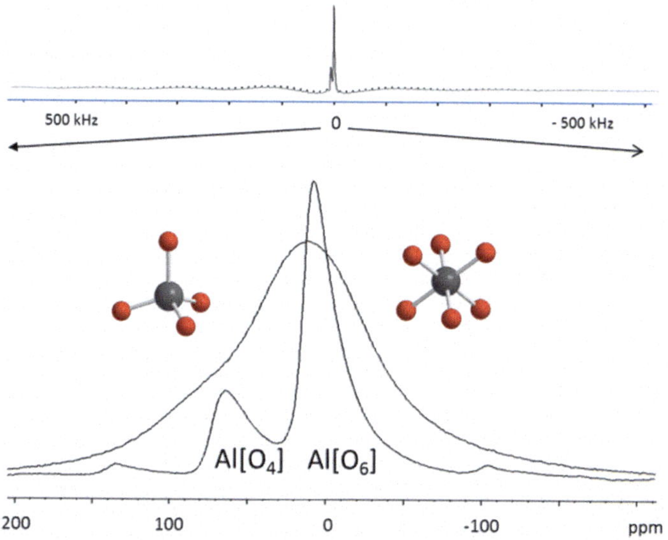

Fig. 12.8 A typical 1D spectrum of a spin 5/2 nuclei. MAS NMR spectrum of $\gamma - Al_2O_3$. The whole range spectrum (upper) with a center expansion (lower) of with (relatively sharp signal) or without (the other broader one) MAS

band's floor, when we wanted to irradiate a pulse to the CT band, it may result in the population transfer from m to $m - 1$ state, which will increase the population of the floor state of the upper neighboring band, and will decrease that of the ceiling state of the lower neighboring band. In addition, even for the static powder samples, a small part of the nuclei, which are observed under ST bands, will have the resonance frequency in the range of the CT band. In the case of MAS experiments, a major number of nuclei will experience the CT band range resonance frequency twice or four times within one rotation. Every electromagnetic wave is distinguished by its frequency but most part of them cannot be distinguished as those coming from a single excitation band. Consequently, all the neighboring band depicted in Fig. 12.9 are irradiated simultaneously leading to a rather complicated situation in that quadrupolar spin. The effect also results in the deformation of the CT band spectrum shape through the intra-spin coherence transfer based on the band energy crossing above noted.

These make the nutation patterns of quadrupolar nuclei complicated. As we have described above, inevitably overlapped ST bands into the CT band and non-linearity of the signal excitation induced by the ceiling-floor mechanism make it difficult to extract quantitative information from the spectrum. In general, the 90° and 180° pulses of these nuclei are determined with a solution sample of a standard material containing those nuclei. When the pulse power is expressed by the nutation frequency with kHz unit, we ordinarily use this nutation rate obtained with a solution sample because it generally shows a simply decaying sine-wave. However, for the solid samples, we cannot get the expected results using the 90° and 180° pulses obtained

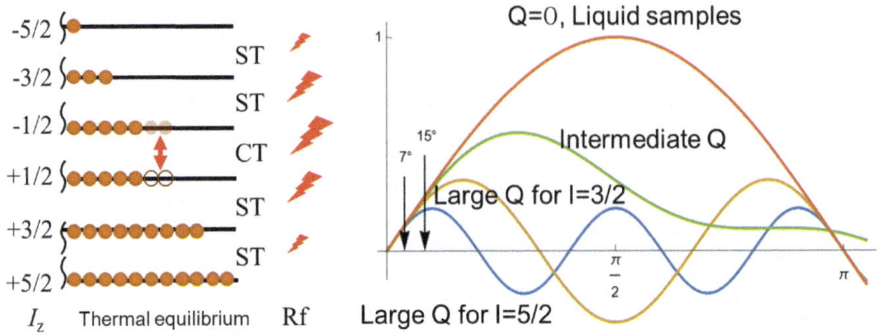

Fig. 12.9 Population determined by Boltzmann low for a $I = 5/2$ nucleus with nutation patterns. When we want to get a CT band signal, we should note that the ST bands are at the same time irradiated which will affect the population of CT bands. This reason and the possible inter-system crossing will make the nutation pattern of the CT band complicated. We can obtain rather better quantitative signal data using very small flip- angle such as 15° or 7°

through such a method. Because the appropriate pulse for excitation pulse differs from one sample to another, a simple method to get a pseudo-quantitative spectrum is to use very short excitation pulses (flip angle $< 15^\circ$ or $< 7^\circ$), although the contamination of ST band signals may not be removed completely [27].

When the second-order quadrupolar coupling is sufficiently small ($C_Q <$ 500kHz), the CT band spectrum can be observed isolated under MAS conditions. For larger C_Q cases, it will be associated with a few numbers of spinning side bands. When that coupling is much larger, even the CT band spectrum cannot be obtained as a well-separated peak on the spectrum. For the samples containing a few numbers of components, we can choose numerical simulation for the analysis and confirmation of the quadrupolar parameters of the components.

Figure 12.10 illustrates the orientation dependence of the quadrupolar interactions on the CT band spectrum with virtual quadrupolar nuclei having 7 MHz of the quadrupolar coupling constant, C_Q, δ_{iso} (70 ppm) and 5 different η's. The ν_0 is 104.6 MHz, which is related to an Al[O_4] site in a silica-alumina compounds with relatively a large C_Q. The correspondence of resonance frequency in the MAS spectrum patterns and the MAS axis orientation dependences relative to the PAS are illustrated by colored spheres.

In the cases which both the quadrupolar interaction and the CSA and the other non-vanishing anisotropic interactions are comparable, numerical simulation will be needed for the analysis of the spectrum. For single crystal samples, the relative orientation between the two PAS of the CS tensor and the EFG tensor is described by a set of Euler angles (α, β, γ), which can be obtained by fitting the simulation data to the experimental ones. However, for the amorphous or mixture solids, these parameters are rarely obtained separately.

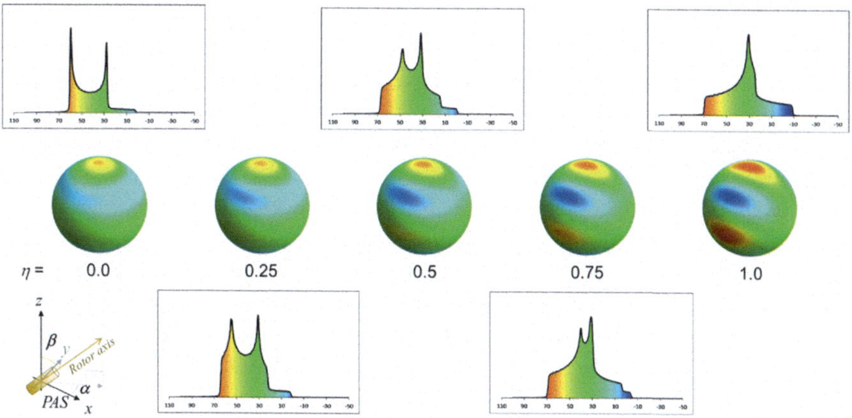

Fig. 12.10 Calculated MAS spectra of CT band region of a virtual spin 5/2 nuclei of ^{27}Al, ν_0 = 104.6 MHz with δ_{iso} = 60 ppm, C_Q = 7MHz having different η, with spherical maps of the frequency-orientation relationship of the MAS axis to the PAS[6]

12.4 Signal Separation Through Quadrupolar Coupling

Most of the spectrometry is being used for the separation of the signals obtained from the objective materials to get information about the amounts and properties of particular ingredients. However, a large second-order quadrupolar interaction of a nucleus often prevents us from observing signals under different environments separately. As a result, there have been long demands for the observation of half-integer quadrupolar nuclei as well-separated, relatively narrow signals. For this purpose, several numbers of methods have been developed that may cancel the second-order quadrupolar broadening in NMR. As systematical methods, double rotation (DOR) [29] and dynamic angle switching (DAS) [29, 30] were implemented both using specifically designed probes. Both of them contain their particular advantages, however, here we have no space to deal with them further. Currently, multi-quantum MAS (MQMAS) is most frequently used for obtaining the quadrupolar parameters and isotopic chemical shifts, and obtaining separated signals depending on these parameters, because this method uses no special probes other than normal MAS probe.

12.4.1 MQMAS

Frydman and his coworker have first proposed MQMAS as a 2D NMR method [31] that can refocus anisotropic quadrupolar interactions using additional F1 dimension (the second Fourier-transformed frequency dimension upon the parameter evolution time, t_1. We use the F2 dimension for the Fourier-transformed frequency dimension

[6] The coloring scheme of the spheres is corrected from the previous edition of this book.

with the observation time, t_2 in accordance with the commonly used convention.) in a 2D map. The word 'multi-quantum' is used first time by a Japanese researcher[7] [32]. Single quantum (SQ) coherence may be a single photon entangled state with a quadrupolar nucleus on its $m = 1/2$ and -1/2 Zeeman states. This coherence can be exclusively used for the observation of the CT band spectra. Similarly, MQ-coherence means that an entangled state between $+m$ and $-m(|m| \leq I)$ states of a quadrupolar spin with an MQ-energized photon that may have been made by $2m$ photons of the incidental electromagnetic wave. Depending on the phase evolution frequency, we call the 'coherence order' of this entangled state, p ($|p| = 2m$, in the present case.). This method is based on the phenomenon that the anisotropy of quadrupolar interaction of MQ coherence is proportional to that of SQ coherence in the frequency domain with a single universal ratio $R(S, p)$, described below

$$R(S, p) = \frac{p\left[36S(S+1) - 17p^2 - 10\right]}{36S(S+1) - 27}, \tag{12.21}$$

which is depending only on the spin quantum number S and the coherence order p. Another important point is that $R(S, p)$ can be selected negative for every combination of S and $|p|$. So, for a quadrupolar nucleus, when it experiences both the SQ and MQ coherences, in the time domain with each the coherence duration ratio of 1 : R, all the nuclear spins under the identical orientation-dependent quadrupolar interactions, however under different orientations to the field, will evolve to refocus at the same time and form an echo signal. This echo is called "isotropic quadrupolar echo". Based on these phenomena, controlling the evolution time of MQ coherence by changing the time between the excitation and conversion pulses, we can observe isotropic quadrupolar echo at a time depending on the C_Q, η and δ_{iso}. The resulted 2D data will be containing the information of $|P_Q|\left(P_Q^2 = C_Q^2\left(1 + \eta^2/3\right)\right)$, called as "quadrupolar parameter" or $|\delta_{QIS}|$, and δ_{iso}, isotropic chemical shift for each of the separated signals.

Number of introductions and reviews of MQMAS have been published. Here we describe the most frequently used z-filtered MQMAS [33, 35] in further detail. This most frequently used method uses symmetrical coherence pathways between $p = 3 \rightarrow 0$ and $-3 \rightarrow 0$ for the 3Q MQMAS. This method is suitable for canceling the phase dispersive component in the finally obtained 2D map.

Z-filtered MQMAS uses three pulses, two of which are at around the maximum strength of that probe; the third one is normally set to be a weak or selective pulse (Fig. 12.11). Under MAS conditions, a strong pulse around 180° is first irradiated for the excitation of multi-quantum coherence. The second strong pulse around 90° is used to convert the MQ coherence into SQ coherence selecting that coherence through applying phase cycling. During the next waiting time, normally around 20 μs, the magnetization to be observed will be settled in the z-direction for avoiding the appearance of dispersive components of the magnetization. The last, a weak or

[7] The name, Dr. Hatanaka is mentioned in [4].

selective pulse will convert the magnetization to the observable SQ coherence and the resulting signal will be observed as echoes with the coherence order $p = -1$. Fourier transformed spectrum will show several numbers of slanted ridge-like peaks in the 2D map because each peak is made by echo signals with those the starting time-shifted sequentially (Fig. 12.10). These can be made parallel to the F2 axis by applying shearing to the 2D spectra. The quadrupolar parameter and the isotropic chemical shifts can be obtained from this finally obtained 2D map using the F1 and F2 values of the center of gravity of each peak by a simple linear combination of these values. For the case of 3Q MQMAS (using coherence between $m = \pm 3/2$), the equations are

$$\delta_{CS} = \frac{10}{27}\delta_2^{CG} + \frac{17}{27}\delta_{ISO}^{CG} \tag{12.22}$$

$$\delta_{QIS} = \frac{17}{27}\left(\delta_2^{CG} - \delta_{ISO}^{CG}\right) \tag{12.23}$$

where δ_2^{CG} and δ_{ISO}^{CG} are the center of gravity of the 2D peak obtained by F2 and F1 values in the sheared 2D map, respectively. From these parameters, the quadrupolar parameter P_Q can also be calculated by the following equations:

$$P_Q^2 = \left(\delta_{ISO}^{CG} - \delta_2^{CG}\right)^2 \cdot F(S) \cdot \nu_0^2 \cdot 10^{-6} \tag{12.24}$$

$$\mathrm{F}(S) = \frac{680\left[2S(2S-1)^2\right]}{81[4S(S+1)-3]} \tag{12.25}$$

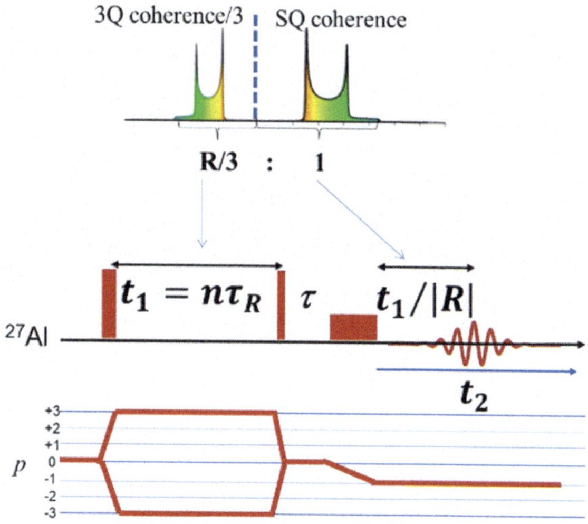

Fig. 12.11 Z-filtered 3Q-MQAS pulse sequence with the time progression of the coherence order and spectrum shapes of 3Q and 1Q coherences are depicted. In the upper part, spectra of virtual 3Q and 1Q signals in the frequency domain are illustrated to easily see that the time revolution of these signals' coherences are kind of mirror image with the negative frequency shrunk by a constant

Here we will show an example of ^{27}Al z-filtered 3Q MQMAS chart. Zeolites and related materials are essential in catalysis and adsorbate materials. MQMAS is frequently used in the research of these materials sciences. Figure 12.12 shows ^{27}Al MQMAS of ALPO-15 taken by using a 4mm MAS rotor spinning at 10 kHz within a MAS probe set in a 400 MHz magnet. You can see five peaks separately observed in the 2D map. In most of silica-alumina or alumino-silicate samples, 2D peaks of ^{27}Al signals of different coordination numbers can be observed separately. In some cases, we can recognize two or three peaks are included within the Al[O_4], Al[O_5] or Al[O_6] sites. From each separated peak, with some help of spectral simulation software, we can obtain a set of quadrupolar parameters, C_Q and η, and an isotopic chemical shift, δ_{iso} for each of the peaks.

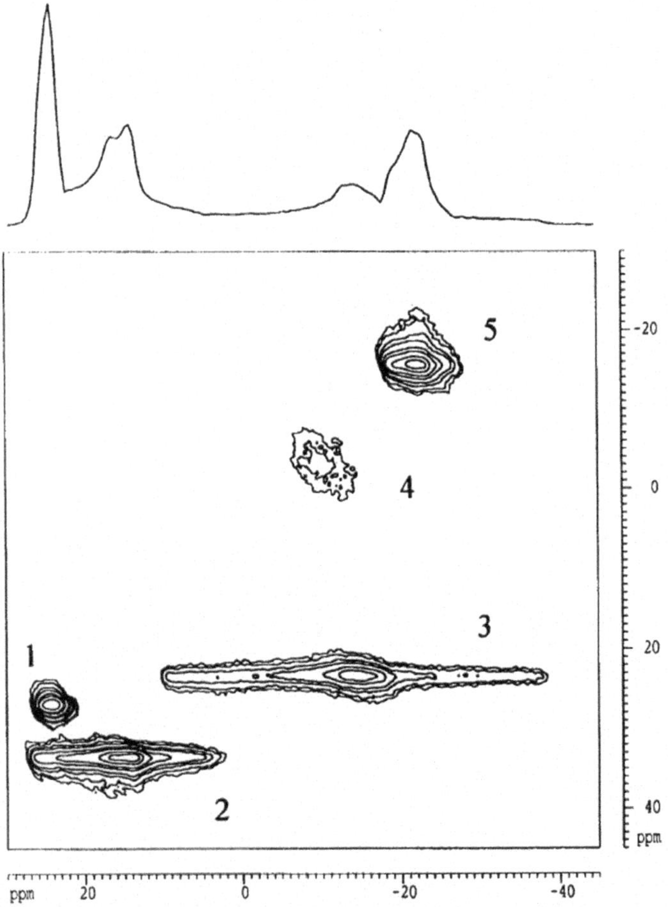

Fig. 12.12 A ^{27}Al z-filtered 3Q MQMAS spectrum of AlPO-14. Two Al[O_4] (1,2), one Al[O_5] (3), and two Al[O_6] (4,5) sites are separately observed in this chart [33] (b). Copyright from Elsevier 4,177,351,452,113

There have been proposed a number of variations based on the same principle [34–38]. Each method has its own advantages and drawbacks. Among them, four pulse sequences, split-t1 type of MQMAS is useful for obtaining data that doesn't need shearing. SPAM may be useful in some cases for its ability to sensitivity enhancement. However in both of the cases, the peak shape might be skewed a little. Note that in the former case, the parameter extraction should be performed in a different way [36].

12.4.2 STMAS

STMAS is a 2D NMR method that provides a spectrum much looks like those spectra obtained by MQMAS, if we look at the result. However, this method is using different coherence pathways, compared to MQMAS, for the echo to be refocused [39]. STMAS uses spinning side bands of satellite transition (ST) bands, gathering them into a single peak in the F1 direction of a 2D map, with an increment of t1 interval synchronized to the rotation period of the MAS rotor. STMAS uses coherent transfer echo that will form after the single pulse excitation of the ST bands and the following conversion pulse applied to the CT band. The initially excited nucleus will experience the two coherent states during the time domain and will make the coherence transfer echoes at a time. The advantage of STMAS compared to MQMAS is mainly in its higher sensitivity [40]. Since MQMAS adopts phase cycling that cancels the main part of the transiently obtained signals, resulting in a relatively lower overall sensitivity. As we have seen in the previous part of this section, the orientation dependence of the second-order quadrupolar interactions with the ST bands is proportional to that of the CT. Contrary to MQMAS, where the orientation dependence of the first order quadrupolar coupling, $Q^{(1)}$ is at any orientation parallel between m and $-m$ states, that of STMAS only vanishes after averaging through the rotation. Sometimes this difference results in a critical defect of the STMAS compared to the MQMAS data of the same sample. This fact shows that the coherence does not connect to each of the states, which are participating in a resonance phenomenon, but to the energy difference of the two Zeeman states. A number of authors are describing the advantage of STMAS to MQMAS on this point [41–43], although STMAS demands severer experimental settings in MAS rotation speed ($\pm$ 2 Hz) and MAS angle (<0.003°) than those demanded by MQMAS. These are needed for preventing loss of refocusing the ST band signals that are necessary for obtaining the coherent transfer echo. In many cases, samples showing fewer ST band signals and their number are not suitable for the STMAS. Since the line shape especially in the F1 axis on the 2D chart will be broadening in those samples. A possible reason that would cause F1 broadening is molecular motion, which is pointed out by an author [44].

12.5 Signal Separation Through Correlations to Other Nuclei

NMR provides us with extremely precise information about the nucleus observed in a particular frequency. What is the most unique property compared to the other spectroscopic method is the ability to correlate information between homo- or hetero nuclei. In this section, we will describe the solid-state NMR using CPMAS- and HMQC-type coherence transfer methods.

Cross-polarization (CP) is a solid-state NMR method using magnetization transfer between two different nuclei under close proximities. When a pair of two $I = 1/2$ spins, identified by s (source) and r (receiver), are simultaneously irradiated and spin-locked at the same nutation rate, $\nu_{1s} = \nu_{1r}$, the relatively less abundant receiver spin with low Larmor frequency will be magnetized by polarization transfer from the other abundant spin with high Larmor frequency, source spin. The condition, $\nu_{1s} = \nu_{1r,}$ is called the Hartman-Hahn matching condition, which allows magnetization transfer between the two nuclei through dipole–dipole interaction between them, and causes a flip-flop between the two spins. CPMAS (meaning CP under magic angle spinning of the sample) is one of the most frequently used solid-state NMR techniques used in the observation of relatively rare nuclei with low Larmor frequency, the magnetization receiver spin S (typically ^{13}C, ^{15}N, etc.,) by using magnetization transfer from abundant nuclei with high Larmor frequency, the magnetization source spin I, such as ^{1}H or ^{19}F. Under the MAS, the Hartman-Hahn matching condition should be modified to

$$\nu_{1S} = \nu_{1I} \pm n\nu_R \tag{12.26}$$

where ν_{1I} and ν_{1S} is the nutation rate of the initially magnetized source spin, and that of the target, receiver nuclear spin. ν_{R} is the rotating rate of the sample and n is 0, 1, or 2. Applying sub-milliseconds to milliseconds of spin-locking pulse to both the ^{1}H channel and X channel simultaneously, magnetization transfer takes place during the pulse through dipole–dipole interaction between the two nuclei being under mutual spatial proximity. The heteronuclear dipolar interaction between spin I and S can be written in the following equation (see also Eq. 12.3)

$$\omega_{IS} = \hbar\gamma_I\gamma_S\frac{1 - 3\cos^2\theta_{IS}}{r_{IS}^3} \tag{12.27}$$

where $\gamma_{I,}\gamma_S$ are the gyromagnetic ratio of spin I and S, and r_{IS}, θ_{IS} are the distance and the relative angle between the two spins, respectively. The magnetization transfer will take place through the flip-flop interaction derived from this interaction.

The magnetization transfer in CPMAS is often treated in a simple thermodynamic transport model that consists of a magnetization buildup time (T_{IS}) of the destination nuclei and the decay of the source, and sometimes, the decay of destination magnetization life time: $T_{1\mathrm{p}}$ (longitudinal relaxation time within a rotating frame).

In some literature, this phenomenological treatment is used for the estimation of a number of nuclei that have long relaxation time, which makes the quantitative estimation through the spectrum difficult. This method will provide good estimation when there are no other nuclei that will not receive any magnetization through CP. Theoretically, the magnetization transfer within a rotating sample including CPMAS is often explained using a tilted rotating frame [45], and in some cases, with Floquet theory [46], which uses large matrix representation containing a number of rotary-assisted states. In any case, CPMAS is thought to be a rather complicated process. The explanation of this phenomenon needs precise estimation of RDC during the ν_1 field irradiated. In some ideal cases, where a dipolar interacting spin-pair is well isolated by the other spins, we can obtain a dipolar coupling constant by performing a Fourier transformation of the build-up curve as the time domain data. The dipolar coupling constant can be obtained from the peak distance that appeared on the resulting pattern with a shape like a Pake doublet [47].

The CPMAS including quadrupolar nuclei can be used as sensitivity enhancement in very limited cases (Fig. 12.13). This is mainly because the spin-lock duration within quadrupolar nuclei is much shorter than that of spin 1/2 nuclei [48]. The sensitivity enhancement through CP can be observed only in the cases of nuclei with sufficiently small quadrupolar couplings. In more general quadrupolar nuclei, a number of states having large anisotropic quadrupolar interaction exist and have intersystem crossings with the central transition band. The spin-lock state, which can be prepared on the CT band coherence between $m = 1/2$ and $m = -1/2$ states, will be easily collapsed by the other interstate transitions within the same nucleus.

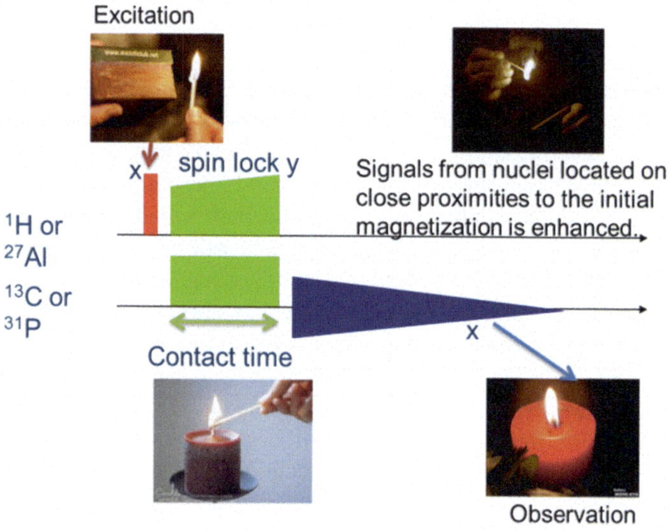

Fig. 12.13 Pulse sequence of CPMAS with schematic pictures of magnetization transfer from the initial magnetization to the destination magnetization

The matching condition of CPMAS with at least one quadrupolar nucleus included is apparently much different from those between spin 1/2 nuclei, because, the nutation of quadrupolar nuclei is much dependent on the C_Q and its orientation. If the C_Q is very small, close to zero, then the nutation rate is the same as that of the liquid state. If the C_Q is sufficiently large, the nutation rate becomes $2I + 1$ times faster than that of the liquid sample. An intermediate quadrupolar nucleus designates an intermediate effective nutation rate. Within this limit, the effective nutation rate of a quadrupolar nucleus varies. Under this situation, the matching condition should be changed to

$$\nu_{1I} = -\nu_{1S}^{eff} + n\nu_R \geq -(2I+1)\nu_{1S} + n\nu_R, \quad n = 1(or\ 2) \tag{12.28}$$

The only matching condition that can be used for the wide range of samples is to set the effective nutation rate of the quadrupolar nuclei, ν_{1S}^{eff}, to $\nu_R - \nu_{1I}$, although in some cases, we can get a CPMAS signal in the condition of $n = 2$, $\nu_{1S}^{eff} = 2\nu_R - \nu_{1I}$. ν_{1S}^{eff} cannot be simply determined in precision by the spin-lock or nutation experiment. In most cases, we should search a range of frequencies to obtain a suitable nutation rate. For the case of magnetization transfer between two quadrupolar nuclei, both the nutation rates become ν_1^{eff}.

The spinlock of quadrupolar nuclei for a duration needed for the magnetization transfer is rather a problem. Alexander Vega introduced an adiabaticity parameter, α, defined by

$$\alpha = \frac{\nu_1^2}{\nu_Q \nu_R} \tag{12.29}$$

which can express the spin-lock efficiencies to the quadrupolar nuclei [49]. When this parameter is much larger than 1, then the spin-lock power is sufficiently strong for effectively delivering spin-locking power into a whole range of the signal width. Contrary, when this parameter is much smaller than 1, then only a limited range of the nucleus within the signal can be spin-locked. The intermediate range, $\alpha \sim 1$, should not be used because, in these regions, the RF field will interfere with quadrupolar interaction and sample rotation to lead to a complex magnetization movement. The shape analysis of the static CP signal obtained by applying this logic to expand the parameter α to be orientation dependent, the spectral patterns of a quadrupolar nucleus taken under different conditions have been well simulated [50].

Even when the sensitivity is much lower compared to the direct excitation of the destination spins, the obtained signals will bear the information that the source spin handed the magnetization to the destination spin via through-space dipolar interactions. Sometimes, the fact, that the magnetization has transferred through-space, may bring us critical information about the relationship between the two nuclei. Or, in other cases, we will be able to observe nuclei more clearly separated through these interactions. 2D spectra using these conditions may bring us the correlation map between the two nuclei if the spectrum is successfully taken. The hetero-nuclear correlation (HETCOR) spectra combined with CPMAS enables us to observe the through-space correlation between the two hetero-nuclei visually in a 2D spectrum.

HMQC (hetero nuclear multiple quantum correlations) is another method to obtain heteronuclear correlations in a 2D spectrum [51]. This method was originally used in liquid-state NMR that extracts heteronuclear interactions through *J*-coupling between them. When we apply a recoupling pulse during the magnetization transfer, this method can be used as the *D*-HMQC method which is very useful for quadrupolar nuclei in general. In solid-state NMR, the coherence time duration (T_2) is not in every case long enough for the magnetization transfer evolution via *J*-coupling. However, high-speed rotation and high magnetic fields together lengthen the available coherence time and meanwhile narrow the matching condition for the magnetization transfer via CP. Both these two conditions will increase the relative importance of HMQC compared to CP-HETCOR used in obtaining hetero-correlation spectra. When the value of *J*-coupling constant is roughly estimated, the only parameter to be adjusted for HMQC experiments is the inversion pulse for the observation nuclei, which makes preparative experiments needed for HMQC easier than those for CP although the signal intensity of HMQC is comparably small or often smaller compared to that of CP, because HMQC actually extracts the dephased part of the signal using subtraction of the signals with and without magnetization transfer using alternation of the pulses on/off's (Fig. 12.14). SEDOR [52], TRAPDOR [53], and REAPDOR [54] are all related methods but with differences in the magnetization transfer methods.

D-R-INEPT (dipolar refocused INEPT) is also developed for taking the correlation between ^{1}H and low-γ quadrupolar nuclei. Nagashima and his collaborators have been developing observation methods of quadrupolar nuclei with DNP and other signal enhancement technologies like as sensitivity-improved new versions of PRESTO or QCPMG [55]. Although they did not get chemical shift correlation spectra between ^{1}H and X directly by this method, they have successfully observed a natural abundance ^{17}O nuclei with or without direct connection to ^{1}H, discriminating them by changing the pulse duration of the R-INEPT sequence. In that document, $^{47,49}Ti$, ^{67}Zn, and ^{95}Mo are also reported with some help of signal enhancements.

We have no space to describe non-half-integer quadrupolar nuclei in this article. Some useful information about quadrupolar nuclei with $I = 1$, ^{2}H, ^{14}N will be described in Chap. 6 in Volume I and Chap. 3 in Volume II. 2D HETCOR experiments

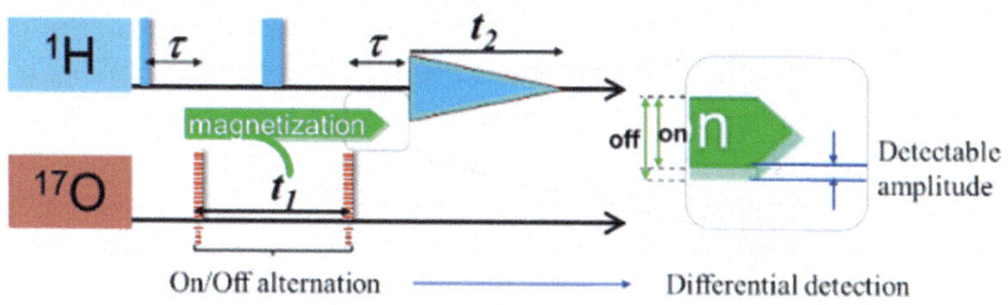

Fig. 12.14 HMQC uses magnetization transfer between ^{1}H and quadrupolar nuclei detecting subtraction of spectra with on/off of two pulses, which promote magnetization transfer

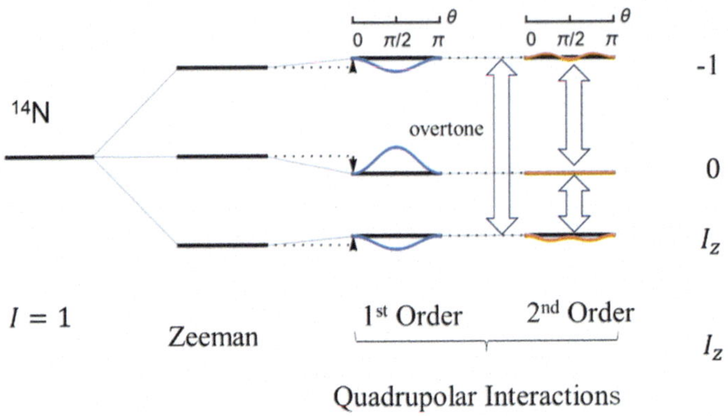

Fig. 12.15 Energy diagram of a ^{14}N ($I = 1$) nucleus, with polar angle (θ) dependence of anisotropies of quadrupolar interactions for each of the Zeeman states (η=0) are illustrated with expanded amplitudes with expansion ratios around 1000, 10000 times, for the 1st and 2nd interactions, respectively.

using *D*-HMQC type correlations are effectively used in ^{1}H, ^{14}N hetero-correlation and ^{14}N-^{14}N homo-correlation spectra [56]. These types of experiments require high-speed rotation and strictly matched magic-angle conditions. For integer spin number nuclei, there are no methods to eliminate the second-order quadrupolar interactions as like MQMAS. For the quadrupolar nuclei of integer spin numbers, unfortunately, there is no method to cancel the second-order quadrupolar interactions (Fig. 12.15). However, although the sensitivity is not every time sufficient, the application of overtone for these nuclei is effective for making clearer correlation peaks on ^{2}H or ^{14}N frequency axes in the 2D map [57].

A number of researchers are combining MQMAS or STMAS with CPMAS, HMQC, or SQ-DQ type correlation spectra for obtaining heteronuclear correlation information with much separated peaks [58, 59]. 1QCP-MQMAS, MQCP-MQMAS are one kind of these methods, which prepare firstly CP magnetization bearing 1Q or MQ coherence selected by phase cycling [60]. As for the 1QCP-MQMAS method, the initial magnetization made on ^{1}H's is first moved to quadrupolar nuclei, and then the resulting magnetization immediately flips back to the z-axis, after a short wait, a normal MQMAS sequence is applied to the quadrupolar nuclei, with ^{1}H decoupling during the MQ and SQ magnetization evolution and the acquisition time [61]. The other method is to obtain the correlation between the isotropically refocused quadrupolar signals with the chemical shift-solved ^{1}H signals. The resulting magnetization is utilized as the initial 3Q magnetization then the second and the third pulses are irradiated consecutively to obtain the MQMAS spectrum [62].

^{27}Al-^{1}H high-resolution, high-sensitivity correlation spectrum has been obtained through STMAS-CP using *w*-PMLG acquisition [63]. Because the T_1 of quadrupolar nuclei is much shorter than that of spin 1/2 nucleus, it is advantageous to choose quadrupolar nuclei as the source magnetization, *w*-PMLG, or *w*-DUMBO acquisition is necessary for obtaining ^{1}H magnetization well separated. Unless such sequence, the resulted ^1H magnetization will be mixed up during the observation period and will return to be observed as broad signals. We will also note that LG-CP should be considered to be used for the ^1H signal separation [64].

12.6 Sensitivity Enhancement Through Population Transfer

For quadrupolar nuclei specifically, methods using the population transfer within a single nucleus can be used for the sensitivity enhancement of the MAS NMR. In this part, we will introduce RAPT (rotation-assisted polarization transfer), FAM (fast amplitude modulation), DFS (double frequency sweep), HS (hyperbolic secant), and WURST (wide-band uniform-rate smooth truncation (WURST). Each of these methods listed here uses a different pulse sequence or a differently shaped pulse for sensitivity enhancement. FAM and DFS can be also used in MQMAS for 3Q-1Q conversion pulses. For the conservation of the $t = 0$ point of the second coherence, WURST has an advantage because it can be made as for having frequency up or down feature with the same rate and a same span that ensure the linear phase calibulation to several numbers of signals in the certain range of frequency domain.

Under MAS conditions, the ST band components of NMR signals will be observed as many spikelet peaks distributed in a wide range within that ST band (usually, MHz's or larger) (see Fig. 12.8). Although these spikelet signals are narrower than that of the CT signal, these signals are positively used in very limited cases, because these spikelet peaks are too small to gain a sufficient *S/N* ratio. The ST band coherence component of magnetization developed in a nucleus will experience one or two cycles of frequency oscillation, which sweeps a certain frequency range, during one MAS cycle. This swept frequency amplitude depends on the relative orientation of the PAS and the MAS axis. The spikelet peaks are emerging from the ST coherence in which the frequency sweep range is sufficiently small that can be averaged by MAS, contrary, some ST coherences experiencing larger frequency change during a MAS rotation period will be dephasing rapidly. The sensitivity enhancement methods listed above are all stimulate the spins under the ST region and sweep them out from ST states (states with $|m| > 1/2$) into the CT state ($|m| = 1/2$) in a particular degree depending on the method and the condition of the spins. Applying these pulses immediately before the first stimulation pulse (90° pulse), we can obtain much larger signals in the CT band in many types of experiments except for the methods using the ST band states, for example, MQMAS or STMAS.

RAPT and FAM are the methods proposed independently at a relatively early time [65, 66] and still developing in each. Both of them use many numbers of short pulse trains with middle strength changing the frequency or amplitude alternatingly, respectively, for the population transfer from the ST to CT states in a quadrupolar nucleus. The pulse trains will provide a frequency distribution in a wide range of frequencies that spread to a large part of the ST bands. The efficiency of population transfer depends on the applied frequency shifts of the pulse train among which, the population transfer will be maximized at the $\pm\ 2\nu_Q$. Because it will vanish at the edge frequency of the ST bands, this observation can be utilized for the estimation of C_Q [67].

DFS, WURST, and HS are the methods using frequency sweeping pulses (more frequently called adiabatic pulses). DFS [68] is based on the amplitude modulation by a frequency-swept sine curve, resulting in the pulse consisting of up-and down-frequency sweeping components of relatively wide ranges in the frequency domain excluding the certain width of the center region so as not to interfere the CT band signals. WURST [69] and HS [70] both use similar phase modulation resulting in one-direction (either up or down) frequency sweeping. Both the two shaped pulses, WURST and HS, have a common phase modulation pattern, a parabola curve rolled up within one phase period resulting an increasing or decreasing frequency change at a constant rate for the pulse that enables a uniform excitation or inversion of spins within a relatively wide range of frequencies. Each shaped pulse is truncated by a different envelope shape. The truncation of the shape is important for cutting off the unexpected frequency extrema. The WURST has an envelope shape closer to rectangular, which can provide a wider range of frequency sweep width by using relatively a weaker pulse than that of HS, and HS may have sharper limits in the frequency domain. The usage of these pulses is different from that of DFS. When these pulses are applied to the population transfer of quadrupolar nuclei, the frequency sweeping range is limited to a frequency span corresponding to one rotor cycle within an ST pulse region located at 150–250 kHz distant frequency higher or lower position from the CT band. Applying 180° pulse to this limited region, during 1–3 rotation cycles, the highest population transfer is achieved so far. Most part of quadrupolar nuclei in ST states will pass through this frequency region, around ν_Q distant from the CT band, 2 or 4 times during one rotor cycle (Fig. 12.16). That will make the CT band population increase.

Recently, a new method that may elucidate the heteronuclear correlation between spin 1/2 and a half-integer quadrupolar spin, PT-*J*-HMQC and PT-*D*-HMQC has been reported [71]. PT means population transfer that is performed by WURST pulses. This method is based on HMQC, additional continuous irradiations, or WRUST pulses to the quadrupolar nuclei during the spin evolution time (Fig. 12.17). Those WURST pulses are applied synchronously to the rotation cycle. During this pulse period, the 3Q-coherence is continuously evolved on the quardrupolar nuclei to realize rapid growth of the correlation peaks through *J*-coupling with the spin 1/2 nuclei (PT-*J*-HMQC). When applying recoupling pulses to the spin 1/2 nuclei, the spin evolution through dipole–dipole interaction can be achieved (PT-*D*-HMQC). In this case, the build-up speed of the destination magnetization is accelerated only a little [72].

WURST can deliver RF power uniformly to a relatively wide range of frequencies. This characteristic is used for the excitation and inversion of signals spread in a relatively wide frequency range. This pulse is recently used in combination with CPMG. For the quadrupolar peak up to 200 kHz of peak width, when the frequency sweeping pulses are applied in both the excitation and inversion pulses, every signal

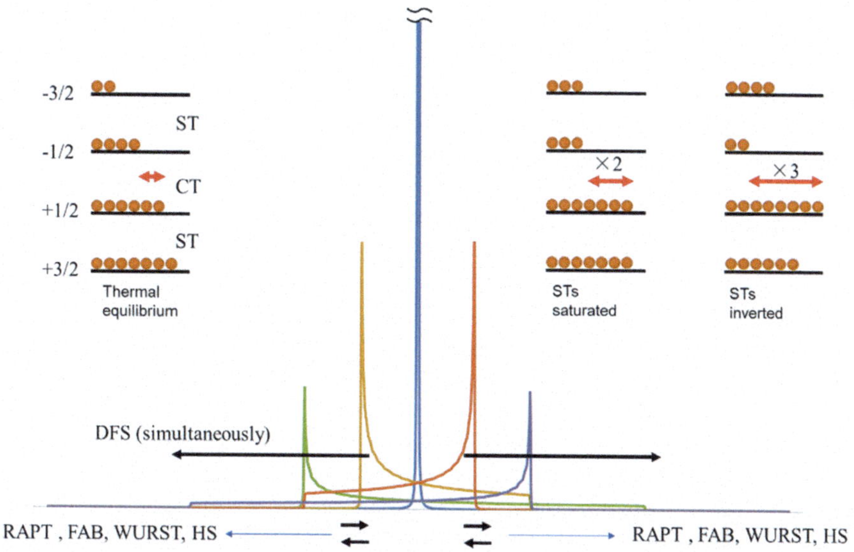

Fig. 12.16 The suitable frequency sweeping range of the NMR spectrum of a $I = 5/2$ nuclei for CT band population enhancement. This frequency range is slowly swept by irradiating weak WURST or HS pulses continuing for one to three rotor cycles. A major part of ST band coherent states is influenced by WURST or HS pulses because most of the coherent states will experience these frequency ranges during MAS

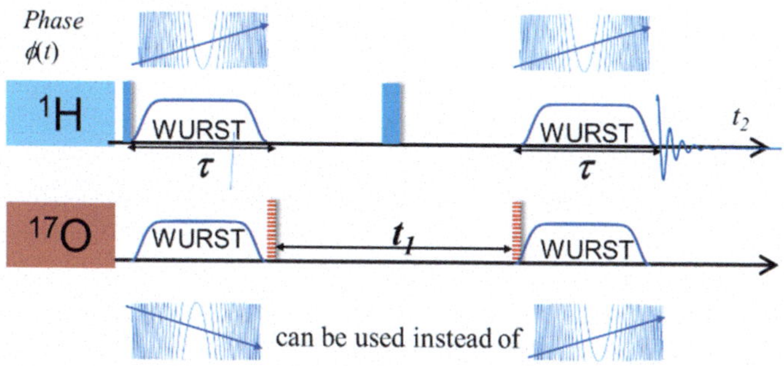

Fig. 12.17 A ^{1}H-^{17}O PT-HMQC sequence for obtaining ^{1}H-^{17}O correlation via polarization transfer. Enhanced correlation spectra can be obtained using the frequency sweeping pulse. Both the upward and downward frequency sweeping WURST pulse phase changing can be used

will be focused correctly [73], then also used for signal acquisition of nuclei having vast C_Q's such as ^{71}Ga, ^{91}Zr, and ^{59}Co with a help of VOCS type experiments [74].

12.7 Confronting the Real World, Real Materials

In solid-state NMR experiments for an objective sample, we should consider many parameters used in that experiment simultaneously. First of all, the rotor size selection is very important. If the larger rotor we use, we can obtain the higher sensitivity [58]. But, when we use the larger rotor, meanwhile we need a larger amount of that sample, and we should give up applying a strong RF field to the sample, and also should we give up rotating the sample under higher rotation speeds. The ν_I homogeneity will also be hindered in the larger rotors. Generally, high-speed rotation and high-power decoupling pulse result in good, however, the temperature rise of the sample will become high which will threaten the sample to potential thermal decompositions. Without saying, the probe specs, especially the endurable voltages and duty cycle limits are more important. Acquisition time and the frequency resolution, sampling time, and the frequency width of the spectrum are in mutually reciprocal relationships, respectively. Especially for the 2D experiments, a larger acquisition time in the t1 (time in the parameter dimension) axis will take much experimental time, hence we often minimize the number of t1 points. To achieve this, we should select a short acquisition time in the t1 domain or a long sampling interval along with t1. Those conditions, when all the acquired data were finally Fourier transformed, will result in the lower resolution or narrower observation window in the F1 domain, respectively. Both of them prevent us from adjusting the phase properly in the F1 domain. Especially if the line width along with the F1 domain is relatively large compared to the spectral width, we should observe in a somewhat wider range than the frequency range in which all the signals are contained for the determination of the phase, especially the first-order phase/frequency coefficient, P1 (comparative to P0) properly. For this purpose, we should take excess numbers of t1 points. All of these trade-off conditions mentioned above are not limited to the experiments for the quadrupolar nuclei, but are common to all the MAS NMR experiments. Thus, when we want to take the NMR of a sample with a sufficiently high sensitivity, many numbers of trade-off factors should be considered in solid-state NMR experiments.

During our daily work using solid-state NMR, we may confront many types of problems to be overcome. First of all, we should pack the sample into the rotor properly. A very small unbalance of the weight distribution in the rotor will collapse the stable rotation of the rotor. The smooth rotation necessary for the observation of MAS NMR signals should be achieved prior to the signal observation. For some types of samples, this will be an extremely difficult task. Consequently, the sample packaging to the rotor is the first important task for obtaining MAS NMR spectra. The dense packing will be better for obtaining a good *S/N* ratio generally, however, in some cases, more sparse packing will be far easier than that of dense packing for obtaining a smooth, high-speed rotation of that rotor. In the latter case, the sample distribution within the rotor will be achieved during the rotation. For taking solid-state NMR, finely grained powder samples will be the most suitable for obtaining good spectra. However, this is not every time possible.

Generally, flowable samples should not be rotated within the rotor, because the convection flow within the rotor induced by the centrifugal force may push off the rotor cap during rotation. There are many ideas and methods, special rotors, and apparatus developed for gels, liquid crystals, liquid-containing solids or solid-containing liquids. A convenient way to take MAS NMR signals of these materials is available by using some tight packing rotor insert that can contain the liquid-like samples within a rotor in the MAS experiment. Though the rotating speed will be limited, if we succeed in sampling, the rotor will start a stable spinning and MAS NMR signals can be taken using that rotor.

In the cases of samples emitting small amounts of solvent vapor or other gases, such samples demand special care. In those cases, using a cap having a central hole should be necessary for letting the gases escape from the rotor.

For the thin film samples, in most cases, the sample will be finely and regularly cut into small pieces. In some particular cases, stacked punched disk packing or scroll-like packing to rotors are successfully used for their MAS NMR experiments. If the very first step of the rotation would help the achievement of a balanced mass distribution through centrifugal force, this kind of packing method would work well, though the initiation of the smooth rotation is not always easy. In some cases, we can observe the chemical shift anisotropy of an oriented sample by comparing the NMR signals obtained with differently packed samples [75].

For the observation of conducting materials, sometimes dilution of the sample with insulation materials for example silica, alumina, etc., should be necessary. When we apply strong RF, the rotation of the sample is possibly interrupted. In those cases, we should dilute the sample further, we should use a weaker pulse power or slower spinning speed, or both.

Deeply dried samples such as silica and alumina, would be inclined to be charged up electrostatically. These samples would not rotate in a strong magnetic field, possibly because of the force between the moving charges and the magnetic field (Lorentz force). In those cases, we need to use some static electricity eliminator during the sample packing within a dry box or similar apparatus.

Especially for the inorganic materials, the preparation, and control of the sample conditions affect the resulting spectrum drastically. Sometimes, we should thoroughly dry up the sample under high-temperature and vacuum conditions, in some of those cases, we often obtain completely different signals depending on the sample preparation conditions. In this section, we will discuss the ^{27}Al NMR of zeolites and related silica-alumina materials in a little more detail. These materials are frequently used as catalysts or adsorbents. The surface conditions of these materials, especially for the amount of adsorbed water on the surface may often affect the activities of these materials. ^{27}Al, $I = 5/2$, ν_0 (^{27}Al) / ν_0 (^{1}H) = 0.26057 and natural abundance 100%. The ^{27}Al NMR spectra obtained in metal oxides such as silica, alumina, zeolites, and silica-alumina ordinary show signals from Al[O_4], Al[O_6] (Fig. 12.8), and less frequently from Al[O_5] sites separately [60]. However, the signal separation patterns of them are very sensitive to the moisture load on the surface of that sample. A simple assumption of the hydration mechanism of these materials will predict that Al[O_5] and then Al[O_6] will appear stepwise along with the hydration of the sample progressing as the $Al[O_4]$ will decrease. However, at least from comparing the signal amounts of dried samples in different conditions, the amount of $Al[O_4]$ and $Al[O_6]$ increased and that of $Al[O_5]$ decreased as the hydration progressed. In addition to that, in some deeply dried silica-alumina samples including zeolites, we often cannot observe a sufficient amount of ^{27}Al signals including the Brønsted acid sites denoted as Al–OH-Si that can be working as a strong acid site. This observation leads us to recognize that there are some 'invisible' aluminum species existing in the deeply dried zeolites or silica-alumina samples. Since the Brønsted acid sites, which can only be observed upon heating of oxonium, $[Al^{-}\text{-O-Si}][H_3O^{+}]$, or ammonium, $[Al^{-}\text{-O-Si}][H_4N^{+}]$ forms of this sites under vacuum, is one of the central importance in the catalytic science, many efforts have been made for the observation of these Brønsted sites. Compared to the ordinary observed C_Q's of $Al[O_4]$ sites (3 ~ 7 MHz), much larger a value, $C_Q = 15.3$ MHz was found for the ^{27}Al signal in the Brønsted acid site in a dry HY zeolite that was measured by TRAPDOR (transfer population double resonance) [61]. In HZSM-5, C_Q ~ 16 MHz is directly observed with a static sample with the help of signal enhancement with DFS [76]. An *ab initio* calculation of a model cluster structure taken from ZSM-5, the $^{27}Al[O_4]$ signal at the Brønsted site possibly exhibits a very large C_Q value (~ 20 MHz) [77]. This is very important knowledge that not every aluminum species can be observed in a NMR chart with a similar order of line width.

In situ experiments are expected to provide very useful information for material development and reaction control in many fields of chemical systems. For this purpose, MAS probes with rotors having gas inlet and outlet, can be operated under sample rotation within the rotor under development.

Hunger and his co-workers first devised a MAS probe in which the rotor was equipped with an in situ gas flow system [64]. For example, a partially protonated Na, HY zeolite (Faujasite) has been investigated using this apparatus. Under normal conditions, it shows both tetrahedral and octahedral Al sites. When the sample was heated and dried under a continuous flow of dry air, almost all the ^{27}Al signals of HY zeolite disappeared. When water-containing air flowed inside the rotor, both the tetrahedral and octahedral coordinated Al species were recovered on the ^{27}Al spectrum. Since not any Al was added or subtracted during these processes, this observation clearly shows that there exist Al species invisible in the NMR chart (Fig. 12.18). The different areas of ^{1}H NMR signal correspond to the water adsorbed. Except for water, the amount of other components within the rotor does not change during this process. However, the structural change in the aluminum coordination environment affected both ^{1}H and ^{27}Al signals. Thus, the word 'invisible' exactly means 'too broad to be observed'.

During the hydration of the sample, a part of adsorbed water molecules may construct the first, second, and higher coordination spheres of the aluminum molecules in the sample. This change may introduce some molecular motions caused by reversible absorption – desorption of water molecules on some local spots and/or the motion of the attached water molecules themselves. The strong dipole moment of water will affect the electrostatic environment of aluminum. These processes may cause motional averaging to the structure of the aluminum coordination sphere and

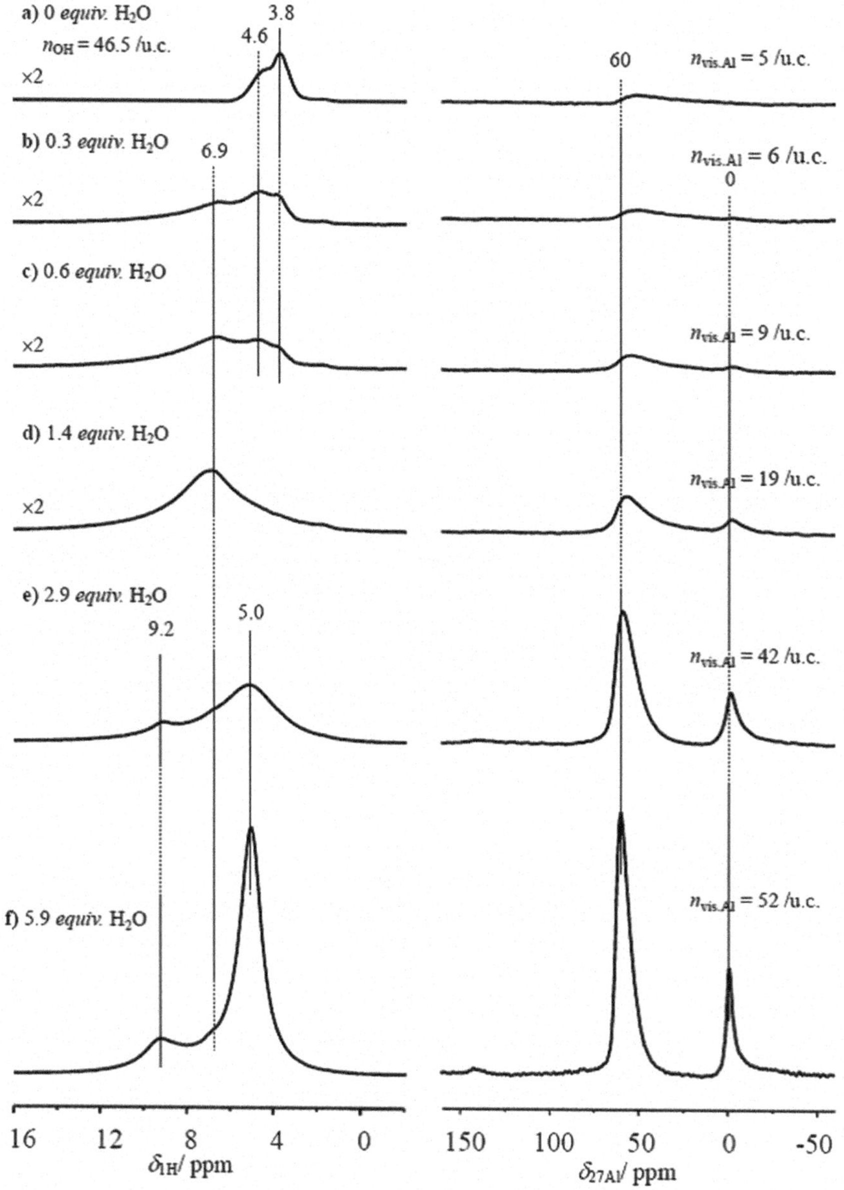

Fig. 12.18 In situ MAS NMR of dried H, Na-Y zeolite upon hydration. [79], as the hydration of the sample progressed, the signal area of proton is increasing and apparently, the signal area of ^{27}Al is also increasing although there are no changes in the Al amount. Copyright permission was obtained personally from Dr. Jian Jiao

make it into a more relaxed one. The local environmental difference between species will shrink within a certain range and enable us to observe the ^{27}Al species as rather simple and much more grouped peaks. Although the above observation poses some limits to the observation of ^{27}Al NMR within simple observation methods adoption, there are still existing other methods to observe such a broad signal, especially in static experiments. In general cases, signals having a width wider than several numbers of rotor cycle frequencies should be observed in static conditions, not by MAS. In those cases, signal separation is mostly done by numerical methods.

As another example, we will describe about ^{17}O NMR briefly. ^{17}O, $I = 5/2$, ν_0 (^{17}O) /ν_0 (^{1}H) = 0.1356 and natural abundance 0.037%. The quadrupole interaction of ^{17}O species is much smaller than those of ^{27}Al species, so the observation of ^{17}O species may be easier if the concentration of ^{17}O is sufficient. Although it may take some cost, the ^{17}O NMR with some isotope enrichment is widely investigated because oxides and hydroxides are important for ceramics, structure materials, electric materials, catalysts, and their supports. In solution NMR, an oxygen species shows chemical shifts from 300 to −100 ppm depending on its electronic states. For example, the oxo species may normally have very high-frequency chemical shifts at around 300 ppm or higher. Alkoxide ions show signals around 200 ppm. Contrary, the oxide species having two coordinated metals are not so high and can be observed ordinarily under 100 to −100 ppm [66]. The lowest frequency species are oxides having oxygens in their crystal's very high symmetrical positions. However, in most of the solid samples, the chemical shifts are no longer discriminative in the 1D solid-state NMR because of the ambiguity brought by large quadrupole couplings of ^{17}O. The typical value of quadrupolar coupling frequencies of ^{17}O is around several MHz that will appear in the spectrum as a peak having several 10's kHz of the total line width. In contrast, the quadrupolar coupling frequencies of ^{17}O can be separated by using MQMAS or STMAS [78].

As an example, we will see zeolite materials here again. In the ^{17}O NMR spectra of zeolite materials, it is of the primary importance to separate the oxygen signals into the kinds of their neighboring two or three atoms. For example, a ^{17}O MAS NMR signal obtained with a ^{17}O-enroched zeolite, separately shows that the Si–O-Si and Si–O-Al signals. The signal from the Brønsted acid site has been also revealed [80]. In some special cases, further separation can also be obtained and the site-dependent signal assignment is partially achieved, which means it will provide a further precision to the site-specific observation of zeolites and related materials [81, 82].

Acknowledgements The author is fully thanking Prof. Naito, the Chief Editor of this book, for his kind and enduring help and encouragement, without his help, I couldn't complete my task. I thank Prof. Takegoshi for his frank and helpful discussion. I also thank Dr. Hayashi who gave me instructions on the solid-state NMR, especially in my launching days. I also thank the committee members of the Japanese Solid-state NMR and Material Forum (no English site, the site only in Japanese, (https://ssnmrforum2021.wixsite.com/ssnmr) and the Nuclear Magnetic Resonance Society of Japan (http://www.nmrj.jp/index.php?page=index-e) and people who have been gathering there and giving me a plenty of fruitful information and discussions. Finally, I would express another deep acknowledgement to the National Institute of Industrial Science and Technology (AIST), and

New Energy and Industry Technology Development Organization (NEDO), and the people who are working there for their financial and everyday support everywhere.

References

1. Man, P.P.: Quadrupole Coupling in NMR. http://www.pascal-man.com
2. Pound, R.V.: Nuclear electric quadrupole interactions in crystals: phys. Rev. **79**, 685–702 (1950)
3. Cohen, M.H., Reif, F.: Quadrupole effects in nuclear magnetic resonance studies of solids. In: Zeitz, F., Turnbull, D. (eds.) Solid State Physics, Advances in Research Applications, vol. 5, pp. 321–455. Academic Press, New York (1957)
4. Abragam, A.: Principles of Nuclear Magnetism. Oxford University Press, London (1961)
5. Slichter, C.P.: Principles of Magnetic Resonance (Springer Series in Solid-State Science 1). Third Enlarged and Updated edn. Springer, New York (1996)
6. Hashi, K., Ohki, S., Matsumoto, S., Nishijima, G., Goto, A., Deguchi, K., Yamada, K., Noguchi, T., Sakai, S., Takahashi, M., Yanagisawa, Y., Iguchi, S., Yamazaki, T., Maeda, H., Tanaka, R., Nemoto, T., Suematsu, H., Miki, T., Saito, K., Shimizu, T.: Achievement of 1020 MHz. J. Magn. Reson. **256**, 30–33 (2015)
7. Nishiyama, Y., Endo, Y., Nemoto, T., Utsumi, H., Yamaguchi, K., Hioka, K., Asakura, T.: Very fast magic angle spinning ^{1}H–^{14}N 2D solid-state NMR: sub-micro-liter sample data collection in a few minutes. J. Magn. Reson. **208**, 44–48 (2011). https://doi.org/10.1016/j.jmr.2010.10.001
8. Mizuno, T., Takegoshi, K.: Development of a cryogenic duplexer for solid-state nuclear magnetic resonance. Rev. Sci. Instrum. **80**, 124072 (2009). https://doi.org/10.1063/1.3263908
9. Hyberts, S.G., Arthanari, H., Wagner, G.: Applications of non-uniform sampling and processing. Top. Curr. Chem. **316**, 125–148 (2011). and references therein
10. Barnes, A.B., De Paëpe, G., van der Wel, P.C.A., Hu, K.-N., Joo, C.-G., Bajaj, V.S., Mak-Jurkauskas, M.L., Sirigiri, J.R., Herzfel, J., Temkin, R.J., Griffin, R.G.: High-field dynamic nuclear polarization for solid and solution biological NMR. Appl. Magn. Reson. **34**(3–4), 237–263 (2008). https://doi.org/10.1007/s00723-008-0129-1
11. Bruker Biospin: DNP NMR. https://www.bruker.com/products/mr/nmr/dnp-nmr/overview.html?gclid=CPuNydr-6tQCFYcDKgod_OwAhg
12. Tolman, J.R., Flanagan, J.M., Kennedy, M.A., Prestegard, J.H.: Nuclear magnetic dipole interactions in field-oriented proteins: information for structure determination in solution. Proc. Natl. Acad. Sci. U.S.A. **92**, 9279–9283 (1995)
13. Hexem, J.G., Frey, M.H., Opella, S.J.: Influence of ^{14}N on ^{13}C NMR spectra of solids. J. Am. Chem. Soc. **103**, 224–226 (1981)
14. Naito, A., Gnanapathy, S., McDowell, C.A.: High resolution solid state ^{13}C NMR spectra of carbons bonded to nitrogen in a sample spinning at the magic angle. J. Chem. Phys. **74**, 5393–5397 (1981)
15. Cavadini, S., Lupulescu, A., Antonijevic, S., Bodenhausen, G.: Nitrogen-14 NMR spectroscopy using residual dipolar splittings in solids. J. Am. Chem. Soc. **128**, 7706–7707 (2006)
16. McManus, J., Kemp-Harper, R., Wimperis, S.: Second-order quadrupolar-dipolar broadening in two-dimensional multiple-quantum MAS NMR. Chem. Phys. Lett. **311**(3), 292–298 (1999)
17. Wi, S.-S., Frydman, L.: Residual dipolar couplings between quadrupolar nuclei in high resolution solid state NMR: description and observations in the high-field limit. J. Chem. Phys. **112**(7), 3248–3261 (2000)
18. Ashbrook, S.E., Wimperis, S.: Second-order cross-term interactions in high-resolution MAS NMR of quadrupolar nuclei. Prog. Nucl. Magn. Reson. Spectrosc. **55**, 160–181 (2009)
19. Hope, M.A., Halat, D.M., Magusin, P.C.M.M., Paul, S., Pengcand, L., Grey, C.P.: Surface selective direct ^{17}O DNMP NMR of CeO_2. Chem. Commun. **53**, 2142–2145 (2017)

20. Kuchel, P.W., Shishmarev, D., Puckeridge, M., Levitt, M.H., Naumann, C., Chapman, B.E.: NMR of $^{133}Cs^{+}$ in stretched hydrogels: one-dimensional, *z*- and NOESY spectra, and probing the ion's environment in erythrocytes. J. Magn. Reson. **261**, 110–120 (2015)
21. Man, P.P., Couty, R., Fraissard, J.: Determination of line intensities of ^{27}Al in Al_2O_3. J. Magn. Reson. **86**, 613–617 (1990)
22. Bryce, D.L., Taulelle, F.: NMR crystallography. Acta. Cryst. **C73**, 126–127 (2017)
23. Kusumi, R., Kadoma, H., Wada, M., Takeda, K., Kimura, T.: In situ solid-state NMR of a magnetically oriented microcrystal suspension. J. Magn. Reson. **309**, 1–6 (2019)
24. Man, P.P.: Quadrupolar interactions in solid-state NMR, general. In: Meyers, R.A. (ed.) Encyclopedia of Magnetic Resonance, pp. 12224–13365. Wiley, Chichester (1990)
25. MacKensie, K.J.D., Smith, M.E.: Multinuclear solid-state NMR of inorganic materials. In: Cahn, R.W. (ed.) Pergamon Materials Series 6th edn. Elsevier, Oxford (2002)
26. Wasylishen, R.E., Ashbrook, S.E., Wimperis, S. (ed.): NMR of Quadrupolar Nuclei in Solid Materials. Weiley, Hoboken (2012)
27. Freude, D.: Quadrupolar nuclei in solid-state nuclear magnetic resonance. In: Meyers, R.A. (ed.) Encyclopedia of Analytical Chemistry, pp. 12188–12224. Wiley, Chichester (2000)
28. De-Lacaillerie, J.B.D., Fretigny, C., Massiot, D.: MAS NMR spectra of quadrupolar nuclei in disordered solids: the Czjzek model. J. Magn. Reson. **192**, 244–251 (2008)
29. Llor, A., Virlet, J.: Chem. Phys. Lett. **152**, 248–253 (1988); Samoson, A., Lippmaa, E., Pines, A.: High resolution solid-state NMR: Averaging of second-order effects by means of a double-rotor. Mol. Phys. **65**, 1013–1018 (1988); Bryce, D.L.: Double-rotation (DOR) NMR spectroscopy: progress and perspectives. Solid State Nucl. Magn. Res. **130**, 101923 (2024)
30. Bax, A., Szeverenyi, N.M., Maciel, G.E.: Chemical shift anisotropy in powdered solids studied by 2D FT NMR with flipping of the spinning axis. J. Magn. Res. **55**, 494–497 (1983); Terao, T, Fujii, T., Onodera, T., Saika, A.: Switching-angle sample-spinning NMR spectroscopy for obtaining powder-pattern-resolved 2D spectra: Measurements of ^{13}C chemical-shift anisotropies in powdered 3,4-dimethoxybenzaldehyde. Chem. Phys. Lett. **107**, 145–148 (1987)
31. Frydman, L., Harwood, J.S.: Isotropic spectra of half-integer quadrupolar spins from bidimensional magic-angle spinning NMR. J. Am. Chem. Soc. **117**, 5363–5364 (1995)
32. Hatanaka, H., Terao, T., Hashi, T.: Excitation and detection of coherence between forbidden levels in three-level spin system by multi-step processes. J. Phys. Soc. Jpn. **39**, 835−836 (1975); Hatanaka, H., Hashi, T.: Coherence-transfer rotary echoes produced by double irradiation in a three-level NMR system. Phys. Rev. **21**, 2677−2680 (1980)
33. Brown, S.P., Heyes, S.J., Wimperis, S.: Two-dimensional MAS multiple-quantum NMR of quadrupolar nuclei. Removal of inhomogeneous second-order broadening. J. Magn. Reson. Ser. **A119**, 280–284 (1996); Amoureux, J.-P., Fernandez, C., Steuernagel, S.Z.: Filtering in MQMAS NMR. J. Magn. Reson. Ser. **A123**, 116–118 (1996)
34. Massiot, D., Touzo, B., Trumeau, D., Coutures, J.P., Virlet, J., Florian, P., Grandinetti, P.J.: Two-dimensional MAS isotropin reconstruction sequences for quadrupolar nuclei. Solid State NMR **6**, 73–83 (1996)
35. Brown, S.P., Wimperis, S.: Two-dimensional MQ-MAS NMR of quadrupolar nuclei. Solid State NMR **6**, 73–83 (1996)
36. Brown, S.P., Wimperis, S.: Two-dimensional MQMAS NMR of quadrupolar nuclei: a comparison of methods. J. Magn. Reson. **128**, 42–61 (1997)
37. Gan, Z.H., Kwak, H.T.: Enhancing MQMAS sensitivity using signals from multiple coherence transfer pathways. J. Magn. Reson. **168**, 346–351 (2004)
38. Amoureix, J.-P., Flambard, A., Delevoye, L., Montagne, L.: A very sensitive high-resolution NMR method for quadrupolar nuclei. Chem. Commun. **27**, 3573–3474 (2005)
39. Gan, Z.: Isotopic NMR spectra of half-integer quadrupolar nuclei using satellite transitions and magic-angle spinning. J. Am. Chem. Soc. **122**, 3242–3243 (2000)
40. Gan Z., Gor'kov, T.A., Samoson, A., Massiot, D.: Seeking higher resolution and sensitivity for NMR of quadrupolar nuclei. J. Am. Chem. Soc. **124,** 5634–5635 (2002)

41. Amoureux, J.-P., Delevoye, L., Fink, G., Taulelle, F., Flambard, A., Montagne, L.: Unified representation of MQMAS and STMAS NMR of half-integer quadrupolar nuclei. Chem. Phys. Lett. **356**, 3200–3201 (2001)
42. Ashbrook, S.E., Wimperis, S.: High-resolution NMR of quadrupolar nuclei in solids: the satellite-transition magic angle spinning (STMAS) experiment. Prog. Nucl. Magn. Reson. Spectrosc. **46**, 52–108 (2004)
43. Takahashi, T., Kanehashi, K., Shimoikeda, Y., Nemoto, T., Saito, K.: Practical comparison of sensitivity and resolution between STMAS and MQMAS for ^{27}Al. J. Magn. Reson. **198**, 228–235 (2009)
44. Antonijevic, S., Ashbrook, S.E., Biedasek, S., Walton, R.I., Wimperis, S., Yang, H.: Dynamics on the microsecond timescale in microporous aluminophosphate ALPO-14 as evidenced by ^{27}Al MQMAS and STMAS NMR spectroscopy. J. Am. Chem. Soc. **128**, 8054–8062 (2006)
45. Takegoshi, K., Nomura, K., Terao, T.: Rotational resonance in the tilted rotating frame. Chem. Phys. Lett. **232**, 424–428 (1995)
46. Marks, D., Vega, S.: A Theory of cross-polarization NMR of non-spinning and spinning samples. J. Magn. Reon. Serie **A118**, 157–172 (1996)
47. Van Rossum, B.-J., de Groot, C.P., Ladizhansky, V., Vega, S., de Groot, H.J.M.: A Method for measuring heteronuclear (^{1}H–^{13}C) distance in high speed MAS NMR spectroscopy. J. Am. Chem. Soc. **122**, 3465–3472 (2000)
48. Amoureux, J.-P., Pruski, M.: Theoretical and experimental assessment of single and multi-quantum cross-polarization in solid-state NMR. Mol. Phys. **100**, 1595–1613 (2002)
49. Vega, A.: CP/MAS of $S = 3/2$ quadrupolar nuclei. Solid State NMR **1**, 17–32 (1992)
50. Hayashi, S., Hayamizu, K.: Line–shapes in CP/MAS NMR of quadrupolar nuclei. Chem. Phys. Lett. **203**, 319–324 (1993); Hayashi, S., Hayamizu, K.: Errata: line-shapes in CP/MAS NMR of quadrupolar nuclei. Chem. Phys. Lett. **205**, 597 (1993)
51. Amoureux, J.-P., Trebosc, J., Wiench, J., Pruski, M. : HMQC and refocused-INEPT experiments involving half-integer quadrupole spins. J. Magn. Reson. **184**, 1–14 (2007); Grey, C.P., Kao, H.-M.: INEPT experiments involving quadrupolar nuclei in solids.: J. Magn. Reson. **133**, 313–323 (1998); Fyfe, C.A., Meyer zu Altenschildesche, H., Wong-Moon, K.C., Grandey, H., Chezeau, J.M.: 1D and 2D solid state NMR investigations of the framework structure of as-synthesized $AlPO_4$–14. Solid State Nucl. Magn. Reson. **9**, 97–106 (1997)
52. Ansermet, J.-P., Slichter, C.P., Sinfelt, J.H.: Solid State NMR techniques for the study of surface phenomena. Prog. Nucl. Mag. Reson. Spectrosc. **22**, 401–421 (1990); van Eck, E.R.H., Veeman, W.S.: The determination of the average 27Al-31P distance in aluminophosphate molecular sieves with SEDOR NMR. Solid State Nucl. Magn. Reson. **1**, 1–4 (1992)
53. van Eck, E.R.H., Janssen, R., Maas, W.E.J.R., Veeman, W.S.: A novel application of nuclear spin-echo double-resonance to aluminophosphates and aluminosilicates. Chem. Phys. Lett. **174**, 428–432 (1990). Grey, C.P., Vega, A.J.: Determination of the quadrupole coupling constant of the invisible aluminium spin in zeolite HY with ^{1}H-^{27}Al TRAPDOR NNR. J. Am. Chem. Soc. **117**, 8232–8242 (1995)
54. Gullion, T.: Measurement of dipolar interactions between spin 1/2 and quadrupolar nuclei by rotational-echo, adiabatic-passage, double-resonance NMR. Chem. Phys. Lett. **246**, 325–330 (1995); Chopin, L., Vega, S., Gullion, T.: A MAS NMR method for measureming ^{13}C-^{17}O distance. J. Am. Chem. Soc. **120**, 4406–4409 (1988); Goldbourt, S., Vega, S., Gullion, T., Vega, A. J.: Interatomic distance measurement in solid-state NMR between a spin-1/2 and a spin-5/2 using a universal REAPDOR curve. J. Am. Chem. Soc. **125**, 11194–11195 (2003)
55. Nagashima, H., Trebosc, J., Kon, Y., Sato, K., Lafon, O., Amoureux, J.-P.: Observation of low-γ quadrupolar nuclei by surface-enhanced NMR spectroscopy. J. Am. Chem. Soc. **142**, 10659–10672 (2020)
56. Pandey, M.K., Amoureux, J.P., Asakura, T., Nishiyama, Y.: Sensitivity enhanced ^{14}N/^{14}N correlation to probe inter-beta-sheet interaction using first magic angle spinning solid-state NMR in biological solids. Phys. Chem. Chem. Phys. **18**, 22583–22589 (2016)
57. Gan, Z., Hung, I., Nishiyama, Y., Amoureux, J.-P., Lafon, O., Nagashima, H., Trebosc, J., Hu, B.: ^{14}N overtone nuclear magnetic resonance of rotating solids. J. Chem. Phys. **149**, 064201 (2018)

58. Pruski, M.K., Lamg, D.R., Fernandez, C., Amoureux, J.P.: Multiple-quantum magic-angle spinning NMR with cross-polarization: spectral editing of high-resolution spectra of quadrupolar nuclei. Solid. State Nucl. Magn. Reson. **7**, 327–331 (1997); Ashbrook, S.E., Wimperis, S.: Multiple-quantum cross-polarization and two-dimensional MQMAS NMR of quadrupolar nuclei. J. Magn. Reson. **147**, 238–249 (2000)
59. Wang, S.H., De Paul, S.M., Bull, L.M.: High-resolution heteronuclear correlation between quadrupolar and spon-1/2 nuclei using multiple-quantum magic-angle spinning. J. Magn. Reson. **125**, 364–368 (1997); Fermandez, C., Morais, C., Rocha, J., Pruski, M.: High-resolution heteronuclear correlation spectra between ^{31}P and ^{27}Al in microporous aliminophophates. Solid State NMR **21**, 61–70 (2002)
60. Siegel, R., Rocha, J., Mafra, L.: Combination STMAS and CRAMPS NMR spectroscopy: high-resolution HETCOR NMR spectra of quadrupolar and ^{1}H nuclei in solids. Chem. Phys. Lett. **470**, 337–341 (2009)
61. Kwak, H.-T., Prasad, S., Clark, T., Grandinetti, P.J.: Enhancing sensitivity of quadrupolar nuclei in solid-state NMR with multiple rotor assisted population transfers. Solid State NMR. **24**, 71–77 (2003)
62. Madhu, P.K., Goldbourt, A., Frydman, L., Vega, S.: Sensitivity enhancement of the MQMAS NMR experiment by fast amplitude modulation of the pulses. Chem. Phys. Lett. **307**, 41–47 (1999)
63. Kwak, H.-T., Prasad, S., Clark, T., Grandinetti, P.J.: Selective suppression and excitation of solid-state NMR resonances based on quadrupole coupling constants. J. Magn. Reson. **160**, 107–113 (2003)
64. Kentgens, A.P.M., Verhagen, R.: Advantages of double frequency sweeps in static, MAS and MQMAS NMR of spin $I = 3/2$ nuclei. Chem. Phys. Lett. **300**, 435–443 (1999)
65. Kupče, Ē, Freeman, R.: Optimized adiabatic pulses for wideband spin inversion. J. Magn. Reson. Ser. A **118**, 299–303 (1996)
66. Siegel, R., Nakashima, T.T., Wasylishen, R.E.: Sensitivity enhancement of NMR spectra of half-integer spin quadrupolar nuclei in solid using hyperbolic secant pulses. J. Magn. Reson. **184**, 83–100 (2007)
67. Wang, Q., Li, Y., Trebosc, J., Lafon, O.,Hu., B., Xu, J., Feng, N., Chen, Q., Amourexu, J.P., Deng, F.: Population transfer HMQC for half-integer quadrupolar nuclei. J. Chem. Phys. **142**, 094201 (2015)
68. Luke, A.: O'Dell: the WURST kind of pulses in solid-state NMR. Solid State Nucl. Magn. Reson. 55–56, 28–41 (2013)
69. O'Dell, L.A., Schurko, R.W.: QCPMG using adiabatic pulses for faster acquisition of ultra-wideline NMR spectra. Chem. Phys. Lett. **464**, 97–102 (2008)
70. Pandey, M.K., Hashi, K., Ohki, S., Nishimjima, G., Matsumoto, S., Noguchi, T., Deguchi, K., Gotoh, A., Shimizu, T., Maeda, H., Takahashi, M., Yanagisawa, T., Suematsu, H., Saito, K., Miki, T., Nishiyama, Y.: 24T high-resolution and–sensitivity solid-state NMR measurements of low-ganmma half-integer quadrupolar nuclei ^{35}Cl and ^{37}Cl. Anal. Sci. **32**(5), 1339–1345 (2016). https://doi.org/10.1246/cl.60850
71. Naito, A.: Structure elucidation of membrane-associated peptide and proteins in oriented bilayers by solid-state NMR spectroscopy. Solid State Nucl. Magn. Reson. **36**, 67–76 (2009)
72. Klinowsky, J.: Solid-state NMR studies of molecular sieve catalysis. Chem. Rev. **91**, 1459–1479 (1991)
73. Grey, C.P., Vega, A.J.: Determination of the quadrupole coupling constant of the invisible aluminium spins in zeolite HY with ^{1}H/^{27}Al TRAPDOR NMR. J. Am. Chem. Soc. **117**, 8232–8242 (1995)
74. Kentgens, A.P.M., Iuga, D., Kalwei, M., Koller, H.: Direct observation of Brønsted acidic site dehydrated zeolites, H-ZSMS using DFS enhanced ^{27}Al MQMAS NMR spectroscopy. J. Am. Chem. Soc. **123**, 2925–2926 (2001)
75. Ehresmann, J.O., Wang, W., Herreros, B., Luigi, D.-P., Venkatraman, T.N., Song, W., Nicholas, J.B., Haw, J.F.: Theoretical and experimental investigation of the effect of proton transfer on the ^{27}Al MAS NMR line shape of zeolite-adsorbate complexes: an independent measure of solid acid strength. J. Am. Chem. Soc. **124**, 10868–10874 (2002)

76. Hunger, M., Horvath, T.: A new MAS NMR probe for in situ investigations of hydrocarbon conversion on solid catalysts under continuous-flow conditions. J Chem. Soc. Chem. Commun. 1423–1424 (1995)
77. Jiano, J.: Chap. 8 Number and nature of framework aluminum atoms in non-hydrated zeolites Y studied by adsorption of probe molecules in quantitative characterization of aluminum in non-hydrated zeolite catalysts by multi-nuclear solid-state NMR spectroscopy. Thesis at Unversitet Stuttgart, (2006) which is presently accessible at https://elib.univ-stuttgart.de/bitstream/11682/824/1/thesis_jiao.pdf
78. Yamada, K.: Chapter 3–recent applications of solid-state ^{17}O NMR. Annual. Rep. NMR Spectrosc. 115–158 (2010)
79. Ashbrook, S.E., Smith, M.E.: Solidstate ^{17}O NMR–an introduction to the background principles and application to inorganic materials. Chem. Soc. Review. **35**, 718–735 (2006)
80. Peng, L., Liu, Y., Kim, N., Readman, J.E., Grey, C.P.: Detection of Brønsted acid sites in zeolite HY with high-field ^{17}O MAS-NMR techniques. Nat. Mater. **4**, 216–219 (2005)
81. Peng, L., Huo, H., Gan, Z., Grey, C.P.: ^{17}O MQMAS NMR studies of zeolite HY. Microporous Mesoporous Mater. **109**, 156–162 (2008)
82. Huo, H., Peng, L., Gan, Z., Grey, C.P.: Solid-state MAS NMR studies of Brønsted acid site in zeolite H-modenite. J. Am. Chem. Soc. **134**, 9708–9720 (2012)

Chapter 13
Protein–Ligand Interactions Studied by NMR

Hidekazu Hiroaki and Daisuke Kohda

Abstract Several solution NMR experiments have become indispensable in academia and industry for studying protein–ligand interactions. In general, solution NMR is superior to other physicochemical methods in terms of its spatial resolution and non-necessity of protein modifications. The applications of solution NMR are loosely classified into two categories: "ligand-based" approach and "protein-based" approach. Many unique experiments have been developed based on the ligand-based approach, including STD, WaterLOGSY, ^{19}F-NMR with R_2 filter, DIRECTION, INPHARMA, ILOE, and trNOE. These experiments frequently comprise the important steps of the drug discovery process, including ligand screening, fragment-based drug discovery (FBDD), pharmacophore mapping, and molecular design. This review provides a practicable classification of these experiments to aid the selection of a suitable experiment depending on the purpose. In contrast to the ligand-based approach, the experiments based on the protein-based approach are rather limited, with the ^{1}H-^{15}N-HSQC-based NMR titration experiment and its variants being preferentially used. This review also discusses several practical aspects of the NMR titration experiment, including data analysis and sample handling.

Keywords WaterLOGSY · INPHARMA · DIRECTION · NMR titration experiment · Fragment-based drug discovery (FBDD)

13.1 Overview

Solution NMR spectroscopy has become an indispensable analytical technique for studying protein–ligand interactions in the last two decades. The detection and quantitative analysis of protein–ligand interactions are important not only for

---

H. Hiroaki
Graduate School of Pharmaceutical Sciences, Nagoya University, Nagoya, Japan
e-mail: hiroaki.hidekazu.j7@f.mail.nagoya-u.ac.jp

D. Kohda (✉)
Medical Institute of Bioregulation, Kyushu University, Fukuoka, Japan
e-mail: dkohda@kyudai.jp

The NMR Society of Japan (ed.), *Experimental Approaches of NMR Spectroscopy II*,
https://doi.org/10.1007/978-981-97-6830-1_13

basic researchers but also for pharmaceutical industries. NMR offers comprehensive information about the molecular interactions in a solution at an atomic resolution.

For this purpose, several NMR experiments have been devised, and they can be loosely classified into two: "ligand-based" approaches and "protein-based" approaches [1–4]. The ligand-based approaches include saturation transfer difference (STD) [5], WaterLOGSY [6], ^{19}F-NMR with R_2 filter [7], DIRECTION [8], INPHARMA [9], inter-ligand nuclear Overhauser experiment (ILOE) [10], and transferred NOE (trNOE) [11]. These experiments focus on the changes in the NMR signals derived from ligands and do not require isotopically labeled protein samples.

The protein-based approaches, in contrast, require isotopically labeled protein samples. After introducing suitable stable isotopes (^{15}N, ^{13}C, or both), heteronuclear correlation spectra, such as heteronuclear single quantum coherence spectroscopy (HSQC) are measured. Since the NMR chemical shift is highly sensitive to the chemical and steric environments surrounding an atom, the changes in cross peaks in the HSQC spectrum reflect a ligand's binding to the target protein. In addition, the key residues relevant to the molecular interactions in the protein can be identified. This residue-wise information is extremely useful for selecting and designing new candidate molecules ("hits") with higher affinity to the target. Consequently, even at the early stage of drug screening projects, protein-based approaches are useful for validating the hit compounds.

This review aims to highlight the merits and demerits of the two approaches according to the purpose of the experiments (Table 13.1) and discuss several practical aspects of the experiments.

13.2 Ligand-Based Approaches

This section focuses on the seven major methods of ligand-based NMR screening: STD, WaterLOGSY, ^{19}F-NMR with R_2 filter, ILOE, INPHARMA, DIRECTION, and trNOE. They do not require isotopically labeled protein samples; usually, a tenfold or more molar excess of a ligand is added to the protein solution. STD, WaterLOGSY, ^{19}F-NMR with R_2 filter, and DIRECTION are one-dimensional (1D), whereas ILOE, INPHARMA, and trNOE are two-dimensional (2D).

13.2.1 Ligand Screening

In recent decades, NMR experiments have been combined with a new drug discovery strategy: "fragment-based drug discovery" (FBDD) [12–14]. It has emerged as one of the trends in lead discovery/generation in pharmaceutical industries. In this strategy, one or several low-molecular-weight compounds—fragments—that weakly bind to the target protein are identified. At the early stage of NMR-assisted FBDD, STD,

Table 13.1 Purpose-oriented classification of various solution NMR experiments that are useful for studying protein–ligand interactions

	Classification	Data acquisition	Number of ligands	Solvent	Protein: ligand ratio	Assignment of ligand signals	Protein sample	Key references
Ligand screening								
STD	Ligand-based	1D	1~several	100% ^2H_2O	1: 20~100	Not required	Unlabeled	5, 15–18, 25
WaterLOGSY	Ligand-based	1D	1~several	5% ^2H_2O–95% ^1H_2O	1: 20~100	Not required	Unlabeled	6, 21–24
CSP/ NMR-titration	Protein-based	2D	1 several	5% ^2H_2O–95% ^1H_2O	1: 0 ~ 20	Not required	$^1H/^{15}N$–labeled, assignment not required	39, 40, 42
Pharmacophore determination								
DIRECTION	Ligand-based	1D	1	100% ^2H_2O	1: 20~100	Required	Unlabeled	8
INPHARMA	Ligand-based	2D	2	100% ^2H_2O	1: 20~100	Required	Unlabeled	9, 31–33
Ligand design								
ILOE	Ligand-based	2D	2	100% ^2H_2O	1: 20~100	Required	Unlabeled	10, 36, 37
trNOE	Ligand-based	2D	1	5% ^2H_2O–95% ^1H_2O or 100% ^2H_2O	1: 20~100	Required	Unlabeled	34, 35
SAR-by-NMR	Protein-based	2D	2	5% ^2H_2O–95% ^1H_2O	1: 0~20	Not required	$^1H/^{15}N$–labeled, assignment required	40
Ligand characterization (K_D determination)								
CSP/ NMR-titration	Protein-based	2D	1	5% ^2H_2O–95% ^1H_2O	1: 0~20	Not required	$^1H/^{15}N$ (^{13}C)–labeled, assignment required	39, 40, 42

(continued)

Table 13.1 (continued)

	Classification	Data acquisition	Number of ligands	Solvent	Protein: ligand ratio	Assignment of ligand signals	Protein sample	Key references
Ligand characterization (determination of the molecular interface)								
CSP/ chemical-shift mapping	Protein-based	2D	1	5% ^2H_2O–95% ^1H_2O	1: 0 ~ 20	Not required	$^1H/^{15}N$ (^{13}C)–labeled, assignment required	41, 42
Cross-saturation	Protein-based	2D	1	80% ^2H_2O–20% ^1H_2O	1: 0 ~ 20	Not required	$^2H/^{15}N$–labeled, assignment required	38

and WaterLOGSY are often used. Weak interactions are usually difficult to detect by methods other than solution NMR.

13.2.1.1 Saturation Transfer Difference (STD)

In STD experiments, the sample is prepared by mixing a target protein (*ca.* 50 μM) and a 10- to 40-fold molar excess of a ligand of interest. To achieve higher spin saturation transfer efficiency, ^2H_2O (99.8%) is recommended as the solvent. STD is the difference between an "ON-resonance saturated" spectrum and an "OFF-resonance saturated" spectrum. To record the former, a selective pulse that excites the protein protons (*i.e.*, the nuclei of ^1H atoms that give ^1H resonances) is applied before an observation 90-degree pulse. After the 90-degree pulse, a weak spin-locking pulse and a water suppression pulse are optionally applied, followed by NMR signal detection. To record the reference "OFF-resonance saturated" spectrum, the same selective excitation pulse—but distant from the protein spectral region—is applied, leaving the protein protons unsaturated.

The STD method has received several technical improvements in the past two decades, including (1) the saturation transfer double difference (STDD) experiment [15], (2) optimized excitation pulse bandwidth (WET-STD) [16], (3) STD with spectral editing (clean-STD) [17], (4) quantitative STD [18], and (5) differential epitope mapping by STD (DEEP-STD) [19, 20]. STDD requires another reference sample (a sample with only the target protein), and to compensate for this inconvenience, it offers a higher signal-to-noise ratio by eliminating the background signals arising from the target protein. STDD was shown to be highly advantageous for target proteins with large molecular sizes. The recently developed DEEP-STD experiments are suited for either epitope mapping or pharmacophore determination at the medium stage of the drug discovery process (discussed later).

13.2.1.2 WaterLOGSY

In WaterLOGSY, another important 1D NMR technique for discriminating whether the ligand of interest binds to the target protein, the solvent, in contrast to STD, is 90% ^1H_2O–10% ^2H_2O. This is because WaterLOGSY is based on the magnetization transfers among the protons in bulk solvent water molecules, the ligand, and the protein. Note that 10% ^2H_2O is not specifically required—it is needed only for the NMR spectrometer to stabilize (*i.e.*, lock) the NMR frequency.

A WaterLOGSY experiment works as follows: At the beginning of the pulse sequence, a selective pulse saturates the protons of the bulk water molecules. Then, the magnetization of the water protons transfers to the protein protons via two mechanisms: (1) The magnetization transfers to the surface protons of the target protein through the nuclear Overhauser effect (NOE) and (2) the exchange between the bulk water protons and the exchangeable protons of the target protein during the mixing time (typically 500 ms). Subsequently, the magnetization of the protein protons is transferred to the ligand protons in the bound state through NOE. A comprehensive review for comparing STD and WaterLOGSY was written by Ramirez et al. [21].

Several improvements in terms of sensitivity have been made to the original WaterLOGSY experiment, including polarization-optimized (PO) WaterLOGSY [22] and the Aroma-WaterLOGSY experiment [23]. The radiation damping process was also incorporated [24], a modification that resulted in 13% sensitivity improvement compared to the PO-WaterLOGSY.

13.2.1.3 STD *versus* WaterLOGSY

Since STD and WaterLOGSY are 1D NMR experiments, they are often used to locate a hit fragment in a mixture with non-hit fragments (up to 10 fragments). In STD experiments, only the hit fragment offers a positive difference signal, and the non-hit fragment signals are eliminated; in WaterLOGSY experiments, the hit fragment offers a positive difference signal, whereas the non-hit fragments offer negative signals. STD is thus simpler than WaterLOGSY. Using a shorter excitation pulse at an optimized exciting frequency, Ley et al. proposed a simple but significant intensity improvement in STD [25]. However, comparing the sensitivity of STD and WaterLOGSY in a realistic drug screening experiment [26], WaterLOGSY was found to be still more sensitive in certain cases. Thus, the comparison of STD (simplicity of the results) and WaterLOGSY (sensitivity) is left for future evaluation.

13.2.1.4 Data Analysis for STD and WaterLOGSY Experiments

STD and WaterLOGSY are commonly employed in combination with FBDD to achieve the first discovery of fragments with weak target-binding affinity. This is particularly crucial in the pharmaceutical industry, which requires the batch processing of the comparison of extensive 1D NMR datasets of numerous compounds. The two experiments, it should be noted, need to prepare for two different spectra—"ON-resonance" and "OFF-resonance"—or for the presence or absence of the target protein. The subsequent processes may involve ranking the compounds based on the 1D NMR data and selecting the candidate fragments. However, to compare and rank the efficiency of these ligand-based approaches for various compounds belonging to different classes, all the 1D NMR spectra must be significantly different from each other. Moreover, owing to experimental errors in, for instance, sample preparation, data acquisition, and data processing, the automated 1D spectral analysis of many compounds derived from automatic NMR measurement is often impractical.

To address these complexities, Vuister et al. developed the CcpNmr AnalysisScreen, a dedicated tool for a high-throughput ligand-based NMR screening project [27]. This application leverages the CcpNmr package as its foundational platform for NMR data processing and offers a suite of functionalities, including the linking of 1D spectra to compounds' chemical formulas, the generation and management of the difference spectra required for STD or WaterLOGSY, and the automated handling of measurement errors through the principal component analysis (PCA) of 1D spectra. Additionally, it facilitates the automatic identification of hit compounds, thus streamlining the FBDD process within the pharmaceutical industry [27].

In 1D NMR experiments that employ the ligand-based approach, when the ratio of compounds to the target protein is changed, the compounds that interact with the protein often exhibit signal broadening. This phenomenon can be leveraged to determine the dissociation constant (K_D) of the interaction between the ligand and the target protein by plotting the reciprocal of the signal linewidth on the x-axis and the ligand concentration on the y-axis [28].

13.2.1.5 Use of ^{19}F-NMR with Fluorinated Ligand Library

As an alternative to STD or WaterLOGSY, the use of ^{19}F-NMR for monitoring ligand–protein interactions for FBDD has been developed [7, 29, 30]. ^{19}F is a highly abundant "NMR-active" nucleus (a natural abundance of 100%), and ^{19}F NMR is highly sensitive (as much as 83% corresponding to ^{1}H-NMR) with narrow linewidths and a large chemical shift divergence. ^{19}F-NMR enables to monitor the interaction between the ligand of interest and the target protein by signal broadening upon binding. ^{19}F-NMR screening is thus useful for FBDD [29].

Recently, Shinya et al. carefully constructed a small fragment library of selected fluorinated compounds [7] that were divided into 22 cocktails: 11 "10-mix" cocktails and 11 "20-mix" cocktails, containing 10–11 or 19–21 compounds, respectively. To use the solution cocktails of more than 20 compounds without overlapping the ^{19}F NMR signals, the cocktails were carefully designed by considering the chemical shifts of the signals. Conducting ^{19}F-NMR screening using R_2-filter experiments, new fragments bound to FKBP12 were discovered [7]. This approach was advantageous, considering the higher throughput of the "cocktail-base" NMR screening than ^{1}H-NMR-based approaches.

13.2.2 Pharmacophore Determination

This section describes the medium-throughput NMR-based drug screening techniques: DIRECTION [8] and INPHARMA [9]. In contrast to STD and WaterLOGSY, these experiments were developed to identify the ligand protons that are in immediate contact with the target protein via NOE observation. Between the lead discovery and lead optimization processes in drug discovery, a core sub-structure common to the related compounds with different affinities for the target protein should be experimentally determined. This sub-structure, which plays a central role in molecular recognition, is called "pharmacophore." The spatial information obtained from the DIRECTION, DEEP-STD, and INPHARMA experiments is particularly useful to define pharmacophores, which results in the further development of a series of derivative compounds to find the best ligand.

13.2.2.1 DEEP-STD

STD NMR, as already mentioned, is a powerful ligand-based tool for screening small molecules and low-molecular-weight fragments whether they interact with a target protein of interest. The recently developed DEEP-STD is a variation of STD spectroscopy for "epitope-mapping" [19]. When applied to small compounds and the target protein, DEEP-STD enables the identification of some key protons forming a contact site with the protein and the estimation of the molecules' pharmacophore. DEEP-STD comprises a series of STD experiments of different saturation radio frequencies and/or different sample solvents (^1H_2O or ^2H_2O). For example, two frequencies—"ON-resonance" for the ligand binding pocket, and "reference (ON-resonance but at different frequency)" as the remote site for the ligand binding pocket—are carefully selected. Then, the results of the two STD experiments with differentially saturated protons of the protein are compared. Alternatively, or in combination, different solvents (either ^2H_2O or ^1H_2O) are used. These data provide the ligand orientation while it is bound to the target protein and help define the pharmacophores.

13.2.2.2 DIRECTION

To classify ligand protons according to their proximity to the target protein, Mizukoshi et al. developed the DIRECTION experiment, a simple and accurate method [8] based on the difference in the protons' longitudinal relaxation rates depending on the saturation state of the protein protons. The experiment involves a series of inversion-recovery 1D NMR experiments to determine the T_1 value of the ligand protons, with and without additional excitation pulse on the target protein protons.

13.2.2.3 INPHARMA

INPHARMA, another important experiment used for pharmacophore mapping (Fig. 13.1) (see review [9, 31]), was designed to identify the spatially equivalent protons between two ligands that share the same binding pocket. The INPHARMA sample thus contains three components: the interest ligand *a*, the reference ligand *b* which is known to bind to the target protein, and the target protein. In the pulse sequence, the excited magnetization of ligand *a* transfers to the proximal protons in the target protein's ligand binding site. During the long mixing time, ligand *a* dissociates, and ligand *b* binds to the binding site. Then, the magnetization of the protein protons transfers to the proximal protons of ligand *b*. Consequently, the spatially equivalent protons of ligand *a* and ligand *b* give a cross peak in the INPHARMA spectrum.

INPHARMA was also shown to be useful for the binding-site identification of a target protein with multiple binding sites. Krimm prepared a set of reference molecules, of which the binding sites on the glycogen phosphorylase were known, and unambiguously assigned the specific binding sites of several compounds [32].

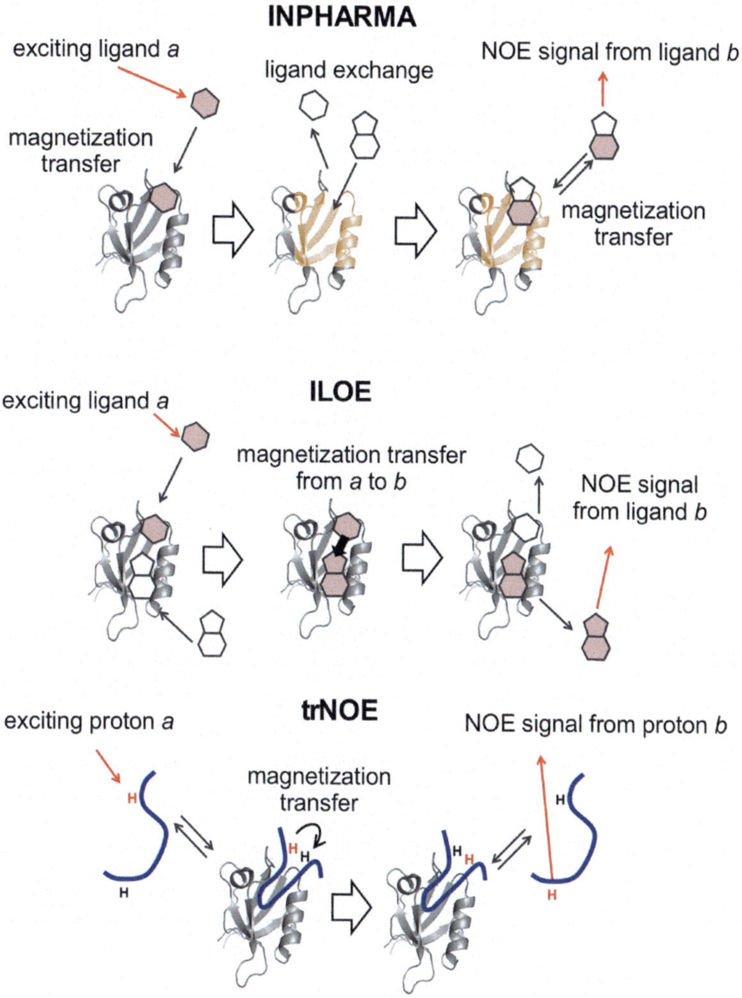

Fig. 13.1 Basic principle of operation of the ligand-based NMR experiments, INPARMA, ILOE, and trNOE for analyzing protein–ligand interactions. Please pay attention to the routes of the magnetization transfer from proton to proton. The colored shapes and sites represent the “heated” ligands and protein parts by NMR pulse irradiation. The red H’s also mean the “heated” protons by irradiation

Skjærven et al. developed a method to build a reliable protein–ligand complex model by docking simulation assisted by INPHARMA-derived NOE constraints [33].

13.2.3 Structural Information-Driven Ligand Design

The NOE is a phenomenon induced by through-space dipole–dipole interactions between two nuclear spins. The NOE effect is measured as the change in the signal intensity of one nuclear spin when another adjacent nuclear spin is saturated by irradiation. The NOE effect is proportional to the inverse of the sixth power of the distance between the two nuclei, thereby providing the distance information between the two nuclei. NOE-based experiments thus offer useful structural restraints for designing novel molecules with a higher affinity to the target protein.

13.2.3.1 Transferred Nuclear Overhauser Effect (trNOE)

Transferred NOE is another NOE-based experiment that helps in designing a new ligand molecule from a linear flexible lead molecule, such as peptides (Fig. 13.1). In a small molecule, the peak intensity of one proton increases when another proximal proton, closer than 4–5 Å, is irradiated by a selective pulse. This phenomenon is referred to as positive NOE. In the presence of a protein, the small molecule is in an exchange equilibrium between the free state and the protein-bound state, which results in an apparent molecular mass increase. In the same NOESY spectrum, the peak intensity now decreases, and a negative NOE can be observed. The negative sign of the NOE confirms that the NOE arises from a fixed conformation of the flexible ligand in the target-bound state. One recent example is the determination of the CCR5-bound conformation of the HIV-1 gp120-derived V2 peptide, which inhibits HIV infection [34]. Recently, Hiroaki et al. also succeeded in determining the bound conformation of glycyrrhizin in the presence of the first PDZ (*P*ostsynaptic Density 95, *D*isc Large, and *Z*onula Occludens-1) domain of mouse ZO-1. They used this PDZ-bound glycyrrhizin conformation to construct the structural model of the ligand–protein complex [35]. In general, designing a new ligand with a fixed conformation is thermodynamically favorable for increasing a flexible ligand's affinity against the target protein by reducing the entropy of the ligand's free state. Thus, the inter-proton distance information of the target-bound conformation supplies an invaluable clue for designing new potent compounds.

13.2.3.2 Inter-Ligand Nuclear Overhauser Effect (ILOE)

The ILOE experiment is a good example of an NOE-based experiment [10] (Fig. 13.1). As described before, the FBDD approach comprises the fragment screening step and the lead generation step from the fragments. NMR-assisted FBDD is a concept in which NMR-derived information is used to promote two FBDD steps: in the first step, several NMR screening experiments are accelerated, including STD, WaterLOGSY, and NMR titration experiments; in the second step, once several fragments are identified, they are linked into one novel molecule.

The ILOE experiment can aid in finding a linking point in fragment-based lead generation, although ILOE is conceptually resembling INPHARMA. Both ILOE and INPHARMA samples contain two fragments *a* and *b*, and the target protein. However, in contrast to INPHARMA, two ligands *a* and *b* in the ILOE sample are assumed to bind side by side to the same binding site. In a 2D NOESY spectrum, an NOE cross peak between the two proximal protons—one in ligand *a* and the other in ligand *b*—is expected to be observed (Fig. 13.1). Two milestone papers were published by Sledz et al. [36] and Rega et al. [37]: Sledz et al. demonstrated that ILOE-based fragment linking obtained a new tight inhibitor that targeted the *M. tuberculosis* pantothenate synthetase (K_D = 880 nM). Rega et al. employed an elegant application of SAR-by-ILOE [37] as an alternative strategy to SAR-by-NMR (described later). They succeeded in developing acylsulfonamide derivatives that targeted the antiapoptic Bcl-1 family proteins.

13.2.3.3 Technical Aspects of trNOE, INPHARMA, and ILOE Experiments

Interestingly, the pulse sequences of the three experiments—trNOE, INPHARMA, and ILOE—are essentially the same as those used for the standard 2D NOESY spectrum. The trNOE sample contains only one ligand, whereas the samples of INPHARMA and ILOE contain, in addition to the target protein, two ligands. The differences in the route of magnetization transfer in the three experiments are illustrated in Fig. 13.1. The major conceptual differences between INPHARMA and ILOE are the binding modes of the two ligands. While the INPHARMA experiment is designed to find the spatially equivalent atoms between the two ligands that compete directly for the same binding site, the ILOE experiment is designed to detect the inter-molecular NOEs between the two ligands that simultaneously bind to the protein side by side. In real experimental setups, the mixing time delay in the NOESY pulse sequence is different. INPHARMA's mixing time is long (500 ms), whereas ILOE has a short mixing time (80 ms). TrNOE typically requires a relatively long mixing time (200–600 ms).

13.3 Protein-Based Approach for Protein–Ligand Interactions

The goals of the protein-based approach include: (1) screening ligands that weakly bind to the target; (2) facilitating the design of new lead compounds for FBDD; (3) identifying the interface on the target protein molecule; (4) detecting the conformational changes of the target protein upon ligand binding; and (5) determining the dissociation constant (K_D). To achieve these goals, the so-called NMR titration experiment—also known as the chemical shift perturbation (CSP) experiment—is most frequently used. For example, goals 1, 2, 4, and 5 are routinely achieved by the NMR titration experiment. Cross saturation experiment is another useful method [38]. An overview of the two protein-based NMR experiments is demonstrated in Fig. 13.2.

The NMR chemical shifts of some (but not all) chemical groups of the target protein are extremely sensitive to the changes in the microscopic environments surrounding the chemical groups. The nuclear spins of protons or other NMR-active nuclei located within or close to the ligand binding site can be used to monitor the ligand's binding to the protein. Comparing the two NMR spectra of the target protein recorded in the absence and presence of the ligand can reveal the protein–ligand interactions. In practice, an NMR titration experiment is a repeated 2D NMR measurement with serial additions of the ligand stock solution to the protein sample.

The NMR titration and the cross saturation experiments are both 2D NMR experiments. The ^{1}H-^{1}H 2D NMR spectrum is usable, but the incorporation of stable isotopes ^{15}N and/or ^{13}C is highly recommended since ^{1}H signals of proteins are frequently heavily overlapped. Note that if heteronuclear ^{1}H-^{15}N or ^{1}H-^{13}C HSQC spectra show a good dispersion of cross peaks, and if at least some cross peaks are separately observed, the signal assignment is not mandatory in case of goal 1 (ligand screening) and goal 5 (K_D determination).

13.3.1 Ligand Screening and NMR-Assisted Fragment-Based Drug Discovery

13.3.1.1 HSQC-Based Ligand Screening

The CSP of amide groups in $^{1}H-^{15}N$ HSQC spectra or methyl groups in $^{1}H-^{13}C$ HSQC spectra are useful for monitoring the protein–ligand interactions. To detect these interactions, a pair of HSQC spectra of the protein recorded in the absence and presence of the ligand is sufficient. To facilitate the screening, a cocktail of several different ligands can be used. In fact, using this cocktail technique, unexpected interactions of non-steroid anti-inflammatory drugs—diclofenac and flufenamic acid—were discovered with the ZO-1 PDZ1 domain [39]. Here, we offer another example of an NMR-titration experiment with ZO-1 PDZ1 and the peptide derived from

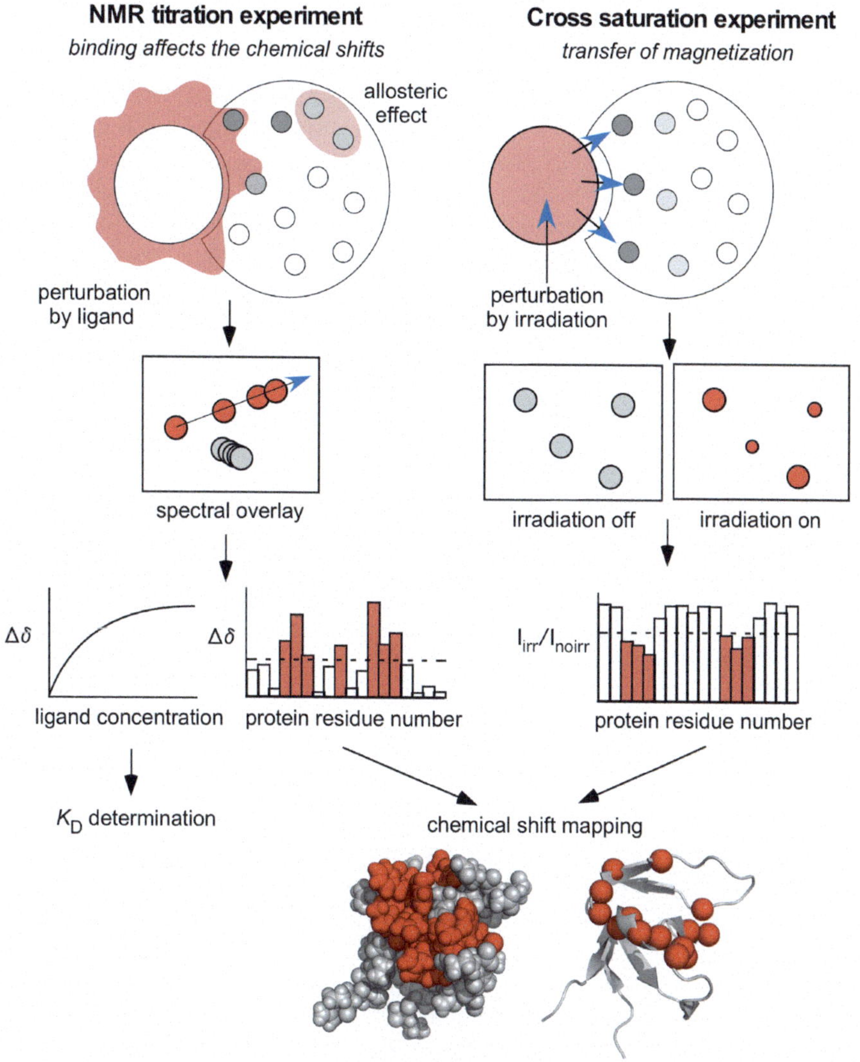

Fig. 13.2 Basic principle of operation of the protein-based NMR experiments, NMR titration (alias, chemical shift perturbation) experiment, and cross-saturation experiment

its physiological ligand claudin-3 (Fig. 13.3). As previously described, these NMR applications were often combined with FBDD approaches and called NMR-assisted FBDD. Moreover, the NMR titration experiment is useful during fragment screening in NMR-assisted FBDD as well as other ligand-based NMR screening experiments, such as STD and WaterLOGSY [14].

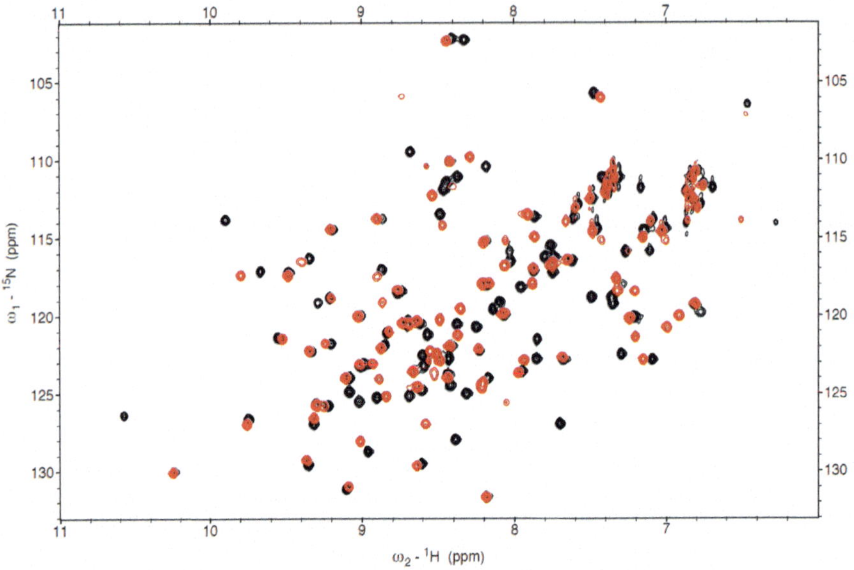

Fig. 13.3 An example of an NMR-titration experiment/chemical shift perturbation (CSP) experiment. The HSQC spectra of 0.1 mM of ^{15}N-labeled mouse ZO-1 PDZ1 domain with (red) or without (black) its physiological-ligand-derived peptide, Pep-Claudin3, were recorded, and the two spectra were overlaid

13.3.1.2 SAR-by-NMR

"SAR-by-NMR," proposed by Fesik et al., is a pioneering work that designed and synthesized a compound with improved affinity from experimentally identified fragments [40]. In brief, it is assumed that two side-by-side ligand binding sites, *A* and *B*, are identified on a ^{15}N-labeled protein, and all the ^{1}H—^{15}N cross peaks, or at least those close to the two binding sites, are assigned. Next, a ligand *a* that binds to *A* is discovered using NMR titration; the chemical structure of the ligand *a* is then optimized to ligand *a'* with higher affinity. The CSP of the protein amide groups provides a guide during the structural optimization process, as it allows to monitor the fitness of the fragment molecules into the binding site *A*. In parallel, a ligand *b* that binds to the site *B* is discovered and optimized into ligand *b'*. Finally, a novel ligand *c* is generated by connecting ligand *a'* and ligand *b'* with an appropriate linker moiety.

13.3.2 Representation and Data Analysis Obtained from CSP Experiments

13.3.2.1 Chemical Shift Mapping

The second important use of the NMR titration experiment is determining the molecular interface of the ligand binding site on the surface of the target protein. If the target protein's three-dimensional structure was determined using X-ray crystallography, solution NMR, or estimated using homology modeling or AI-based prediction, then the nuclei exhibiting large chemical shift changes are mapped onto the surface of the target protein structure. Needless to say, the specific signal assignment of the amide groups or methyl groups in the corresponding HSQC spectrum must be accomplished beforehand. It might seem logical to perform mapping on an atom-to-atom basis, but residue-basis coloring is better for easy interpretation—that is, all atoms belonging to the residue bearing the amide or methyl group can be colored for better visual effect. This is referred to as "chemical shift mapping" (Fig. 13.2).

Before mapping, the chemical shift changes are normalized according to the following equation:

$$\Delta\delta = [(\Delta\delta_{1H})^2 + (\Delta\delta_{15N}/w)^2]^{1/2} \tag{13.1}$$

where $\Delta\delta_{1H}$ and $\Delta\delta_{15N}$ are the chemical shift changes of the ^{1}H and ^{15}N chemical shifts of an amide N–H group, respectively, and w is the normalization weight. Usually, $w = 5\sim7$ is used to balance the different ranges of the ^{1}H and ^{15}N chemical shift distributions of the amide cross peaks.

Based on the amplitude of the normalized chemical shift change, $\Delta\delta$, each residue is painted in different colors. The threshold value of the coloring is determined arbitrarily. Williamson et al. offered a reasonable recommendation for determining the coloring threshold [41, 42]:

(1) Calculate the average, $\Delta\delta_{ave}$, and the standard deviation, σ, for all the groups.
(2) Omit the groups of changes larger than $\Delta\delta_{ave} + 3\sigma$, and re-calculate a new average, $\Delta\delta_{ave}$', and the standard deviation, σ', using the rest of the groups.
(3) Color the residues with $\Delta\delta$ larger than $\Delta\delta_{ave}$' + σ'.

In this protocol, step 2 helps in avoiding the deleterious effects of groups with accidentally large chemical shift changes. Alternatively, the authors recommended avoiding only those groups on the surface-exposed residues.

A flaw of quantitative CSP analysis is the lack of an obvious correlation between the amplitude of CSP and the strength of the interaction. It is not always true that the residues with large CSPs are critical in molecular recognition. Positive and negative chemical shift changes are induced by various factors, and two opposing factors may cancel out to accidentally give nearly zero values. In such a case, the amino acid residue important for the molecular interactions exhibits insignificant CSP. In another case, the residues located far from the ligand binding site may exhibit large CSPs by undergoing allosteric conformational changes upon ligand binding. One must thus recognize this limitation and be extremely careful about interpreting the chemical shift mapping. The chemical shift mapping problems can be largely resolved by performing magnetization saturation transfer from a ligand to a highly deuterated protein sample [38, 43]. This method is called the cross-saturation experiment.

13.3.2.2 HADDOCK Modeling

One important goal of protein–ligand interaction NMR experiments is to determine the 3D structure of the protein–ligand complex, which is useful for further rational drug design applications. Although CSP information is mainly derived from either the backbone amide signals or the sidechain methyl signals, it is still useful for this purpose. Indeed, it can be used to validate the structural model of the complex obtained by any docking simulation program. AutoDockFilter is a famous example of this "post-filtering" strategy [44]. However, the HADDOCK program seems more powerful and comprehensive for building a reliable protein–ligand complex model with a user-friendly interface [45]. The program uses CSP data to generate mutually ambiguous interaction restraints, which are then used for the docking simulation. HADDOCK, it should be noted, can accept some other types of experimental NMR information, such as data derived from STD or cross-saturation experiments.

13.3.2.3 K_D Determination Using NMR Titration

Here, we introduce a basic mathematical analysis to determine K_D values using the NMR titration experiment. In a system in which the ligand-free state and the bound state exist in a two-state exchange equilibrium, and in which association and dissociation are fast enough on the chemical shift time scale, all the cross peaks move smoothly during ligand titration. Typically, the analysis is focused on a single cross peak that exhibits the largest CSP upon ligand binding in the ^1H-^{15}N (or ^1H-^{13}C) HSQC spectrum.

In a simple two-state equilibrium and a single binding site model, K_D is defined using the following equation:

$$K_D = R \cdot L / RL \tag{13.2}$$

where R and L are the concentrations of the free protein (or free receptor) and the free ligand, respectively. RL represents the concentration of the complex. The total concentration of the protein and the ligand, R_{total} and L_{total}, are given by the following equations:

$$R_{total} = R + RL \tag{13.3}$$

$$L_{total} = L + RL \tag{13.4}$$

The normalized chemical shift difference, Δδ, and the maximum chemical shift change, $\Delta\delta_{max}$, are connected to each other as follows:

$$\Delta\delta = \Delta\delta_{max} \cdot (1/\, 2R_{total}) \left[(R_{total} + L_{total} + K_D) - \{(R_{total} + L_{total} + K_D)^2 - 4\, R_{total} \cdot L_{total}\}\right]^{1/2} \tag{13.5}$$

The normalized chemical shift change, Δδ, is plotted against the total concentration of the ligand, L_{total}, to give a saturation curve. A non-linear least square fitting algorithm, with a given R_{total}, determines the K_D and $\Delta\delta_{max}$ values. In some cases, several cross-peaks can be collectively used to determine a single K_D value and a set of $\Delta\delta_{max}$ values by a global fitting algorithm. Of course, one must assume that these cross-peaks monitor the same binding process. A suitable range of K_D values for the NMR titration experiment is between 1 mM and 1 μM. If the affinity is weaker than $K_D = 1$ mM, the titration experiment requires an excessive amount of a ligand and is thus unfeasible. If the affinity is stronger than $K_D = 1$ μM, the protein concentration must be reduced, which renders the NMR measurement insensitive. Unfortunately, due to few titration data points, usually five to ten, the precision of the K_D determination is not high. Nevertheless, NMR-based K_D determination has tremendous merits, including (1) the unnecessity of the immobilization of the protein and the ligand to solid supports, beads, or sensor chip surfaces; (2) the unnecessity of the modification of the protein with any fluorescent dyes; and (3) the opportunity to directly observe the exchange equilibrium of protein–ligand interactions in a solution.

13.3.2.4 The "Per-Residue Plots" Analysis of CSP Data to Distinguish the Different Binding Modes of Ligands

Though chemical shift mapping is an intuitive and straightforward method to visualize CSP data and illustrate an overview of the ligand binding site, it can offer limited information regarding ligands' binding modes. To overcome this disadvantage, the so-called "per-residue plots" analysis of CSP data was used, with Riedinger et al. successfully utilizing it to distinguish the two different binding modes of 10 MDM2 inhibitors [46]. They focused on a limited number of residues surrounding the binding pocket of the N-terminal p53 binding region of MDM2. This method is remarkably different from chemical shift mapping, as it (1) focuses on a limited number of backbone amide signals and (2) uses ^{15}N and ^{1}H CSPs with signs separately instead of normalized amide group chemical shifts. Regarding the results, the signed CSPs of 10 compounds were graphically clustered into three groups. The different clusters corresponded to the different ligand binding orientations to the same ligand binding pocket. This method is thus particularly useful for classifying several ligand candidates according to their binding mode.

13.3.2.5 Hierarchical Cluster Analysis (HCA) of CSP Data

Similar to "per-residue plots" analysis, Hiroaki et al. recently employed hierarchical cluster analysis (HCA) to analyze several non-steroid anti-inflammatory drugs (NSAIDs) that bound to ZO-1-PDZ1 [47]. HCA is a common statistical method used to analyze the similarity of data and clustering them. In theory, CSP data may contain two distinct pieces of information: affinity and the key residues of ligand binding. Hiroaki et al. focused on highlighting the variety of key residues corresponding to different ligands. To this end, the cosine similarity of CSP data for clustering instead of Euclidean distances was prepared and subjected to HCA. Finally, based on the differences between the key binding residues, the nine NSAIDs were classified into two clusters. While the "per-residue plots" analysis requires pre-defining the residues to be analyzed, HCA does not demand such a prerequisite and is thus widely applicable.

13.3.2.6 Principal Component Analysis (PCA) of CSP Data

Most of the common CSP-data-representation methods—chemical shift mapping, HADDOCK modeling, "per-residue plots" analysis, and HCA—may have a limitation of application that researchers often tend to ignore: these methods can be applied only when the target protein's structure is relatively rigid and the changes in the chemical shift of the signals are mainly attributable to the ligand's spatial proximity and not to structural fluctuations. Regarding flexible targets, such as intrinsically disordered proteins (IDPs), this basic premise does not fit. In such a case, we recommend using PCA instead of CSP mapping [48]. Sakurai et al. pioneered the use of PCA for CSP experiments, demonstrating its advantages while studying the folding intermediate of βMG [49, 50]. Hiroaki et al. applied a similar approach to visualize the CSP data of Aβ(1–42) and several fibril-formation inhibitors [48]. Some compounds induced chemical shift changes on many residues on Aβ(1–42), either directly or indirectly. In addition, since Aβ(1–42) is known to be disordered in an aqueous condition, chemical shift mapping was not applicable. These works demonstrated PCA's power in visualizing the similarities and differences in heteronuclear single quantum coherence (HSQC) chemical shift information in reduced dimensions. The various data-representation methods of CSP experiments are summarized in Fig. 13.4.

13.4 Experimental Aspects of the NMR Study of Protein–Ligand Interactions

13.4.1 Stable Isotope Labeling

Most protein-based approaches for protein–ligand interactions rely on 2D heteronuclear chemical shift correlation experiments, such as ^{1}H-^{15}N HSQC and ^{1}H-^{13}C HSQC. Usually, the cross peaks in a 2D ^{1}H-^{1}H NMR spectrum of the target protein are heavily overlapped. Such signal overlapping can be solved by introducing a second heteronuclear dimension. The natural abundances of ^{15}N and ^{13}C are 0.4% and 1.1%, respectively. In the protein-based approach, the first step is to decide the isotope labeling strategy of the target protein. The most cost-effective isotope labeling is ^{15}N-uniformly labeling using the *Escherichia coli* expression system. *E. coli* cells grow in minimal media containing 2 to 4 g/L of glucose ($^{12}C_6$ or $^{13}C_6$ if needed) as the sole carbon source and 0.5 to 1 g/L of ^{15}N-ammonium chloride as the sole nitrogen source. Trace amounts of vitamins, minerals, and nucleosides are often supplemented. Tip: Do not use milliQ water when preparing minimal media. Tap water is a better choice for growing *E. coli* cells in minimal media.

Other microorganism-based recombinant protein expression systems for isotope labeling have also been developed, including *Brevibacillus choshinensis* [51], *Pichia pastoris* [52, 53], *Schizosaccharomyces pombe* [54], and *Kluyveromyces lactis* [55]. Since these microorganisms can grow in minimal media that contain simple nitrogen-source compounds, they are cost-effective. If the target protein is difficult to produce in these microorganism-based expression systems, cell cultures of higher eukaryotes can be used. Regarding insect cells, the baculovirus-Sf9 insect cell [56] as well as the *Drosophila* S2 cell [57] expression system were employed to prepare the isotope-labeled protein samples. Regarding plant cells, Ohki et al. established an expression system using tobacco BY-2 cells [58], demonstrating the preparation of sufficient amounts of ^{15}N-labeled proteins, including dihydrofolate reductase, chicken calmodulin, porcine protein kinase C-dependent protein phosphatase-1 inhibitor, and bovine pancreatic trypsin inhibitor. $K^{15}NO_3$ and $^{15}NH_4{}^{15}NO_3$ were used as nitrogen sources. Regarding mammalian cells, Werner et al. first reported the isotope labeling of GPCRs using HEK293 cells [59]. Sastry et al. successfully prepared the isotopically labeled outer domain of the HIV-1 gp120 glycoprotein using HEK293 cells and an adenovirus expression vector [60]. The commercially available NMR-ready synthetic medium for mammalian cells, CGM-6000 (Cambridge Isotope Laboratory), was used. However, the main problem of these methods is cost. Recently, Egorova-Zachernyuk et al. reported a cost-effective recipe for ^{15}N-labeling using HEK293 cells with yeast and algal auto-lysates [61].

Alternatively, several cell-free protein synthesis systems and many commercially available cell-free kits have been developed (reviewed in [62]). In several structural genomics projects, *E. coli* cell-free systems and wheat germ cell-free systems were used for highly automated protein production. Depending on the degree of purification of the cell-free extracts, some of these have limits concerning their use for amino acid-specific labeling, as the cell-free extracts contain various metabolic enzymes in addition to the entire machinery of protein production.

13.4.2 Tips for the NMR Titration Experiment

Chemical shifts are extremely sensitive to environmental factors, such as pH and salt concentrations. Thus, the accidental changes in chemical shift values due to the unwanted solvent effects of the ligand solution must be considered during the NMR titration experiments. If such unwanted solvent effects occur, the chemical shift changes cannot fit into a saturation curve shape. As a control, using titrations with the same solvent without any ligands is helpful. Ideally, the ligand solution should have, through dialysis or ultrafiltration, identical solvent components in the same concentrations as the protein solution. In practice, the volume change during titration should be, preferably, less than 10% and the volume effect on the concentrations of the components must be corrected during data analysis. To satisfy this prerequisite, the concentration of the ligand solution must be strong enough to reduce the volume change. Alternatively, the protein solution and the ligand solution can be premixed according to the preplanned ratios in separate NMR tubes. A relatively large amount of the ligand is necessary, but the volume change does not matter. The latter method is also useful when ligand solubility is limited by the saturating concentration. In drug screenings, candidate drug compounds are frequently hydrophobic; consequently, dimethyl sulfoxide or other organic solvents must be added to the water solution. Obviously, to prevent the side effects of this addition, another titration experiment must be performed. In the case of synthesized peptides, trifluoroacetic acid (TFA) is present in the lyophilized powder of the peptide and must be neutralized before use. However, due to the medium size of peptides, neutralization by dialysis or ultrafiltration is sometimes difficult.

A normal 5-mm diameter NMR tube requires a sample volume of about 0.4 mL. A special micro NMR tube (Shigemi microtubes; https://www.shigemi.co.jp/english) is useful for minimizing the sample volume (Fig. 13.5). In normal uses, an inner glass rod is inserted into an outer glass tube. A typical working sample volume is about 0.12 to 0.15 mL. For titration experiments, the inner glass rod is sometimes impracticable, but the outer glass NMR tube can be still used to reduce the sample volume to 0.25 mL. Adding small volumes of the ligand solution is extremely difficult for beginners. First, the head of an NMR tube is tilted to the side to the brink of spilling. Then, with a micropipette, a small volume of the ligand solution is added to the solution near the mouth of the NMR tube. Alternatively, a pair of two capillary glass tubes (Drummond) tandemly connected by a silicon tube can be used to directly blow out a small volume of the ligand solution into the protein solution at the bottom of the NMR tube. It is convenient to use a wind-up type centrifuge apparatus, one specially tailored for an NMR tube to collect the solution on the wall to the bottom of the NMR tube.

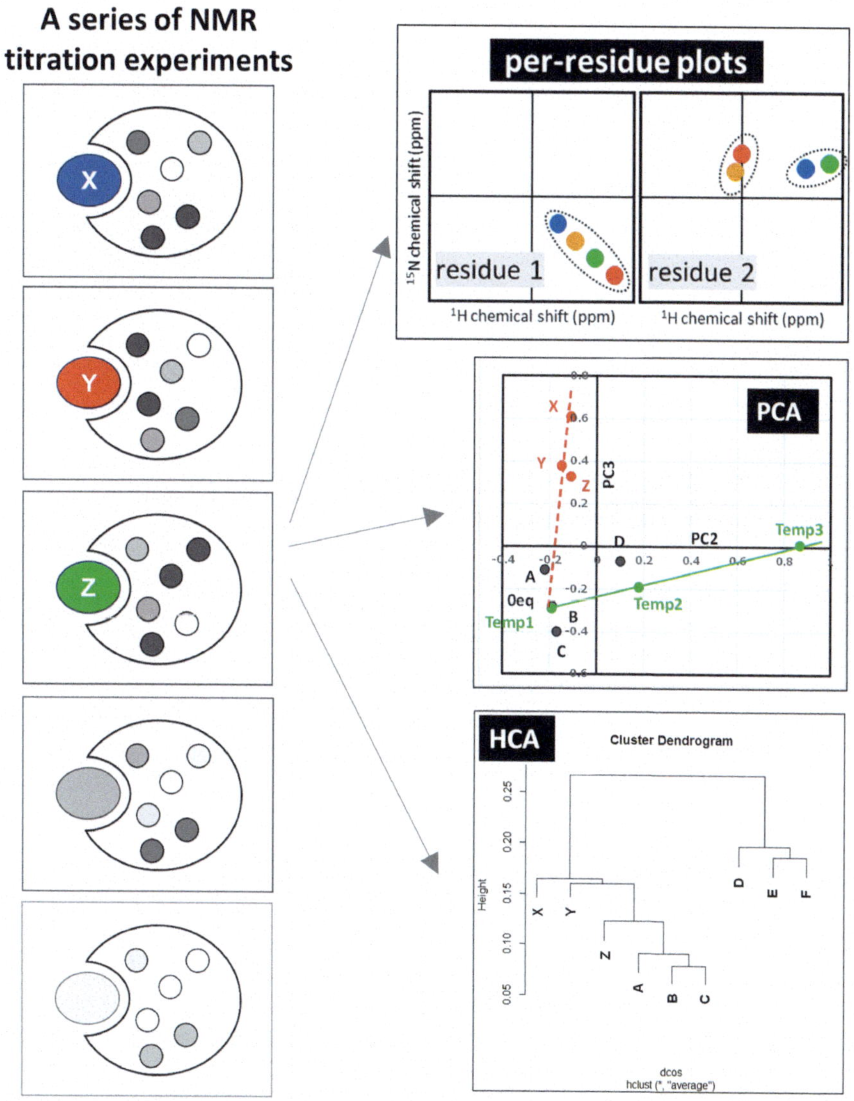

Fig. 13.4 Variety of visualization methods for the results of chemical shift perturbation experiments for understanding protein–ligand interactions. The results of NMR titration experiments with ligands X, Y, Z, and so on (left) are analyzed by "per-residue plot" (right top), principal component analysis (PCA, right middle), and hierarchical cluster analysis (HCA, right bottom), respectively. In each visualization method, ligands are classified into some clusters based on the variety of contact residues

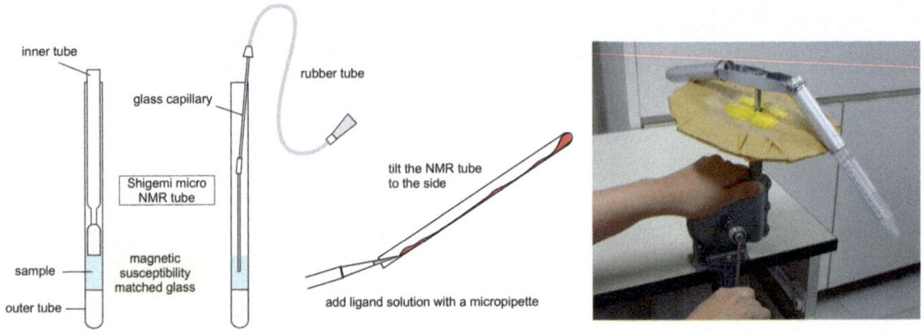

Fig. 13.5 Sample handling tips for NMR titration study. The left panel shows two ways to add a small volume of ligand solution in an NMR tube. The right panel shows a wind-up type centrifuge apparatus for the collection of the solution on the wall to the bottom of the NMR tube

13.5 Other Complementary Methods for Studying Protein–Ligand Interactions

In recent decades, various non-NMR methods combined with FBDD have become indispensable core technologies in early-stage drug discovery (see reviews [63, 64]), especially in the development of protein–protein interaction modulators [65]: surface plasmon resonance (SPR), differential scanning fluorometry (DSF), and isothermal titration calorimetry (ITC) [66, 67]. Employing these methods complementarily is always useful for confirming ligand binding and the determined K_D values.

13.6 Perspective

This review summarized the recent solution NMR techniques for protein–ligand interactions (Table 13.1). Solution NMR is an indispensable tool for detecting and analyzing protein–ligand interactions regardless of various ranges of their dissociation constant, K_D. The versatility of NMR spectroscopy can influence the entire drug development process, from upstream basic research to downstream clinical applications. Further development and improvement of solution NMR will create new trends in pharmaceutical applications, FBDD, the development of protein–protein interaction inhibitors, and peptide-mimetics strategy.

References

1. Pellecchia, M., Sem, D.S., Wüthrich, K.: NMR in drug discovery. Nat. Rev. Drug Discov. **1**, 211–219 (2002)

2. Pellecchia, M., et al.: Perspectives on NMR in drug discovery: a technique comes of age. Nat. Rev. Drug Discov. **7**, 738–745 (2008)
3. Cala, O., Guillière, F., Krimm, I.: NMR-based analysis of protein-ligand interactions. Anal. Bioanal. Chem. **406**, 943–956 (2014)
4. Hiroaki, H.: Recent applications of isotopic labeling for protein NMR in drug discovery. Expert Opin. Drug Discov. **8**, 523–536 (2013)
5. Viegas, A., Manso, J., Nobrega, F.L., Cabrita, E.J.: Saturation-transfer difference (STD) NMR: a simple and fast method for ligand screening and characterization of protein binding. J. Chem. Edu. **88**, 990–994 (2011)
6. Dalvit, C., Fogliatto, G., Stewart, A., Veronesi, M., Stockman, B.: WaterLOGSY as a method for primary NMR screening: practical aspects and range of applicability. J. Biomol. NMR **21**, 349–359 (2001)
7. Shinya, S., et al.: ^{19}F Chemical library and ^{19}F-NMR for a weakly bound complex structure. RSC Med. Chem. **13**, 1100–1111 (2022)
8. Mizukoshi, Y., Abe, A., Takizawa, T., Hanzawa, H., Fukunishi, Y., Shimada, I., Takahashi, H.: An accurate pharmacophore mapping method by NMR spectroscopy. Angew. Chem. Int. Ed. Engl. **51**, 1362–1365 (2012)
9. Sánchez-Pedregal, V.M., Reese, M., Meiler, J., Blommers, M.J.J., Griesinger, C., Carlomagno, T.: The INPHARMA method: protein-mediated interligand NOEs for pharmacophore mapping. Angew. Chemie. **117**, 4244–4247 (2005)
10. Becattini, B., Culmsee, C., Leone, M., Zhai, D., Zhang, X., Crowell, K.J., Rega, M.F., Landshamer, S., Reed, J.C., Plesnila, N., Pellecchia, M.: Structure-activity relationships by interligand NOE-based design and synthesis of antiapoptotic compounds targeting bid. Proc. Natl. Acad. Sci. U. S. A. **103**, 12602–12606 (2006)
11. Ni, F.: Recent developments in transferred NOE methods. Prog. Nucl. Magn. Reson. Spectrosc. **26**, 517–606 (1994)
12. Rees, D.C., Congreve, M., Murray, C.W., Carr, R.: Fragment-based lead discovery. Nat. Rev. Drug Discov. **3**, 660–672 (2004)
13. Ciulli, A., Williams, G., Smith, A.G., Blundell, T.L., Abell, C.: Probing hot spots at protein-ligand binding sites: a fragment-based approach using biophysical methods. J. Med. Chem. **49**, 4992–5000 (2006)
14. Orita, M., Warizaya, M., Amano, Y., Ohno, K., Niimi, T.: Advances in fragment-based drug discovery platforms. Expert Opin. Drug Discov. **4**, 1125–1144 (2009)
15. Claasen, B., Axmann, M., Meinecke, R., Meyer, B.: Direct observation of ligand binding to membrane proteins in living cells by a saturation transfer double difference (STDD) NMR spectroscopy method shows a significantly higher affinity of integrin alpha(IIb)Beta3 in native platelets than in liposomes. J. Am. Chem. Soc. **127**, 916–919 (2005)
16. Furihata, K., Shimotakahara, S., Shibusawa, Y., Tashiro, M.: An effective pulse sequence for detecting a ligand binding with a protein receptor using a WET sequence and the repeated Z-Filters. Anal. Sci. **26**, 1107–1110 (2010)
17. Bergeron, S.J., Henry, I.D., Santini, R.E., Aghdasi, A., Raftery, D.: Saturation transfer double-difference NMR spectroscopy using a dual solenoid microcoil difference probe. Magn. Reson. Chem. **46**, 925–929 (2008)
18. Kemper, S., Patel, M.K., Errey, J.C., Davis, B.G., Jones, J.A., Claridge, T.D.W.: Group epitope mapping considering relaxation of the ligand (GEM-CRL): including longitudinal relaxation rates in the analysis of saturation transfer difference (STD) experiments. J. Magn. Reson. **203**, 1–10 (2010)
19. Monaco, S., Tailford, L.E., Juge, N., Angulo, J.: Differential epitope mapping by STD NMR spectroscopy to reveal the nature of protein-ligand contacts. Angew. Chem. Int. Ed. Engl. **56**, 15289–15293 (2017)
20. Monaco, S., Walpole, S., Doukani, H., Nepravishta, R., Martínez-Bailén, M., Carmona, A.T., Ramos-Soriano, J., Bergström, M., Robina, I., Angulo, J.: Exploring multi-subsite binding pockets in proteins: DEEP-STD NMR fingerprinting and molecular dynamics unveil a cryptic subsite at the GM1 binding pocket of cholera toxin B. Chem. – Eur. J. **26**, 10024–10034 (2020)

21. Ramirez, B.E., Antanasijevic, A., Caffrey, M.: Ligand screening using NMR. In: Anderson, W.F. (ed.) Structural Genomics and Drug Discovery: Methods and Protocols. Methods in Mol. Biol. 1140 305–313 2014
22. Gossert, A.D., Henry, C., Blommers, M.J.J., Jahnke, W., Fernández, C.: Time efficient detection of protein-ligand interactions with the polarization optimized PO-WaterLOGSY NMR experiment. J. Biomol. NMR **43**, 211–217 (2009)
23. Hu, J., Eriksson, P.O., Kern, G.: Aroma WaterLOGSY: a fast and sensitive screening tool for drug discovery. Magn. Reson. Chem. **48**, 909–911 (2010)
24. Sun, P., Jiang, X., Jiang, B., Zhang, X., Liu, M.: Biomolecular ligands screening using radiation damping difference WaterLOGSY spectroscopy. J. Biomol. NMR **56**, 285–290 (2013)
25. Ley, N.B., Rowe, M.L., Williamson, R.A., Howard, M.J.: Optimising selective excitation pulses to maximise saturation transfer difference NMR spectroscopy. RSC Adv. **4**, 7347–7351 (2014)
26. Antanasijevic, A., Ramirez, B., Caffrey, M.: Comparison of the sensitivities of WaterLOGSY and saturation transfer difference NMR experiments. J. Biomol. NMR **60**, 37–44 (2014)
27. Mureddu, L.G., Ragan, T.J., Brooksbank, E.J., Vuister, G.W.: CcpNmr AnalysisScreen, a new software programme with dedicated automated analysis tools for fragment-based drug discovery by NMR. J. Biomol. NMR **74**, 565–577 (2020)
28. LeBlanc, R.M., Mesleh, M.F.: A drug discovery toolbox for nuclear magnetic resonance (NMR) characterization of ligands and their targets. Drug Discov. Today Technol. **37**, 51–60 (2020)
29. Dalvit, C.: Ligand-and substrate-based ^{19}F NMR screening: principles and applications to drug discovery. Prog. Nucl. Magn. Reson. Spectrosc. **51**, 243–271 (2007)
30. Dalvit, C., Vulpetti, A.: Ligand-based fluorine NMR screening: principles and applications in drug discovery projects. J. Med. Chem. **62**, 2218–2244 (2019)
31. Carlomagno, T.: NMR in natural products: understanding conformation, configuration, and receptor interactions. Nat. Prod. Rep. **29**, 536–554 (2012)
32. Krimm, I.: INPHARMA-based identification of ligand binding site in fragment-based drug design. MedChemComm **3**, 605–610 (2012)
33. Skjærven, L., Codutti, L., Angelini, A., Grimaldi, M., Latek, D., Monecke, P., Dreyer, M.K., Carlomagno, T.: Accounting for conformational variability in protein-ligand docking with NMR-guided rescoring. J. Am. Chem. Soc. **135**, 5819–5827 (2013)
34. Cimbro, R., et al.: Tyrosine-sulfated V2 peptides inhibit HIV-1 infection via coreceptor mimicry. EBioMedicine **10**, 45–54 (2016)
35. Hibino, E., Goda, N., Hisada, M., Tenno, T., Hiroaki, H.: Direct inhibition of the first PDZ domain of ZO-1 by glycyrrhizin is a possible mechanism of tight junction opening of Caco-2 cells. Food Func. **13**, 1953–1964 (2022)
36. Sledz, P., Silvestre, H.L., Hung, A.W., Ciulli, A., Blundell, T.L., Abell, C.: Optimization of the interligand overhauser effect for fragment linking: application to inhibitor discovery against mycobacterium tuberculosis pantothenate synthetase. J. Am. Chem. Soc. **132**, 4544–4545 (2010)
37. Rega, M.F., Wu, B., Wei, J., Zhang, Z., Cellitti, J.F., Pellecchia, M.: SAR by interligand nuclear overhauser effects (ILOEs) based discovery of acylsulfonamide compounds active against Bcl-XL and Mcl-1. J. Med. Chem. **54**, 6000–6013 (2011)
38. Shimada, I.: NMR techniques for identifying the interface of a larger protein-protein complex: cross-saturation and transferred cross-saturation experiments. Methods Enzymol. **394**, 483–506 (2005)
39. Tenno, T., Goda, N., Umetsu, Y., Ota, M., Kinoshita, K., Hiroaki, H.: Accidental interaction between PDZ domains and diclofenac revealed by NMR-assisted virtual screening. Molecules **18**, 9567–9581 (2013)
40. Shuker, S.B., Hajduk, P.J., Meadows, R.P., Fesik, S.W.: Discovering high-affinity ligands for proteins: SAR by NMR. Science **274**, 1531–1534 (1996)
41. Schumann, F.H., Riepl, H., Maurer, T., Gronwald, W., Neidig, K.-P., Kalbitzer, H.R.: Combined chemical shift changes and amino acid specific chemical shift mapping of protein-protein interactions. J. Biomol. NMR **39**, 275–289 (2007)

42. Williamson, M.P.: Using chemical shift perturbation to characterise ligand binding. Prog. Nucl. Magn. Reson. Spectrosc. **73**, 1–16 (2013)
43. Shimada, I., Ueda, T., Matsumoto, M., Sakakura, M., Osawa, M., Takeuchi, K., Nishida, N., Takahashi, H.: Cross-saturation and transferred cross-saturation experiments. Prog. Nucl. Magn. Reson. Spectrosc. **54**, 123–140 (2009)
44. Stark, J., Powers, R.: Rapid protein-ligand costructures using chemical shift perturbations. J. Am. Chem. Soc. **130**, 535–545 (2008)
45. Van Zundert, G.C.P., Rodrigues, J.P.G.L.M., Trellet, M., Schmitz, C., Kastritis, P.L., Karaca, E., Melquiond, A.S.J., Van Dijk, M., De Vries, S.J., Bonvin, A.M.J.J.: The HADDOCK2.2 web server: user-friendly integrative modeling of biomolecular complexes. J. Mol. Biol. **428**, 720–725 (2016)
46. Riedinger, C., et al.: Analysis of chemical shift changes reveals the binding modes of isoindolinone inhibitors of the MDM2-P53 interaction. J. Am. Chem. Soc. **130**, 16038–16044 (2008)
47. Tenno, T., Kataoka, K., Goda, N., Hiroaki, H.: NMR-guided repositioning of non-steroidal anti-inflammatory drugs into tight junction modulators. Int. J. Mol. Sci. **22**, 2583 (2021)
48. Iwaya, N., Goda, N., Matsuzaki, M., Narita, A., Shigemitsu, Y., Tenno, T., Abe, Y., Hoshi, M., Hiroaki, H.: Principal component analysis of data from NMR titration experiment of uniformly ^{15}N labeled amyloid beta (1–42) peptide with osmolytes and phenolic compounds. Arc. Biochem. Biophys. **690**, 108446 (2020)
49. Sakurai, K., Goto, Y.: Principal component analysis of the PH-dependent conformational transitions of bovine beta-lactoglobulin monitored by heteronuclear NMR. Proc. Natl. Acad. Sci. U.S.A. **104**, 15346–15351 (2007)
50. Sakurai, K., Maeno, A., Lee, Y.H., Akasaka, K.: Conformational properties relevant to the amyloidogenicity of β2-microglobulin analyzed using pressure- and salt-dependent chemical shift data. J. Phys. Chem. B **123**, 836–844 (2019)
51. Tanio, M., Tanaka, T., Kohno, T.: ^{15}N isotope labeling of a protein secreted by *Brevibacillus choshinensis* for NMR study. Anal. Biochem. **373**, 164–166 (2008)
52. Rodriguez, E., Krishna, N.R.: An economical method for ^{15}N/^{13}C isotopic labeling of proteins expressed in *Pichia pastoris*. J. Biochem. **130**, 19–22 (2001)
53. Sugiki, T., Ichikawa, O., Miyazawa-Onami, M. Shimada, I., Takahashi H.: Isotopic labeling of heterologous proteins in the yeast *Pichia pastoris* and *Kluyveromyces lactis*. In: Shekhtman, A., Burz, D.S. (ed.) Protein NMR Techniques. Methods Mol Biol. **831**, 19–36 (2012)
54. Takegawa, K., Tohda, H., Sasaki, M., Idiris, A., Ohashi, T., Mukaiyama, H., Giga-Hama, Y., Kumagai, H.: Production of heterologous proteins using the fission-yeast (*Schizosaccharomyces pombe*) expression system. Biotechnol. Appl. Biochem. **53**, 227–235 (2009)
55. Sugiki, T., Shimada, I., Takahashi, H.: Stable isotope labeling of protein by *Kluyveromyces lactis* for NMR study. J. Biomol. NMR **42**, 159–162 (2008)
56. Strauss, A., Bitsch, F., Fendrich, G., Graff, P., Knecht, R., Meyhack, B., Jahnke, W.: Efficient uniform isotope labeling of Abl kinase expressed in baculovirus-infected insect cells. J. Biomol. NMR **31**, 343–349 (2005)
57. Meola, A., Deville, C., Jeffers, S.A., Guardado-Calvo, P., Vasiliauskaite, I., Sizun, C., Girard-Blanc, C., Malosse, C., van Heijenoort, C., Chamot-Rooke, J., Krey, T., Guittet, E., Pêtres, S., Rey, F.A., Bontems, F.: Robust and low cost uniform ^{15}N-labeling of proteins expressed in *Drosophila* S2 cells and *Spodoptera frugiperda* Sf9 cells for NMR applications. J. Struct. Biol. **188**, 71–78 (2014)
58. Ohki, S., Dohi, K., Tamai, A., Takeuchi, M., Mori, M.: Stable-isotope labeling using an inducible viral infection system in suspension-cultured plant cells. J. Biomol. NMR **42**, 271–277 (2008)
59. Werner, K., Richter, C., Klein-Seetharaman, J., Schwalbe, H.: Isotope labeling of Mammalian GPCRs in HEK293 cells and characterization of the C-terminus of bovine rhodopsin by high resolution liquid NMR spectroscopy. J. Biomol. NMR **40**, 49–53 (2008)
60. Sastry, M., Xu, L., Georgiev, I.S., Bewley, C.A., Nabel, G.J., Kwong, P.D.: Mammalian production of an isotopically enriched outer domain of the HIV-1 Gp120 glycoprotein for NMR spectroscopy. J. Biomol. NMR **50**, 197–207 (2011)

61. Egorova-Zachernyuk, T.A., Bosman, G.J.C.G.M., DeGrip, W.J.: Uniform stable-isotope labeling in Mammalian cells: formulation of a cost-effective culture medium. Appl. Microbiol. Biotechnol. **89**, 397–406 (2011)
62. Takeda, M., Kainosho, M.: Cell-free protein production for NMR studies. Methods Mol. Biol. **831**, 71–84 (2012)
63. Lepre, C.A.: Practical aspects of NMR-based fragment screening. Methods Enzymol. **493**, 219–239 (2011)
64. Harner, M.J., Frank, A.O., Fesik, S.W.: Fragment-based drug discovery using NMR spectroscopy. J. Biomol. NMR **56**, 65–75 (2013)
65. Scott, D.E., Ehebauer, M.T., Pukala, T., Marsh, M., Blundell, T.L., Venkitaraman, A.R., Abell, C., Hyvönen, M.: Using a fragment-based approach to target protein-protein interactions. ChemBioChem **14**, 332–342 (2013)
66. Wielens, J., Headey, S.J., Rhodes, D.I., Mulder, R.J., Dolezal, O., Deadman, J.J., Newman, J., Chalmers, D.K., Parker, M.W., Peat, T.S., Scanlon, M.J.: Parallel screening of low molecular weight fragment libraries: do differences in methodology affect hit identification? J. Biomol. Screen. **18**, 147–159 (2012)
67. Mashalidis, E.H., Śledź, P., Lang, S., Abell, C.: A three-stage biophysical screening cascade for fragment-based drug discovery. Nat. Protoc. **8**, 2309–2324 (2013)

Chapter 14
Protein Structure and Dynamics Determination by Residual Anisotropic Nuclear Spin Interactions

Shin–ichi Tate

Abstract Proteins in a weakly aligned state restore the anisotropic nuclear spin interactions for dipoles and chemical shifts, which vanish in rapidly tumbling molecules in an isotropic solution. The partially restored anisotropic nuclear spin interactions, also known as 'residual anisotropic spin interactions', observed in a protein offer extra structural information to improve its spatial structure and better understand its structural dynamics over a wider time range than nuclear spin relaxations. The successful utilization of residual anisotropic nuclear spin interactions relies on properly aligning a protein to a degree that balances sensitivity and the magnitude of residual anisotropy. Various types of aligning media are devised to gain residual anisotropy for proteins. The integration of residual anisotropic interactions with the structural data from other methods enabled the investigation of a large amplitude conformational change in a protein, as seen in domain rearrangement. These methods include small-angle X-ray scattering (SAXS), paramagnetic relaxation enhancement (PRE), and pseudo-contact shifts (PCS). Conventional NMR techniques, which rely on short-range structural information from NOEs and scalar nuclear spin couplings, have failed to elucidate such conformational changes of large magnitudes. This chapter covers the residual anisotropy in solution NMR, from the basic theory of residual anisotropic nuclear spin interactions to their application in analyzing protein structure and dynamics.

Keywords RDC · RCSA · TROSY · Structure refinement · Structure dynamics

S. Tate (✉)
Department of Mathematical and Life Sciences, School of Science, Hiroshima University, Hiroshima, Japan
e-mail: tate@hiroshima-u.ac.jp

Research Center for the Mathematics On Chromatin Live Dynamics (RcMcD), Hiroshima University, Hiroshima, Japan

International Institute for Sustainability with Knotted Chiral Meta Matter (WPI-SKCM2), Hiroshima University, Hiroshima, Japan

The NMR Society of Japan (ed.), *Experimental Approaches of NMR Spectroscopy II*,
https://doi.org/10.1007/978-981-97-6830-1_14

14.1 Why Do We Need Residual Anisotropic Nuclear Spin Interactions in Solution NMR?

A large number of high-resolution crystal structures of proteins are available, and the number continues to grow. Many people expected that the extensive collection of protein structures would help understand the relationship between protein structure and function, but this has not been the case. The crystal structures and solution structures of proteins determined by NMR reveal only a few possible structural states. The accurate views of protein structure dynamics are essential to deepen our understanding of how proteins work. The functionally relevant protein dynamics may involve large-scale motions associated with domain rearrangement. (Fig. 14.1).

The high-resolution crystal structures provide us with a comprehensive understanding of how proteins and ligands interact. Machine learning on the interactions in the Protein Data Bank (PDB) database helps improve the accuracy of computer-based predictions for molecular recognition by proteins [1]. Using chemical shift perturbations to identify the interacting surface in proteins binding to a target, we can model the structure of the protein-target molecule complex. This NMR-based protein complex modeling requires analyzing the potential chemical bonding network formed by the residues on the identified protein surface with the target molecule through informatics analysis of the structural databases [2]. The approaches that rely on the database assume that the protein structure does not change significantly from the structure determined in the crystal. This naïve assumption is not always valid for proteins in binding to targets. Many examples demonstrate that proteins change their structures upon binding to targets in solution; the bound forms are significantly different from their apo-forms determined in crystals [3]. The structural change in a

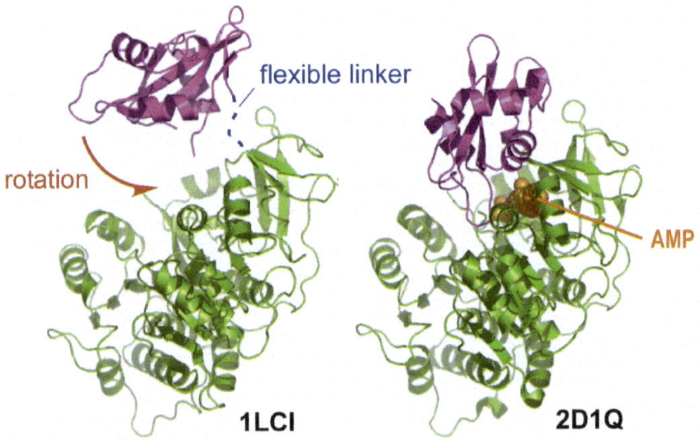

Fig. 14.1 Protein domain rearrangement observed for Luciferase upon binding to ATP. PDB codes 1LCI and 2D1Q for apo- and AMP-bound forms, respectively. The ATP binding significantly changes the relative domain orientation between the large and small domains of Luciferase

protein is evident in cases where the protein consists of domains connected by flexible linkers or intrinsically disordered regions (IDRs) [4]. The substantial domain rearrangement in a multi-domain protein upon binding to a target is functionally relevant [5, 6].

The analysis of the structural change of a protein, which involves large amplitude motion as seen in domain rearrangement, has been challenging to analyze using conventional NMR methods that depend on the short-range structural parameters obtained from NOEs and scalar couplings [7, 8]. Almost two decades ago, solution NMR revolutionized the analysis of protein structure dynamics by introducing anisotropic nuclear spin interactions to overcome the existing limitation [9, 10]. The anisotropic nuclear spin interactions are inherent physical properties related to nuclear dipolar interactions and magnetic shielding observed as chemical shift anisotropy (CSA). The anisotropic nuclear spin interactions, however, are not observed in proteins when they are in an isotropic solution. In contrast, proteins in an anisotropic medium, as in the solution containing magnetically aligned liquid crystals, can restore the anisotropic nuclear spin interactions [11]. The anisotropic nuclear spin interaction provides tensor values that describe the orientation angles and magnitudes of proteins in relation to the external magnetic field. [11]. The tensor values can help determine the relative domain orientation in a protein with multiple domains [12]. The introduction of anisotropic nuclear spin interactions made it feasible to analyze large-amplitude structural changes in proteins using solution NMR, something that was not achievable with conventional NMR techniques. The anisotropic nuclear spin interactions observed for proteins in anisotropic environments are here referred to as "*residual*" anisotropies, as a portion of these anisotropies remain uncancelled in the measurements.

The residual anisotropies can determine the relative domain orientations but cannot provide the distances between the domains. The combined use of other methods is essential to explore the domain dynamics of a protein. This involves determining the relative orientation and distances of the protein's domains. We may use electron spin resonance (ESR), fluorescence resonance energy transfer (FRET), and X-ray small angle scattering (SAXS) along with the paramagnetic relaxation enhancements (PRE) and pseudo contact shifts (PCS) in NMR, in addition to the residual anisotropies. By combining the other structural analyses mentioned earlier, the residual anisotropies have significantly expanded the applications of NMR, enabling a more precise understanding of protein structure dynamics across various time scales [13].

In this chapter, I will begin with a short overview of the fundamental theories for residual anisotropies, and then use them to examine protein motions with significant amplitudes. I will also mention the TROSY-based approach for determining the molecular structure of high-molecular-weight proteins with residual anisotropies. This method has expanded the size limitations for using residual anisotropies in protein structure analyses [14, 15].

14.2 Theoretical Backgrounds of the Residual Anisotropies

Conventional NMR structure analysis based on the short-range spin interactions does not provide the global structural information required to determine protein molecular shape or morphology. The residual anisotropies, which become apparent when a protein is weakly aligned, provide the orientation angles of the protein relative to the magnetic field. This helps determine the protein's morphology.

Two different types of residual anisotropies can be observed for the amide ^{1}H-^{15}N spin pair on a peptide plane in proteins; one is the nuclear spin dipolar-dipolar interaction, and the other is anisotropic shielding against the external magnetic field, which is referred to as chemical shift anisotropy (CSA) (Fig. 14.2).

The observed anisotropies do not appear on an NMR spectrum for proteins in an isotropic solution because each peptide plane in a rapidly tumbling protein experiences all magnetic field orientations, which cancels out the anisotropies. On the other hand, if a protein is dissolved in an aligning medium, a fraction of proteins will exhibit a specific orientation with respect to the magnetic field depending on its molecular shape [15]. As a result, the anisotropies will partially recover.

In the presence of the magnetically aligned bicelle in a protein solution, the protein is trapped in a confined space formed by a cluster of aligning bicelles. The protein still moves around inside a space, but it often hits the bicellar wall, which hinders certain orientations of the protein and leads to incomplete cancellation of the anisotropies [16, 17]. This incomplete cancellation of the protein anisotropies in a bicellar solution results in the '*residual*' anisotropies.

In a properly prepared aligning liquid crystalline solution, there is sufficient space for proteins to move freely. Proteins still give high-resolution NMR signals under these conditions. In a higher concentration of liquid crystals, the tumbling of the

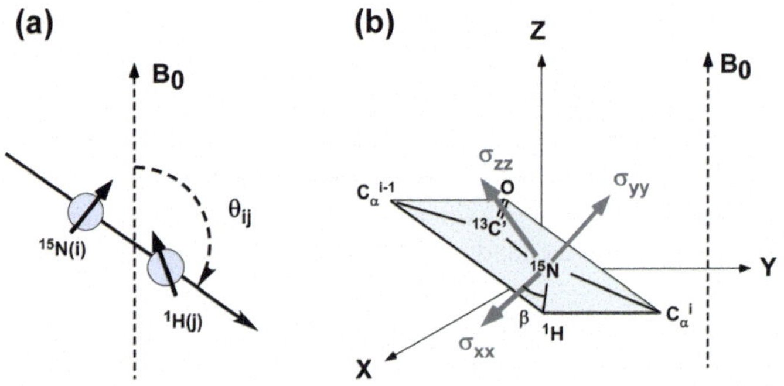

Fig. 14.2 Two anisotropic nuclear spin interactions observed for the peptide bond. **a** Dipolar interaction between ^{1}H and ^{15}N nuclei in the amide bond. The anisotropy for the nuclear spin interaction is defined by the orientation of the NH bond against the magnetic field, B_0. **b** Chemical shift anisotropy (CSA) for ^{15}N nuclear spin in the peptide bond. The shielding anisotropy for ^{15}N spin against the external magnetic field, B_0, is described by a tensorial value

protein becomes severely limited, leading to broader NMR signals and reduced spectral resolution and sensitivity. In experiments to observe residual anisotropies, we carefully adjust the alignment conditions to restore significant anisotropies while maintaining narrow spectral lines on an NMR spectrum. This type of alignment is referred to as 'weak' alignment [11].

Anisotropic dipolar couplings are observed in solid-state NMR, providing through-space interactions between two nuclear spins. The weak alignment condition has allowed the use of anisotropic dipolar couplings in solution NMR, known as residual dipolar couplings (RDCs). The common through-space nuclear dipolar interactions in solution NMR are nuclear Overhauser effects (NOEs) detected between protons within 5 Å [18]. NOEs observed in solution NMR provide distance information between protons, but they do not provide any orientation angles between the proton pairs. The strength of anisotropic dipolar couplings is determined in the alignment tensor frame that defines the molecular orientation to the magnetic field, B_0. The dipolar couplings, therefore, give relative angles of the connecting vector for two interacting nuclear dipoles and the external magnetic field (Fig. 14.2a).

Like the RDCs, the protein under weak alignment produces residual CSAs (RCSAs) that are observed as chemical shift changes in an anisotropic state, distinct from those in the isotropic state [19]. The CSA is another anisotropic nuclear spin interaction observed in solid-state NMR, but it is not observed for proteins in isotropic solution. Amide nitrogen (^{15}N) and carbonyl carbon (^{13}C') in a peptide plane have significant CSAs (Fig. 14.2b). The chemical shifts of these nuclei change when they are in an anisotropic state compared to when they are in an isotropic state. The difference in the chemical shifts corresponds to RCSA. The RCSA values of these nuclei depend on the orientation of the peptide plane against the magnetic field, B_0.

If the peptide plane aligns the s_{zz} axis of amide ^{15}N CSA parallel to the magnetic field, B_0, the chemical shift of the ^{15}N nucleus exhibits a high-field shift (Fig. 14.2b). When the peptide plane aligns the s_{xx} axis of the amide ^{15}N parallel to the magnetic field, the ^{15}N chemical shift induces a low-field shift (Fig. 14.2b). The RCSA of the amide 15N nucleus thus determines the orientation of a peptide plane against the magnetic field. A set of RCSAs of the amide ^{15}N nuclei in a protein determines the molecular orientation of the protein relative to the magnetic field.

The combined effects of RDC and RCSA are noticeable in a ^{1}H-^{15}N TROSY correlation spectrum of the protein in a weakly aligning state [14, 20, 21]. The residual anisotropy in a TROSY spectrum, Dd_{TROSY}, includes the RDC from ^{1}H-^{15}N spin pair (Fig. 14.2a) and the RCSA from ^{15}N CSA in a peptide plane (Fig. 14.2b), which provides more robust structural information for determining the peptide plane orientation to the magnetic field, B_0, compared to relying solely on RDC and RCSA. The combinatorial use of the residual anisotropies is especially beneficial when applied to proteins over 100 kDa. This is because the alignment tensor is determined based on the chemical shift change, Dd_{TROSY}, observed in TROSY that provides the highest sensitivity for ^{1}H-^{15}N correlation spectra for high molecular weight proteins [21].

14.2.1 Residual Dipolar Coupling, RDC

Residual dipolar coupling (RDC) gives the direction of the internuclear vector connecting a pair of nuclei against the external magnetic field, B_0. In the peptide bond of a protein isotopically enriched with ^{15}N, the RDC observed for the amide ^{15}N nucleus defines the angle between the NH bond vector and a magnetic field (Fig. 14.2a).

The RDCs for amide ^{1}H-^{15}N spin pairs are of common choice in protein structure analysis because of the good spectral dispersion in the ^{1}H-^{15}N correlation spectrum. The RDCs are measured from ^{1}H-^{15}N spectra collected without ^{1}H decoupling during ^{15}N nuclear spin evolution time, t_1 (Fig. 14.3). In practice, the IPAP HSQC is used to prevent signal overlaps that diminish the accuracy in reading signal positions [22].

On a ^{1}H-couple ^{1}H-^{15}N HSQC spectrum, each amide bond gives a pair of signals split along the ^{15}N dimension (Fig. 14.3). The split for a protein in an isotropic condition represents a single bond scalar coupling between ^{1}H and ^{15}N nuclei, $^{1}J_{NH}$, which is typically 93 Hz (Fig. 14.3a). The corresponding splits observed for a protein in a weakly aligning state vary residue by residue (Fig. 14.3b). The changes in $^{1}J_{NH}$ values come from the overlaid contribution of the RDCs that become apparent in a weakly aligned state. As described above, the RDC varies according to the NH

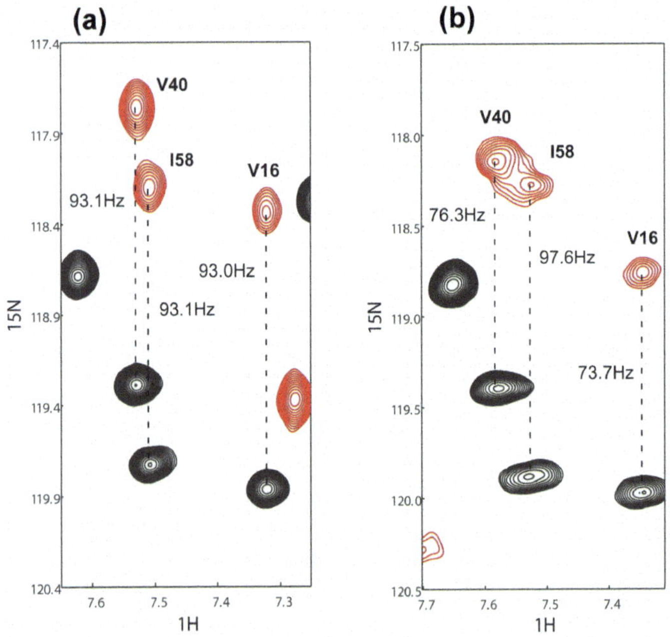

Fig. 14.3 RDC measurement with ^{1}H-coupled HSQC. Isotropic sample **a** and aligned sample **b**. Split with for each paired signal corresponds to $^{1}J_{NH}$. The apparent $^{1}J_{NH}$ is changed for the aligned sample. The difference in $^{1}J_{NH}$ is defined as $^{1}J_{NH}$ (align) – $^{1}J_{NH}$ (isotropic) gives RDC

bond vector orientation to the magnetic field, the observed change in $^1J_{NH}$ value is different in each residue. The RDC is defined as $^1J_{NH}$ (aligned state) - $^1J_{NH}$ (isotropic state).

14.2.2 Theoretical Description on RDC

We focus only the amide ^{1}H-^{15}N spin pair in the following theoretical description of the RDC. In the following formulation, the indices *i* and *j* are used for ^{1}H and ^{15}N nuclei, respectively. The dipolar Hamiltonian to describe the dipolar coupling is written as below:

$$H_{ij}^D(t) = -\left(\frac{\mu_0}{4\pi}\right)\frac{\gamma_i\gamma_j h}{2\pi^2 r_{ij}^3} I_{iz} I_{jz} P_2(\cos\Theta(t)) \tag{14.1}$$

where m_0 is the permeability in a vacuum, *h* is Planck's constant, r_{ij} is the distance between ^{1}H and ^{15}N atoms, g_i and g_j are the gyromagnetic ratios for ^{1}H and ^{15}N nuclear spin, respectively. I_{iz} and S_{jz} are the spin angular-momentum operators for ^{1}H and ^{15}N nuclear spins, respectively. The angular part of the dipolar Hamiltonian is described by the second-rank Legendre function, $P_2(\cos\Theta(t))$, here $\Theta(t)$ is a time-dependent angle between the magnetic field, B_0, and the internuclear vector in the NH bond (Fig. 14.4a). In determining molecular orientation by the RDC, the bond libration effect is negligible due to its faster motion over the protein tumbling that occurs on the nsec time scale. In considering the fast bond libation that happens in the psec time range, the time-averaged effective inter-nuclear distance is used for ^{1}H-^{15}N bond length, r_{ij}; which distance was estimated by NMR to be 1.04 Å, slightly longer than the static bond length 1.02 Å [23].

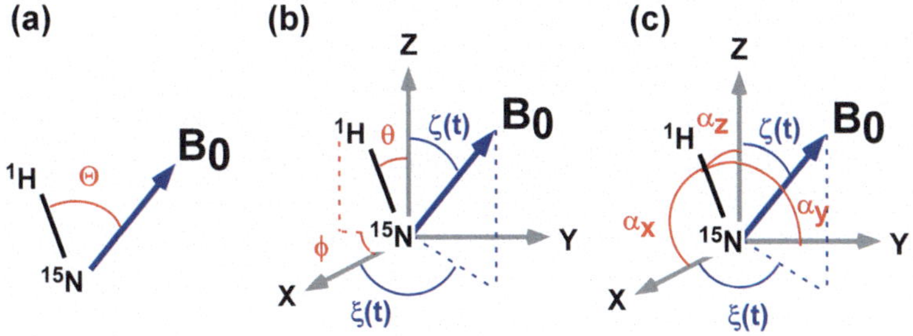

Fig. 14.4 Three different representations for NH bond vector against a magnetic field. **a** Angle dependency of the ^{1}H-^{15}N RDC is described by Θ, the angle between the NH bond vector and a magnetic field. **b** NH bond vector on a molecular coordinate system, whose direction is defined by polar angles (θ, ϕ). Molecular reorientation relative to a magnetic field is described by the time-dependent polar angles $(\zeta(t), \xi(t))$. **c** Direction cosines recast the NH bond vector to three molecular axes

The experimentally measured RDCs, $D_{ij}^{res.}$, is represented as the time-averaged value for the dipolar Hamiltonian, Eq. (14.1).

$$D_{ij}^{res.}(t) = -\left(\frac{\mu_0}{4\pi}\right)\frac{\gamma_i\gamma_j h}{2\pi^2 r_{ij}^3}\langle P_2(\cos\Theta(t))\rangle \tag{14.2}$$

The angle bracket denotes the time-averaged value. When a protein rapidly tumbles in an isotropic solution, no residual anisotropy remains, thus leading to $\langle P_2(\cos\Theta(t))\rangle = 0$. In an anisotropic condition, the molecular tumbling incompletely vanishes the time-averaged term, giving $\langle P_2(\cos\Theta(t))\rangle \neq 0.$. The magnitude of the RDC depends on the orientation of the NH bond vector against the magnetic field (Fig. 14.4a) and the order of the aligning magnitude of the protein in a liquid crystalline medium, which is described by the term $\langle P_2(\cos\Theta(t))\rangle$ in Eq. (14.2).

When using the RDCs as structural parameters, we need to specify the direction of each NH bond vector in the molecular coordinate frame. This frame is a Cartesian coordinate system used to describe all atomic positions in a protein.

The orientation of an NH bond vector in a protein is described by polar angles (*q*, *f*) in the molecular coordinate system (Fig. 14.4b). The time-dependent polar angles describe the orientation of the magnetic field in the molecular coordinate system, $(\zeta(t), \xi(t))$ (Fig. 14.4b): the time-dependent polar angles describe the directional change of the magnetic field sensed by the tumbling protein.

The Legendre function $\langle P_2(\cos\Theta(t))\rangle$ is expanded by the spherical harmonics according to the spherical harmonic addition theorem [24]:

$$P_2(\cos\Theta(t) = \frac{4\pi}{5}\sum\nolimits_{q=-2}^{2} Y_{2q}^*(\theta,\varphi)\langle Y_{2q}(\zeta(t),\xi(t))\rangle \tag{14.3}$$

where the Y_{2q}'s are the normalized spherical harmonics. Using the relations in Eq. (14.3), the RDC ($D_{ij}^{res.}(t)$ in Eq. (14.2)) is expressed with the spherical harmonics:

$$D_{ij}^{res.}(t) = -\left(\frac{\mu_0}{4\pi}\right)\frac{\gamma_i\gamma_j h}{2\pi^2 r_{ij}^3}\frac{4\pi}{5}\sum\nolimits_{q=-2}^{2} Y_{2q}^*(\theta,\varphi)\langle Y_{2q}(\zeta(t),\xi(t))\rangle \tag{14.4}$$

In Eq. (14.4), the five time-averaged spherical harmonics $\langle Y_{2q}(\zeta(t),\xi(t))\rangle$ with q = -2, -1, 0, 1, 2 define the orientation angles and their magnitudes for protein. Recasting these five terms makes a more intuitively acceptable representation to describe the aligning state of the protein, which is Saupe's order matrix. The use of Saupe's order matrix reforms Eq. (14.4) as below:

$$D_{ij}^{res.}(t) = -\left(\frac{\mu_0}{4\pi}\right)\frac{\gamma_i\gamma_j h}{2\pi^2 r_{ij}^3}\frac{4\pi}{5}\sum\nolimits_{k,l=x,y,z} S_{kl}\cos(\alpha_k)\cos(\alpha_l) \tag{14.5}$$

where S_{kl} is an element in Saupe's order matrix. The Saupe order matrix is a traceless, symmetric 3×3 matrix: it comprises five independent elements as described below.

Saupe's order matrix defines each bond vector orientation by direction cosines against the molecular coordinate axes with angles a_x, a_y, and a_z (Fig. 14.4c). Five elements in the Saupe order matrix are described by the time-averaged spherical harmonics $\langle Y_{2q}(\zeta(t), \xi(t))\rangle$ as described below:

$$S_{xz} = \sqrt{\frac{3}{8}}\sqrt{\frac{4\pi}{5}}(\langle Y_{21}(\zeta(t), \xi(t))\rangle - \langle Y_{2-1}(\zeta(t), \xi(t))\rangle)$$

$$S_{yz} = i\sqrt{\frac{3}{8}}\sqrt{\frac{4\pi}{5}}(\langle Y_{21}(\zeta(t), \xi(t))\rangle + \langle Y_{2-1}(\zeta(t), \xi(t))\rangle)$$

$$S_{xy} = i\sqrt{\frac{3}{8}}\sqrt{\frac{4\pi}{5}}(\langle Y_{22}(\zeta(t), \xi(t))\rangle - \langle Y_{2-2}(\zeta(t), \xi(t))\rangle)$$

$$S_{xx} - S_{yy} = \sqrt{\frac{3}{2}}\sqrt{\frac{4\pi}{5}}(\langle Y_{22}(\zeta(t), \xi(t))\rangle + \langle Y_{2-2}(\zeta(t), \xi(t))\rangle)$$

$$S_{zz} = \sqrt{\frac{4\pi}{5}}\langle Y_{20}(\zeta(t), \xi(t))\rangle \tag{14.6}$$

The Saupe order matrix describes the orientation angles and the orders of the aligning magnitude of protein against the magnetic field, B_0. Diagonalization of the Saupe order matrix gives the alignment tensor. The alignment tensor consists of the principal values that represent the orders of the aligning magnitudes along principal axes and the tensorial axes defined in the molecular coordinate system, whose angles define the orientation of protein to the magnetic field.

The Z-axis of the alignment tensor is defined as the most ordered axis, and the others are defined according to the absolute magnitudes of the principal values $|S_{zz}| \geq |S_{yy}| \geq |S_{xx}|$. An asymmetry, $\eta = (S_{xx} - S_{yy})/S_{zz}$, , represents the deviation from the axially symmetric ordering. The orientation of the alignment tensor axes (the principal axes of the alignment tensor) are represented in the molecular coordinate system by the Euler angles.

In the alignment tensor axes system (alignment tensor frame), the RDC, $D_{ij}^{res.}$, is recast in a more simplified and intuitively acceptable form, in which two new parameters, A_a and A_r, are introduced. A_a and A_r are made of the non-zero average spherical harmonics:

$$\langle Y_{20}(\zeta(t), \xi(t))\rangle = \sqrt{\frac{5}{4\pi}}S_{zz} = \sqrt{\frac{5}{4\pi}}A_a$$

$$\langle Y_{2\pm2}(\zeta(t), \xi(t))\rangle = \sqrt{\frac{5}{24\pi}}(S_{xx} - S_{yy}) = \sqrt{\frac{15}{32\pi}}A_r \tag{14.7}$$

Using the parameters defined in Eq. (14.7), $D_{ij}^{res.}$, is transformed into the formulation that is readily applied to the structure analysis with the RDCs:

$$D_{ij}^{res.} = -\left(\frac{\mu_0}{4\pi}\right)\frac{\gamma_i\gamma_j h}{2\pi^2 r_{ij}^3}\left[A_a\left(3\cos^2\theta - 1\right) + \frac{2}{3}A_r\sin^2\theta\cos 2\varphi\right] \quad (14.8)$$

where the polar angles (q, f) define the orientation of the NH bond vector in the alignment tensor frame. A_a and A_r are the axial and rhombic components of the ordering magnitudes, respectively. The orientation of the alignment tensor is described in the molecular coordinate frame with the Euler angles (a, b, g).

14.2.3 Residual Chemical Shift Anisotropy, RCSA

The other anisotropy associated with the amide ^{15}N nucleus comes from the anisotropic nuclear shielding or chemical shift anisotropy, CSA (Fig. 14.2b). CSA is not observed for a protein in an isotropic solution. The peptide plane in a protein that rapidly tumbles in solution experiences changes in direction completely relative to the magnetic field, which averages over the CSA and results in an isotropic chemical shift, d_{iso}. In a solid state, where each molecule aligns in a specific direction to the magnetic field, we can see three different components of the anisotropic chemical shifts for the ^{15}N nucleus in a peptide plane, d_{xx}, d_{yy}, and d_{zz}. These components reprint the directional differences in the magnetic shielding (Fig. 14.2b) [27]. The CSA is partially restored in a weakly aligned protein because the anisotropy is not fully averaged. This incomplete averaging results in a change in the chemical shift from its isotropic shift, d_{iso}, along ^{15}N axis. This is the residual CSA (RCSA) (Fig. 14.5) [19].

The ^{15}N RCSA is defined as $Dd^{15}N$ = $d^{15}N$(aligned) – $d^{15}N$(isotropic). The magnitude of RCSA varies according to the peptide plane orientation to the magnetic field (Fig. 14.2b). In the peptide bond, ^{13}C' also has also substantial CSA, and the RCSA for ^{13}C' can be used similarly as in the ^{15}N RCSA application [28].

The RCSA values, $Dd^{15}N$, relate the CSA tensor in each ^{15}N nucleus to the molecular coordinate frame with Saupe's order matrix elements S_{ij}:

$$\Delta\delta^{15}\mathrm{N} = \frac{2}{3}\sum\nolimits_{i=x,y,z}\sum\nolimits_{j=x,y,z}\sum\nolimits_{k=x,y,z} S_{ij}\cos(\theta_{ik})\cos\left(\theta_{jk}\right)\delta_{kk} \quad (14.9)$$

Here, cos(q_{ij}), denotes the direction cosine between the *i*-axis in the molecular coordinate system and *the j*-axis in the CSA tensor principal axis. The principal value for the CSA tensor axis k is denoted by d_{kk} (kk = xx, yy, and zz), the magnetic shielding magnitude along each principal axis (Fig. 14.2).

The alignment tensor is obtained through diagonalizing Saupe's order matrix, whose matrix elements are determined with a set of $Dd^{15}N$ data using Eq. (14.9). In determining Saupe's order matrix, the pre-defined CSA tenor parameters including b angle to define the CSA tensor position in a peptide plane and two principal values d_{zz} and d_{xx} with assuming the isotropic shift $d_{iso} = (d_{xx} + d_{yy} + d_{zz})/3 = 0.0$: the s_{yy} axis is normal to the peptide plane with the angle b = 19.6 ± 2.5 degrees

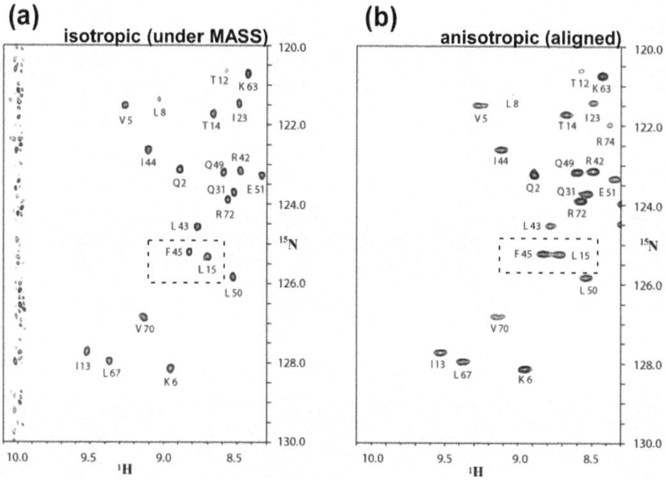

Fig. 14.5 ^{15}N chemical shift change induced by a weak alignment. ^{1}H-^{15}N HSQC spectra are compared for the sample of ^{15}N labeled ubiquitin dissolved in 7.5% (w/v) DMPC/DHPC/CTAB ternary bicelle solution between isotropic and aligned states: (a) under the magic angle sample spinning condition, giving an isotropic spectrum, (b) the aligned condition. Chemical shift changes along the ^{15}N axis caused by a weak alignment are clearly observed, for example, the signals in the dotted square. These spectral changes come from the incomplete cancellation of the ^{15}N chemical shift anisotropy (CSA) effect under a weak aligning condition. The change along the ^{15}N axis is the value for the residual CSA, RCSA

(Fig. 14.2b) [29]. In determining the alignment tensor with the RDCs, the only pre-defined parameter is the libration average bond length, e.g., $< r_{NH} > = 1.04$ Å. In the case of the RCSA, the tensor determination is more pre-knowledge demanding.

The ^{15}N CSA tensor value depends on the local backbone torsion angles [29]. Various methods, including solid-state and solution NMR, have reported the ^{15}N CSA tensor values in protein [19, 27, 29–32]. These data gave consensus ^{15}N CSA tensor parameters for the residues in each type of secondary structure, such as a-helix, b-sheet, and others. In determining the alignment tensor with the RCSAs, the secondary structure-specific ^{15}N CSA tensor parameters will be used as pre-defined parameters. Using the secondary structure-specific ^{15}N CSA tensors gives the alignment tensors that can reproduce the experimentally observed Dd^{15}N values within the experimental errors.

At the end of this section, I briefly describe how the ^{15}N CSA tensor parameters are determined by solution NMR, in which the weak alignment is essentially used.

We reported the ^{15}N CSA tensor determined for the protein in a weakly aligning state in solution, including a magnetically aligned bicelle [19]. In our approach, we applied magic-angle sample spinning (MASS) to collect the RCSA data, Dd^{15}Ns. The MASS to the magnetically aligned bicelles eliminates its magnetically induced ordering, thus, the protein in the bicellar solution completely averages out the anisotropies, while the protein stays in a weakly aligning state unless the MASS is applied [19]. As seen in Eqs. (14.5) and (14.9), the RDCs and RCSAs are represented by Saupe's order matrix. In the same sample, Saupe's order matrix elements, S_{ij}, are shared in the formula for RDC and RCSA as Eqs. (14.5) and (14.9), respectively. Once the Saupe order matrix is determined by the RDCs using the relation in Eq. (14.5), which is also directly used in Eq. (14.9),

We can observe the RCSA, Dd^{15}N, values from the ^{15}N chemical shift changes between the signals for the protein in a weakly aligned state and the same sample under the MASS (Fig. 14.5). MASS is essential to measure the RCSA. Chemical shifts are quite sensitive to the chemical environment around the protein, including co-existing solutes. In accurately measuring RCSAs, we need to keep the sample conditions the same to limit the causes for the change in chemicals to the alignment of proteins. Applying the MASS to the weakly aligned sample fulfills this requirement in measuring the RCSAs; the MASS eliminates the residual anisotropies with keeping the chemical environment the same.

With the RCSAs, Dd^{15}Ns, collected with the MASS and the Saupe order matrix determined by the RDCs, we can determine the ^{15}N CSA tensors with the relation in Eq. (14.9). Because the ^{15}N CSA tensor consists of five independent elements, determining the ^{15}N CSA tensor needs at least five Dd^{15}Ns. In using the single aligned sample, we only determine the averaged ^{15}N CSA tensor over the residues of at least five in the sample. Because of this requirement, the residues in the same type of secondary structure are used to give the secondary structure specific ^{15}N CSA tensor [19].

The residue-specific ^{15}N CSA tensor can be determined once we can collect the RCSAs and RDCs for the same sample in at least five different aligning states, meaning the protein has different orientation angles to the magnetic field. Those samples are not readily accomplished by merely changing the aligning media. However, the amino acid changes to loops in a protein are demonstrated to change the aligning states in the same bicelle medium significantly. With the structurally innocent mutants, Bax and co-workers determined the residue-specific ^{15}N CSA tensors in the protein GB3 based on the RDCs and RCSAs for the protein in six different aligning states [29]. The RCSAs, Dd^{15}N values, for the protein in each aligning state were collected with the MASS. The Saupe order matrices determined from the RDCs were subjected to determine the residue-specific ^{15}N CSA tensor with six different sets of the RCSAs using Eq. (14.9). The results have shown that the ^{15}N CSA tensor shows only a moderate degree of variation from the average value, but it has larger magnitudes for the residue in a-helix (Ds = -173 ± 7 ppm) than b-sheet (Ds = -162 ± 6 ppm) residues [29]; Ds is defined as $(s_{yy} - s_{xx})/s_{zz}$ [33]. The results ensure that the use of secondary structure-specific ^{15}N CSA tensor is practically acceptable in the structure analysis using the RCSA.

14.3 Practical Procedures to Use the Residual Anisotropies

Measuring the residual anisotropies demands a weak alignment of protein in solution to attain the observable changes by anisotropies while keeping the spectral resolution high enough. The alignment order required for measuring residual anisotropies is approximately 10^{-3}, giving around 20 Hz in maximum absolute magnitude for amide ^{1}H-^{15}N RDC. To prepare a weak alignment, we usually use the liquid crystalline medium to assist the protein in aligning to the magnetic field. Most proteins have limited inherent magnetic susceptibility. Therefore, proteins do not spontaneously align even under a strong NMR magnetic field. Some of the proteins, like heme-containing proteins, align without the aid of the magnetically aligning co-solute: They have substantial magnetic susceptibility in a heme group [34]. The double-stranded DNAs show spontaneous magnetic alignment in the NMR magnet [35]. In this section, I will describe typical examples of preparing for weakly aligning samples to measure residual anisotropies.

14.3.1 Magnetically Aligning Liquid Crystalline Media

Magnetically ordering liquid crystalline media are commonly used in collecting the residual anisotropies. Discoidal phospholipid assembly, bicelle, is one of the prevailingly used materials to prepare for weakly aligning protein [36]. The discoidal bicelle is composed of a mixture of dimyristoylphosphatidylcholin (DMPC) and dihexanoylphosphoatyldylcholine (DMPC) in a ratio of 3:1. The phospholipid binary mixture forms lipid bilayers disks with 30 nm ~ 40 nm in diameter. Bicelle has substantial anisotropy in magnetic susceptibility that allows it to spontaneously align under the NMR magnetic field: the normal of the discoidal bicellar surface orients perpendicular to the magnetic field (Fig. 14.6a).

In measuring the residual anisotropies, an appropriate amount of bicelle is put into the protein solution to establish an ideal degree of aligning state; the higher concentration goes to higher alignment [17]. In an NMR magnet, the associated bicelles transit into a liquid crystalline phase, in which the discoidal molecules are ordered to arrange their bicellar normally perpendicular to the magnetic field. Bicelle has a discoidal shape with a flat surface (Fig. 14.6a). In the liquid crystalline phase of the bicelles, proteins are entrapped in the cavities inside the bicellar liquid crystalline. Proteins in cavities bump to the bicellar walls and experience steric limits for their tumbling. Under such conditions, proteins imperfectly average the anisotropic interactions to make the residual anisotropies observable [16, 37].

The aligning magnitude is tuned by the bicelle concentration; higher bicellar concentration will produce a dense liquid crystalline phase, thus achieving a higher aligning magnitude. On the other hand, bicelle requires a minimal concentration to maintain the liquid crystalline phase; in lower bicellar concentrations under the critical limit, bicellar solution shows phase separation to be back in a micellar state.

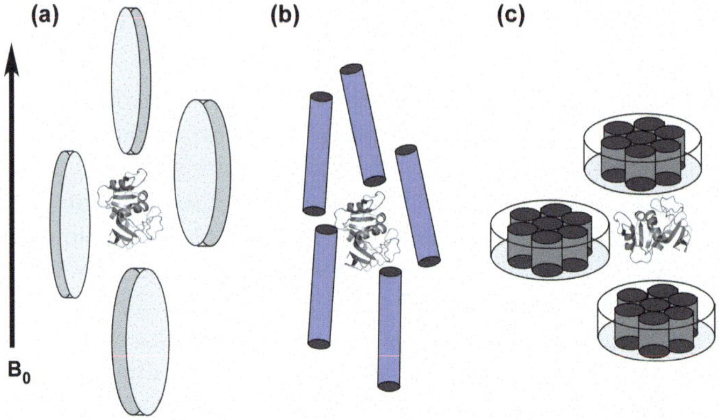

Fig. 14.6 Weak alignment made by using various media. **a** Discoidal shape phospholipid bicelle medium, **b** rod-like filamentous phage, and **c** Purple membrane. Because of significant anisotropic magnetic susceptibility, the molecules spontaneously align in a magnetic field. Proteins within the aligning media are aligned through the collisional interaction or electrostatic interactions with the aligning molecules

The bicelle made of DMPC/DHPC binary phospholipid has a neutral charge on its surface, and it induces residual anisotropy by bumping proteins to the bicellar walls. In using the charged bicelle made with doping charged molecule added in the binary bicelle, the aligning property of proteins changes because of the doped charges on the bicellar surface; the electrostatic interactions between the charged bicellar wall and protein surface charges alter the aligning states of the molecule. The addition of SDS, sodium dodecyl sulfate, into binary phospholipid, makes the bicelle negatively charged, while the doping CTAB, cetyl trimethyl ammonium bromide, makes it positively charged. Change in the aligning states of proteins improves the quality of the structural parameters derived from the residual anisotropies.

Bicelle, as an aligning medium, has practical limitations in its application to protein research. One is the limited temperature range for keeping the bicelles in a liquid crystalline state; typically, it ranges from 27 °C to 45 °C. Some proteins may be denatured in such higher temperatures. Bicelle becomes unstable in acidic or alkali conditions. The liquid crystalline state of the bicelle is only stable under neutral conditions. This also limits the sample conditions to prepare for protein samples. Additionally, the stability of the liquid crystalline state of bicelle is sensitive to salt conditions. We, therefore, need to care about the solution conditions to achieve the appropriate aligning condition. In some cases, bicellar media cannot be used for the residual anisotropy measurement. This is why various aligning media have been reported to overcome the limitations in the residual anisotropy experiments.

14.3.2 Naturally Occurring Materials that Spontaneously Align in a Magnetic Field

Spontaneously aligning molecules are also useful as aligning media for observing residual anisotropy. Rod-like filamentous phage and purple membrane are representative examples of this application.

Filamentous phage, made of a rod-like coat protein, is a well-known example of a naturally occurring molecule used for weak alignment [38]. Because of the highly anisotropic shape of the filamentous phage, it spontaneously aligns in a high magnetic field while keeping its rod axis parallel to a magnetic field (Fig. 14.6b). Filamentous phage has a negatively charged surface. Thus, it induces protein alignment through electrostatic interactions with protein; the aligning mechanism differs from that of the binary phospholipid bicellar medium. The concentration of the phage suspension in the protein solution can tune the aligning order.

The purple membrane has a two-dimensional crystal lattice structure of bacteriorhodopsin (bR) that is rich in a-helices (Fig. 14.6c). Purple membrane constitutes ordered a-helices and thus it has intrinsic high anisotropic magnetic susceptibility. Because of the high anisotropic magnetic susceptibility, the purple membrane spontaneously aligns in a high magnetic field.

A suspension of purple membrane is added to a protein solution to weakly align the protein. The concentration of purple membrane in the solution tunes the aligning magnitude [39]. Protein in the solution with the purple membrane aligns through electrostatic interactions on the negatively charged surface of the aligned purple membrane.

Purple membrane and filamentous phage cannot be applied to basic proteins having positively charged surfaces: The basic proteins tightly adsorb onto the negatively charged surfaces through strong electrostatic interactions, which broadens the NMR signals to prohibit high-resolution spectral analyses.

Other spontaneous aligning media are reported to compensate for the limits of the other aligning media, which include collagen gels and DNA nanotubes [40–42].

14.3.3 Compressed Acryl Amide Gel

The other method uses anisotropically compressed acrylamide gel. Acrylamide hydrogel has cavities to entrap proteins; the cavity size changes according to the ratios of the composing chemicals, acrylamide, and bis-acrylamide. Acrylamide forms a linear polymer chain and bis-acrylamide bridges acrylamide polymers. The increased ratio of bis-acrylamide to acrylamide generates a smaller cavity [43].

The acrylamide gel has a spatially isotropic cavity. Protein soaked in the gel retains isotropic tumbling and shows no residual anisotropies on its NMR spectrum. Weak alignment of protein in the gel is induced by the spatially anisotropic cavities in the gel through stretching or compressing the gels.

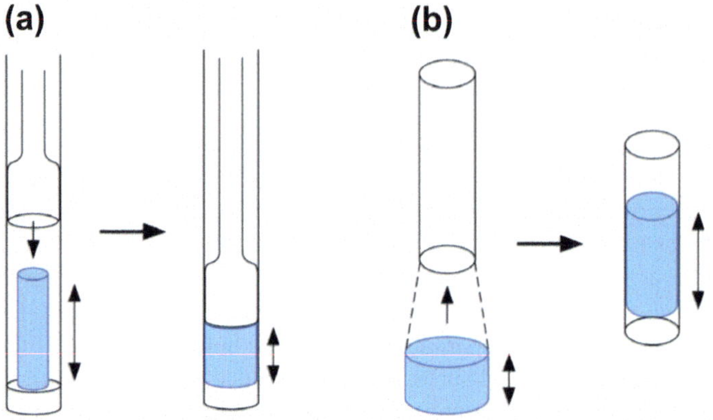

Fig. 14.7 Anisotropically compressed gels to achieve weak alignment. **a** Vertically compressed gel. **b** Laterally compressed, thus, stretched gel. A cast gel chip (blue) is inserted into an NMR glass tube. In the vertical compression, the cast gel is pressed by a glass rod in an NMR tube. These compressions induce spatially anisotropic cavities within the acrylamide gel

In measuring the residual anisotropy with acrylamide gels, a cast gel containing a protein solution is deformed inside an NMR sample tube (Fig. 14.7). There are two ways to make anisotropically deformed gel: one is to press the gel along the NMR tube (compressed gel) and the other is squeeze it in the lateral dimension; thus, it becomes stretched vertically (stretched gel) (Fig. 14.7).

In order to vertically compress the acrylamide gel, the gel is cast to have a shorter diameter than the inner diameter for the NMR sample tube. After inserting the gel, it was vertically compressed with a grass rod in a Shigemi NMR tube [44, 45] (Fig. 14.7a). On the other hand, the stretched gel is made from cast gel, which has a slightly larger diameter than the inner diameter of the NMR sample tube; the gel is inserted into the tube using a tapered device [46] (Fig. 14.7b). Because of the different pressed directions, these two preparations induce different protein orientations to each other.

The charge doping in the anisotropically compressed gel is also possible [47]. The negative charge is doped by replacing a part of acrylamide with acrylate, while a positive charge is introduced by adding DADMAC, diallyl dimethylammonium chloride, in the casting gel chip. The charge doping in the hydrogel changes aligning property through the additive electrostatic interaction between protein and the gel becomes active.

Acrylamide gel is neutral in charge, and thus, it aligns protein through collisional interaction between protein and anisotropically distributing acryl amide granules that exclude protein invasion within the compressed gel. The non-charged acrylamide gel is readily used for any type of protein, independent of the surface charge of the protein. In working on the protein having a significant surface charge, we carefully choose the doping to the hydrogel to avoid adsorbing the protein to the gel by electrostatic interaction to keep the spectral resolution.

The aligning order of the magnitude is tuned by changing the concentration of acrylamide and/or the ratio of the composing chemicals and also the diameter of the cased gel chip. In addition, the extent of the charge doping may have to be considered in some cases. There are more parameters to adjust than that for the liquid crystalline media. We, therefore, need some trial experiments to gain the optimal gel condition to have the anisotropic spin interactions observable in enough spectral resolution. It is noted that the lateral compression using the so-called Bax device enables rather uniform compression to give clear NMR signals for measuring the residual anisotropies [48]. The vertical compression is hard to gain the uniformly compressed gel, which sometimes gives seriously distorted NMR signals to ruin the spectral resolution.

The great advantage of using the compressed gel is in the chemical stability of acrylamide gel. In using acrylamide gel for a weak alignment, we do not need to worry about the sample conditions including temperature, salt concentration, and solution pH. Under any sample conditions, we can align protein with the compressed gel. It is a keen difference from the limitation posed by the other aligning materials including liquid crystalline media and the naturally occurring materials. The compressed gel is used as a universal media for weakly aligning protein. Especially for the membrane proteins dissolved in detergents, the lipid-bicelle is out of choice and the spontaneous aligning molecules are also limited in use. In such cases, the use of acrylamide gel is the preferential choice of the aligning media to use the residual anisotropies [49].

14.3.4 Protein Structures in Weakly Aligned Media

Adequate preparation of a weakly aligning protein sample is crucial for NMR to accurately determine domain orientations in a protein, as it requires global structural information from residual anisotropies. Because protein is dissolved in an artificial anisotropic medium like magnetically ordering liquid crystalline solution, one may worry about the structural changes that might occur under such unusual conditions: you may suspect if the domains in a protein in an aligning medium orient the same as in an isotropic solution.

The mechanisms for aligning protein in magnetically aligned bicelle solution and rod-like filamentous phage suspension are well understood [16]. The observed RDCs can be reproduced using the simulation according to restrained protein tumbling in the confined space formed by the aligning medium. The successful reproduction of the RDCs for the protein aligned by bicelle or filamentous phage supports that the proposed theory describing the mechanisms for achieving a weak protein alignment in such anisotropic media is valid.

According to the theory, the observed residual anisotropies can be established by a small fraction of protein experiencing rotational restriction through the collision between the protein and the magnetically aligning molecule. The charged filamentous phage, for example, restricts the rotation of protein near the medium through electrostatic interactions. The fraction represents approximately 0.1% of the total number of observed protein molecules in the sample. The interactions between the aligning medium and the protein do not actually force the protein to align with the medium's surface; they simply prevent some of the protein from tumbling near the medium. In other words, interacting with aligning media should not alter the protein conformation or domain arrangement.

The bicellar medium is in a liquid crystalline state above its phase-transition temperature. At lower temperatures, it forms a micelle solution that does not align in a magnetic field and therefore does not cause protein alignment. If there are no significant spectral changes of protein in the micelle solution, we can eliminate the possibility of non-specific contacts causing a structural alteration in the protein.

In the case of the charged filamentous phage, the increased concentration of cations, which shield the surface charge on the medium, impairs the ordering of the media in a magnetic field. The anisotropic spin interactions of the protein in the phage suspension decrease in magnitude as the cation concentration rises. If the spectral changes show a linear dependency on the cation concentration, we can exclude the structural changes caused by the interaction with the medium, filamentous phage in this case.

The use of compressed or stretched acrylamide gel enables a straightforward comparison of spectral characteristics between samples in reference gel (non-compressed or non-stretched gel) and the isotropic solution without acrylamide gel. If there is no spectral change observed in the comparison, it means that the protein structure remains unchanged in acrylamide gel.

Under weak alignment conditions, the spectral changes caused by interacting with the media are almost negligible. This ensures that no obvious structural changes occur to the protein, even in the case of proteins with multiple domains.

14.4 RDC-Based Domain Orientation Analysis

In this section, we will describe the experimental procedure for determining the domain orientation of a multi-domain protein, from the RDC data collection to the structure determination. I will also discuss the limitation of the RDC-based analyses to highlight the need for our TROSY-based DIORITE (Determination of Induced ORIentation by Trosy Experiment) approach.

14.4.1 Collecting the RDC Data

The RDC is measured from a pair of ^{1}H-coupled HSQC spectra for the samples in isotropic and anisotropic states. The ^{1}H-coupled HSQC spectrum displays a pair of split peaks along the ^{15}N axis for each ^{1}H-^{15}N correlation. The increased number of peaks on a ^{1}H-coupled HSQC spectrum can lead to more signal overlaps, making it difficult to accurately read the peak positions. To avoid this issue, an NMR technique is used to separate the up-field and down-field components into distinct 2D spectra, IPAP-HSQC [22]. The IPAP-HSQC experiment provides two separate spectra with in-phase and anti-phase doubles, respectively. The addition or subtraction of the spectra will produce two separate 2D spectra that only display the up-field or down-field components of each doublet. This signal separation reduces signal overlap on a ^{1}H-coupled HSQC spectrum and keeps the spectral resolution to the same extent as in the original HSQC spectrum (Fig. 14.3).

The IPAP-HSQC spectra reveal the separation gap between the up-field and down-field components, providing the apparent $^1J_{NH}$ values for isotropic and aligned samples. Consequently, the RDC is determined through the change in the measured apparent $^1J_{NH}$ values between isotropic and anisotropic states.

14.4.2 Domain Orientation Analysis Based on the RDCs

Here, we describe the analysis of the domain orientation of proteins based on the RDCs. The domain orientation analysis should be done for the protein with a known structure determined by X-ray. The main focus of the analysis is to explore how the domain reorients when interacting with another protein or ligand. In these cases, each domain is assumed to keep the same structure as in a crystal.

The alignment tensor for a weakly aligned protein is determined based on the RDCs and its structural coordinate, Eq. (14.5): direction cosines are calculated from the structure coordinate. Because the Saupe order matrix has five independent elements, we need more than five RDC data to determine the Saupe order matrix for the corresponding part. The singular value decomposition (SVD) of the matrix containing the relations for the observed RDCs will yield the Saupe order matrix [25, 26, 50]. Diagonalizing the Saupe order matrix provides the orientation of the alignment tensor frame in relation to the molecular coordinate system and the magnate of the orders along each principal axis. As described in the theory part, the alignment tensor frame orientation is defined by the Euler angles (α, β, γ).

We are considering a protein with two domains here. We assume that the high-resolution structure of each domain is available, and each domain structure retains the structure as in the crystal. Based on the collected RDCs, the alignment tensor for each domain is determined independently using the procedure described above. As schematically shown in Fig. 14.8, the determined tensor frames for each domain are used as a guide to define the orientation of the domain: one domain coordinate is rotated to align its tensor frame with the other (Fig. 14.8). It is noted here, the RDCs do not provide any distance information between the domains. If the segment between domains is highly flexible, additional distance restraints may be needed to construct the complete structure. These restraints can be obtained from other experiments, such as paramagnetic relaxation enhancement (PRE) and/or small angle X-ray scattering (SAXS) [51–53].

The alignment tensor determined by a set of RDCs has four potential orientations. This is because the inversion around each principal axis yields the same RDC values.

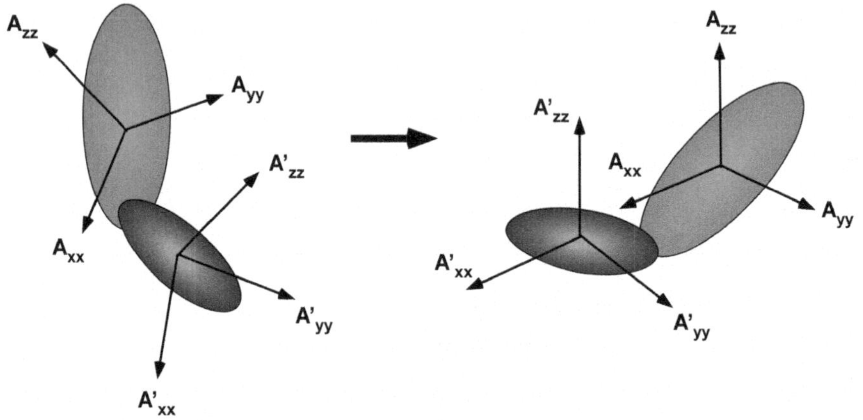

Fig. 14.8 Schematic representation for the procedure to determine the relative domain orientation based on the alignment tensors for each domain. Sets of RDC data determine the alignment tensor for each domain, independently. The domain orientation of a protein is established by rotating one domain coordinate to make an overlay of its tensor frame on the other

Therefore, the structural inversion around the principal axis cannot be discriminated experimentally. To solve the uncertainty in the orientation angle, we will incorporate extra aligning conditions using various aligning substances like charged bicelle or charged acrylamide gel. In domain orientation analysis, structural restrictions like the linker length between domains or possible clashes between domains can help identify a particular domain orientation, even when using a single set of RDCs obtained from the protein in one alignment state [34, 54].

The alignment tensor's magnitudes along each principal axis indicate the level of alignment present. If they differ across domains, it may indicate that each one rotates differently. Comparing the alignment orders between the domains in a protein offers valuable insights into how the protein domain functions [34].

14.4.3 Significance of the Domain Orientation Analysis by RDCs

The domain orientation analysis for protein in solution provides valuable insight, even when its high-resolution X-ray structure is available. Some cases demonstrate the different domain arrangements between solution and crystalline structures. The RDC analysis on maltose binding protein (MBP) in the complex with b-cyclodextrin has revealed a different domain orientation in the solution compared to that in the crystal [55]. This may indicate that the crystal contact causes a subtle change in the orientation of the domains. Bacteriophage T4 lysozyme in solution has a more open conformation compared to the crystal structure, which was also analyzed using

RDCs [56]. This observation seems to align with the steric requirements for ligand binding.

These instances demonstrate how RDC-based domain analysis serves as a valuable complement to X-ray crystallography in determining the relative domain orientation of protein domains, defining the protein's overall morphology. The RDC-based analysis plays a significant role in exploring how the domain rearranges upon binding to another protein or ligand. In cases where X-ray is unable to solve the complex structures, the RDCs in a solution state can provide the structure in the complex with the target molecule. This approach does not demand tedious and time-consuming NOE analysis as required in conventional NMR structure determination but only requires the backbone resonance assignments and a set of IPAP-HSQC spectra. The structure determination is much more efficient than the conventional NMR structure determination.

14.4.4 Molecular Size Limitation in the RDC-Based Approach

The use of RDCs in domain orientation analysis has become widely acknowledged as a valuable tool for elucidating the overall structure of proteins in solution, complementing X-ray crystallography. This approach, however, has severe size limitations. Here, this obstacle in the RDC application is discussed.

The size limitation arises from the rapid transverse relaxation rate of one of the split components observed on a 2D IPAP-HSQC spectrum. The high-filed component shows a faster transverse relaxation rate than the other. This component has an even faster relaxation rate than the HSQC counterpart. This is due to the cross-correlated relaxation interference to the amide ^{15}N spin relaxation process; for the high-field component, the cross-correlated relaxation process additively affects, while for the low-field component, the interference reduces its relaxation rate. The transverse relaxation process of the HSQC signal is not affected by the interference.

When measuring the RDCs with IPAP-HSQC for high molecular weight proteins, the high-field components of each amide spin pair will broaden and significantly reduce the signal intensities, making them unobservable. Observing the high-field component becomes more challenging in an aligned state. The addition of dipolar interactions appearing in the weakly aligned protein further broadens the high-field components. For proteins larger than 20 kDa, it is often difficult to observe the high-field component in an aligned state, making it impossible to measure the RDCs. The RDC-based domain orientation analysis with IPAP-HSQC is practically limited to around 20 kDa.

Simulation of the line broadening on each component of a double as per the molecule weight is shown in Fig. 14.9. In this figure, the low-field component with a longer transverse relaxation time is named TROSY component, and the other with a shorter transverse relaxation time is named anti-TROSY component. The

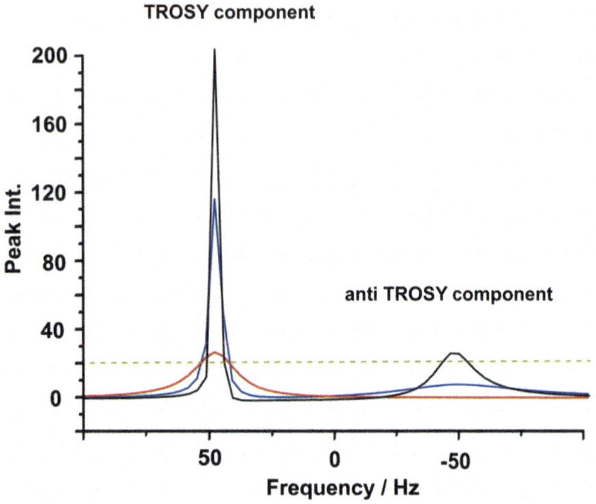

Fig. 14.9 Simulation of the molecular size dependency of the TROSY and anti-TROSY components observed on an IPAP-HSQC spectrum. The data represent a slice peak along the ^{15}N axis. This simulation assumed a 750 MHz experiment. Black, blue, and red lines are the simulated peaks for the sizes 20 kDa, 150 kDa, and 800 kDa, respectively. The dotted line is assumed the expected noise level

slower transverse relaxation associated with the low-field component is attributed to the mechanism implemented in TROSY (Transverse Relaxation Optimized SpetroscopY) [21]. In the simulation, it was shown that the anti-TROSY component experiences significant broadening, even for medium-sized proteins, such as 20 kDa.

The difference in line widths between the TROSY and anti-TROSY components will become significant for higher molecular weight proteins. As seen in the simulation, proteins over 150 kDa give severely broadened anti-TROSY signals, making it already hard to observe. An 800 kDa protein does not give measurable anti-TROSY signals. The size limitation of the RDC-based approach is clearly demonstrated in this simulation.

It should be noted that the TROSY component can maintain observable signal intensity even for an 800 kDa protein (Fig. 14.9). This motivated us to devise an approach to determine an alignment tensor only using TROSY components.

14.4.5 Existing Remedy for Overcoming the Size Limitation in the RDC-Based Approach

Alternative remedies have been proposed to surpass the size restriction of the RDC-based approach. They all rely on the TROSY technique.

The difference in the transverse relaxation rates between the TROSY (low-field) and anti-TROSY (high-field) components split along the ^{15}N axis is explained by relaxation interference. The same effect is present in the split signals along the ^{1}H dimension. When observing the ^{1}H-^{15}N single bond correlation spectrum without decoupling during t_1 and t_2 durations, each spin pair produces a quartet on the spectrum, resulting in split signals in both the ^{1}H and ^{15}N dimensions. The pure TROSY signal is the one having the longest transverse relaxation time among the quartets. By using the protein labeled with ^{15}N and ^{2}H, we reduce unwanted relaxation processes by breaking the ^{1}H-^{1}H dipolar interaction network in a protein, thereby enhancing the TROSY effect. This enables the observation of the ^{1}H-^{15}N correlation spectrum for proteins over 100 kDa [21].

One suggested solution is the combinatorial use of TROSY and HSQC. The difference between TROSY and HSQC signals along the ^{15}N axis yields half of the RDC value. As discussed earlier, the transverse relaxation rate of the HSQC signal is slower than that of the anti-TROSY component in an IPAP-HSQC spectrum. Therefore, the size limitation problem should be alleviated by replacing the anti-TROSY signal with the HSQC counterpart. For an 81.4 kDa protein, the transverse relaxation times for the TROSY, anti-TROSY, and HSQC signals are reported to be 65 ms, 10 ms, and 30 ms, respectively [57]. When comparing the transverse relaxation times of anti-TROSY and HSQC signals, the combinatorial use of TROSY and HSQC signals helps alleviate the molecule size limitation of the RDC-based approach.

Another remedy is to use of J-scaled TROSY, also known as J-enhanced (JE) TROSY [58, 59]. In the J-scaled TROSY experiment, the signal positions change depending on the $^{1}J_{NH}$-evolution encoded in the chemical shift evolution period. Standard TROSY experiment allows full $^{1}J_{NH}$-evolution evolution during t_1, thus giving the TROSY signals that are shifted from the corresponding HSQC signals by $^{1}J_{NH}$ / 2. The J-scaled TROSY experiment reduces the $^{1}J_{NH}$-evolution in the chemical shift evolution period by adding a defined duration to decouple ^{1}H-^{15}N nuclear spin pairs. The J-scaled TROSY signal positions, therefore, become close to the corresponding HSQC signals depending on the applied decoupling duration. The chemical shift difference between the TROSY signal and the corresponding J-scaled TROSY signal with a defined $^{1}J_{NH}$ decoupling period allows the calculation of the $^{1}J_{NH}$ value. The J-scaled TROSY method does not rely on signals with shorter transverse relaxation times, like anti-TROSY and HSQC signals. This feature allows measuring RDCs for high-molecular-weight proteins.

The signals on the J-scaled TROSY spectrum are more intense than those on the HSQC spectrum and their corresponding anti-TROSY components, but weaker than their 'pure' TROSY counterparts. This is because the J-scaling period mixes the TROSY and anti-TROSY coherence to make the resultant signals broadened. Increasing the J-scaling results in a significant chemical shift change of the J-scaled TROSY signal compared to the corresponding 'pure' TROSY signal. This leads to more accurate RDC values, but it also causes smaller signal intensities for the J-scaled TROSY. We need to choose a compromised J-scaling factor between the signal intensity and the accuracy for the observed RDCs.

The above approaches expanded the RDC application up to 50 kDa protein [60]. However, it is also reported that the rapid transverse relaxation of the non-TROSY component proves to be an obstacle in measuring the RDCs for an 81.4 kDa protein [57]. Using the full advantage of the TROSY signals is expected to expand the RDC application to higher-molecular weight proteins. Our DIORITE method is one of those that could be used for that purpose.

14.5 Alignment Tensor Determination Using Only TROSY

As mentioned earlier, the problem of molecular size limitation in RDC-based domain orientation analysis has not been fully resolved, despite some successful results from existing approaches. Most biologically interesting multi-domain proteins or their complexes tend to be over 100 kDa. The current approaches are not believed to be suitable for application to proteins such as biologically significant proteins. This is because they do not fully take advantage of TROSY spectroscopy, which provides the signals with the longest transverse relaxation time to enable the NMR signals observable for the high-molecular-weight proteins. Our approach, DIOIRTE, utilizes 'pure' TROSY spectra to take full advantage of the signals with the longest transverse relaxation time. This may offer significant advantages over the existing methods when applied to the high-molecular-weight proteins. In this section, we will describe the theoretical aspects of the TROSY-based alignment tensor determination. This method allows the analysis of domain orientation for larger proteins.

14.5.1 Alignment-Induced TROSY Shift Changes

The TROSY shift changes along the ^{15}N axis when the protein is transferred from isotropic to anisotropic states. This induced TROSY shift change in an anisotropic solution contains two anisotropic spin effects observed in a peptide plane; they are the ^{1}H-^{15}N dipolar interaction and ^{15}N CSA. As depicted in Fig. 14.10, the induced TROSY shift change, $\Delta\delta_{TROSY}$, contains half of RDC and the full RCSA effects. In a Cartesian representation using the Saupe order matrix, the following relation should fold:

$$\begin{aligned}\Delta\delta_{TROSY} &= -\frac{1}{2}\left(\frac{\mu_0}{4\pi}\right)\frac{\gamma_i\gamma_j h}{2\pi^2 r_{ij}^3}\frac{4\pi}{5}\sum_{k,l=x,y,z} S_{kl}\cos(\alpha_k)\cos(\alpha_l) \\ &+\frac{2}{3}\sum_{k,l=x,y,z}\sum_{j=x,y,z}\sum_{k,l=x,y,z} S_{kl}\cos\left(\alpha_{kj}\right)\cos(\alpha_l j)\delta_{jj} \\ &= \sum_{k,l=x,y,z} S_{kl}\left\{\frac{1}{2}D_{NH}^0\cos(\alpha_k)\cos(\alpha_l)+\frac{2}{3}\Delta_{kl}\right\}\end{aligned} \tag{14.10}$$

$$\Delta_{kl}+\sum_{k,l=x,y,z}\sum_{j=x,y,z} S_{kl}\cos\left(\alpha_{kj}\right)\cos(\alpha_l j)\delta_{jj} \tag{14.11}$$

where $\cos(\alpha_k)$ is the direction cosine for the NH bond vector relative to the molecular axis k (k = x, y, z). D_{NH}^0 is the static dipolar coupling constant, which equals 23.0 and 21.7 kHz for assumed NH bond lengths 1.02 Å and 1.04 Å, respectively; in solution, the effective NH bond length is estimated to be 1.04 Å, which values include the bond libration effects. The term $\cos\left(\theta_{kj}\right)$ is the direction cosine of the CSA principal

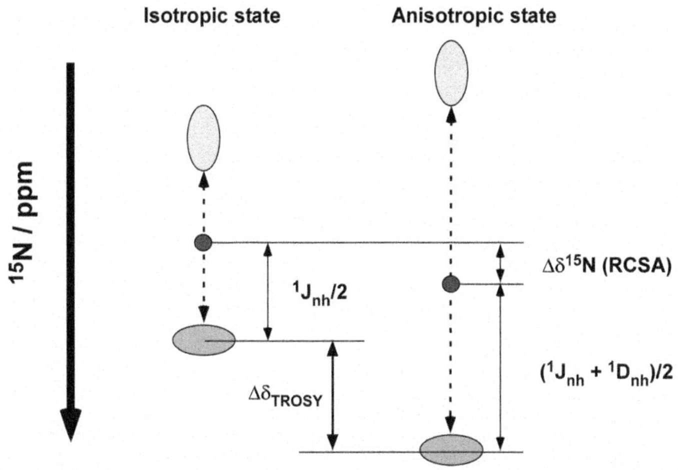

Fig. 14.10 Schematic drawing of the relationship among the signals for ^{1}H-^{15}N doublet in a ^{1}H coupled HSQC spectra observed for protein in isotropic and aligned states. $\Delta\delta_{TROSY}$ contains the contributions from half of RDC and full RCSA

axis *j* relative to the molecular axis *k*. The principal value of the CSA tensor along *the* *j*-axis is denoted as δ_{jj}. As done for the RDC, the SVD calculation to the equations Eq. (14.10) for the residues in a protein gives the Saupe order matrix. At least, five $\Delta\delta_{TROSY}$ data are required to determine the Saupe order matrix constituted by five independent elements. The alignment tensor is obtained through the diagonalization of the Saupe order matrix.

DIORITE is the algorism to determine the alignment tensor solely from TRSOY spectra. DIORITE can accurately determine an alignment tensor when used with high-molecular-weight proteins over 50 kDa. DIORITE utilizes two anisotropic spin interactions: RDC and RCSA. As described in the theory part, RCSA contains the tensorial information about the peptide plane orientation against a magnetic field, while RDC gives the bond vector orientation. The RDC value remains constant when the peptide plane is rotated along the NH bond axis, while the RCSA undergoes significant changes. DIORITE including the RCSA effect, thus, can determine the peptide plane orientation. DIORITE is more informative than the RDC when working on the ^{1}H-^{15}N spin pairs.

14.5.2 CSA Tensor Parameters Used in DIORITE

Determining the alignment tensor for DIORITE requires the ^{15}N CSA tensor for each residue. In the RCSA section, obtaining an accurate ^{15}N CSA tensor for each residue in a protein is not easy, as it is highly dependent on local structure, such as backbone torsion angle, hydrogen bonding, and other factors. In the DIORITE analysis, the ^{15}N CSA tensor for the residue must be pre-defined as input parameters.

In the domain orientation analysis with DIORITE, we use a high-resolution domain coordinate from X-ray. The ^{15}N CSA tensors determined experimentally for small proteins in solution have demonstrated that the ^{15}N CSA tensor value is primarily dependent on the backbone torsion angle. Based on this correlation, correlation, we adopt the secondary structure-specific ^{15}N CSA tensor values in the practical analysis with DIORITE.

The quality of the back-calculated $\Delta\delta_{TROSY}$ values was assessed on ubiquitin. The $\Delta\delta_{TROSY}$ values calculated with the secondary structure-specific ^{15}N CSA tensors and those with the residue-specific tensors were compared (Fig. 14.11); the residue-specific ^{15}N CSAs used here were determined based on a set of spin relaxation analyses by Bodenhausen and co-workers [61]. The use of the secondary structure-specific ^{15}N CSA tensor consistently yields results within an acceptable error range compared to those obtained with residue-specific values (Fig. 14.11). The analysis of DIORITE using the secondary structure-specific ^{15}N CSA tensors can be a practical approach for exploring the orientation of domains in a protein.

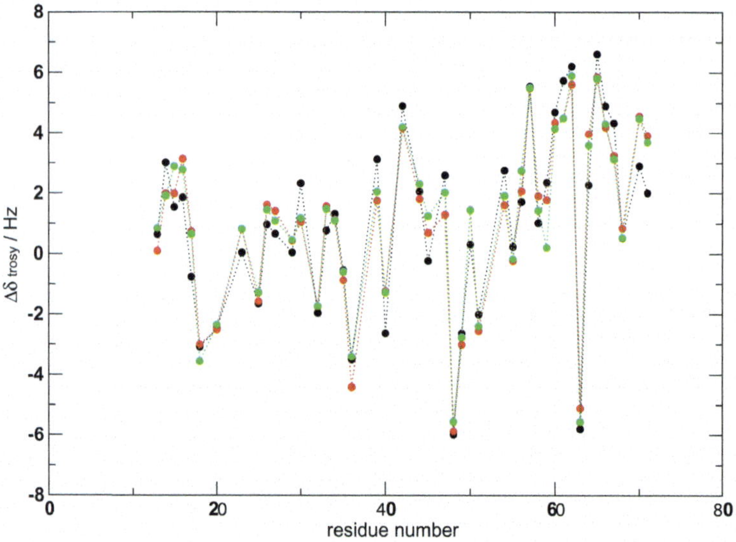

Fig. 14.11 Comparison of the back-calculated $\Delta\delta_{TROSY}$ obtained by DIORITE algorism using the secondary structure specific and the residue-specific ^{15}N CSA tensors. Red and green circles are the values with secondary structure-specific and residues-specific tensors, respectively. Black circles are the observed $\Delta\delta_{TROSY}$

14.5.3 DIORITE Analysis Using Different Magnetic Field Strengths

The TROSY effect reduces the line width along the ^{15}N dimension at a higher magnetic field. The original paper on TROSY estimated that the optimal frequency for obtaining the narrowest lines is around 1 GHz [62]. In this section, we will discuss the best magnetic-field strength for the DIORITE analysis.

The TROSY shift change, $\Delta\delta_{TROSY}$, induced in an anisotropic state shows significant magnetic-field dependency because of the field-dependent change in the RCSA. The magnitude of RCSA depends on the magnetic-field strength, while the RDC keeps the same value in different magnetic-field. The RCSA alters $\Delta\delta_{TROSY}$ ss value in different magnetic fields.

In a peptide plane, the least shielded ^{15}N CSA tensor axis, δ_{xx}, is close to the NH bond; the angle β ranges from 15 to 20 degrees (Fig. 2b). The small b angle results in the cancellation between the RDC and RCSA values making the observed $\Delta\delta_{TROSY}$ value smaller than the corresponding RDC: the RDC and RCSA exhibit opposite signs. Therefore, $\Delta\delta_{TROSY}$ should be roughly equivalent to that half of RDC minus RCSA in their absolute values. The value $\Delta\delta_{TROSY}$ tends to be approximately one-third of the RDC in an absolute magnitude.

The cancellation between the two effects depends non-linearly on the magnetic field strength. By using ubiquitin, we simulated the room mean square (rms) values at different magnetic-field strengths as indicated by the proton resonance frequencies (Fig. 14.12). As seen in the simulation, the rms $\Delta\delta_{TROSY}$ value, which shows the variation of $\Delta\delta_{TROSY}$ value in a protein, decreases up to 800 MHz and then increases according to the field strength (Fig. 14.12). In lower magnetic-field strength, where the RCSA contribution is small and the $\Delta\delta_{TROSY}$ is roughly approximated by a half of RDC. The increasing RCSA, which has an opposite sign to RDC, reduces the $\Delta\delta_{TROSY}$ value. However, the RCSA becomes dominant in $\Delta\delta_{TROSY}$ value in a higher magnetic field, which may explain the profile (Fig. 14.12).

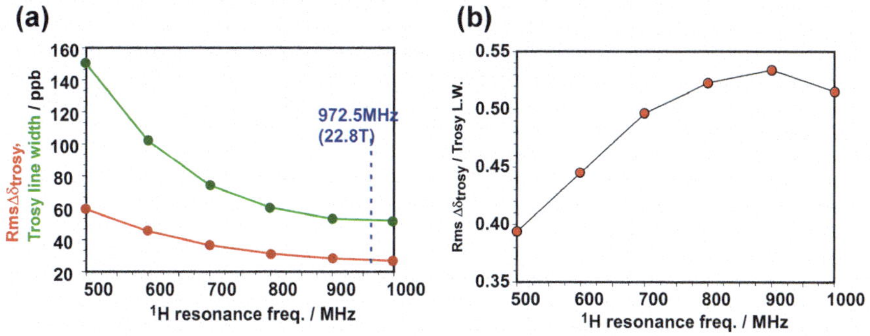

Fig. 14.12 Field dependency of the TROSY line width and the root-mean-square (rms) $\Delta\delta_{TROSY}$. **a** TROSY line width (green) and rms $\Delta\delta_{TROSY}$ (red) according to the magnetic field strength **b** Field strength dependency of the effective resolution in measuring $\Delta\delta_{TROSY}$

The condition for weak alignment is carefully tuned to prevent signal broadening. The alignment order is typically tuned to around 10^{-3}, giving the maximal absolute magnitude of RDC around 20 Hz. Under the condition, $\Delta\delta_{TROSY}$ is expected to give about 6 Hz. Even under the 800 MHz where the $\Delta\delta_{TROSY}$ value is minimal, the expected $\Delta\delta_{TROSY}$ is 5.5 Hz. The accuracy in reading peak positions at a signal-to-noise ration of 40:1 is estimated to be 0.6 Hz. Therefore, the experimental error in the observed $\Delta\delta_{TROSY}$ should be 0.8 Hz. Accordingly, the expected $\Delta\delta_{TROSY}$ is well resolved within the peak reading errors. It is noted that the reading error in the RDC should be 1.2 Hz because of twice the subtraction required.

The resolution of the TROSY spectrum also needs to be taken into account when discussing the performance of DIORITE at different magnetic-field strengths. Using the average ^{15}N CSA tensor value estimated from ubiquitin, we plotted the dependencies of ^{15}N TROSY line width and rms $\Delta\delta_{TROSY}$ value against proton resonance frequencies. The values are in ppb (parts per billion) units instead of Hz, allowing for comparison in the context of spectral resolution (Fig. 14.13a). In this simulation, the optimal frequency to minimize the ^{15}N line width of the observed TROSY signal is 972.5 MHz, close but slightly different from the value estimated in the original paper. When comparing the line narrowing of TROSY signal, the rms $\Delta\delta_{TROSY}$ value shows less dependency on the magnetic-field strength. The effective accuracy in observing $\Delta\delta_{TROSY}$ value should be estimated by the value $\Delta\delta_{TROSY}$ divided by the ^{15}N line width of TROSY signal (Fig. 14.13b). In this simulated profile, the optimal magnetic-field strength for observing $\Delta\delta_{TROSY}$ is around 900 MHz for ^{1}H resonance frequency. Therefore, the magnetic-field strength for the maximal TROSY effect is almost optimal for the DIORITE analysis. DIORITE analysis can fully benefit from the TROSY effect on a 900 MHz NMR spectrometer, which is now readily available on the market.

14.5.4 Practical Aspects of the DIORITE Data Collection

DIORITE uses TROSY chemical shift changes $\Delta\delta_{TROSY}$ induced by a weak alignment of protein [63, 64]. In general, the chemical shift is highly sensitive to the solution conditions, such as temperature, sample concentration, pH, and ionic strength. In measuring $\Delta\delta_{TROSY}$, we need to isolate the factors that affect the chemical shift, focusing particularly on the alignment effects. To achieve this, an anisotropically compressed polyacrylamide gel is the best medium for aligning protein. The anisotropically compressed gel, also known as stretched gel, is created by inserting a cast gel chip with a slightly larger diameter than the inner diameter of the NMR tube. The gel chip, with the same diameter as the NMR tube, is not compressed after insertion, allowing the isotropic cavity to be maintained. The protein in this uncompressed gel is not aligned and does not exhibit any anisotropic spin interactions. The sample, therefore, can be used as the reference TROSY spectrum for measuring $\Delta\delta_{TROSY}$. Because the uncompressed gel has the same acrylamide composition as

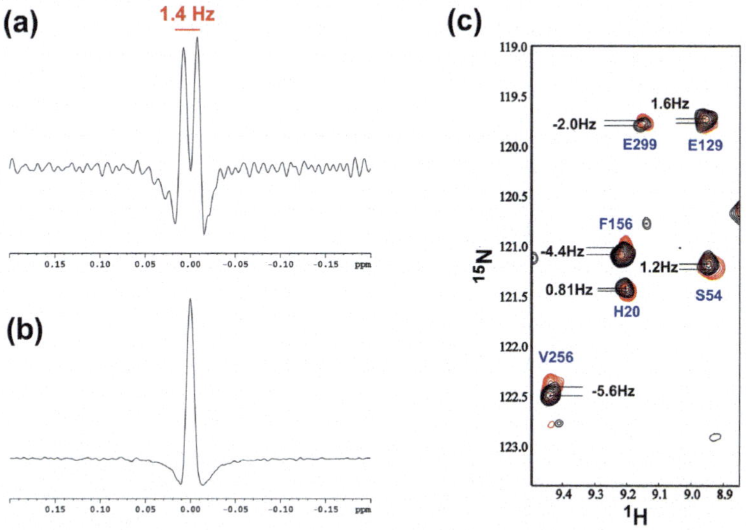

Fig. 14.13 TROSY shift changes induced by weak alignment by using compressed acrylamide gel. TROSY shift changes are measured from the reference positions collected for the sample in an uncompressed gel that is made of the same chemical composition to that used for compressed gel. **a** D_2O residual quadrupole split observed in a compressed gel, **b** the residual quadrupole split is not observed in an uncompressed gel, showing that the gel can be used for the reference. **c** TROSY shift changes, Dd_{TROSY}, observed for a 38 kDa protein (MBP); a pair of TROSY spectra collected in compressed (red) and uncompressed gels (black)

the compressed gel, the proteins in this gel experience the same chemical environments. This guarantees that the observed TROSY shift changes solely result from the alignment effects. It is difficult to obtain reference data using other alignment media, such as bicelle and filamentous phage. They are not suitable media for DIORITE experiments.

In measuring $\Delta\delta_{TROSY}$ values, we use a set of TROSY spectra for the aligned and isotropic samples (Fig. 13c). The formation of the uneven cavity in the stretched gel can be observed by the splitting of the signal from HOD. The water molecule's deuteron within an uneven cavity displays a doublet, indicating residual quadrupolar coupling of the deuteron (Fig. 14.13a). The stronger the anisotropy in a cavity, the larger the split. The singlet water deuteron signal in the reference gel confirms the isotropic cavity in the gel (Fig. 14.13b).

In measuring an NMR spectrum, the signal from water deuterons is used as a frequency lock to establish the frequency standard. The resonance positions observed are biased due to the split of a deuteron signal in an aligned state; one of the doublet envelopes is used for the frequency lock, therefore biasing the signals by half of the split width. This offset should be subtracted from each signal position on the TROSY spectrum for proteins in an aligned state.

14.5.5 DIORITE Analyses on MBP in Different Ligand-Bound States

Here, we demonstrate the applications of the DIORITE analysis. Maltose binding protein (MBP) has two domains. From a series of X-ray analyses, MBP is known to show domain reorientation upon binding to ligand, and the domain rotation angle depends on the size of the ligand molecule.

MBP binds to b-cyclodextrin (b-CD) comprising seven glucose units, and it shows a slight change in the relative domain orientation from that in apo form as demonstrated by X-ray analysis. MBP also binds to different types of cyclodextrins comprising of the different number of glucose units; a-CD (six glucose units), g-CD (eight glucose units). We expected to see the domain rotation angle changes according to the size of the three types of CDs.

Using the anisotropically compressed gel (stretched gel) and uncompressed reference gel, we collected a pair of TROSY spectra for each MBP in the complex with a-, b-, and g- CD. Using the apo-form MBP X-ray coordinate, we analyzed the relative domain orientation using DIORITE; the model structure was constructed based on the alignment tensors individually determined for N- and C-terminal domains. On each complex, the back-calculated and observed $\Delta\delta_{TROSY}$ values were well correlated, suggesting the alignment tensors for each domain were well determined (Fig. 14.14).

The DIORITE analyses on the MBP complexes demonstrated that for the smallest ligand, a-CD, the domain rotation angle was significantly larger than that for the b-CD complex structure. On the other hand, MBP in the complex with g-CD retains almost the same domain orientation relative to the b-CD complex; the size over the b-CD does not change the domain rotation angle.

The example analyses showed that the MBP significantly reoriented in response to the ligand size. It is worth mentioning that the DIORIET analysis is highly effective in detecting ligand-dependent domain reorientation. This can be achieved by a pair of TROSY spectra for the sample in both aligned and isotropic states.

14.6 Conclusion

X-ray structures continue to accumulate exponentially every year. The large compilation of protein coordinates is a driving force for conducting protein research, utilizing the protein coordinates at hand. This chapter delves into the use of protein structure data from X-ray crystallography to assist in protein structure analysis in solution with new NMR approaches.

There has been a lot of discussion on how important it is to determine the structure of a solution using NMR. Most of the structures for single-domain proteins or isolated domains have only minor structural differences compared to the corresponding structures solved by X-ray. This reduces the significance of NMR structure

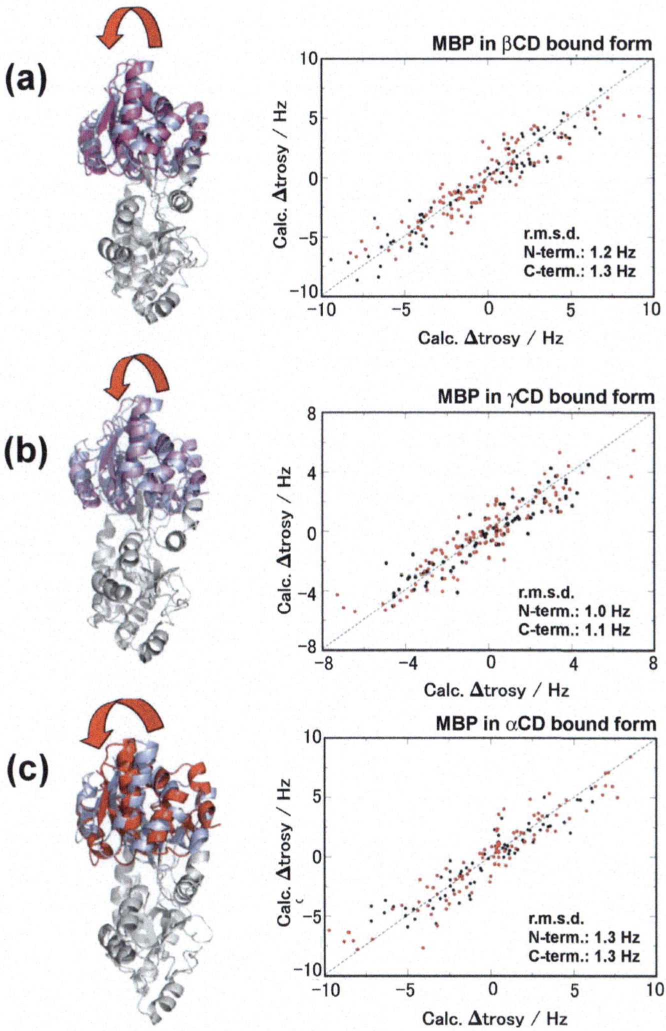

Fig. 14.14 Domain rotation angle change according to the ligand size in maltose binding protein (MBP). **a** b-CD, **b** g-CD, and **c** a-CD complex states. The hinge rotation angles induced by ligand binding were 13 deg., 14 deg. And 22 deg. for the cases of b-CD, g-CD, and a-CD, respectively

analysis, except when crystallization is difficult. When an X-ray structure is available for a protein that behaves as an independent structural unit, solution structure determination by NMR is typically not carried out, since the crystal structure is not expected to significantly differ from the solution structure. In most cases, the size limitation in NMR structure determination prevents the analysis of solution structures, even when it is necessary. NMR solution structure analysis is recognized as a useful method, but it's only applicable in limited cases.

The structure analysis of multiple domain proteins seems to present a different situation. The distinction in protein domain orientations between solution and crystalline states has been evidenced by a few examples. The number of examples is increasing because it is becoming evident that many proteins have domains connected by flexible or unstructured linkers, as indicated by the sequence. In proteins, the domain arrangement is likely defined artificially by crystal packing, and may not accurately represent the protein's morphology in a solution state.

The NMR techniques using weak alignment have made it possible to directly determine the relative domain orientation in solution, something that conventional NMR methods have never achieved. Some variations of the methods were introduced in this review, along with their practical limitations. Our developed DIORITE approach has a significant advantage in analyzing the orientation of domains over the existing methods, especially when applied to higher molecular weight proteins. The domain orientation analysis by DIORITE will enhance the X-ray structure-assisted investigation of protein morphological changes in solution, which are linked to functional activity.

The rearrangement of a protein's structure, or its morphological change, when it binds to a ligand or interacts with another protein, becomes crucial in discussing the regulation of protein function, especially after obtaining the high-resolution crystal structure in a certain state, such as the apo form. The combination of DIORITE with X-ray structure data can provide detailed insight into how proteins function by altering their morphology in solution. NMR is now becoming a complementary technique to X-ray crystallography for analyzing the structure of multi-domain proteins in solution.

References

1. Morris, G. M. et al.:AutoDock4 and AutoDockTools4: Automated docking with selective receptor flexibility. J. Comput. Chem. **30**, 2785–2791 (2009)
2. Dominguez, C., Boelens, R., Bonvin, A.M.: HADDOCK: a protein-protein docking approach based on biochemical or biophysical information. J. Am. Chem. Soc. **125**, 1731–1737 (2003)
3. de Vries, S.J., van Dijk, M., Bonvin, A.M.: The HADDOCK web server for data-driven biomolecular docking. Nat. Protoc.Protoc. **5**, 883–897 (2010)
4. Fuxreiter, M.: Fuzziness: linking regulation to protein dynamics. Mol. Biosyst.Biosyst. **8**, 168–177 (2012)
5. Evenäs, J., Mittermaier, A., Yang, D., Kay, L.E.: Measurement of 13Cα−13Cβ dipolar couplings in 15N,13C,2H-Labeled proteins: application to domain orientation in maltose binding protein. J. Am. Chem. Soc. **123**, 2858–2864 (2001)
6. Evenas, J. et al.:Ligand-induced structural changes to maltodextrin-binding protein as studied by solution NMR spectroscopy. J. Mol. Biol. **309**, 961–974 (2001)
7. Cavanagh, J., Fairbrother, W.J., Palmer III, A. G., Skelton, N.J.: Protein NMR Spectroscopy. 411–531 (Academic Press, Inc., 1996)
8. Wuthrich, K.: NMR of Proteins and Nucleic Acids. (John Wiley & Sons, Inc., 1986)
9. Tjandra, N., Tate, S., Ono, A., Kainosho, M., Bax, A.: The NMR structure of a DNA Dodecamer in an Aqueous Dilute Liquid Crystalline Phase. J. Am. Chem. Soc. **122**, 6190–6200 (2000)
10. Tjandra, N., Bax, A.: Direct measurement of distances and angles in biomolecules by NMR in a dilute liquid crystalline medium. Science **278**, 1111–1114 (1997)

11. Bax, A.: Weak alignment offers new NMR opportunities to study protein structure and dynamics. Protein Sci. **12**, 1–16 (2003)
12. de Alba, E., Tjandra, N.: Residual dipolar couplings in protein structure determination. Methods Mol. Biol. **278**, 89–106 (2004)
13. Delaforge, E. et al.:Large-Scale Conformational Dynamics Control H5N1 Influenza Polymerase PB2 Binding to Importin α. J. Am. Chem. Soc. (2015)
14. Tate, S.-i.:Anisotropic nuclear spin interactions for the morphology analysis of proteins in solution by NMR spectroscopy. Anal. Sci. **24**, 39–50 (2008)
15. Prestegard, J.H., Bougault, C.M., Kishore, A.I.: Residual dipolar couplings in structure determination of biomolecules. Chem. Rev. **104**, 3519–3540 (2004)
16. Zweckstetter, M.: NMR: prediction of molecular alignment from structure using the PALES software. Nat. Protocols **3**, 679–690 (2008)
17. Ottiger, M., Bax, A.: Characterization of magnetically oriented phospholipid micelles for measurement of dipolar couplings in macromolecules. J. Biomol. NMR **12**, 361–372 (1998)
18. Wüthrich, K.: Ch. 7, 117–129 (John & Wiley & Sons, Ltd, 1986)
19. Kurita, J., Shimahara, H., Utsunomiya-Tate, N., Tate, S.: Measurement of 15N chemical shift anisotropy in a protein dissolved in a dilute liquid crystalline medium with the application of magic angle sample spinning. J. Magn. Reson. **163**, 163–173 (2003)
20. Tate, S.-I., Shimahara, H., Utsunomiya-Tate, N.: Molecular-orientation analysis based on alignment-induced TROSY chemical shift changes. J. Magn. Reson. **171**, 284–292 (2004)
21. Pervushin, K., Riek, R., Wider, G., Wuthrich, K.: Attenuated T2 relaxation by mutual cancellation of dipole-dipole coupling and chemical shift anisotropy indicates an avenue to NMR structures of very large biological macromolecules in solution. Proc. Natl. Acad. Sci. U.S.A. **94**, 12366–12371 (1997)
22. Ottiger, M., Delaglio, F. & Bax, A.: Measurement of J and dipolar couplings from simplified two-dimensional NMR spectra. J. Magn. Reson. (San Diego, Calif.: 1997) **131**, 373–378 (1998)
23. Yao, L., Vögeli, B., Ying, J., Bax, A.: NMR Determination of Amide N−H Equilibrium bond length from concerted dipolar coupling measurements. J. Am. Chem. Soc. **130**, 16518–16520 (2008)
24. Varshalovich, D.A.M., A.N., Khersonskii,V.K.: Quantum theory of angular momentum. (World Scientific Publishing Co. Pte. Ltd., 1998)
25. Saupe, A.: Recent results in the field of liquid crystals. Angew. Chem. Int. Ed. Engl. **7**, 97–112 (1968)
26. Saupe, A., Englert, G.: High-Resolution nuclear magnetic resonance spectra of orientated molecules. Phys. Rev. Lett. **11**, 462–464 (1963)
27. Hiyama, Y., Niu, C.H., Silverton, J.V., Bavoso, A., Torchia, D.A.: Determination of 15N chemical shift tensor via 15N–2H dipolar coupling in Boc-glycylglycyl[15N glycine]benzyl ester. J. Am. Chem. Soc. **110**, 2378–2383 (1988)
28. Lundstrom, P., Hansen, D.F., Kay, L.E.: Measurement of carbonyl chemical shifts of excited protein states by relaxation dispersion NMR spectroscopy: comparison between uniformly and selectively (13)C labeled samples. J. Biomol. NMR **42**, 35–47 (2008)
29. Yao, L., Grishaev, A., Cornilescu, G., Bax, A.: Site-specific backbone amide (15)N chemical shift anisotropy tensors in a small protein from liquid crystal and cross-correlated relaxation measurements. J. Am. Chem. Soc. **132**, 4295–4309 (2010)
30. Fushman, D., Tjandra, N., Cowburn, D.: An Approach to direct determination of protein dynamics from 15N NMR relaxation at multiple fields, independent of variable 15N Chemical shift anisotropy and chemical exchange contributions. J. Am. Chem. Soc. **121**, 8577–8582 (1999)
31. Boyd, J., Redfield, C.: Characterization of 15N chemical shift anisotropy from Orientation-Dependent changes to 15N chemical shifts in dilute bicelle solutions. J. Am. Chem. Soc. **121**, 7441–7442 (1999)
32. Cornilescu, G., Bax, A.: Measurement of proton, nitrogen, and carbonyl chemical shielding anisotropies in a protein dissolved in a dilute liquid crystalline phase. J. Am. Chem. Soc. **122**, 10143–10154 (2000)

33. Harris, R.K., Becker, E.D., Cabral de Menezes, S.M., Goodfellow, R., Granger, P.:NMR nomenclature: nuclear spin properties and conventions for chemical shifts. IUPAC Recommendations 2001. In: International Union of Pure and Applied Chemistry. Physical Chemistry Division. Commission on Molecular Structure and Spectroscopy. Magn. Reson. Chem. **40**, 489–505 (2002)
34. Tolman, J.R., Flanagan, J.M., Kennedy, M.A., Prestegard, J.H.: NMR evidence for slow collective motions in cyanometmyoglobin. Nat. Struct. Biol. **4**, 292–297 (1997)
35. Williams, D.C., Cai, M., Clore, G.M.: Molecular basis for synergistic transcriptional activation by Oct1 and Sox2 revealed from the solution structure of the 42-kDa Oct1·Sox2·Hoxb1-DNA ternary transcription factor complex. J. Biol. Chem. **279**, 1449–1457 (2004)
36. Ottiger, M., Bax, A.: Bicelle-based liquid crystals for NMR-measurement of dipolar couplings at acidic and basic pH values. J. Biomol. NMR **13**, 187–191 (1999)
37. Berlin, K., O'Leary, D.P., Fushman, D.: Improvement and analysis of computational methods for prediction of residual dipolar couplings. J. Magn. Reson. **201**, 25–33 (2009)
38. Hansen, M.R., Mueller, L., Pardi, A.: Tunable alignment of macromolecules by filamentous phage yields dipolar coupling interactions. Nat. Struct. Biol. **5**, 1065–1074 (1998)
39. Sass, J., et al.: Purple membrane induced alignment of biological macromolecules in the magnetic field. J. Am. Chem. Soc. **121**, 2047–2055 (1999)
40. Bellot, G., McClintock, M.A., Chou, J.J., Shih, W.M.: DNA nanotubes for NMR structure determination of membrane proteins. Nat. Protocols **8**, 755–770 (2013)
41. Douglas, S.M., Chou, J.J., Shih, W.M.: DNA-nanotube-induced alignment of membrane proteins for NMR structure determination. Proc. Natl. Acad. Sci. **104**, 6644–6648 (2007)
42. Ma, J., Goldberg, G.I., Tjandra, N.: Weak alignment of biomacromolecules in collagen gels: an alternative way to yield residual dipolar couplings for NMR measurements. J. Am. Chem. Soc. **130**, 16148–16149 (2008)
43. Naghash, H.J., Okay, O.: Formation and structure of polyacrylamide gels. J. Appl. Polym. Sci.Polym. Sci. **60**, 971–979 (1996)
44. Ishii, Y., Markus, M.A., Tycko, R.: Controlling residual dipolar couplings in high-resolution NMR of proteins by strain induced alignment in a gel. J. Biomol. NMR **21**, 141–151 (2001)
45. Tycko, R., Blanco, F.J., Ishii, Y.: Alignment of Biopolymers in Strained Gels: A New Way To Create Detectable Dipole−Dipole Couplings in High-Resolution Biomolecular NMR. J. Am. Chem. Soc. **122**, 9340–9341 (2000)
46. Chou, J., Gaemers, S., Howder, B., Louis, J., Bax, A.: A simple apparatus for generating stretched polyacrylamide gels, yielding uniform alignment of proteins and detergent micelles*. J. Biomol. NMR **21**, 377–382 (2001)
47. Meier, S., Häussinger, D., Grzesiek, S.: Charged acrylamide copolymer gels as media for weak alignment. J. Biomol. NMR **24**, 351–356 (2002)
48. Liu, Y., Prestegard, J.H.: A device for the measurement of residual chemical shift anisotropy and residual dipolar coupling in soluble and membrane-associated proteins. J. Biomol. NMR **47**, 249–258 (2010)
49. Opella, S.J., Marassi, F.M.: Structure determination of membrane proteins by NMR spectroscopy. Chem. Rev. **104**, 3587–3606 (2004)
50. Losonczi, J. A., Andrec, M., Fischer, M.W. Prestegard, J.H.:Order matrix analysis of residual dipolar couplings using singular value decomposition. Journal of magnetic resonance (San Diego, Calif.: 1997) **138**, 334–342 (1999)
51. Clore, G.M., Schwieters, C.D.: Theoretical and computational advances in biomolecular NMR spectroscopy. Curr. Opin. Struct. Biol. **12**, 146–153 (2002)
52. Kozak, M., Lewandowska, A., Ołdziej, S., Rodziewicz-Motowidło, S., Liwo, A.: Combination of SAXS and NMR techniques as a tool for the determination of peptide structure in solution. J. Phys. Chem. Lett.s **1**, 3128–3131 (2010)
53. Grishaev, A., Tugarinov, V., Kay, L.E., Trewhella, J., Bax, A.: Refined solution structure of the 82-kDa enzyme malate synthase G from joint NMR and synchrotron SAXS restraints. J. Biomol. NMR **40**, 95–106 (2008)

54. Tolman, J.R., Al-Hashimi, H.M., Kay, L.E., Prestegard, J.H.: Structural and dynamic analysis of residual dipolar coupling data for proteins. J. Am. Chem. Soc. **123**, 1416–1424 (2001)
55. Skrynnikov, N.R., et al.: Orienting domains in proteins using dipolar couplings measured by liquid-state NMR: differences in solution and crystal forms of maltodextrin binding protein loaded with beta-cyclodextrin. J. Mol. Biol. **295**, 1265–1273 (2000)
56. Goto, N.K., Skrynnikov, N.R., Dahlquist, F.W., Kay, L.E.: What is the average conformation of bacteriophage T4 lysozyme in solution? A domain orientation study using dipolar couplings measured by solution NMR. J. Mol. Biol. **308**, 745–764 (2001)
57. Tugarinov, V., Kay, L.E.: Quantitative NMR studies of high molecular weight proteins: application to domain orientation and ligand binding in the 723 residue enzyme malate synthase G. J. Mol. Biol. **327**, 1121–1133 (2003)
58. Kontaxis, G., Clore, G.M., Bax, A.: Evaluation of cross-correlation effects and measurement of one-bond couplings in proteins with short transverse relaxation times. J. Magn. Reson. **143**, 184–196 (2000)
59. Bhattacharya, A., Revington, M., Zuiderweg, E.R.P.: Measurement and interpretation of 15N–1H residual dipolar couplings in larger proteins. J. Magn. Reson. **203**, 11–28 (2010)
60. Jain, N.U., Noble, S., Prestegard, J.H.: Structural characterization of a mannose-binding protein-trimannoside complex using residual dipolar couplings. J. Mol. Biol. **328**, 451–462 (2003)
61. Loth, K., Pelupessy, P., Bodenhausen, G.: Chemical Shift Anisotropy Tensors of Carbonyl, Nitrogen, and Amide Proton Nuclei in Proteins through Cross-Correlated Relaxation in NMR Spectroscopy. J. Am. Chem. Soc. **127**, 6062–6068 (2005)
62. Pervushin, K., Riek, R., Wider, G., Wuthrich, K.: Attenuated T2 relaxation by mutual cancellation of dipole-dipole coupling and chemical shift anisotropy indicates an avenue to NMR structures of very large biological macromolecules in solution. Proc Natl Acad Sci U S A **94**, 12366–12371 (1997)
63. Tate, S.: Anisotropic nuclear spin interactions for the morphology analysis of proteins in solution by NMR spectroscopy. Anal. Sci. **24**, 39–50 (2008)
64. Tate, S., Shimahara, H. & Utsunomiya-Tate, N.:Molecular-orientation analysis based on alignment-induced TROSY chemical shift changes. J. Magn. Reson. (San Diego, Calif.: 1997) **171**, 282–291 (2004).

The manufacturer's authorised representative in the EU is Springer Nature Customer Service Centre GmbH, Europaplatz 3, 69115 Heidelberg, Germany. If you have any concerns regarding our products, please contact ProductSafety@springernature.com

Printed and bound by CPI Group (UK) Ltd, Croydon, CR0 4YY
01/10/2026
02225969-0016